Single Cell Oils
Microbial and Algal Oils

Second Edition

T0348615

Single Cell Oils
Microbial and Algal Oils

Second Edition

Editors
Zvi Cohen
Colin Ratledge

AOCS
PRESS

Urbana, Illinois

AOCS Press, Urbana, IL 61802

Library of Congress Cataloging-in-Publication Data

Single cell oils : microbial and algal oils / editors, Zvi Cohen, Colin Ratledge.—2nd ed.
   p. cm.
  Includes bibliographical references and index.
  ISBN 978-1-893997-73-8 (alk. paper)
  1. Single cell lipids. I. Cohen, Zvi, Ph. D. II. Ratledge, Colin.

TP248.65.S55S54 2010
664'.3—dc22

                  2009047325

Contents

Preface to the Second Edition

The pace of developments in the field of microbial oils and fats—the Single Cell Oils of the title—over the past 4 to 5 years has surprised us both. So much so, that a new edition of this book has had to be prepared to keep readers up to date. Of considerable interest and, to an extent, to our amazement have been the many commercial proposals since 2005 for developing microbial oils as sources of biofuels, particularly using photosynthetic algae. Some of these ideas appear very serious indeed with developments and investments being announced at very frequent intervals. Reports of these activities are now included in this new edition, together with considerations for developing the potential of bacteria for biofuel production. Biofuels can undoubtedly be produced from microalgae but the cost is highly problematic. Even with additional research and perhaps genetic manipulation to improve productivities and lower costs, we remain somewhat skeptical about the feasibility of producing such biofuels at a cost sufficiently low to be competitive with petroleum oils. However, our skills in gazing into the crystal ball to divine the future are somewhat limited. No one can tell what will be the price of petroleum oils in the years to come, let alone guess what levels of subsidies some governments might be prepared to pay for the dubious honor of being able to claim that they are using 'microbial or algal oils for fuel'. But, and we should not grumble too much, therefore, these activities are acting as tremendous stimuli for developing new ideas and interests in this subject. Indeed, this is the very type of stimulus that the entire subject of microbial oils has been lacking since its inception in the early decades of the 20th century.

Biofuels apart, what now can be regarded as the more 'traditional' areas for the uses of microbial oils—the production of the very long-chain polyunsaturated fatty acids (VL-PUFAs) for infant, adult, and animal nutrition—have also seen considerable advancement over the past 5 years. The application of genetic techniques for cloning and transforming oleaginous microorganisms into the production of novel PUFAs, or for producing ones previously not easily available either from animals or plants, has been successful; and chapters on the molecular breeding of yeasts and molds are now included in this second edition. The days of producing what have previously been described as 'designer' oils are clearly now with us. In addition, we have seen a much wider appreciation of the value of VL-PUFAs as essential dietary components by the general public and, from this, a realization that supplies of these oils from the traditional marine sources may not be sustainable nor, for many groups of people, acceptable, because of their animal origins. The presence of various pollutants that are concentrated in marine lipids further exacerbate the situation. Thus, there is now much more impetus from many companies around the world for producing and exploiting microbial oils. These oils have gone from being academic curiosities to being minor commodity oils in their own right within a very short space of time.

Of course, everyone—even the editors—admit that microbial oils, particularly using nonphotosynthetic microorganisms, are expensive to produce because of the need for large-scale fermentation systems that involve the purchase of even larger amounts of sugar as fermentation feedstocks. If the key fatty acids could be produced by genetically modified plants, this would be a much cheaper route for obtaining them. But, in spite of well more than a decade of devoted research by a myriad of plant geneticists and even plant breeders, it has proved virtually impossible to produce plants that yield useful amounts of the key VL-PUFAs. Again, our crystal ball does not allow us to see the future in this area with any accuracy but it seems more than likely that at least another decade will have to pass before we see,the emergence of any plant-derived VL-PUFA. Thus, meanwhile, and perhaps in perpetuity, *Single Cell Oils* will command a key place in the marketplace. They may be expensive but they are produced by reliable, safe technologies that are not subject to the vagaries of the weather (or national politics)' do not involve the use of pesticides, insecticides, and herbicides; and are true organically derived materials. We believe that their position in the marketplace is now secure.

Zvi Cohen
Colin Ratledge
August 2009

Single cell oils (SCO) have come of age. They have become accepted biotechnological products fulfilling key roles in the supply of the major very long chain polyunsaturated fatty acids (PUFA), now known to be essential for infant nutrition and development. But their acknowledgment as being potential sources of oils and fats has been a slow process. Many critics in the early years of SCO doubted whether they could ever be produced at a reasonable price; even if they could, there were grave doubts as to whether SCO would be accepted by the general public. This was in spite of the "general public" having no apparent objection to consuming bacteria and yeasts as part of their everyday diet in the form of yogurts, cheeses, beers, and sourdough breads. When the product is good, the public will buy it; when the product is essential, the public will line up to buy it; and when our babies need the product, the line is likely to be a very long one indeed. SCO are the edible oils extracted from micro-organisms—the single-celled entities that are at the bottom of the food chain. The best producers with the highest oil contents are various species of yeasts and fungi with several key algae also able to produce high levels of nutritionally important PUFA. Interest in SCO, as they have now become known, stretches back for over a century. Attempts have been made to harness the potential of various organisms, especially during the two world wars, in order to produce much needed oils and fats. Attempts have also been made to produce substitute materials for some of the major oilseed crops and even to produce a superior type of cocoa butter material. But it has been their potential to produce PUFA that has now galvanized the current interest in these SCO as oils rich highly desirable fatty acids essential for our well being and not readily available either from plants or animals. This monograph has arisen from a symposium organized by David Kyle for the American Oil Chemists' Society in May 2003 that covered many of the ongoing projects in this area. It echoes two earlier conferences of the AOCS, the first in 1982 in Toronto and the second in Chicago in 1992, also organized by David Kyle. Over the intervening years, the position of SCO has become much more secure. Processes that were just "twinkles in the eye" in 1992 now exist as commercial realities; SCO production processes occur not only in the United States, but also in Europe, Japan, and China. Interest in them is widespread and the prospects of producing a complete range of PUFA is within our grasp. Whether the next decade or so will see SCO being overtaken by oils coming from genetically engineered plants, as has been predicted by some, will remain a tantalizing prospect. The future, as always, will be awaited with interest. In the meantime, SCO are here and available.

Zvi Cohen
Colin Ratledge
January 2005

PART 1

Introduction and Overview

1

Single Cell Oils for the 21st Century

Colin Ratledge

Department of Biological Sciences, University of Hull, Hull, HU6 7RX, UK

Introduction

Single cell oils (SCOs) are defined as the edible oils obtainable from single celled microorganisms that are primarily yeasts, fungi (or molds), and algae (Ratledge, 1976). Bacteria, which are also single cell microorganisms, do not usually accumulate edible oils but instead tend to produce other storage materials, such as poly-beta-hydroxybutyrate. However, the possible uses of bacteria for producing lipids in sufficient quantities to be commercially useful are covered elsewhere in this book (see Chapter 14), but these uses are entirely for lipids as biofuels and would not be suitable for edible purposes. The edible oils that are produced by eukaryotic microorganisms (i.e., those that have a defined nucleus and are, therefore, distinct from prokaryotic bacteria having no defined nucleus) are similar in type and composition to those oils and fats from plants or animals. This chapter aims to provide an introductory overview to SCO and to show that current interest in their production and use has developed out of a long history of exploitating microorganisms as sources of oils and fats. Without these early endeavors, it is quite possible that none of the commercial SCO products currently on the market would have been developed as the basic understanding behind the exploitation of microbial oils would have delayed for several decades.

The key event that changed microbial oils from being more or less academic curiosities 20–30 years ago (see Ratledge, 1992) to becoming important nutraceuticals for adults, as well as infants, was the emergence of overwhelming evidence for the dietary significance of very long chain, polyunsaturated fatty acids (VL-PUFAs), coupled with the realization that there was no adequate or safe source from plants or animals. What were originally considered to be unusual microorganisms have turned out to be extraordinarily important, since they are still the only realistic sources of these oils. The belief that genetically modified plants would be created in a few short years to supply these essential VL-PUFAs and that SCOs would quickly be superseded by GM-oils has not been substantiated. Such sources of PUFAs appear as remote as ever, and, perhaps somewhat hopefully, microbial oils may be expected to continue to be of economic importance for several decades to come.

The Early Years

There has been interest in microbial lipids for over 130 years (Nageli & Lowe, 1878) and in exploiting them as alternative sources of oils and fats for human consumption since the early decades of the 20th century. Paul Lindner, working in Berlin, Germany, appears to have been the first person to develop a small-scale process to make fat using a species of yeast then called *Endomyces vernalis* and currently known as *Trichosporon pullulans* (Lindner, 1922; Woodbine, 1959). The development of microorganisms as a source of oils and fats continued to escalate during the first four decades of the last century with a number of groups in various countries studying not only the process of lipid biosynthesis but also the factors influencing its accumulation. These early endeavors in microbial oil production were reviewed in considerable depth by Woodbine (1959), and this review remains possibly the most thorough one available covering the world-wide development of microbial oil production from its very inception through the mid-1950s.

The problem, though, was that the oils and fats produced by oleaginous species of yeasts and fungi (the groups of microbes that are the highest producers are referred to as the "oleaginous" species) were not significantly different from the oils and fats obtained from plant seeds. Since these microorganisms had to be grown in culture media that contained glucose or sucrose as a source of carbon, which was derived from agricultural crops, the cost of turning one agricultural commodity into another (i.e., turning sugar into oil) was never going to be economically feasible as the cost of sugar is never more than about a quarter of the price of most commodity plant oils, such as corn oil, soybean oil, and rapeseed (Canola) oil. Moreover, it is not a question of turning one ton of sugar into one ton of oil. Microorganisms are not that efficient; it takes about 5 tons of sugar to make 1 ton of oil (see Ratledge & Wynn, 2002). Therefore either some zero-cost carbon source had to be found or oils had to be identified that exceeded the prices of the usual commodity oils by a considerable margin in order for SCOs to become economically viable propositions. There was no value in developing SCO processes that simply produced facsimiles of existing commodity plant oils.

In spite of these obvious economic limitations, considerable work on the production of microbial oils took place from the 1920s through to the late 1950s. This work laid the foundation for understanding lipid production in microorganisms. In brief, it was established that:

- The number of microorganisms capable of accumulating oil to more than 20% of their biomass weight was relatively small in comparison with the total number of species.
- Oil-accumulating microorganisms were mainly species of yeast and fungi; only a few bacteria produced much extractable edible oil. The oil produced by these

yeasts and fungi was, like plant oils, mainly composed of triacylglycerols having component fatty acids (FA) that were, in almost every case, similar to what had already been recognized in plant oils.

- Some algae were recognized as producing fairly high amounts of lipid, but this lipid tended to be more complex than that produced by yeasts and fungi, as it included the lipids of the photosynthetic apparatus; nevertheless they still contained the same FAs that occurred in plant oils. Some PUFAs were observed to be similar to those found in fish oils.

- Oil accumulation in oleaginous microorganisms could be increased by starving the cells of a supply of nitrogen or a nutrient other than carbon. The cells responded to deprivation of nitrogen in the growth medium by entering into a lipid storage phase in which excess carbon, still present in the medium, was converted preferentially into storage lipid (triacylglycerols) materials. If the cells were subsequently returned to a situation in which the missing nutrient was now made available, the oil reserves could be mobilized and re-channelled into cellular materials. Lipid accumulation was, therefore, a stress-induced response with the oil being an intracellular storage material that could reintroduce carbon and energy into the metabolic processes of the cell when the prevailing conditions were appropriate.

These views were upheld by extensive investigations, involving both biochemical and molecular studies, carried out on the mechanism of lipid accumulation in oleaginous microorganisms (see, for example, Ratledge & Wynn, 2002 for review).

A typical profile for the accumulation of lipid in an oleaginous microorganism is shown in Fig. 1.1. This figure shows that lipid accumulation in a microbial cell begins when nitrogen is exhausted from the medium. The medium, therefore, has to be formulated with a high C:N ratio to ensure that nitrogen is exhausted while other nutrients, including carbon, remain in excess. In practice, the C:N ratio is about 40–50:1, although the optimum ratio needs to be determined for each individual organism. To produce the greatest number of cells, the concentration of nitrogen and carbon may need to be increased while keeping them in the same proportion; this enables the balanced growth phase to continue until it reaches the maximum biomass density that the fermentor can sustain before the lipid accumulation phase begins. In practice, it is probably advantageous to cultivate cells using NH_4OH as a pH titrant in the fermentor so that cells can grow at their fastest rate without any nutrient limitation and then, when cell density is at its highest optimum level, switching to NaOH as a titrant, while ensuring that the supply of carbon is still in excess. This switch in titrant results in immediate N exhaustion in the culture medium so that lipid accumulation commences when the greatest possible biomass has been achieved (see Ganuza et al., 2008). There are occasionally suggestions that lipid accumulation may not require exhaustion of nitrogen from the growth medium and that the process of oleaginicity is growth-associated in some microorganisms—see, for example, Eroshin et al., 2000, 2002. In other words, lipid accumulates in the cells as they continue to multiply.

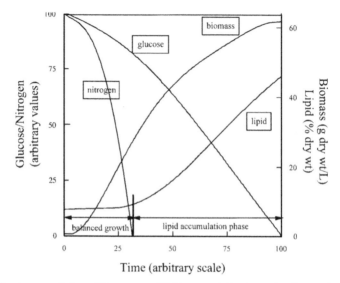

Fig. 1.1. Idealized representation of the process of lipid accumulation in an oleaginous microorganism. The composition of the culture medium is formulated so that the supply of nitrogen, which is usually an ammonium salt, is growth limiting. After its exhaustion, cells do not multiply any further, but they continue to assimilate glucose (the usual carbon feedstock). This glucose is then channeled into the synthesis of storage lipid (triacylglycerol) within the cells. The extent of lipid accumulation depends upon the individual microorganism—lipid contents may vary between 20 and 70% of the biomass.

However, when the growth patterns of these organisms are closely examined, it is apparent that these organisms, which are, in fact, very few, are slow-growers. They clearly have some metabolic impairment. In such cells, glucose is being assimilated into the cells faster than it can be converted into cell biomass and, consequently, its excess is then diverted into the synthesis of a storage material, such as a lipid.

Although attempts were made in Germany during World War II to produce microbial fat to supplement the meager supplies that could be obtained from conventional sources (mainly animal fat with a little plant oil), these efforts were limited. However, some oil-rich biomass production was achieved with fungi. The fungi, mainly *Oidium lactis* (now *Geotrichum candidum*), were grown on waste lactose (from a cheese creamery) or agricultural waste materials (Bunker, 1945, 1946, 1963; Ledingham et al., 1945), and the resultant cells seem to have been fed mainly to army horses after being molded into bricks using hay and straw (Bunker, 1946). Some biomass may have been included in soups and sausages for human consumption, but they were mainly viewed as a protein supplement rather than a source of fat. Although feeding the oil-rich fungi to army horses may sound rather trivial, the German army during this period had approximately one million horses to support and, using unconventional sources of feed material, was clearly considered reasonable. There does not, however, seem to have been any extraction of the oils and, even with the amount of effort that had gone into developing these processes, the arrival of the first SCO was still some way off.

The development of efficient large-scale production of microbial oils (and, indeed, all microbial products) was limited by the availability of appropriate large-scale fermentors necessary to produce the biomass (microbial cells) at high densities (over 50 g dry wt/L). Laboratory-scale fermentors were relatively unheard of until the 1950s, and industrial-scale stirred tank fermentors were rare. We see this lack of technology evidenced by the UK's need to transfer the technology for penicillin production in the early 1940s (which had used static cultivation of *Penicillium chyrsogenum* in adapted hospital bed-pans) to the US, which had the only accessible stirred tank bioreactors in the world. This technological deficit was a clear limitation not only to microbial oil production but also to almost all other microbial products that needed aerated, submerged cultivation systems. Some fermentors existed in many countries to produce beer and related materials, but these were for the anaerobic production of microbial products and had no facilities for aeration or stirring. Moreover, most were open vessels and, therefore, were prone to airborne contamination, which would have been disastrous for any process seeking to produce food-grade materials.

The major stimulus for developing large-scale fermentation technology and, from it, the production of laboratory-scale fermentor units that would enable the necessary experimental work to be carried out, was probably the advent of single cell protein (SCP) production in the late 1950s. Several petroleum companies—principally BP Ltd. of the UK—began to explore the conversion of *n*-alkanes, unwanted waste materials from the initial phase of fractionating petroleum oil, into edible biomass. Yeasts (especially *Yarrowia lipolytica*—formerly known as *Candida lipolytica*) were found that could grow rapidly on the alkanes, but stirred and aerated fermentors were essential to achieve optimal conversion. The ensuing biomass was rich in protein (about 50% w/w) and proved to be a useful major feed material for animals. As the manufacturers felt a little uneasy about describing their product as "microbial protein," the name SCP was coined as an appropriate euphemism to disguise the origins of the material.

This period of SCP production ended because of unfavorable economics in 1975 with the price escalation of crude petroleum oil and the maintenance of the low price of soybean meal—the major competitor of the SCP—in the USA. But, at the end of this period, the world had developed systems for submerged microbial cultivation to an unparalleled degree. Biotechnology had arrived, and not just for SCP production. Production of antibiotics, amino acids, and organic acids, such as citric acid, now used sophisticated, stirred tank fermentation technology which had replaced the cultivation of microorganisms in static cultures that primarily used shallow tray systems.

With new technology becoming widely available (not forgetting that the availability of laboratory-scale fermentors at a reasonable cost allowed research to be carried out at the 1–2 L level), interest in producing microbial oils once more re-emerged in the mid-1960s (Kessell, 1968; Ratledge, 1968). However, enthusiasm for producing such products had largely waned since plant seed oils were now inexpensive and there was no prospect of producing oils from microbial sources that could rival their price. There were, though, some prospects of producing one or two "rare" microbial oils (Shaw 1965,1966a, 1966b) that were not readily available from conventional plant sources, but these ideas were still embryonic and lacked focus because the market for such materials was very uncertain. The examination of microorganisms carried out by Shaw (1966b) focused on identifying possible sources of arachidonic acid (ARA; 20:4n-6)—not for use in human nutrition (for which nothing was known at that point), but as a chicken-flavor material! Only after this work had been done did he realize that chicken flavor was not due to ARA, but instead to some entirely unrelated compound. The work of Shaw (1965,1966a), however, proved invaluable for identifying microorganisms that might be used for the production of various long chain PUFAs.

The other main development that occurred in the early 1960s, which was of considerable importance for the study of microbial oils, was the development of gas chromatography. Previously, FA analysis had been laborious and tedious. It also required relatively large amounts of material. Gas chromatography altered all this; one could very quickly analyze a number of oils and fats for their component FA and use just milligram amounts of material. The stage was, therefore, set for a reexamination of microorganisms as potential sources of oils and fats; this shift can be seen in the seminal work of Bob Shaw, mentioned above.

Developments in the Last Quarter of the 20th Century

Work in the author's laboratory (Gill et al., 1977; Botham & Ratledge, 1979; Boulton & Ratledge, 1983; Evans & Ratledge, 1985) consolidated the mechanism of oil accumulation in yeasts being grown in laboratory fermentors, using both batch and continuous fermentations. This work also established that the approximate conversion efficiency of the starting substrate (glucose) to the product (triacylglycerol oil) was maximally 22% (w/w) under optimized growth conditions. But, in spite of understanding a great deal about the process of lipid accumulation,,there was no clear target for which the microbial oils could be considered for commercial development. It was then brought to the author's attention that there might be a small niche market for an oil rich in γ-linolenic acid (GLA, 18:3n-6).

A Process for GLA Production

In the mid-1970s, GLA was only available as a minor component (about 9% of the total FA) in the oil of evening primrose (*Oenothera biennsis*), but this oil was considered, by virtue of its GLA content, efficacious to relieve many symptoms , even for the treatment of multiple sclerosis—a claim that has since been discounted. At the time, evening primrose oil commanded a price of about $50 per kg when most commodity plant seed oils were fetching less than a hundredth of this price. The prospects of a commercially viable SCO were instantly evident since the work of Shaw (1965,

1966a, 1966b) had established the consistent occurrence of GLA in a group of lower fungi known as the Zygomycetes.

Research carried out in the author's laboratory established that several members of this group were appropriate for producing an oil rich in GLA. Further examination of the short list of candidate microorganisms using pilot-scale submerged fermentation technology established the most suitable one; commercialization of the process followed this selection with the first oil being produced in 1985 (Ratledge, 2006).

The very first SCO was thus produced using *Mucor circinelloides* (also known as *Mucor javanicus*). It was grown in large-scale fermentors of 220 m^3 (65000 US gallons) run by J. & E. Sturge Ltd. at Selby, North Yorkshire, UK, who normally used their skills in fermentation technology to produce citric acid from another fungus, *Aspergillus niger*. The oil was sold under the trade name of "Oil of Javanicus" and also as "GLA-Forte," a name used by one retailer of the oil. It achieved some limited penetration of the over-the-counter food supplement market. The process closed down in 1990, primarily due to a change in company ownership (to Rhone-Poulenc Ltd.), but also because of falling prices in evening primrose oil and the advent of borage oil as a cheaper alternative source of GLA. By this time, about 25-30 tons of the SCO had been produced. Each fermentor run produced about 10 tons of biomass from which about 2 to 2.5 tons of oil could be extracted. An account of this SCO-GLA process, including details from the necessarily extensive screening and selection program, can be reviewed (Ratledge, 2006).

Although the SCO-GLA was superior to evening primrose oil in all respects—higher content of GLA (Table 1.A), higher stability to oxidation, absence of high

Table 1.A. Fatty Acid Profiles of Fungi and Plants Used Commercially for γ-Linolenic Acid (GLA) Production.

		Relative % (w/w) of Major Fatty Acids								
	Oil Content (% w/w)	16:0	16:1	18:0	18:1	18:2 (n-6)	18:3 (n-6) GLA	18:3 (n-3)	20:1	22:1
Mucor circinelloides[a]	25	22	1	6	40	11	**18**	—	—	—
Mortierella isabellina[b]	~50	27	1	6	44	12	**8**	—	0.4	—
Mortierella ramanniana[b]	~40	24	—	5	51	10	**10**	—	—	—
Evening primrose	16	6	—	2	8	75	**8–10**	0.2	0.2	—
Borage	30	10	—	4	16	40	**22**	0.5	4.5	2.5
Blackcurrant	30	6	—	1	10	48	**17**	13	—	—

[a]Oil of Javanicus.

[b]Production organisms used by Idemitsu Co. Ltd, Japan. Oil contents of cells uncertain but approximate levels indicated.

levels of competing FAs, such as linoleic acid, and a much lower content of herbicide, insecticide and pesticide residues—the fungal oil was difficult to sell to a public (mainly in the UK and some other European countries) that wanted evening primrose oil. An oil that was superior to evening primrose oil but was not called "evening primrose oil" was largely ignored by the public, even though marketing publicity carefully eschewed mentioning the microbial/fungal origins of "Oil of Javanicus."

Although this first SCO failed to bring in a reasonable profit for the producers, it was, nevertheless, a significant milestone in the development of SCO. Its arrival encouraged other companies in other countries to explore the possibilities of using microorganisms as sources of similar, and even more expensive, oils and fats; however, the targeting of potential oils for niche markets was still critical. A process similar to the GLA-SCO process in the UK was developed by Idemitsu Kosan Co. Ltd., Tokyo, Japan, using *Mortierella isabellina* and possibly also *Mort. ramanniana* as the production organisms (Nakahara et al., 1992). The oils produced, however, had a much lower GLA content than the oil produced by *Mucor circinelloides* (Table 1.A), although each fungus had about twice the oil content of the *Mucor* cells. Sales of these oils to the Japanese domestic market began in 1988 but probably ceased shortly afterwards; certainly no PUFA oil, let alone an SCO-GLA, appears in the current portfolio of company products.

A Process for a Cocoa Butter Equivalent Fat

Some interest was developed in the early 1980s for the possible production of a cocoa butter equivalent (CBE) fat using yeasts (Moreton, 1988; Davies, 1988, 1992; Smit et al., 1992). Yeasts, unlike many molds and fungi, tend to produce only limited amounts of PUFA, and some strains can have relatively high contents of stearic acid (18:0). For a successful CBE, it is necessary to have a oil or fat that has roughly equivalent amounts of stearate, oleate, and palmitate all accommodated on the same triacylglycerol molecule preferably as *sn*-1 stearoyl, *sn*-2 oleoyl, *sn*-3 palmitoyl glycerol (Table 1.B). The main obstacle in achieving this goal was how to increase the rather low content of stearic acid in yeast fat to reach at least 25%. This level was initially attained using an inhibitor of the delta 9-desaturase that converts stearic acid into oleic acid (Moreton, 1988). However, the inhibitor used, stearidonic acid, was too expensive and would have substantially increased the price of the final product. In an alternative approach, mutants of a yeast, *Candida curvata* (now, *Cryptococcus curvatus*), were created that had diminished activities of this desaturase (Smit et al., 1992) and thus produced considerably increased amounts of stearic acid without having to use an expensive inhibitor (Table 1.B). These mutants,however, were not entirely stable when used in large-scale fermentors. Consequently, it was preferred to use the original wild-type yeast, which already had a higher natural level of stearic acid (15% of the total fatty acids) than most other yeasts, as the possible production organism (Davies, 1988,1992). The key procedure used to increase the level of stearic acid was

Table 1.B. Fatty Acid Profiles of Cocoa Butter Equivalent-Single Cell Oils (CBE-SCO): Microbial Oils Used as a CBE Compared with Cocoa Butter.

	Relative % (w/w) of major fatty acids					
	16:0	18:0	18:1	18:2	18:3 (n–3)	24:0
Cryptococcus curvatus Wt[a]	30	15	45	5	0.5	2
C. curvatus NZ[b]	18	24	48	3	1	2
C. curvatus R26-20[c]	15	47	25	8	2	—
C. curvatus R25-75[c]	33	25	33	7	1	—
C. curvatus F33.10[c]	24	31	30	6	—	4
Yeast isolate K7-2[d]	26	25	38	6	1	1
Cocoa butter	23–30	32–37	30–37	2–4	—	—

[a]Wt, wild type yeast (original strain).
[b]NZ, strain used in New Zealand. See Davies (1988, 1992).
[c]Mutant strains produced with partial deletions of Δ9-desaturase. See Smit *et al.* (1992)
[d]Isolated in New Zealand. See Davies (1988, 1992).

to maintain a very low aeration rate in the fermentors during lipid accumulation so that the activities of the desaturases were then limited by the supply of oxygen, which is the essential co-substrate for their activity. Davies and his colleagues were able to carry out a one-off, large-scale run at 250 m³, which established the feasibility of the process (Davies, 1992).

In spite of achieving a good quality CBE (Table 1.B), which could be incorporated into chocolate at the permitted level of 5% of the total fat, thereby giving improved characteristics over the use of a conventional plant-derived CBE (R.J. Davies, personal communication), the yeast process was abandoned because it was not sufficiently cost-effective. The cost of cocoa butter, which had been up to about $8000 per ton when the research work began, had fallen by the late 1980s to about $3000 per ton. Since a CBE could only fetch about 60% of this price, this fluctuation left insufficient profit for the process to proceed beyond the pilot-scale level despite the process using a virtually zero-cost feedstock - lactose. Lactose arises as a byproduct of cheese creamery processes in New Zealand where there is so much of it that there are severe problems ensuring its environmentally friendly disposal!

Also taken into consideration when deciding to abandon this yeast CBE-SCO project was uncertainty about the chocolate industry using the product in confectionery products (e.g., for cakes, toppings, etc.) rather than in chocolate for direct consumption. Unease about using a "microbial fat" in chocolate products that depend very much on marketing images for high sales was a telling factor. Thus, with its market uptake uncertain; the presence of adequate, alternative sources of other CBEs (namely, from palm oil fractionation); and the apparent low profitability of the microbial process, another SCO program was then terminated.

SCO for the 21st Century

The Quest for a Docosahexaenoic Acid-Rich SCO

Having established that microorganisms could produce high quality oils and fats at the highest levels of purity and safety—though, admittedly, at a price—it was then a question of identifying which, if any, possible market might be exploited by these materials. Top consideration had to be given to oils that would be appropriate for human consumption rather than for animals, because these would be the markets able to command the highest prices, as had already been seen with the GLA-SCO product. At the same time, oils that could not be readily obtained from plant or animal sources would give additional advantage to a microbial route of production as the ensuing oil would be free from serious competition. With these general considerations in mind, work on the nutritional benefits and effects of the VL-PUFAs found in fish oil was of major importance.

There has been a steady investigation of the possible dietary benefits of fish oil since the pioneering work of Sinclair in the 1940s (see Ewin, 2001). However, the findings that received international recognition arose from reports of Danish scientists investigating the reasons why cardiovascular problems seemed non-existent, or to affect significantly fewer people, in Greenland Eskimos compared with other populations and in spite of the very high intake of fat by this group (Haraldsson & Hjaltason, 2001). A low incidence of heart disease in other fish-eating populations—e.g., the Norwegians and the Japanese—also helped to focus attention on the importance of docosahexaenoic acid (DHA; 22:6n-3) and eicosapentaenoic acid (EPA; 20:5n-3) as the two major PUFAs of fish oils. By the 1980s, the importance of both EPA and DHA for human nutrition was established, and then, in the 1990s, the particularly beneficial effects of DHA during pregnancy for the nutrition of premature and full-term babies began to appear (see Haraldsson & Hjaltason, 2001; Huang & Sinclair, 1998). The presence of DHA and arachidonic acid (ARA) in the mother's milk and their occurrence as the major FA of brain lipids and retinal membrane lipids reinforced the concept that it would be highly beneficial if both of these FA could be included in the diet of pregnant women and in formulas designed for the neonatal baby. A more complete account of the nutritional advantages of DHA and ARA is covered in Chapters 16–19.

Since it is DHA, rather than EPA, that is considered important for infant nutrition, this finding meant that fish oils were not entirely satisfactory sources, because all these oils contained both FAs in roughly equal proportions (see Breivik, 2007); EPA was not, however, a "neutral" material that could be taken along with DHA. It appeared to interfere metabolically with the efficacy of DHA uptake and its incorporation into brain and retinal lipids and, thus, was contra-indicated (Craig-Schmidt & Huang, 1998). Tuna oil, though, appeared to be an exception when compared to most other fish oils; it has a DHA to EPA ratio of 4:1 (Haraldsson & Hjaltason, 2001), which is about the same ratio as in a mother's milk. But tuna oil was clearly in short, if not diminishing, supply and, in any case, it did still have some EPA. The only solution that seemed appropriate was to embark on a very expensive process to fractionate DHA from fish oil. This route would require several steps culminating in the use of preparative level, high performance liquid chromatography (HPLC) which, by its very nature, was prohibitively expensive. No other source of DHA seemed available to nutritionists during the early 1990s.

Nutritionists, however, are not microbiologists and tend not to bother about microbial lipids or to know much about their composition, except for the recognition that some marine microorganisms do contain DHA, but usually in association with EPA. It did not seem apparent to any nutritionist in the late 1980s that microorganisms could be the key to providing a supply of DHA. It took someone who was aware of both the need for a good supply of DHA-rich oil and, simultaneously, had knowledge of the FA composition of key microorganisms to put literally two and two together and identify a potentially useful microbial source of DHA. This discovery was a major breakthrough and was pioneered by David Kyle, who launched a company, Martek Corp., in the late 1980s that focused exclusively on developing a process using *Crypthecodinium cohnii* as the organism of choice for DHA production. *Crypthecodinium cohnii* was already known as a producer of an oil that contained no PUFA other than DHA (Harrington & Holz, 1969; Beach & Holz, 1973), but it was not apparent that it could be grown in very large scale fermentors to produce sufficient biomass to warrant considering it as a commercial source of oil. Kyle and his colleagues, in a remarkably short period of time, demonstrated that this production was feasible, and they went on to produce this oil which has since had a major impact on the infant nutrition market. A detailed account of the current process for producing DHASCO™ using *C. cohnii* is given in Chapter 6. Suffice to say, for the purposes of this overview, this oil now enjoys worldwide sales and its production is measured in thousands of tons per year, making it the world's most valuable SCO.

An ARA-Rich SCO

DHA, as indicated previously, was not the only FA that appeared to be important to infant nutrition. The other FA was ARA (for review, see Craig-Schmidt & Huang, 1998). By a happy coincidence, a microbial source of ARA was already known from the work of Shimizu in Japan (Chapter 2) using the Zygomycetes fungus, *Mortierella alpina* (Yamada et al., 1987, 1992). However, the use of this oil for infant nutrition had not been considered, and the opportunity of exploiting this technology independently from the Japanese work was then undertaken, again by Martek Corp.

Martek, it has to be said, was the only company that recognized that the infant formula market needed a secure and safe supply of VL-PUFAs and appreciated that this demand could not be met by using fish oils. Microbial oils were the only realistic source of such VL-PUFAs. The foresight of this company was due to the recognition that both DHA and ARA could have huge markets even if only10% of all newborn babies were fed on enriched infant formulas and even if only 1% of the weight of the formula was DHA/ARA. Multiply 0.1% of all infant formulas that are produced in the

USA and Europe (not to mention in the 100+ other countries in which the product is now sold), and the potential of SCO for this market can be quickly appreciated.

Microbial oils rich in both of these FAs (ARA and DHA) are now the main SCOs in current production. Both processes were developed by Martek, though the one for ARA was further developed by DSM (Chapter 5) working under license from Martek. Both processes began at a commercial level in the 1990s (Kyle, 1996,1997), and both look as if they will continue their expansion during the second decade of this century. In all probability, they will continue to dominate the market for both DHA and ARA for some time to come, as it is highly unlikely that the demand for these VL-PUFA will diminish. Indeed, all the indications are that the demand for both FA will continue to grow until, possibly, there will be no infant formulas produced in Western countries that will be without both materials. The only change would be the rather unlikely event of a significantly higher proportion of mothers choosing to breast-feed their babies rather than opting to formula-feed.

The FA profiles of the SCO in current commercial production are given in Table 1.C.

Table 1.C. Fatty Acid Profiles (given as rel. % w/w) of SCO in Current Production[a].

A. Arachidonic Acid-SCO Processes Using *Mortierella alpina* strains

	14:0	16:0	18:0	18:1	18:2	18:3 (n–6)	20:3 (n–6)	20:4 (n–6)	22:0	24:0
DSM process[b]	0.4	8	11	14	7	4	4	49	—	1
Wuhan Alking process[c]	0.2	6.3	2.2	3.7	4.0	1.6	—	70.2	2.7	5.3

B. Docosahexaenoic Acid -SCO processes

	12:0	14:0	16:0	16:1	18:0	18:1	18:2	18:3 (n–3)	20:3 (n–6)	22:5 (n–6)	22:6 (n–3)
Martek process[d] (DHASCO™)	4	20	18	2	0.4	15	0.6	—	—	—	39
OmegaTech process[e] (DHASCO-S)	—	13	29	12	1	1	2	3	1	12	25
Nutrinova process[f] (DHA-FNO)	—	3	31	—	1	—	—	—	—	11	45

[a]For other abbreviations see Table 1.B.
[b]See chapter 5.
[c]See *Yuan et al.* (2002).
[d]Uses *Crypthecodinium cohnii*, see Chapter 6.
[e]Uses *Schizochytruim* sp., see Chapter 4.
[f]Uses *Ulkenia* sp., see Kiy et al. (2005); now operated by Lonza Ltd, Switzerland. Oil sold as DHA Functional Nutrition Oil.

Other Sources of PUFA-SCO

DHA-Rich Oils

Not unexpectedly, once the DHA-SCO and ARA-SCO oils were announced, other possible microbial sources of these materials were examined. An account is given in Chapter 4 of the process developed by Omega Tech Inc., Boulder CO, to produce an oil rich in DHA using a species of *Schizochytrium* (Barclay et al., 1994). Briefly, the oil produced was not a "DHA-only" oil, but it had about 15% of the DHA content as docosapentaenoic acid (DPA; 22:5n-6). This latter FA, although not of the same n-3 family of FA as DHA, was metabolically neutral and did not detract from the efficacy of uptake of DHA into key brain lipids; it did not though add to the DHA content of the oil and, to this extent, diminished the overall efficiency of DHA production in the organism. However, by the time this process was fully launched, the market for a DHA-only oil had been established by the *Crypthecodinium* oil, and its position has proved unimpeachable. The *Schizochytrium* oil should, nevertheless, be considerably less expensive than the former oil as this organism grows about four times faster than *C. cohnii* (Ganuza et al., 2008) and to very high cell densities—cell dry weight values of over 200 g/L, attained after 72 hours of growth, have been attained (Bailey et al., 2003).

The marketing of the *Schizochytrium* oil, unlike the *Crypthecodinium cohnii* oil that is used exclusively in the infant formula market (see Section), is focused primarily on its use in adult nutrition (see Chapter 17). In addition, there may be benefits for including it in animal and fish feeding materials (Abril & Barclay, 1998; see also, Chapters 18 and 19). The opportunity to incorporate either the oil or the biomass containing the oil into feed material for farmed fish could have huge implications. Currently, about 5 tons of fishmeal are needed to bring one ton of fish to maturity in these fish-farms—the exact amount varies with the species of fish being farmed. This process is clearly non-sustainable, and alternatives to fishmeal are being actively sought. Since the key ingredient of fishmeal for growth and development, especially in the very earliest stages of fish growth (for larvae and fish fry), is the VL-PUFA, then an alternative source of DHA would be extremely attractive. Although the costs of producing *Schizochytrium* biomass (for fish feeding, one need not extract the oil, but, instead, the whole biomass can be used) are considerably less than producing *Crypthecodinium* biomass, this biomass still appears much more costly than fishmeal itself. Nevertheless, it is a sustainable source of DHA. If *Schizochytrium* biomass/ DHA turns out not to be, ultimately, too prohibitive in price, governments or regulatory agencies in some countries or regions may choose to ban fishmeal, or at least to place a moratorium on its use, in favor of a sustainable, alternative source of VL-PUFA for fish feeding.

A further reason for a move away from using fishmeal for fish feeding is the presence of various residues of man-made chemicals that have entered the world's oceans and seas. These toxins include dioxins and polychlorobiphenyls (PCB), as well as organo-heavy metals, which include mercury compounds. The presence of such materials is already too high to allow fish oils to be given as dietary supplements to infants in the USA.

Further markets for SCOs are also likely to be developed for other food uses and for which either the oil itself or the biomass could be used. *Schizochytrium* biomass, containing high levels of DHA, has already been successfully incorporated into poultry feed to produce DHA-enriched eggs, which have been a minor marketing success in the form of DHA-Gold eggs (Abril & Barclay, 1998). DHA-enriched milk and milk-derived products (e.g., cheeses, yogurts, etc.) and other food products are obvious extensions of this concept. It may be expected that over the next decade or two there will be a growing appreciation of the need for PUFAs, such as DHA and, perhaps, ARA, in adult nutrition, in addition to their use in infant foods. The development of a whole range of DHA-supplemented foods—from enriched breads, cereal foods, and margarines to salad dressings—is then entirely feasible. The use of oils or biomass from organisms, such as *Schizochytrium*, is then bound to rise, and rise quite sharply, should these predictions be fulfilled.

It is also evident that other microbial sources of DHA are already being developed and considered as additional commercial sources of DHA-rich oils and DHA-rich biomasses to meet expected increases in the market for PUFAs. Processes using marine organisms, referred to variously as *Ulkenia* or *Labyranthula*, are under development in Japan (Tanaka et al., 2003; Yokochi et al., 2002). Lonza Ltd., in Switzerland, has developed a full-scale process for DHA production using an *Ulkenia* sp. that has been reviewed by Kiy et al. (2005)—see also Table 1.C. This oil is sold both as Lonza DHA FNO (Functional Nutrition Oil) and as Lonza DHA CL (Clear Liquid). There is, thus, an apparent profusion, and, perhaps, also confusion, of taxonomic names in this area. Because of the relative novelty of these DHA-producing organisms, there is still discussion about their various names; for example, the genus *Schizochytrium* is now considered to comprise three different genera: *Schizochytrium* itself, *Aurantiochytrium*, and *Oblongichytrium* (Yokoyama & Honda, 2007). These groups of organisms, together with the previously mentioned ones, would all be included in the broader taxonomic order of the Labyrithulomycetes (see Leander & Porter, 2001). All labyrinthula are then of current interest because of their abilities to produce DHA (Raghukumar, 2008; Jakobsen et al., 2008). Most species would seem to produce both DHA and DPA (see Table 1.C), but some isolates have been reported that produce only DHA (Kumon et al., 2006); there is even one that produces the usually minor fatty acid, DPA, as its sole PUFA (Kumon et al., 2003).

ARA-Rich Oils

Alternative microbial sources of ARA are also being sought. A process for ARA production in China using a new strain of *Mortierella alpina* has been reported (Yuan et al., 2002). This process appears to operate at the 50-100 ton level (50,000-100000 L). The process was originally developed by Wuhan Alking Bioengineering Co., Ltd., but is now a joint venture with Cargill Ltd., which is known as Cargill Alking Bioengineering (Wuhan) Co., Ltd. The percent of ARA in the oil was originally given as 73%—see Table 1.C—but more recent indications suggest that the level of ARA is probably not in excess of 50%. Other companies in China, such as Franken Biochemical Co. Ltd, Shandong, and Deyang Huatai Biopharmaceutical Resources Co. Ltd, may also be involved in ARA production. Work also appears to be ongoing in identifying new organisms of interest for ARA production: a new strain of *Mortierella alliacea* has been reported with contents of ARA similar to those found in *M. alpina* of over 40% (Aki et al., 2001), and recent work (reviewed in Chapter 10) has found a new phototrophic algae, *Parietochloris incisa,* that has the highest content of ARA of any phototrophically grown alga, at nearly 60% of the total FA.

The overall activity in these areas to identify new and, possibly, improved sources of DHA and ARA implies considerable economic potential in these processes. The lucrative nature of the markets will, therefore, continue to attract further interest from established biotechnology companies, and, perhaps, even pharmaceutical companies, all wishing for a share of the revenue.

PUFA-SCO for Clinical Applications

Clinical applications of the VL-PUFA seem, at the moment, to be restricted to the use of EPA, rather than DHA or ARA, which are now regarded as nutraceuticals or dietary supplements. Currently, EPA—usually as its ethyl or methyl ester - is produced by fish oil fractionation and HPLC technology making the final product extremely expensive. There are, however, possible emerging microbial sources of it and methods for its production as an SCO are under active consideration in a number of laboratories around the world (see Chapters 3, 8, and 10).

The potential market size for EPA is difficult to estimate as there is no major, cheap supply of it currently available, and the present market size is, therefore, limited by the high cost of the material. Possible clinical uses of this PUFA are given in greater detail in the introductory material of Chapters 3 and 8. EPA oils appear to alleviate or even cure various illnesses, which include schizophrenia, bipolar disorder, certain cancers, Alzheimer's disease, and atherosclerosis, all of which are currently treated by expensive pharmaceutical drugs. EPA may also be useful in the prevention of coronary events in hypercholesterolaemic patients (Yokoyama et al., 2007). It is worth pointing out that pharmaceutical companies have a vested interest in maintaining the status quo, as they are the sole providers of the expensive medications for the treatment of most of these illnesses and disorders. Pharmaceutical companies will not encourage clinical and medical practitioners to prescribe, or suggest consumption of, a simple, over-the-counter FA that would be much cheaper than a pharmaceutical drug. The efficacy of an EPA-oil, in comparison with a pharmaceutical drug, for the treatment or alleviation of any disorder has yet to be proven. If the treatment of the various disorders that have been mentioned, especially ones involving brain disorders, is going to be the major future market for EPA oils, then the stringency of testing will have to be increased enormously as the claims being made for its efficacy would go beyond the simple one of nutritional benefit. EPA oils would need to be as rigorously tested through appropriate clinical trials as any pharmaceutical drug that they were seeking to replace. Perhaps, for this reason, the acceptance of EPA oils for medical treatments will be slow as the cost of carrying out the necessary clinical trials may be beyond the financial resources of most non-pharmaceutical companies.

There is no doubt, though, that EPA is a very useful anti-inflammatory compound and can be given safely to many types of patients presented with a variety of disorders (see also, Chapter 8), possibly under the guise of a beneficial dietary supplement. Nevertheless, problems about its source remain. Fish oils, for reasons discussed previously, are unlikely to be a satisfactory long-term source of EPA; alternative microbial sources would seem, therefore, to be the preferred option. Up until recently, these sources have appeared to be mainly photosynthetic algae (Wen & Chen, 2005; Ward & Singh, 2005), although it has been known for some time that filamentous fungi, such as *Mortierella alpina* and *Mort. elongata*, can, if grown at low temperatures (12° C), produce EPA as the major PUFA (Shimizu et al., 1988,1989; Bajpai et al.,1992). However, cultivation of these fungi in this way is not a realistic proposition for commercial production; not only would considerable refrigeration costs be incurred to cool the fermentors, but also the process itself is very lengthy (up to 2–3 weeks).

Current developments are therefore seeking to produce EPA using organisms that can grow heterotrophically (i.e., using a fixed carbon source, such as glucose), quickly (3–5 days), and at 28–32° C. Such microorganisms could include some algae (see Chapter 8), but the highest chance of success would appear to be through genetic modification of known oleaginous organisms. An account of the work carried out to modify the yeast, *Yarrowia lipolytica*, in EPA production is given in detail in Chapter 3. This work has been highly successful in attaining its objective although the task of converting the yeast, which normally produces linoleic acid (18:2) as its principal unsaturated fatty acid, into EPA (20:5) production has involved the insertion of a large number of genes into the yeast DNA. An alternative approach has been indicated by the recent work of Ando et al. (2009): *Mort. alpina*, which normally produces ARA (see above), has been genetically modified by the insertion of a single gene (coding for an omega-3 desaturase) so that ARA (20:4) is now converted into EPA (20:5) by the addition of one further double bond at the omega-3 position. The key question, in both of these cases, is to ask whether genetically modified (GM) oils will be accepted by the general public; this question also applies to any GM plant oil that might be produced in the future (see below).

Other PUFA-SCOs suitable for clinical use and, perhaps, for dealing with specific metabolic disorders await market opportunities. Prospects of producing a variety of other PUFAs besides DHA, ARA, and EPA are discussed in Chapter 2; various mutants of *Mort. alpina* have been produced that synthesize useful amounts of stearidonic acid (18:4,n-3), dihomo-γ-linolenic acid (20:3n-6), eicosatrienoic acid (also known as Mead acid; 20:3n-9), eicosatetraenoic acid (20:4n-3), and several other PUFAs. Of these products, only stearidonic acid can be obtained from a plant source (*Echium*). In addition to these FA, DPA (22:5 n-6), which occurs in the DHA-rich oils from *Schizochytrium* spp., is thought to be produced by Nagase-Suntory Co., Ltd. in Japan. Whether this is produced as a by-product of the fractionation of *Schizochytrium* oil or is produced using a specific organism, such as the novel labyrinthulid isolate that was recently reported to produce DPA as its sole VL-PUFA (Kumon et al., 2003), is unknown. The isolate mentioned, though, needs to be cultivated on soybean oil to achieve DPA production; growth on linoleic acid or α-linolenic acid gives very low yields.

The applications and potential market sizes of these unusual oils, either for treatment of various disorders or as dietary supplements, remains uncertain. For the first time, however, sufficient amounts of them can be produced by microbial technology that allows their evaluation to take place. It would not be too surprising if one or more of these VL-PUFAs might be found to be beneficial for the treatment of certain conditions, in which case, their large-scale production would quickly follow.

SCO and Competition from Genetically Modified (GM) Plant Oils

While future prospects for the continued production of various SCOs look extremely strong, there is the undoubted prospect that one or more of the current SCOs may be produced in plants at some future stage. Genetic manipulation of plants for the improvement of certain characteristics has long been underway, and there are now several major industrial companies engaged in attempting to clone key genes in agronomically important plants to convert the existing FA of the oilseeds into ARA, EPA, or DHA. Since none of these FA occurs in an agricultural crop, it is necessary for genes coding for various FA desaturases and elongases (Fig. 1.2) to be taken from microorganisms (as these are the only convenient source of such genes) and inserted into the plant's DNA. These then have to be expressed (i.e., made to work), the resultant proteins have to be catalytically active (i.e., made to do the same job that they did in the original microorganism), but, in addition, they need to work only in the plant seed and, moreover, work only at the time of oil accumulation in the seed. Thus, the right genes have to be in the right place and work at the right time. If the new PUFA were produced throughout the entire plant—in the leaves, stem, and roots—the plant would probably be unable to grow properly and would, most likely, be unable to stand upright!

There are, therefore, an enormous number of problems to be overcome for the successful genetic engineering of VL-PUFAs in plants. Even finding the right genes in a microorganism is not an easy task. As David Kyle points out in Chapter 20, the enzyme reactions to be carried out are complex: both the desaturation and elongation reactions require more than one protein (Fig. 1.2). The simplest solution seems to be for geneticists to try to clone an entire gene sequence from a microorganism; this sequence will then be able to serve as code for the entire set of proteins needed for the synthesis of the new FA. Current progress in these areas has been frustratingly slow because of these difficulties. Further details concerning these problems are presented in Chapter 20 and in a number of recent reviews written on this topic (see, for example, Damude & Kinney, 2007, 2008; Graham et al., 2007).

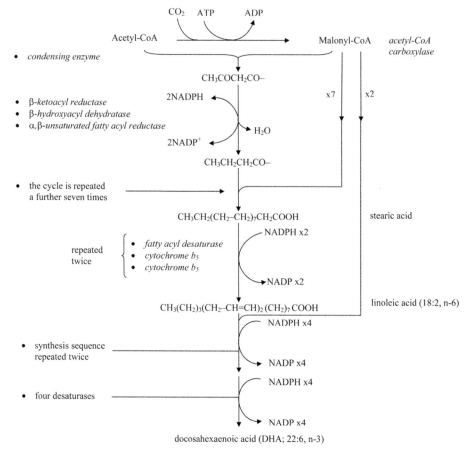

Fig. 1.2. Biosynthesis routes of two fatty acids—linoleic acid (18:2n-6) and docosahexaenoic acid (DHA, 22:6n-3)—showing the requirements for reduced nicotinamide adenine dinucleotide phosphate (NADPH) as the bio-reductant and adenosine triphosphate as the energy supply. The NADPH requirement for DHA biosynthesis is 80% greater than that needed by linoleic acid biosynthesis. (This calculation assumes that DHA is being synthesized by a conventional eukaryotic fatty acid synthetase with accompanying desaturases—but, see Metz et al., 2002, 2009 and Ratledge, 2004 for more details.) In GM plants designed for DHA production, it is not clear how this additional supply of reductant will be provided or even if it can be provided without detriment to the well-being of the plant itself. Abbreviations: ADP, adenosine diphosphate; ATP, adenosine triphosphate; and NADP, nicotinamide adenine dinucleotide phosphate.

There is, however, an additional problem concerning the successful production of GM plants that can synthesize significant amounts of VL-PUFA: the considerable metabolic cost to the plant for producing these materials. All FA are chemically reduced entities containing many methylene (-CH$_2$-) groups; for every acetate group that is used in synthesizing a FA chain, two molecules of the reductant, reduced nicotinamide adenine dinucleotide phosphate (NADPH), are needed to reduce the acetyl group (-CH$_2$CO-) into -CH$_2$CH$_2$- (Fig. 1.2). In addition, for every double bond that is introduced into the FA molecule via a FA desaturase, a further mol of NADPH is needed. Thus, to make one mol of a FA, such as linoleic acid (18:2), 18 mol of NADPH are needed (2×8 for each condensation reaction + 1 for each double bond) (Fig. 1.2). A further 8 mol of adenosine triphosphate (ATP) (as an energy source) are also needed for each condensation reaction, since ATP is needed for the conversion of acetyl-CoA into malonyl-CoA, which is used at each stage of the FA biosynthesis sequence.

For the biosynthesis of DHA, the demand for NADPH rises to 26 mol/mol DHA, and a total of 10 mol of ATP are also needed. [This assumes that DHA is to be synthesized in a GM plant via a conventional eukaryotic FA synthetase system and accompanying desaturases and that it will be produced by the extension and desaturation of the existing FA of the plant. However, there is also the polyketide synthase (PKS) system that produces DHA in *Schizochytrium* that is distinct from the conventional fatty acid synthase (Metz et al., 2002); if this system could be cloned into a plant, it would decrease the NADPH requirement to 20 mol/mol DHA, but the demand for ATP would be unchanged. This cloning would create another problem to be solved as the PKS system generates free fatty acids (Metz et al., 2009) and not thioesters of fatty acids.] DHA synthesis may, thus, require an 80% increase in the supply of NADPH compared to the amount needed for linoleic acid synthesis. It is by no means clear where this extra NADPH is to come from; in microorganisms it comes by the malic enzyme reaction (see Zhang et al., 2007):

$$\text{Malate} + \text{NADPH} \rightarrow \text{Pyruvate} + \text{CO}_2 + \text{NADP}^+$$

The same may well apply to plants in which case it will be necessary to clone additional genes to produce this extra reducing power to achieve high levels of DNA synthesis. Ultimately, of course, the NADPH and ATP have to be generated from the photosynthetic reactions of the plant; most plants that grow in temperate climates are energy-limited by virtue of the availability of sunlight. A GM plant synthesizing appreciable amounts of DHA will be even more energy-limited and, consequently, could not grow to the same extent as the original plant that produces linoleic acid. Energy will have to be taken away from other essential reactions of the cells, and this subtraction will deplete the overall vigor of the plant. Alternatively, the plant may grow normally but then not divert sufficient energy into lipid synthesis so that the production of DHA would fall below expectations. It may, therefore, be necessary to grow a GM crop for DHA production under conditions of high light intensity so that energy production (ATP) and NADPH generation will not be limiting.

All of these factors add up to create an enormous genetic engineering task. Knowing whether a GM plant will be produced within the next 20 years that can synthesize useful amounts of DHA and the other PUFA is, of course, crystal ball-gazing. Several major industrial companies, including BASF, Monsanto, and DuPont, have extensive research activities in this area. From their point of view, it is perhaps not a question of "if" PUFA-GM crops can be produced, but rather "when" they will be created. But we may add a cautionary note: attaining high-yielding, VL-PUFA-producing crops still seems to be as far away as ever. In the late 1990s, it was suggested that such GM crops were only a decade away; a decade later, we are still waiting and still being told to wait another decade! Will they ever arrive?

When (or if) these GM crops are produced, we must ask the very relevant ethical question as to whether the public will accept such materials? Already there is a considerable groundswell of public opinion throughout Europe against all GM crops and this adverse opinion, which is not founded on the basis of any rational scientific argument, is unfortunately spreading in North America, as well as other parts of the world. By the time successful PUFA-GM crops have been created, governments of many countries may have banned their use for human food. On the other hand, let us hope that reason may prevail over prejudiced irrationality.

We stand, thus, at the brink of many exciting developments and some dilemmas. For at least the next two decades, it is more than likely that the supply of key VL-PUFAs (DHA, ARA, and EPA) will be met almost entirely from microbial sources. Work to create GM crops has a long way to go before the first plants are produced that can synthesize useful amounts of a VL-PUFA; it would then probably take a minimum of 8–10 years for the first of these experimental plants to be grown commercially once all the difficulties in achieving high levels of gene expression have been solved.

Predicting future scientific successes (and failures) is a fool's game; after all, who could have predicted the current success of SCO at the beginning of the 1990s (Ratledge, 1992, 1995)? SCO have been successful because they are a product that is not obtainable from other sources and fulfills a primary demand for materials essential for the development and well being of infants, as well as adults. They are a product whose time has come. How long they remain as the prime sources of these materials is, of course, completely uncertain. Should viable, robust GM crops be created that can be cultivated successfully—assuming that their cultivation will be permitted—then there is no doubt that these products will be the cheapest source of desirable PUFA-oils. There is no way that a microbial process requiring a fixed source of carbon and a considerable input of energy to run the large-scale production fermentors could compete with the lower costs of production that correspond to an agricultural oilseed crop. However, there is still a very long way to go to achieve a GM crop that will produce the necessary amount of an VL-PUFA. As always, we look to the future with considerable interest.

References

Abril, R.; W.R. Barclay. Production of docosahexaenoic acid-enriched poultry eggs and meat using an algae-based feed ingredient. *The Return of omega-3 Fatty Acids into the Food Supply I. Land-Based Animal Food Products and Their Health Effects*, Simopoulos, A.P.;, Ed.; S. Karger, Basel, Switzerland, 1998; pp 77–88.

Aki, T.; Y. Nagahata; K. Ishihara; Y. Tanaka; T. Morinaga; K. Higashiyama; K. Akimoto; S. Fujikawa; S. Kawamoto; S. Shigeta; et al. Production of arachidonic acid by filamentous fungus, *Mortierella alliacea* strain YN-15, *J. Am. Oil Chem. Soc.* **2001**, *78*, 599–604.

Ando, A.; Y. Sumida; H. Negoro; D.A. Suroto; J. Ogawa; E. Sakuradani; S. Shimizu. Establishment of *Agrobacterium tumefaciens*-mediated transformation of an oleaginous fungus *Mortierella alpina* 1S-4 and its application for eicosapentaenoic acid-producer breeding. *Appl. Environ. Microbiol.* **2009**, doi:10.1128/AEM.00648-09.

Bailey, R.B.; D. DiMasi; J.M. Hanson; P.J. Mirrasoul; C.M. Ruecher; G.T. Veeder; T. Kaneko; W.R. Barclay. U.S. Patent 6,607,900, **2003**.

Bajpai, P.; P. Bajai; O.P. Ward. Optimization of culture conditions for production of eicosapentaenoic acid by *Mortierella elongata* NRRL 5513. J. *Ind. Microbiol.* **1992**, *9*, 11–18.

Barclay, W.R.; K.M. Meager; J.R. Abril. Heterotrophic production of long chain omega-3 fatty acids utilizing algae and algae-like microorganisms. *J. Appl. Phycol.* **1994**, *6*, 123–129.

Beach, D.H.; G.G. Holz. Environmental influences on the docosahexaenoate content of the triacylglycerols and phosphatidylcholine of a heterotrophic, marine dinoflagellate, *Crypthecodinium cohnii. Biochim. Biophys. Acta* **1973**, *316*, 56–63.

Botham, P.A.; C. Ratledge. A biochemical explanation for lipid accumulation in *Candida* 107 and other oleaginous micro-organisms. *J. Gen. Microbiol.* **1979**, *114*, 361–375.

Boulton, C.A.; C. Ratledge. Use of transition studies in continuous cultures of *Lipomyces starkeyi*, an oleaginous yeast, to investigate the physiology of lipid accumulation. *J. Gen. Microbiol.* **1983**, *129*, 2863–2869.

Breivik, H. Ed. *Long-chain omega-3 specialty oils.* The Oily Press: Bridgwater, UK, 2007.

Bunker, H.J. Fodder yeast plants at I.G. Farbenindustrie, Wolfen. *C.I.O.S. Report* **1945**, *Item 22*, File 29-4. HMSO: London.

Bunker, H.J. The wartime production of food yeast in Germany. *Proc. Soc. Appl. Bacteriol.* **1946**, *1*, 10–14.

Bunker, H.J. Microbial food. *Biochemistry of Industrial Micro-organisms*; Rainbow, C.; A.H. Rose, Eds.; Academic Press: London, 1963; pp 34–67.

Craig-Schmidt, M.C.; M.C. Huang. Interaction of n-6 and n-3 fatty acids: implications for supplementation of infant formula with long-chain polyunsaturated fatty acids. *Lipids in Infant Nutrition*; Huang, Y.-S.; A.J. Sinclair, Eds.; AOCS Press:, Champaign, IL, 1998; pp 63–84.

Damude, H.G.; A.J. Kinney. Engineering oilseed plants for a sustainable, land-based source of long chain polyunsaturated fatty acids. *Lipids* **2007**, *42*, 179–185.

Damude, H.G.; A.J. Kinney. Engineering oilseeds to produce nutritional fatty acids. *Physiol. Plantarum* **2008**, *32*, 1–10.

Davies, R.J. Yeast oil from cheese whey—process development. *Single Cell Oil*; Moreton, R.S., Ed.; Longman-Wiley: London and New York, 1988; pp 99–145.

Davies, R.J. Scale up of yeast technology. *Industrial Applications of Single Cell Oils*; Kyle, D.J.; C. Ratledge, Eds.; AOCS: Champaign, IL, 1992; pp 196–218.

Eroshin, V.K.; A.D. Satroutdinov; E.G. Dedyukhina; T.I. Chistyakova. Arachidonic acid production by *Mortierella alpina* with growth-coupled lipid synthesis. *Proc. Biochem.* **2000**, *35*, 1171–1175.

Eroshin, V.K.; E.G. Dedyukhina; A.D. Satroutdinov; T.I. Chistyakova. Growth-coupled lipid synthesis in *Mortierella alpina* LPM 301, a producer of arachidonic acid. *Mikrobiologiya/Microbiology* **2002,** *71,* 200–204 (English version).

Evans, C.T.; C. Ratledge. The physiological significance of citric acid in the control of metabolism in lipid-accumulating yeasts. *Biotech. Gen. Eng. Rev.* **1985,** *3,* 349–375.

Ewin, J. *Fine Wines & Fish Oil: The Life of Hugh Macdonald Sinclair.* Oxford University Press: Oxford, UK, 2001.

Ganuza, E.; A.J. Anderson; C. Ratledge. High-cell-density cultivation of Schizochytrium sp. in an ammonium/pH-auxostat fed-batch system. *Biotechnol. Lett.* **2008,** *30,* 1559–1564.

Gill, C.O.; M.J. Hall; C. Ratledge. Lipid accumulation in an oleaginous yeast, *Candida* 107, growing on glucose in single-stage continuous culture. *Appl. Environ. Microbiol.* **1977,** *33,* 231–239.

Graham, I.A.; T. Larson, J.A. Napier. Rational metabolic engineering of transgenic plants for biosynthesis of omega-3 polyunsaturates. *Curr. Opin. Biotechnol.* **2007,** *18,* 142–147.

Haraldsson, G.G.; B. Hjaltason. Fish oils as sources of important polyunsaturated fatty acids. *Structured and Modified Lipids*; Gunstone, F.D., Ed.; Marcel Dekker: New York, 2001; pp 313–350.

Harrington, G.W.; G.G. Holz. The monoenoic and docosahexaenoic fatty acids of a heterotrophic dinoflagellate. *Biochim. Biophys. Acta* **1968,** *164,* 137–139.

Huang, Y.-S.; A.J. Sinclair, Eds. *Lipids in Infant Nutrition.* AOCS Press: Champaign, IL, 1998.

Jakobsen A.N.; I.M. Aasen; K.D. Josefsen; A.R Strom. Accumulation of docosahexaenoic acid-rich lipid in thraustochyrid *Aurantiochytrium* sp. strain T66: effects of N and P starvation and O_2 limitation. *Appl. Microbiol. Biotechnol.* **2008,** *80,* 297–306.

Kessell, R.H.J. Fatty acids of *Rhodotorula gracilis*: fat production in submerged culture and the particular effect of pH value. *J. Appl. Bact.* **1968,** *31,* 220–231.

Kiy, T.; M. Rusing; D. Fabritius. Production of docoahexaenoic acid by the marine microalga *Ulkenia* sp. *Single Cell Oils,* 1ˢᵗ ed.; Cohen, Z.; C. Ratledge, Eds.; AOCS Press: Champaign, IL, 2005; pp 99–106.

Kumon, Y.; R. Yokoyama; T. Yokochi; D. Honda; T. Nakahara. A new labyrinthulid isolate, which solely produces n-6 docosapentaenoic acid. *Appl. Microbiol. Biotechnol.* **2003,** *63,* 22–28.

Kumon, Y.; R. Yokoyama; Z. Haque; T. Yokochi; D. Honda; T. Nakahara. A new labyrinthulid isolate that produces only docosahexaenoic acid. *Mar. Biotechnol.* **2006,** *8,* 170–177.

Kyle, D.J. Production and use of a single cell oil which is highly enriched in docosahexaenoic acid. *Lipid Technol.* **1996,** *8,* 107–110.

Kyle, D.J. Production and use of a single cell oil highly enriched in arachidonic acid. *Lipid Technol.* **1997,** *9,* 116–121.

Leander, C.A.; D. Porter. The Labyrinthulomycota is comprised of three distinct lineages. *Mycologia* **2001,** *93,* 459–464.

Ledingham, G.A.; D.H.F. Clayson; A.K. Balls. Production of *Oidium lactis* on waste sulphite liquor. *B.I.O.S. Final Report* **1945,** *236,* Report III, 31–44. His Majesty's Stationery Office: London.

Lindner, P. Das Problem der Biologischen Fettbildung und Fettgewinnung. *Z. Angew. Chem.* **1922,** *35,* 110–114.

Metz, J.G.; P. Roessler,; D. Facciotti; et al. Production of polyunsaturated fatty acids by polyketide synthases in both prokaryotes and eukaryotes. *Science,* **2001,** *293,* 290–293.

Metz, J.G.; J. Kuner; B. Rosenzweig; J.C. Lippmeier; P. Roessler; R. Zirkle. Biochemical characterization of polyunsaturated fatty acid synthesis in *Schizochytrium:* release of the products as free fatty acids. *Plant Physiol. Biochem.* **2009,** *47,* 472–478.

Moreton, R.S. Physiology of lipid accumulating yeasts. *Single Cell Oil;* Moreton, R.S., Ed.; Longman-Wiley: London and New York, 1988; pp 1–32.

Nageli, C.; C. Loew. Ueber die Chemische Zusammensetzung der Hefe. *Ann. Chem.* **1878,** *193,* 322–348.

Nakahara, T.; T. Yokocki; Y. Kamisaka; O. Suzuki. Gamma-Linolenic acid from genus Mortierella. *Industrial Applications of Single Cell Oils;* Kyle, D.J.; C. Ratledge, Eds.; AOCS: Champaign, IL, 1992; pp 61–97.

Raghukumar, S. Thraustochytrid marine poroists: production of PUFAs and other emerging technologies. *Mar. Biotechnol.* **2008,** *10,* 631–640.

Ratledge, C. Growth of moulds on a fraction of n-alkanes predominant in tridecane. *J. Appl. Bact.* **1968,** *31,* 232–240.

Ratledge, C. Microbial production of oils and fats. *Food from Waste;* G.G. Birch; K.J. Parker; J.T. Worgan, Eds.; Applied Science Publishers: UK, 1976; pp 98–113.

Ratledge, C. Microbial lipids: commercial realities or academic curiosities. *Industrial Applications of Single Cell Oils;* Kyle, D.J.; C. Ratledge, Eds.; AOCS: Champaign, IL, 1992; pp 1–15.

Ratledge, C. Single cell oils—have they a biotechnological future? *Trends Biotech.* **1995,** *11,* 278–284.

Ratledge, C. Fatty acid biosynthesis in microorganisms being used for single cell oil production. *Biochemie* **2004,** 86, 807–815.

Ratledge, C. Microbial production of γ-linolenic acid. *Handbook of Functional Lipids;* Akoh, C., Ed.; CRC Press LLC: Boca Raton, FL, 2006; pp 19–45.

Ratledge, C.; J.P. Wynn. The biochemistry and molecular biology of lipid accumulation in oleaginous microorganisms. *Adv. Appl. Microb.* **2002,** *51,* 1–51.

Shaw, R. The occurrence of gamma-linolenic acid in fungi. *Biochim. Biophys. Acta* **1965,** *98,* 230–237.

Shaw, R. The polyunsaturated fatty acids of microorganisms. *Adv. Lipid Res.* **1966a,** *4,* 107–174.

Shaw, R., The fatty acids of phycomycete fungi, and the significance of the γ-linolenic acid component,. *Comp. Biochem. Physiol.* **1966b,** *18,* 325–331.

Shimizu, S.; H. Kawashima; Y. Sinmen; K. Akimoto; H. Yamada. Production of eicosapentaenoic acid by *Mortierella alpina* 1S-4. *J. Am. Oil Chem. Soc.* **1989,** *66,* 237–241.

Shimizu, S.; Y. Shinmen; H. Kawashima; K. Akimoto; H. Yamada. Fungal mycelia as a novel source of eicosapentaenoic acid: activation of enzyme(s) involved in eicosapentaenoic acid production at low temperature. *Biochem. Biophys. Res. Commun.* **1988,** *150,* 335–341.

Smit, H.; A. Ykema; E.C. Verbree; I.I.G.S. Verwoert; M.M. Kater. Production of cocoa butter equivalents by yeast mutants. *Industrial Applications of Single Cell Oils;* Kyle, D.J.; C. Ratledge, Eds.; AOCS: Champaign, IL, 1992; pp 185–195.

Tanaka, S.; T. Yaguchi; S. Shimizu; T. Sogo; S. Fujikawa. U.S. Patent 6,509,179, **2003.**

Ward, O.P.; A. Singh. Omega-3/5 fatty acids: alternative sources of production. *Proc. Biochem.* **2005,** *40,* 3627–3652.

Wen, Z.; F. Chen. Prospects for eicosapentaenoic acid production using microorganisms. *Single Cell Oils,* 1ˢᵗ edn.; Cohen, Z.; C. Ratledge, Eds.; AOCS Press, 2005; pp 138–160.

Woodbine, M. Microbial fat: Microorganisms as potential fat producers. *Prog. Ind. Microb.* **1959,** *1,* 179–245.

Yamada, H.; S. Shimizu; Y. Shinmen. Production of Arachidonic Acid by *Mortierella* fungi. *Agric. Biol. Chem.* **1987,** *51,* 785–790.

Yamada, H.; S. Shimizu; Y. Shinmen; K. Akimoto; H. Kawashima; S. Jareonkitmongkol. Production of dihomo-γ-linolenic acid, arachidonic acid and eicosapentaenoic acids by filamentous fungi. *Industrial Applications of Single Cell Oils*; Kyle, D.J.; C. Ratledge, Eds.; AOCS: Champaign, IL, 1992; pp 118–138.

Yokochi, T.; T. Nakahara; M. Yamaoka; R. Kurane. U.S. Patent 6,461,839, **2002.**

Yokoyama, R.; D. Honda. Taxonomic rearrangement of the genus *Schizochytrium* sensu lacto based on morphology, chemotaxonomic characteristics, and 18S rRNA gene phylogeny (Thraustochytriacease, Labyrinthulomycetes): emendatin for *Schizocytrium* and erection of *Aurantiochytrium* and *Oblongichytrium* gen. nov. *Mycoscience,* **2007,** *48,* 199–211.

Yokoyama, M.; H. Origasa; M. Matsuzaki; Y. Matsuzawa; Y. Saito; Y. Ishikawa; S. Oikawa; J. Sasaki; H. Hishida; T. Itakura; et al. Effects of eicosapentaenoic acid on major coronary events in hypercholesterolaemic patients (JELIS): a randomised open-label, blinded endpoint analysis. *Lancet* **2007,** *369,* 1090–1098.

Yuan, C.; J. Wang; Y. Shang; G. Gong; J. Yao; Z. Yu. Production of arachidonic acid by *Mortierella alpina* I_{49}-N_{18}. *Food Technol. Biotechnol.* **2002,** *40,* 311–315.

Zhang, Y.; I.P. Adams; C. Ratledge. Malic enzyme: the controlling activity for lipid production? Overexpression of malic enzyme in *Mucor circinelloides* leads to a 2.5-fold increase in lipid accumulation. *Microbiology* **2007,** *153,* 2013–2025.

PART 2

Production of Single Cell Oils Using Heterotrophically Grown Microorganisms

2

Arachidonic Acid-Producing *Mortierella alpina:* Creation of Mutants, Isolation of the Related Enzyme Genes, and Molecular Breeding

Eiji Sakuradani,[a] Akinori Ando,[b] Jun Ogawa,[b] and Sakayu Shimizu[a]
[a]Division of Applied Life Sciences, Graduate School of Agriculture, Kyoto University, Kitashirakawa-oiwakecho, Sakyo-ku, Kyoto 606-8502, Japan
[b]Research Division of Microbial Sciences, Kyoto University, Kitashirakawa-oiwakecho, Sakyo-ku, Kyoto 606-8502, Japan

Introduction

Polyunsaturated fatty acids (PUFAs) play important roles not only as structural components of membrane phospholipids but also as precursors of the eicosanoids of signaling molecules, including prostaglandins, thromboxanes, and leukotrienes, which are essential for all mammals. Fish oils, animal tissues, and algal cells are conventional, relatively rich sources of C20 PUFAs, which are not present in plants. For practical purposes, however, these conventional sources are not satisfactory with regard to either the lipid contents or the PUFA contents of the resultant lipids. To find more suitable sources of PUFAs in microbes, the first attempts at PUFA production with γ-linolenic acid (GLA, 18:3n-6) as the target were performed in the UK (Ratledge, 1992) and Japan (Suzuki et al., 1981), with *Mucor* fungi being used. Since then, various PUFAs have been studied with the aim of effective production. For example, arachidonic acid (AA, 20:4n-6), dihomo-γ-linolenic acid (DGLA, 20:3n-6), and Mead acid (MA, 20:3n-9) are now commercially produced by using *Mortierella* fungi (Certik et al., 1998; Certik & Shimizu, 1999; Sakuradani et al., 2005b; Shimizu & Yamada, 1990; Yamada et al., 1992), and docosahexaenoic acid (22:6n-3), docosapentaenoic acid (22:5n-6), and eicosapentaenoic acid (EPA, 20:5n-3) by using marine microorganisms, Labyrinthulae, and microalgae (Certik & Shimizu, 1999; Kyle et al., 1992; Nakahara et al., 1996; Raghukumar, 2008; Ratledge, 2004; Singh & Ward, 1997; Spolaore et al., 2006; Yazawa et al., 1992).

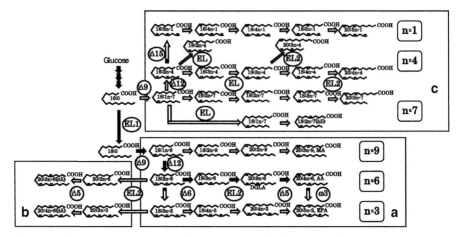

Fig. 2.1. Pathways for the biosynthesis of PUFAs in *Mortierella alpina* 1S-4 and its mutants.

Although success in this area over the last 25 years has generated much interest in the development of microbial fermentation processes, manipulation of the lipid compositions of microorganisms requires new biotechnological strategies to obtain high yields of the desired PUFAs.

The genus *Mortierella* has been shown to be one of the promising single cell oil (SCO) sources rich in various types of C20 PUFAs (Amano et al., 1992; Shimizu & Jareonkitmongkol, 1995), after several *Mortierella* strains were reported to be potential producers of AA in 1987 (Totani & Oba, 1987; Yamada et al., 1987). In particular, several *Mortierella alpina* strains have been extensively studied for the practical production of AA (Shinmen et al., 1989). Some of them are now used for the commercial production of SCO rich in AA. Among them, *M. alpina* 1S-4 has a unique ability to synthesize a wide range of fatty acids and has several advantages not only as an industrial strain but also as a model for lipogenesis studies. The biosynthetic pathways for n-9, n-6, and n-3 PUFAs in *M. alpina* 1S-4 are shown in Fig. 2.1a. The main product of the strain, AA, is synthesized through the n-6 pathway, which involves Δ12 and Δ6 desaturases, elongase (EL2), and Δ5 desaturase. Depending on the conditions, the total amount of AA varies between 3 and 20 g/L (30–70% of the total cellular fatty acids), with 70–90% of the AA produced being present as triacylglycerols (Higashiyama et al., 1998, 2002; Shimizu et al., 2003b).

Here, we describe recent progress in the breeding of commercially important arachidonic acid-producing *M. alpina* strains, particularly approaches for creating desaturase and elongase mutants with unique pathways for PUFA biosynthesis involving conventional chemical mutagenesis and modern molecular genetics. Such mutants are useful not only for the regulation and overproduction of valuable PUFAs but also as excellent models for the elucidation of fungal lipogenesis.

Derivation of Mutants from *M. alpina* 1S-4

A wide variety of mutants defective in desaturases ($\Delta 9$, $\Delta 12$, $\Delta 6$, $\Delta 5$, and n-3) or elongase (EL1) or mutants with enhanced desaturase activities ($\Delta 6$ and $\Delta 5$) have been derived from *M. alpina* 1S-4 by treating the parental spores with *N*-methyl-*N'*-nitro-*N*-nitrosoguanidine (Jareonkitmongkol et al., 1992c). In addition, diacylglycerol-accumulating mutants and several lipid-excretive ones have been isolated by the same method. They are valuable both as producers of useful PUFAs (novel or already existing) and for providing valuable information on PUFA biosynthesis in this fungus (Certik et al., 1998). The main features of these mutants grown on glucose and the biosynthesis of various types of PUFAs by them are outlined below (see also Fig. 2.2).

Δ9 Desaturase-Defective Mutants

Stearic acid (18:0) is the main fatty acid in the mycelial oil (up to 40%) produced by these mutants (Jareonkitmongkol et al., 2002). However, $\Delta 9$ desaturase is not

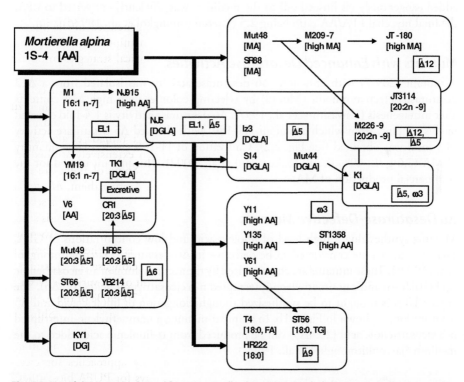

Fig. 2.2. List of the mutants derived from *Mortierella alpina* 1S-4. Symbols in squares indicate apparent mutation sites. Fatty acids in square brackets are major fatty acids produced by the mutants. Abbreviations: arachidonic acid, AA; diacylglycerol, DG; dihomo-γ-linolenic acid, DGLA; free fatty acid, FA; Mead acid, MA; and triacylglycerol, TG.

completely blocked. A total blockage would be lethal since low activity of the enzyme is necessary for introduction of the first double bond at the ninth carbon (from the carboxyl end) of the fatty acid chain to maintain cell viability (see next section).

Δ12 Desaturase-Defective Mutants

The attributes of Δ12 desaturase-defective mutants include the absence of n-6 and n-3 PUFAs and high levels of n-9 PUFAs, such as oleic acid (18:1n-9), octadecadienoic acid (18:2n-9), eicosadienoic acid (20:2n-9), and MA, in their mycelia (Jareonkit-mongkol et al., 1992a). Cultivation of these mutants under the optimal conditions yields a unique oil rich in large quantities of MA. However, the addition of either n-6 or n-3 fatty acids causes a rapid decrease in n-9 fatty acid formation by these mutants and an increase in the AA or EPA level, respectively, because of the substrate specific-ity of Δ6 desaturase, which prefers linoleic acid, α-linolenic acid, and oleic acid, in that order (Jareonkitmongkol et al., 1993d). Therefore, the same mutants can be used for the production of an EPA-rich oil with a low AA level. α-Linolenic acid, when added exogenously (as linseed oil) to the medium, was efficiently converted to EPA, the final mycelial EPA/AA ratio being 2.5 (Jareonkitmongkol et al., 1993d).

Mutants with Enhanced Desaturase Activities

A mutant (209-7) with enhanced Δ6 desaturase activity was isolated from a Δ12 desaturase-defective mutant (Mut48) by selecting colonies with high MA contents after mutagenesis (Kawashima et al., 1997). Δ6 desaturase activity is 1.4-fold elevated in this mutant, from which a mutant (JT-180) with elevated Δ5 desaturase activity (3.3-fold) was obtained (see Fig. 2.2). Cultivation of JT-180 yields a large quantity of MA (2.6 g /L, 49% in oil) (Sakuradani et al., 2002). This mutant is used for the commercial production of MA.

Δ6 Desaturase-Defective Mutants

Mutants synthesizing a high level of linoleic acid and low concentrations of GLA, DGLA, and AA are considered to be defective in Δ6 desaturase (Jareonkitmongkol et al., 1993c). These mutants are characterized by the accumulation of an eicosadienoic acid (20:2n-6) and a nonmethylene-interrupted n-6 eicosatrienoic acid (20:3Δ5). The latter PUFA is thought to be synthesized though elongation of linoleic acid and Δ5 desaturation, as shown in Fig. 2.1b. In a similar manner, a nonmethylene-interrupted n-3 eicosatrienoic acid (20:4Δ5) can be produced from α-linolenic acid added to the medium (Jareonkitmongkol et al., 1993c).

Δ5 Desaturase-Defective Mutants

The fatty acid profiles of these mutants are characterized by a high DGLA level and a reduced concentration of AA (Jareonkitmongkol et al., 1993b). Production of DGLA by these mutants is advantageous because it does not require inhibitors and the yield is relatively high (4.1 g/L, 42% in oil; AA content, <1%) (Jareonkitmongkol et al., 1992b, 1993a, 1993b; Kawashima et al., 2000). One of these mutants is used for the commercial production of DGLA.

Double Mutants Defective in Both Δ12 and Δ5 Desaturases

These mutants accumulate 20:2n-9 in large quantities (Kamada et al., 1999); however, the conversion of oleic acid (endogenously produced from glucose) to 20:2n-9 hardly proceeds when the medium contains α-linolenic acid, which is preferentially converted to 20:4n-3 (Kawashima et al., 1998).

n-3 Desaturase-Defective Mutants

These mutants are unable to synthesize n-3 PUFAs when grown at low temperatures (<20°C) (Jareonkitmongkol et al., 1994; Sakuradani et al., 2004a). The wild strain usually gives the highest AA yield at 20°C, although a part of the AA formed is further converted to EPA and the resultant oil includes a small amount of EPA (ca 3% in oil). Therefore, these mutants are superior to the wild strain when SCO with a relatively higher content of AA is needed (Sakuradani et al., 2004a).

Elongase (EL1 for the Conversion of Palmitic Acid to Stearic Acid)-Defective Mutants

The fatty acid profiles of the mutants with low EL1 activity are characterized by high levels of palmitic acid (16:0) and palmitoleic acid (16:1n-7) with small amounts of various kinds of n-7 and n-4 PUFAs, which are not detectable in the wild strain. The total content of these PUFAs in the oil reaches about 30%. These PUFAs are thought to be derived from the palmitoleic acid accumulated through the n-7 and n-4 pathways, respectively, as shown in Fig. 2.1c (Sakuradani et al., 2004c; Shimizu et al., 2003b). In a similar manner, n-1 PUFAs can be produced from n-1 hexadecaenoic acid (16:1n-1) added to the medium (see Fig. 2.1c) (Shimizu et al., 2003a, 2003b). Therefore, palmitoleic acid corresponds to oleic acid, and the n-7, n-4, and n-1 pathways to the n-9, n-6, and n-3 pathways, respectively, in the wild strain. In Fig. 2.1c, it is not clear whether EL1 or EL2 elongates unsaturated C16 fatty acids to the corresponding C18 fatty acids, because EL1 activity is not completely blocked.

Diacylglycerol-Accumulating Mutant

Triacylglycerols in *M. alpina* 1S-4 account for 90% of the total lipids, whereas mutant KY1 derived from the wild strain accumulates 30% diacylglycerols in the total lipids. This mutant may be defective in an acyltransferase involved in the conversion of diacyglycerols to triacylglycerols. KY1 is expected to be a producer of diacylglycerols rich in C20 PUFAs (Sakuradani et al., 2004b).

Lipid-Excretive Mutants

In a process for the formation of lipid particles, it is hypothesized that triacylglycerols accumulate between the membrane bilayer of the endoplasmic reticulum, and then formed particles containing triacylglycerols are released into the cytosol. The lipid particles are thought to consist of a great number of triacylglycerols and to be overlaid with a lipid monolayer. Mutant V6, excreting lipid particles outside of its filaments, was first isolated on the screening of mutants (Ueda et al., 2001). While this mutant shows the same lipid productivity and fatty acid composition as the wild strain, the excreted lipids are accumulated in this medium. The morphological form of V6 cultivated on a solid medium is very different from that of the wild strain: many lipid particles containing triacylglycerols are observed on the surface of V6 mycelia. On investigating the effects of stabilizers of osmotic pressure, compounds related to the cell wall synthesis, and inhibitors of cell wall synthesis on the lipid excretion by V6, we observed repression of lipid excretion and the formation of intact mycelia in V6. We assume that V6 excretes accumulated lipids from its mycelia due to a defective cell wall structure caused by a mutation in the metabolic pathways for cell wall synthesis (Ando et al., 2006). On the other hand, when V6 is cultivated in a liquid medium, the extracellular production of lipids increases with the cultivation period. Although extracellular production depends on cultural conditions, it reaches 10-40% of the total lipids. The extracellular lipids are not completely separated from water in the liquid medium but are dispersed uniformly as small particles of about 2 μm in diameter. This phenomenon indicates that extracellular lipid particles are overlaid by a surface-active membrane. The polar lipids comprising the particle membranes were confirmed to be phospholipids, sterol derivatives, and cerebrosides (Ioka et al., 2003). In addition, some lipid-excretive mutants were newly isolated from desaturase/elongase-defective mutants. The lipid-excreting mechanism and the characteristic of the extracellular lipid particles remain unclear.

Analysis of Enzyme Genes Involved in PUFA Biosynthesis in *M. alpina*

The genes encoding fatty acid desaturases and elongases for C20 PUFA biosynthesis and enzymes in the electron transport system for fatty acid desaturation from *M. alpina* 1S-4 (shown in Table 2.A) were characterized, as described below.

Δ9 Desaturase

The Δ9 desaturase (Δ9-1) in *M. alpina* 1S-4 has a cytochrome, b_5-like domain linked to its carboxyl terminus, as also seen for yeast Δ9 desaturase (Sakuradani et al., 1999c). *Mortierella* Δ9-1 exhibits 45% and 34% amino acid sequence similarity with those of *Saccharomyces cerevisiae* and rat, suggesting that the *Mortierella* Δ9-1 is a membrane-bound protein of the acyl-CoA type. The *Mortierella* Δ9-1 genomic gene has only one intron, with a GC end occurring at the 5' terminus instead of a GT end, which

Table 2.A. Isolation of Enzyme Genes Involved in PUFA Biosynthesis by *M. alpina* 1S-4.

Type	Isozyme	Substrate	Product
Δ9	Δ9-1	18:0	18:1n-9
	Δ9-2	18:0	18:1n-9
	Δ9-3 (ω9)	24:0, 26:0	24:1n-9, 26:1n-9
Δ12		18:1n-9	18:2n-6
Δ6	Δ6-1	18:2n-6	18:3n-6
	Δ6-2	18:2n-6	18:3n-6
Δ5		20:3n-6	20:4n-6
ω3		n-6 PUFA	n-3 PUFA
EL1		16:0	18:0
EL2		18:3n-6	20:3n-6
MAELO		20:0, 22:0	22:0, 24:0
CbR	CbR-1	—	—
	CbR-2	—	—
Cyt. b_s		—	—

is generally found in the introns of eukaryotic genes. A full-length cDNA clone was expressed under the control of the *amyB* promoter in *Aspergillus oryzae*, resulting in drastic changes in transformant cell fatty acid composition; the presence of palmitoleic acid and oleic acid increased significantly with accompanying decreases in those of palmitic acid and stearic acid.

Multiple Δ9-desaturases (Δ9-2 and Δ9-3) other than Δ9-1 are present in *M. alpina* ATCC 32222 (MacKenzie et al., 2002) and *M. alpina* 1S-4 (Abe et al., 2006). Both Δ9-1 and Δ9-2 desaturate stearic acid to oleic acid, whereas Δ9-3 desaturates a very long saturated fatty acid (26:0) to the corresponding monosaturated fatty acid (26:1n-9) (MacKenzie et al., 2002). The genes encoding Δ9-2 and Δ9-3 are not transcribed as much as that encoding Δ9-1 in the wild strain.

Δ12 and ω3 Desaturases

A cDNA for Δ12 desaturase has been cloned from *M. alpina* 1S-4 (Sakuradani et al., 1999a) based on the sequence information of ω3 desaturase genes (from *Brassica napus* and *Caenorhabditis elegans*), which are involved in the desaturation of linoleic acid to α-linolenic acid. The amino acid sequence of the Δ12 desaturase shows 43.7%

identity, the highest match, with that of a microsomal Δ12 desaturase (from *Glycine max*, soybean), whereas it exhibits 38.9% identity with that of a microsomal ω3 desaturase (from soybean). The *Mortierella* Δ12 desaturase was confirmed to be involved in the desaturation of oleic acid to linoleic acid by its expression in both *S. cerevisiae* and *A. oryzae*. The cDNA expression in *A. oryzae* allowed a fungal transformant to accumulate linoleic acid corresponding to more than 70% of the total fatty acids. The *M. alpina* 1S-4 Δ12 desaturase is the first example of a cloned, non-plant Δ12 desaturase.

M. alpina 1S-4 is capable of producing not only AA but also EPA at a cultural temperature below 20°C. Based on the conserved sequence information on *M. alpina* 1S-4 Δ12 desaturase and *Saccharomyces kluyveri* ω3 desaturase, the ω3 desaturase gene from *M. alpina* 1S-4 was cloned (Sakuradani et al., 2005a). Homology analysis with protein databases revealed that the amino acid sequence showed 51% identity, at the highest, with that of the *M. alpina* 1S-4 Δ12 desaturase, whereas it exhibited 36% identity with that of the *S. kluyveri* ω3 desaturase. Analysis of the fatty acid composition of the yeast transformant expressing the *Mortierella* ω3 desaturase gene demonstrated that C18 and C20 n-3 PUFAs were accumulated through conversion of exogenous C18 and C20 n-6 PUFAs. The substrate specificity of the *M. alpina* 1S-4 ω3 desaturase differs from those of the known fungal ω3 desaturases from *S. kluyveri* (Oura & Kajiwara, 2004) and *Saprolegnia diclina* (Pereira et al., 2004). Plant, cyanobacterial, and *S. kluyveri* ω3 desaturases desaturate C18 n-6 PUFAs, *Saprolegnia diclina* ω3 desaturase desaturates C20 n-6 PUFAs, and *Caenorhabditis elegans* ω3 desaturase prefers C18 n-6 PUFAs to C20 n-6 PUFAs as substrates (Meesapyodsuk et al., 2000). The substrate specificity of *M. alpina* 1S-4 ω3 desaturase is rather similar to that of *C. elegans* ω3 desaturase, but the *M. alpina* ω3 desaturase could more effectively convert AA into EPA when expressed in the yeast. The *M. alpina* 1S-4 ω3 desaturase is the first known fungal desaturase that uses both C18 and C20 n-6 PUFAs as substrates.

Δ5 and Δ6 Desaturases

A cDNA encoding Δ5 desaturase has been isolated from two *M. alpina* strains (CBS 210.32 and ATCC 32221), and its function was confirmed by its expression in *S. cerevisiae* and canola *Brassica napus* (Knutzon et al., 1998; Michaelson et al., 1998). Expression in transgenic canola seeds resulted in the production of unique PUFAs, taxoleic acid (Δ5,9-18:2), and pinolenic acid (Δ5,9,12-18:3), which are the Δ5 desaturation products of oleic acid and linoleic acid, respectively. The deduced amino acid sequence of the Δ5 desaturase from the CBS strain exhibits 22% identity with that of the Δ6 desaturase from the cyanobacterium *Synechocystis* and 20% identity with that of the borage Δ6 desaturase. It also contains a cytochrome, b_5-like domain at the N-terminus, like the above Δ6 desaturases.

A cDNA encoding Δ6 desaturase (Δ6-1) has been cloned from *M. alpina* 1S-4 (Sakuradani et al., 1999b) based on the sequence information of Δ6 desaturase genes (from borage and *Caenorhabditis elegans*), which are involved in the desaturation of linoleic acid to GLA. The predicted amino acid sequence shows similarities with those sequences of the above Δ6 desaturases and contains a cytochrome, b_5-like domain at the N-terminus, this structure being different from the cases of *M. alpina* Δ9-1 and yeast Δ9 desaturase, which have the corresponding domain at the C-terminus. The full-length cDNA clone was expressed under the control of the *amyB* promoter in *A. oryzae*, resulting in the accumulation of GLA to the level of 25.2% of the total fatty acids. Coexpression of both Δ12 and Δ6 desaturase cDNAs from *M. alpina* ATCC32221 in *S. cerevisiae* was also shown to cause the accumulation of GLA (Huang et al., 1999). A second Δ6 desaturase (Δ6-2) is present in *M. alpina* 1S-4 (Sakuradani et al., 2003). The amino acid sequence homology between Δ6-1 and Δ6-2 is very high (92%). The genes are transcribed at different levels in *M. alpina* 1S-4. Usually, the Δ6-1 gene is transcribed much more highly (2- to 17-fold) than the Δ6-2 one.

Elongases

A cDNA clone with a 957-nucleotide open reading frame encoding 318 amino acid residues, which exhibits 25% identity with the ELO2 protein (EL2) responsible for the elongation of saturated and monosaturated fatty acids, has been isolated from *M. alpina* ATCC 32221 and expressed in *S. cerevisiae* (Parker-Barnes et al., 2000). Coexpression of this cDNA with that of *M. alpina* Δ5 desaturase in the yeast resulted in the conversion of exogenously added GLA to AA via DGLA, as well as the conversion of n-3 octadecatrienoic acid (18:3n-3) to EPA, suggesting that EL2 plays a critical role in the elongation of both n-6 and n-3 C18 PUFAs to the corresponding C20 PUFAs, respectively. This is the first example of a cloned elongase.

M. alpina 1S-4 has not only EL2 but also two elongase genes (MAELO and EL1). Although a homologous MAELO gene had already been isolated from *M. alpina* ATCC 32221, its function had not yet been identified. The MAELO gene from *M. alpina* 1S-4 was confirmed to encode a fatty acid elongase by its expression in yeast *S. cerevisiae* (Sakuradani et al., 2008). Analysis of the fatty acid composition of the yeast transformant revealed the accumulation of 22:0, 24:0, and 26:0 fatty acids. On the other hand, the MAELO gene-silenced strain obtained on RNA interference (RNAi) exhibited low contents of 20:0, 22:0, and 24:0 fatty acids and a high content of stearic acid (18:0) when compared with those in the wild strain. The enzyme encoded by the MAELO gene was demonstrated to be involved in the biosynthesis of 20:0, 22:0, and 24:0 fatty acids in *M. alpina* 1S-4.

The *M. alpina* EL1 gene was confirmed to encode a fatty acid elongase by its expression in yeast *S. cerevisiae*, resulting in the accumulation of 18-, 19-, and 20-carbon monounsaturated fatty acids and eicosanoic acid. The EL1 yeast transformant efficiently elongated exogenous 9-hexadecenoic acid, 9,12-octadecadienoic acid, and 9,12,15-octadecatrienoic acid. The EL1 gene-silenced strain obtained from *M. alpina* 1S-4 exhibited a low content of octadecanoic acid and a high content of hexadecanoic acid when compared with those in the wild strain. The enzyme encoded by the EL1 gene was demonstrated to be involved in the conversion of hexadecanoic acid to octadecanoic acid—its main role in *M. alpina* 1S-4.

NADH-Cytochrome b_5 Reductase and Cytochrome b_5

A cDNA clone with an open reading frame encoding 298 amino acid residues, which shows marked sequence similarity to NADH-cytochrome b_5 reductases (CbRs) from other sources (*S. cerevisiae*, bovine, human, and rat), was isolated from *M. alpina* 1S-4 (Sakuradani et al., 1999d). The expression of the full-length cDNA in *A. oryzae* resulted in an increase (4.7-fold) in ferricyanide-reduction activity involving the use of NADH as an electron donor in its microsomes. This *Mortierella* CbR has been purified, with a 645-fold increase in the NADH-ferricyanide reductase specific activity. The purified CbR preferred NADH to NADPH as an electron donor. The second CbR gene of the same fungus has also been characterized (Certik et al., 1999). Its cDNA and predicted amino acid sequence exhibit about 70% similarity to those of the first one.

Cytochrome b_5 has been purified from *Mortierella hygrophila* IFO 5941 and characterized in some detail (Kouzaki et al., 1995). Both the cytochrome b_5 genomic gene and cDNA from *M. alpina* 1S-4 have been cloned (Kobayashi et al., 1999). The amino acid sequence of *M. alpina* 1S-4 cytochrome b_5 exhibits 48%, 40%, and 39% identity with those of rat, chicken, and yeast, respectively, over 100 amino acids of the N-terminus. By contrast, there is no significant identity among these sequences over 30 amino acids of the C-terminus. The soluble form of the cytochrome b_5 reached 16% of the total soluble protein in *Escherichia coli*. The holo-cytochrome b_5 accounted for 8% of the total cytochrome b_5 in the transformants.

Identification of Mutation Sites in Desaturase-Defective Mutants

Based on the information from the enzyme genes involved in PUFA biosynthesis, the mutation sites of desaturase-defective mutants have been analyzed. The gene mutations in various mutants are summarized in Table 2.B.

The nucleotide sequences of the genomic genes encoding three Δ9 desaturases in Δ9 desaturase-defective mutants (T4, ST56, and HR222) were determined, resulting in identification of mutation sites in only Δ9-1 among these mutants (Abe et al., 2006). The G nucleotides at the 408th, 794th, and 1080th positions from the start codon in the Δ9-1 genes of HR222, T4, and ST56 were replaced with A nucleotides, respectively. These mutations caused amino acid replacement of P136Stop, G265D, and P360Stop in the Δ9-1 of HR222, T4, and ST56, respectively. Transcriptional analysis revealed that the genes encoding Δ9-2 and Δ9-3 are not transcribed as much as those encoding Δ9-1 in the wild strain, whereas the Δ9-2 gene is transcribed effectively at the same level as or a higher level than the Δ9-1 gene in the mutants defective

Table 2.B. Summary of Mutations Found in Desaturase-Defective Mutants.

Type	Mutant	Parent	Accumulated fatty acid	Gene mutation	Amino acid replacement
Δ9-1	T4	1S-4	18:0	G794A	G265D
	ST56	Y61	18:0	G1080A	P360Stop
	HR222	1S-4	18:0	G408A	P136Stop
Δ12	Mut48	1S-4	18:1n-9, 20:3n-9	C497T	P166L
	SR88	1S-4	18:1n-9, 20:3n-9	C346T	H116Y
Δ6-1	Mut49	1S-4	18:2n-6, 20:3Δ5	GT-Terminal disruption in 2nd intron	Frame shift
	ST66	1S-4	18:2n-6, 20:3Δ5	C1124A	W314Stop
	YB214	1S-4	18:2n-6, 20:3Δ5	C1169A	G390D
	HR95	1S-4	18:2n-6, 20:3Δ5	G924A	T375K
Δ5	S14	1S-4	20:3n-6	New AG-terminal construction in 1st intron	Frame shift
	Iz3	1S-4	20:3n-6	G566A	G189E
	M226-9	Mut48	20:2n-9	G903A	W301Stop
	Mut44	1S-4	20:3n-6	None	None
ω3	Y11	1S-4	AA	G837A	W232Stop
	K1	Mut44	DGLA	G1299A	W386Stop

in Δ9-1 (Abe et al., 2006). On the other hand, Δ9-3 mRNA is hardly synthesized in either the wild strain or the mutants. These results suggest that Δ9-2 acts as Δ9-1 in these mutants.

The mutation sites on the Δ12 desaturase gene in *Mortierella alpina* Δ12 desaturase-defective mutants SR88, TM912, and Mut48 accumulating Mead acid were identified (Sakuradani et al., 2009a). Each mutation resulted in an amino acid replacement (H116Y and P166L) in the Δ12 desaturase gene from SR88 and Mut48, respectively. JT-180 (Δ12 desaturation-defective and a Δ6 and Δ5 desaturation-enhanced mutant) exhibited the same mutation in the Δ12 desaturase gene as Mut48. On the other hand, there was no mutation site in the Δ6 and Δ5 desaturase-encoding genes or their promoter regions.

The mutation sites in the ω3 desaturase genes of two ω3 desaturase-defective mutants (Y11 and K1) were identified (Sakuradani et al., 2009b). The mutation of the ω3 desaturase gene in Y11 and K1 resulted in an amino acid replacement

(W232Stop and W386Stop), respectively, which caused a lack of ω3 desaturase activity in these mutants.

Multiple Δ6 desaturase (Δ6-1 and Δ6-2) genes of Δ6 desaturase-defective mutants (Mut49, ST66, YB214, and HR95) were analyzed, resulting in identification of mutation sites in only Δ6-1—i.e., not in Δ6-2 (Abe et al., 2005a). The C nucleotides at the 1124th and 1169th positions and the G nucleotide at the 942nd position from the start codon in the Δ6-1 desaturase genes of ST66, YB214, and HR95 were replaced with A nucleotides, respectively. These mutations caused amino acid replacement of T375K, G390D, and W314Stop in Δ6-1 of HR95, YB214, and ST66, respectively. The G nucleotide constituting the GT terminal of the Δ6-1 2nd intron in Mut49 was replaced with an A nucleotide, resulting in a frameshift mutation.

The Δ5 desaturase genes of *M. alpina* 1S-4 mutants (Mut44, S14, Iz3, and M226-9) were also analyzed (Abe et al., 2005b). The *M. alpina* 1S-4 Δ5 desaturase genomic gene consists of 2,319 bp of nucleotides with seven introns. The G nucleotides at the 566th and 903rd positions from the start codon in the Δ5 desaturase genes of Iz3 and M226-9 were replaced with A nucleotides, respectively. These mutations caused amino acid replacement of G189E and W301Stop in the Δ5-desaturases of Iz3 and M226-9, respectively. The C nucleotide at 10 bases upstream from the AG terminal of the first intron was replaced with an A nucleotide in the Δ5 desaturase gene of S14. The Δ5 desaturase cDNA of S14 was 8 bp longer than that of the wild strain, as judged on comparison of the cDNAs. We assumed that the new A nucleotide created by the mutation was recognized as a component of the new AG terminal of the first intron and thereby caused a frameshift mutation. The Δ5 desaturase gene and its promoter region of Mut44, which exhibits low Δ5 desaturase activity, had no nucleotide replacement, suggesting that there is a specific factor that influences Δ5 desaturase activity.

Genetic Manipulation of *M. alpina* Strains

Transformation systems for *M. alpina* strains were reported by two laboratories (Mackenzie et al., 2000; Takeno et al., 2004a, 2004b). The first system utilizes the gene for resistance to an antibiotic, hygromycin. By using a homologous histone H4 promoter, *M. alpina* CBS 224.37 was transformed successfully to become hygromycin-resistant. Genetically stable transformants require inclusion of a homologous ribosomal DNA region in the vector to promote chromosomal integration.

The second system involves uracil auxotrophs that lack orotate phosphoribosyl transferase activity (OPRTase) derived from *M. alpina* 1S-4 (Takeno et al., 2004a). *M. alpina* uracil auxotrophs grow on a medium containing 0.5 mg 5-fluoroorotic acid (5-FOA)/ml as a growth inhibitor, with the addition of a little uracil. Such uracil auxotrophs obtained seem to lack OPRTase or orotidine-5'-monophospate decarboxylase (OMPdecase) activity, according to a general pyrimidine nucleotide biosynthetic pathway; however, all of the auxotrophs were proved to be OPRTase-defective mutants on comparative assaying of the two enzyme activities. The *ura3* and *ura5* genes encoding OMPdecase and OPRTase, respectively, were isolated from *M. alpina* 1S-4 on the basis of the information on the amino acid sequences of the corresponding enzymes from other organisms. The uracil auxotrophs were proved to each have a point mutation in their *ura5* genes, but not in their *ura3* genes.

Transformation with a vector containing the homologous *ura5* gene as a marker, produced by microprojectile bombardment, was successfully performed (Takeno et al., 2004b); other methods frequently used for transformation, such as ones involving protoplasting, lithium acetate, and electroporation, did not give satisfactory results because of the difficulty of effective protoplast formation with the use of general and commercial lytic enzymes, such as chitinase, chitosanase, and glucanase, for cell walls. As a result, transformants were obtained at a transformation frequency of 0.4 transformants/μg vector DNA. In order to isolate stable transformants, all transformants were inoculated into a medium containing 5-FOA/uracil, stable ones that never grew on the medium being selected. More than 90% of the transformants could grow on the medium, and these were proved to be the unstable ones. The stable transformants exhibited OPRTase activity comparable to that of the wild strain and retained the *ura5* gene originating from the transformation vector regardless of the culture conditions, while the unstable ones easily lost the marker gene under uracil-containing conditions.

By using this transformation system, the expression of the gene encoding EL2, which has been shown biochemically to be the limiting step for AA biosynthesis (Wynn and Ratledge, 2000), constructed in a similar cassette on expression of the homologous *ura5* gene, was successfully performed (Takeno et al., 2005b). The resultant transformants yielded more AA than the wild strain, as shown in Table 2.C.

Agrobacterium tumefaciens-mediated transformation (ATMT) of *M. alpina* 1S-4 was performed (Ando et al., 2009b). The *ura5* gene was used as a selectable marker under the control of homologous fungal promoter histone H4.1 and heterologous fungal terminator *trpC*t in the transfer-DNA (T-DNA) region. The frequency of transformation reached more than 400 transformants/108 spores. Southern blot analysis revealed that most of the integrated T-DNA appeared as a single copy and was found at a random position in the chromosomal DNA. Additionally, the mitotic stability of the transformants was much higher (60-80%) compared with that obtained on the previous biolistic transformation (10%). Moreover, all transformants exhibited excellent growth.

M. alpina 1S-4 was endowed with zeocin resistance by the integration of a zeocin-resistance gene at the rDNA locus of genomic DNA (Takeno et al., 2005c). Twenty mg/ml zeocin completely inhibited the germination of *M. alpina* 1S-4 spores, and decreased the rate of growth of fungal filaments, to some extent. It was suggested that the preincubation period and temperature had a great influence on the transformation efficiency. Overexpression of the gene encoding EL2 in the wild 1S-4 by using this zeocin system led to high AA production. In addition, the fungicide carboxin completely inhibited the hyphal growth and spore germination of *M. alpina* 1S-4 (Ando et al., 2009a). The *sdhB* gene encoding the iron-sulfur (Ip) subunit of the succinate dehydrogenase (SDH, EC 1.3.99.1) complex was cloned from *M. alpina* 1S-4. The

Table 2.C. PUFA production by *M. alpina* Transformants.

Accumulated PUFA	Host[a]	Target gene[b]	Method[c]	Note
AA	JT-180	Δ12	OE	Higher accumulation of AA in JT-180 than wild strain 1S-4
AA	1S-4	EL2	OE	High AA production (4.4 g/l)
20:3n-6(Δ5), 20:2n-6	1S-4	Δ6	Ri	Accumulation of 20:3n-6(Δ5) and 20:2n-6
EPA	1S-4	ω3	OE	High EPA production (0.8 g/l, 30%)
20:4n-3	S14	ω3	OE	High 20:4n-3 production (1.8 g/l, 35%)
MA	1S-4	Δ12	Ri	Accumulation of n-9 PUFAs
16:0, 16:1n-7	1S-4	EL1	Ri	Accumulation of 16:0 and 16:1n-7
n-4/n-7 PUFA	M1	MAELO	Ri	Accumulation of n-4/n-7 PUFAs and decrease of n-6 PUFAs
n-7 PUFA	M1	Δ12	Ri	Accumulation of n-7 PUFAs and decrease of n-4 PUFAs
18:0, PUFA	1S-4	MAELO	Ri	No accumulation of 22:0 and 24:0 and a small increase of 18:0 and the following n-6 PUFAs
22:4n-6, 22:5n-3	1S-4	PavELO, ω3	OE	Detection of a small amount of 22:4n-6 and 22:5n-3 in wild strain 1S-4

[a]JT-180, Δ12 desaturase activity-defective mutant; M1, EL1 elongase activity-defective mutant; S14, Δ5 desaturase activity-defective mutant.

[b]The genes, except for *PavELO*, are derived from *M. alpina* 1S-4. The MAELO elongase catalyzes the elongation of long chain saturated fatty acids of more than C18 length in *M. alpina* 1S-4. PavELO is a fatty acid elongase that converts C20 PUFAs to C22 PUFAs in marine microalga *Pavlova* sp.

[c]OE, overexpression; Ri, RNA interference.

deduced amino acid sequence of SdhB from *M. alpina* 1S-4 showed high similarity to SdhB sequences from other organisms. The mutated *sdhB* (*CBXB*) gene encodes a modified SdhB with an amino acid substitution (a highly conserved histidine residue within the third cysteine-rich cluster of SdhB replaced by a leucine residue) and is known to confer carboxin resistance. Transformants obtained with a plasmid containing the *CBXB* gene from *M. alpina* 1S-4 exhibited carboxin resistance. These genes for zeocin and carboxin resistance are, thus, useful as selective markers for the transformation of the oleaginous fungus *M. alpina* 1S-4 and its mutants.

Gene disruption in *M. alpina* 1S-4 was performed by a RNAi method with double strand RNA (Takeno et al., 2005a). The Δ12 desaturase gene-silenced strains accumulate 18:2n-9, 20:2n-9, and MA, which are not detected in either the control strain or wild type strain 1S-4. The fatty acid composition of stable transformants was similar to that of Δ12 desaturation-defective mutants previously identified (Table 2.C). Thus, RNAi could be used to alter the types and relative amounts of fatty acids produced by commercial strains of this fungus without mutagenesis or other permanent changes in the genetic background of the producing strains.

Transformation systems for useful mutants derived from *M. alpina* 1S-4 were developed in the same way as for the wild strain: the spores of uracil auxotrophs isolated from the mutants were treated by microprojectile bombardment. Thus far, the productivities of various PUFAs in *M. alpina* 1S-4 and the derived mutants on overexpression or RNAi of the enzyme genes involved in PUFA biosynthesis have been improved, as shown in Table 2.C. Δ12 desaturation-defective mutant JT-180 exhibits enhanced Δ5- and Δ6-desaturase activities, as described above. On overexpression of the Δ12-desaturase gene in JT-180, it accumulated a higher amount of AA than the wild strain. Overexpression of the ω3-desaturase gene in the wild strain and S14 (Δ5 desaturation-defective mutant) led to the high production of EPA (0.8 g/L, 30% of the total fatty acids) and 20:4n-3 (1.8 g/L, 35%), which usually comprise about 10% of the total fatty acids of the wild strain and S14 cultivated at low temperature (<20°C), respectively. The transformant of EL1-defective mutant M1 obtained on RNAi of the Δ12-desaturase gene accumulated n-7 PUFAs and decreased n-4 PUFAs. The M1 transformant obtained on RNAi of the MAELO gene accumulated n-4/n-7 PUFAs and decreased n-6 PUFAs, which suggests that MAELO is involved in elongation of not only long chain saturated fatty acids such as 20:0 and 22:0, but also 16:0. Overexpression of both the PavElO (involved in the conversion of C20 to C22 PUFA in marine microalga *Pavlova* sp.) and ω3-desaturase genes led to the formation of C22 PUFAs (22:4n-6 and 22:5n-3) in the wild strain. Such breeding of useful mutants isolated previously, as well as the wild strain, improved the productivity of PUFAs.

Conclusion

The unique transforming enzyme systems of *M. alpina* 1S-4 and its mutants can be used to obtain various PUFAs of the n-9, n-6, n-3, n-7, n-4, and n-1 series from glucose or precursor fatty acids added to the medium. They also make it possible to control the fatty acid profiles of fungal mutants and to regulate the flow of glucose or exogenous fatty acids to obtain a desired PUFA. Because of the simplicity of their metabolic systems, these mutants are potentially ideal models for the elucidation of fungal lipogenesis. The present study on *M. alpina* and its mutants was focused on the molecular engineering of these enzyme systems and has pioneered the improvement of PUFA productivity. The breeding of mutants and transgenic strains may make it possible to effectively produce desired PUFAs.

References

Abe, T.; E. Sakuradani; T. Asano; H. Kanamaru; Y. Ioka; S. Shimizu. Identification of mutation sites on Δ6 desaturase genes from *Mortierella alpina* 1S-4 mutants. *Biosci. Biotechnol. Biochem.* **2005a,** *69(5),* 1021–1024.

Abe, T.; E. Sakuradani; T. Ueda; S. Shimizu. Identification of mutation sites on Δ5 desaturase genes from *Mortierella alpina* 1S-4 mutants. *J. Biosci. Bioeng.* **2005b,** *99(3),* 296–299.

Abe, T.; E. Sakuradani; T. Asano; H. Kanamaru; S. Shimizu. Functional characterization of Δ9 and ω9 desaturase genes in *Mortierella alpina* 1S-4 and its derivative mutants. *Appl. Microbiol. Biotechnol.* **2006,** *70(6),* 711–719.

Amano, H.; Y. Shinmen; K. Akimoto; H. Kawashima; T. Amachi; S. Shimizu; H. Yamada. Chemotaxonomic significance of fatty acid composition in the genus *Mortierella* (Zygomycetes, Mortierellaceae). *Micotaxonomy* **1992,** *94(2),* 257–265.

Ando, A.; S. Takeno; M. Ochiai; H. Kawashima; E. Sakuradani; J. Ogawa; S. Shimizu. Analysis of the enzymes involved in the synthesis and degradation of cell walls in arachidonic acid-producing fungus *Mortierella alpina* 1S-4. JSBBA Conference; The Japan Society for Bioscience, Biotechnology, and Agrochemistry: Kyoto, 2006; p 245.

Ando, A.; E. Sakuradani; K. Horinaka; J. Ogawa; S. Shimizu. Transformation of an oleaginous zygomycete Mortierella alpina 1S-4 with the carboxin resistance gene conferred by mutation of the iron-sulfur subunit of succinate dehydrogenase. *Curr. Genet.* **2009a,** *55(3),* 349–356.

Ando, A.; Y. Sumida; H. Negoro; D.A. Suroto; J. Ogawa; E. Sakuradani; S. Shimizu. Establishment of *Agrobacterium tumefaciens*-mediated transformation of an oleaginous fungus *Mortierella alpina* 1S-4 and its application for eicosapentaenoic acid-producer breeding. *Appl. Environ. Microbiol.* **2009b,** *75(17),* 5529-5535.

Certik, M.; E. Sakuradani; S. Shimizu. Desaturase-defective fungal mutants: useful tools for the regulation and overproduction of polyunsaturated fatty acids. *Trends Biotechnol.* **1998,** *16(12),* 500–505.

Certik, M.; E. Sakuradani; M. Kobayashi; S. Shimizu. Characterization of the second form of NADH-cytochrome b_5 reductase gene from arachidonic acid-producing fungus *Mortierella alpina* 1S-4. *J. Biosci. Bioeng.* **1999,** *88(6),* 667–671.

Certik, M.; S. Shimizu. Biosynthesis and regulation of microbial polyunsaturated fatty acid production. *J. Biosci. Biotechnol.* **1999,** *87(1),* 1–14.

Higashiyama, K., T. Yaguchi; K. Akimoto; S. Fujikawa; S. Shimizu. Enhancement of arachidonic acid production by *Mortierella alpina*. *J. Am. Oil Chem. Soc.* **1998,** *75(11),* 1501–1505.

Higashiyama, K.; S. Fujikawa; E. Park; S. Shimizu. Production of arachidonic acid by *Mortierella* fungi. *Biotechnol. Bioprocess Eng.* **2002,** *7(5),* 252–262.

Huang, Y.S.; S. Chaudhary; J.M. Thurmond; E.G. Bobik; L. Yuan; G.M. Chan; S.J. Kirchner; P. Mukerji; D.S. Knutzon. Cloning of Δ12- and Δ6-desaturases from *Mortierella alpina* and recombinant production of γ-linolenic acid in *Saccharomyces cerevisiae*. *Lipids* **1999**, *34(7)*, 649–659.

Ioka, Y.; T. Ito; E. Sakuradani; J. Ogawa; K. Aoki; K. Yamamoto; S. Shimizu. Identification of components of a lipid particle produced by a lipid-excretive mutant derived from *Mortierella alpina* 1S-4. JSBBA Conference; The Japan Society for Bioscience, Biotechnology, and Agrochemistry: Tokyo, 2003; p 16.

Jareonkitmongkol, S.; H. Kawashima; S. Shimizu; H. Yamada. Production of 5,8,11-*cis*-eicosatrienoic acid by a Δ12-desaturase-defective mutant of *Mortierella alpina* 1S-4. *J. Am. Oil Chem.* **1992a**, *69(9)*, 939–944.

Jareonkitmongkol, S.; H. Kawashima; H.; N. Shirasaka; S. Shimizu; H. Yamada. Production of dihomo-γ-linolenic acid by a Δ5-desaturase-defective mutant of *Mortierella alpina* 1S-4. *Appl. Environ. Microbiol.* **1992b**, *58(7)*, 2196–2200.

Jareonkitmongkol, S.; S. Shimizu; H. Yamada. Fatty acid desaturation defective mutants of an arachidonic acid-producing fungus, *Mortierella alpina* 1S-4. *J. Gen. Microbiol.* **1992c**, *138*, 997–1002.

Jareonkitmongkol, S.; H. Kawashima; S. Shimizu. Inhibitory effects of lignan compounds on the formation of arachidonic acid in a Δ5-desaturase-defective mutant of *Mortierella alpina* 1S-4. *J. Ferment. Technol.* **1993a**, *76(5)*, 406–407.

Jareonkitmongkol, S.; E. Sakuradani; S. Shimizu. A novel Δ5-desaturase-defective mutant of *Mortierella alpina* 1S-4 and its dihomo-γ-linolenic acid productivity. *Appl. Environ. Microbiol.* **1993b**, *59(12)*, 4300–4304.

Jareonkitmongkol, S.; S. Shimizu; H. Yamada. Occurrence of two nonmethylene-interrupted fatty acids in a Δ6-desaturase-defective mutant of the fungus *Mortierella alpina* 1S-4. *Biochim. Biophys. Acta* **1993c**, *1167(2)*, 137–141.

Jareonkitmongkol, S.; S. Shimizu; H. Yamada. Production of an eicosapentaenoic acid-containing oil by a Δ12 desaturase-defective mutant of *Mortierella alpina* 1S-4. *J. Am. Oil Chem. Soc.* **1993d**, *70(2)*, 119–123.

Jareonkitmongkol, S.; E. Sakuradani; S. Shimizu. Isolation and characterization of an ω3-desaturation-defective mutant of an arachidonic acid-producing fungus, *Mortierella alpina* 1S-4. *Arch. Microbiol.* **1994**, *161(4)*, 316–319.

Jareonkitmongkol, S.; E. Sakuradani; S. Shimizu. Isolation and characterization of a ω9-desaturation-defective mutant of an arachidonic acid-producing fungus, *Mortierella alpina* 1S-4. *J. Am. Oil Chem. Soc.* **2002**, *79(10)*, 1021–1026.

Kamada, N.; H. Kawashima; E. Sakuradani; K. Akimoto; J. Ogawa; S. Shimizu. Production of 8,11-*cis*-eicosadienoic acid by a Δ5 and Δ12 desaturase-defective mutant derived from the arachidonic acid producing fungus *Mortierella alpina* 1S-4. *J. Am. Oil Chem. Soc.* **1999**, *76(11)*, 1269–1274.

Kawashima, H.; M. Nishihara; Y. Hirano; N. Kamada; K. Akimoto; K. Konishi; S. Shimizu. Production of 5,8,11-eicosatrienoic acid (Mead acid) by a Δ6 desaturation activity-enhanced mutant derived from a Δ12 desaturase-defective mutant of an arachidonic acid-producing fungus, *Mortierella alpina* 1S-4. *Appl. Environ. Microbiol.* **1997**, *63(5)*, 1820–1825.

Kawashima, H.; E. Sakuradani; N. Kamada; K. Akimoto; K. Konishi; J. Ogawa; S. Shimizu. Production of 8,11,14, 17-*cis*-eicosatetraenoic acid (20:4ω3) by a Δ5 and Δ12 desaturase-defective mutant of an arachidonic acid-producing fungus *Mortierella alpina* 1S-4. *J. Am. Oil Chem. Soc.* **1998**, *75(11)*, 1495–1500.

Kawashima, H.; K. Akimoto; K. Higashiyama; S. Fujikawa; S. Shimizu. Industrial production of dihomo-γ-linolenic acid by a Δ5 desaturase-defective mutant of *Mortierella alpina* 1S-4 fungus. *J. Am. Oil Chem. Soc.* **2000**, *77(11)*, 1135–1138.

Knutzon, D.S.; J.M. Thurmond; Y.S. Huang; S. Chaudhary; E.G. Bobik; G.M. Chan; S.J. Kirchner; P. Mukerji. (1998) Identification of Δ⁵-desaturase from *Mortierella alpina* by heterologous expression in bakers' yeast and canola. *J. Biol. Chem.* **1998**, *273(45)*, 29360–29366.

Kobayashi, M.; E. Sakuradani; S. Shimizu. Genetic analysis of cytochrome b_5 from arachidonic acid-producing fungus, *Mortierella alpina* 1S-4: cloning, RNA editing and expression of the gene in *Escherichia coli*, and purification and characterization of the gene product. *J. Biochem. (Tokyo)* **1999**, *125(6)*, 1094–1103.

Kouzaki, N.; H. Kawashima; M.C.M. Chung; S. Shimizu. Purification and characterization of two forms of cytochrome b_5 from an arachidonic acid-producing fungus, *Mortierella hygrophila*. *Biochim. Biophys. Acta* **1995**, *1256(3)*, 319–326.

Kyle, D.J.; V.J. Sicotte; J.J. Singer; S.E. Reeb. Bioproduction of docosahexaenoic acid (DHA) by microalgae. *Industrial Applications of Single Cell Oils*; D.J. Kyle, C. Ratledge, Eds.; AOCS Press: IL, 1992; pp 287–300.

Mackenzie, D.A.; P. Wongwathanarat; A.T. Carter; D.B. Archer. Isolation and use of a homologous histone H4 promoter and a ribosomal DNA region in a transformation vector for the oil-producing fungus *Mortierella alpina*. *Appl. Environ. Microbiol.* **2000**, *66(11)*, 4655–4661.

MacKenzie, D.A.; A.T. Carter; P. Wongwathanarat; J. Eagles; J. Salt; D.B. Archer. A third fatty acid Δ9-desaturase from *Mortierella alpina* with a different substrate specificity to ole1p and ole2p. *Microbiology* **2002**, *148(6)*, 1725–1735.

Meesapyodsuk, D.; D.W. Reed; C.K. Savile; P.H. Buist; S.J. Ambrose; P.S. Covello. Characterization of the regiochemistry and cryptoregiochemistry of a *Caenorhabditis elegans* fatty acid desaturase (*FAT-1*) expressed in *Saccharomyces cerevisiae*. *Biochemistry* **2000**, *39(39)*, 11948–11954.

Michaelson, L.V.; C.M. Lazarus; G. Griffiths; J.A. Napier; A.K. Stobart. Isolation of a Δ⁵-fatty acid desaturase gene from *Mortierella alpina*. *J. Biol. Chem.* **1998**, *273(30)*, 19055–19059.

Nakahara, T.; Y. Yokochi; T. Higashihara; S. Tanaka; T. Yaguchi; D. Honda. Production of docosahexaenoic and docosapentaenoic acid by *Schizochytrium* sp. isolated from Yap Islands. *J. Am. Oil Chem. Soc.* **1996**, *73(11)*, 1421–1426.

Oura, T.; S. Kajiwara. *Saccharomyces kluyveri FAD3* encodes an ω3 fatty acid desaturase. *Microbiology* **2004**, *150(6)*, 1983–1990.

Parker-Barnes, J.M.; T. Das; E. Bobik; A.E. Leonard; J.M. Thurmond; L.T. Chaung; Y.S. Huang; P. Mukerji. Identification and characterization of an enzyme involved in the elongation of n-6 and n-3 polyunsaturated fatty acids. *Proc. Natl. Acad. Sci. (USA)* **2000**, *97(15)*, 8284–8289.

Pereira, S.L.; Y.S. Huang; E.G. Bobik; A.J. Kinney; K.L. Stecca; J.C.L. Packer; P. Mukerji. A novel ω3-fatty acid desaturase involved in the biosynthesis of eicosapentaenoic acid. *Biochem. J.* **2004**, *378(2)*, 665–671.

Raghukumar, S. Thraustochytrid Marine Protists: production of PUFAs and Other Emerging Technologies. *Mar. Biotechnol. (NY)* **2008**, *10(6),* 631–640.

Ratledge, C. Microbial lipids: Commercial realities or academic curiosities. *Industrial Applications of Single Cell Oils*; D.J. Kyle, C. Ratledge, Eds.; AOCS Press: IL, 1992; pp 1–15.

Ratledge, C. Fatty acid biosynthesis in microorganisms being used for Single Cell Oil production. *Biochimie* **2004**, *86(11),* 807–815.

Sakuradani, E.; M. Kobayashi; T. Ashikari; S. Shimizu. Identification of Δ12-fatty acid desaturase from arachidonic acid-producing *Mortierella* fungus by heterologous expression in the yeast *Saccharomyces cerevisiae* and the fungus *Aspergillus oryzae. Eur. J. Biochem.* **1999a**, *261(3),* 812–820.

Sakuradani, E.; M. Kobayashi; S. Shimizu. Δ6-Fatty acid desaturase from an arachidonic acid-producing *Mortierella* fungus: gene cloning and its heterologous expression in a fungus, *Aspergillus. Gene* **1999b**, *238(2),* 445–453.

Sakuradani, E.; M. Kobayashi; S. Shimizu. Δ⁹-Fatty acid desaturase from an arachidonic acid-producing fungus: unique gene sequence and its heterologous expression in a fungus, *Aspergillus. Eur. J. Biochem.* **1999c**, *260(1),* 208–216.

Sakuradani, E.; M. Kobayashi; S. Shimizu. Identification of an NADH-cytochrome b_5 reductase gene from an arachidonic acid-producing fungus, *Mortierella alpina* 1S-4, by sequencing of the encoding cDNA and heterologous expression in a fungus, *Aspergillus oryzae. Appl. Environ. Microbiol.* **1999d**, *65(9),* 3873–3879.

Sakuradani, E.; N. Kamada; Y. Hirano; M. Nishihara; H. Kawashima; K. Akimoto; K. Higashiyama; J. Ogawa; S. Shimizu. Production of 5,8,11-cis-eicosatrienoic acid by a Δ5 and Δ6 desaturation activity-enhanced mutant derived from a Δ12 desaturation activity-defective mutant of *Mortierella alpina* 1S-4. *Appl. Microbiol. Biotechnol.* **2002**, *60(3),* 281–287.

Sakuradani, E.; S. Shimizu. Gene cloning and functional analysis of a second Δ6-fatty acid desaturase from an arachidonic acid-producing *Mortierella* fungus. *Biosci. Biotechnol. Biochem.* **2003**, *67(4),* 704–711.

Sakuradani, E.; Y. Hirano; N. Kamada; M. Nojiri; J. Ogawa; S. Shimizu. Improvement of arachidonic acid production by mutants with lowered ω3-desaturation activity derived from *Mortierela alpina* 1S-4. *Appl. Microbiol. Biotechnol.* **2004a**, *66(3),* 243–248.

Sakuradani, E.; T. Kimura; J. Ogawa; S. Shimizu. Accumulation of diacylglycerols by a mutant derived from an arachidonic acid-producing fungus, *Mortierella alpina* 1S-4. SBJ Annual Meeting; The Society for Biotechnology, Japan: Nagoya, 2004b; p 155.

Sakuradani, E.; M. Naka; H. Kanamaru; Y. Ioka; M. Nojiri; J. Ogawa; S. Shimizu. Novel biosynthetic pathways for n-4 and n-7 polyunsaturated fatty acids in a mutant of an arachidonic acid-producing fungus, *Mortierella alpina* 1S-4. The 95th AOCS Annual Meeting and Expo; The American Oil Chemists' Society: Cincinnati, USA, 2004c; p 22.

Sakuradani, E.; T. Abe; K. Iguchi; S. Shimizu. A novel fungal ω3-desaturase with wide substrate specificity from arachidonic acid-producing *Mortierella alpina* 1S-4. *Appl. Microbiol. Biotechnol.* **2005a**, *66(6),* 648–654.

Sakuradani, E.; S. Takeno; T. Abe; S. Shimizu. Arachidonic acid-producing *Mortierella alpina*: creation of mutants and molecular breeding. *Single Cell Oils*; Z. Cohen, C. Ratledge, Eds.; AOCS Press: IL, 2005b; pp 21–35.

Sakuradani, E.; S. Murata; H. Kanamaru; S. Shimizu. Functional analysis of a fatty acid elongase from arachidonic acid-producing *Mortierella alpina* 1S-4. *Appl. Microbiol. Biotechnol.* **2008,** *81(3),* 497–503.

Sakuradani, E.; T. Abe; K. Matsumura; A. Tomi; S. Shimizu. Identification of mutation sites on Δ12 desaturase genes from *Mortierella alpina* 1S-4 mutants. *J. Biosci. Bioeng.* **2009a,** *107(2),* 99–101.

Sakuradani, E.; T. Abe; S. Shimizu. Identification of mutation sites on ω3 desaturase genes from *Mortierella alpina* 1S-4 mutants. *J. Biosci. Bioeng.* **2009b,** *107(1),* 7–9.

Singh, A.; O.P. Ward. Microbial production of docosahexaenoic acid (DHA, C22:6). *Adv. Appl. Microbiol.* **1997,** *45,* 271–312.

Shimizu, S.; H. Yamada. Production of dietary and pharmacologically important polyunsaturated fatty acids by microbiological processes. *Comments Agric. Food Chem.* **1990,** *2(3),* 211–235.

Shimizu, S.; S. Jareonkitmongkol. *Mortierella* species (fungi): production of C20 polyunsaturated fatty acids. *Biotechnology in Agriculture and Forestry (Medical Plants VIII)*; Y.P.S. Bajaj, Ed.; Springer-Verlag: Berlin, 1995; 33; pp 308–325.

Shimizu, S.; J. Ogawa; E. Sakuradani. Metabolic engineering for oleaginous microorganisms. JSBBA Conference; The Japan Society for Bioscience, Biotechnology, and Agrochemistry: Tokyo, 2003a; p 371.

Shimizu, S.; E. Sakuradani; J. Ogawa. Production of functional lipids by microorganisms: arachidonic acid and related polyunsaturated fatty acids, and conjugated fatty acids. *Oleoscience* **2003b,** *3(3),* 129–139.

Shinmen, Y.; S. Shimizu; H. Yamada. Production of arachidonic acid by *Mortierella* fungi: selection of a potent producer and optimization of culture conditions for large-scale production. *Appl. Microbiol. Biotechnol.* **1989,** *31,* 11–16.

Spolaore, P., C. Joannis-Cassan; E. Duran; A. Isambert. Commercial applications of microalgae. *J. Biosci. Bioeng.* **2006,** *101(2),* 87–96.

Suzuki, O.; T. Yokochi; T. Yamashina. Studies on production of lipids in fungi (II). Lipid compositions of six species of *Mucorales* in *Zygomycetes. J. Jpn. Oil Chem. Soc.* **1981,** *30,* 863–868.

Takeno, S.; E. Sakuradani; S. Murata; M. Inohara-Ochiai; H. Kawashima; T. Ashikari; S. Shimizu. Cloning and sequencing of the *ura3* and *ura5* genes, and isolation and characterization of uracil auxotrophs of the fungus *Mortierella alpina* 1S-4. *Biosci. Biotechnol. Biochem.* **2004a,** *68(2),* 277–285.

Takeno, S.; E. Sakuradani; S. Murata; M. Inohara-Ochiai; H. Kawashima; T. Ashikari; S. Shimizu. Establishment of an overall transformation system for an oil-producing filamentous fungus, *Mortierella alpina* 1S-4. *Appl. Microbiol. Biotechnol.* **2004b,** *65(4),* 419–425.

Takeno, S.; E. Sakuradani; S. Murata; M. Inohara-Ochiai; H. Kawashima; T. Ashikari; S. Shimizu. Molecular evidence that the rate-limiting step for the biosynthesis of arachidonic acid in *Mortierella alpina* is at the level of an elongase. *Lipids* **2005a,** *40(1),* 25–30.

Takeno, S.; E. Sakuradani; A. Tomi; M. Inohara-Ochiai; H. Kawashima; T. Ashikari; S. Shimizu. Improvement of the fatty acid composition of an oil-producing filamentous fungus, *Mortierella alpina* 1S-4, through RNA interference with Δ12-desaturase gene expression. *Appl. Environ. Microbiol.* **2005b,** *71(9),* 5124–5128.

Takeno, S.; E. Sakuradani; A. Tomi; M. Inohara-Ochiai; H. Kawashima; S. Shimizu. Transformation of oil-producing fungus, *Mortierella alpina* 1S-4, using Zeocin, and application to arachidonic acid production. *J. Biosci. Bioeng.* **2005c**, *100(6)*, 617–622.

Totani, N.; A. Oba. The filamentous fungus *Mortierella alpina*, high in arachidonic acid. *Lipids* **1987**, *22(12)*, 1060–1062.

Wynn, J.P.; C. Ratledge. Evidence that the rate-limiting step for the biosynthesis of arachidonic acid in *Mortierella alpina* is at the level of the 18:3 to 20:3 elongase. *Microbiology* **2000,** *14(9)6*, 2325–2331.

Ueda, T.; E. Sakuradani; J. Ogawa; S. Shimizu. Extracellular production of lipids and analysis of the lipid-excretive mechanism in a lipid-excretive mutant derived from arachidonic acid–producing fungus *Mortierella alpina* 1S-4. JSBBA Conference; The Japan Society for Bioscience, Biotechnology, and Agrochemistry: Kyoto, 2001; pp 372.

Yamada, H.; S. Shimizu; Y. Shinmen. Production of arachidonic acid by *Mortierell elongata* 1S-5. *Agric. Biol. Chem.* **1987,** *51(3)*, 785–790.

Yamada, H.; S. Shimizu; Y. Shinmen; K. Akimoto; H. Kawashima; S. Jareonkitmongkol. Production of dihomo-γ-linolenic acid, arachidonic acid, and eicosapentaenoic acid by filamentous fungi. *Industrial Applications of Single Cell Oils*; D.J. Kyle, C. Ratledge, Eds.; AOCS Press: IL, 1992; pp 118–138.

Yazawa, K.; K. Watanabe; C. Ishikawa; K. Kondo; S. Kimura. Production of eicosapentaenoic acid from marine bacteria. *Industrial Applications of Single Cell Oils*; D.J. Kyle, C. Ratledge, Eds.; AOCS Press: IL, 1992; pp 29–51.

Note: Our recent research led to increased EPA production by overexpression of ω3 desaturase gene through ATMT. The EPA content in the transformant obtained reached a maximum of 40% of the total fatty acids and 0.44 g/L after cultivation at 28°C for 2 days and then 12°C for 16 days.

3

Metabolic Engineering of an Oleaginous Yeast for the Production of Omega-3 Fatty Acids

Quinn Zhu, Zhixiong Xue, Naren Yadav, Howard Damude, Dana Walters Pollak, Ross Rupert, John Seip, Dieter Hollerbach, Daniel Macool, Hongxiang Zhang, Sidney Bledsoe, David Short, Bjorn Tyreus, Anthony Kinney, and Stephen Picataggio

Biochemical Sciences and Engineering, Central Research and Development, E.I. du Pont de Nemours and Company, Wilmington, DE 19880

Introduction

Numerous clinical studies have demonstrated that the omega-3 fatty acids in fish oil significantly reduce the risk of cardiovascular disease in adults (Benatti et al., 2004; Breslow, 2006; Jacobson, 2008; Lee et al., 2008; Leaf, 2008). These results have prompted health organizations and government agencies worldwide to establish dietary guidelines that include fish as a regular part of a healthy diet. The major omega-3 fatty acids in fish oil are eicosapentaenoic acid (EPA, 20:5n-3) and docosahexaenoic acid (DHA, 22:6n-3). Co-application of EPA and DHA for prevention and treatment of other diseases—such as Alzheimer's disease, anorexia nervosa, arthritis, asthma, burns, Crohn's disease, cystic lung fibrosis, inflammatory bowel disease, rheumatoid arthritis, obesity, osteoporosis, type II diabetes, ulcerative colitis, and osteoarthritis—have been summarized, in detail, in several reviews (Belluzzi et al., 1996; Teitelbaum & Walker, 2001; McColl, 2003; Freedman et al., 2004; Salari et al., 2008). Recent studies have showed that EPA and DHA are also useful for the prevention and treatment of attention deficit/hyperactivity disorder and mental health disorders, such as anxiety and depression (Buydens-Branchey et al., 2008; Freeman et al., 2008)

Medical research is beginning to reveal the molecular pharmacology of EPA and DHA. Mechanistic studies have shown that these fatty acids may reduce serum triglyceride levels by activating the nuclear peroxisome proliferator-activated receptors (PPARs) that regulate lipid traffic, storage, and metabolism in various tissues. EPA and DHA inhibit conversion of the omega-6 fatty acid, arachidonic acid (ARA, 20:4n-6), to eicosanoids that mediate the inflammatory response. EPA, in particular, leads to the synthesis of "less-inflammatory" prostaglandins, leukotrienes and thromboxanes (Kinsella et al., 1990; Funk, 2001). In addition, EPA and DHA are precursors to distinct E-series and D-series resolvins that induce a broad anti-inflammatory response (Hong et al., 2003; Arita et al., 2005). Both EPA and DHA inhibit the inflammatory response and promote an anti-inflammatory response. The role of these fatty acids in the prevention and treatment of chronic diseases is gaining attention.

EPA and DHA are important fatty acids for human growth and development. Our bodies do not synthesize sufficient omega-3 fatty acids to derive their health benefits; therefore, we must obtain them in our diet because humans lack the delta12- and delta15-desaturases that provide the essential linoleic acid (LA, 18:2n-6) and alpha-linolenic acid (ALA, 18:3n-3), precursors for EPA and DHA biosynthesis (Fig. 3.1). In addition, the delta6- and delta5-desaturases are inefficient and convert less than 1% of dietary ALA to EPA and DHA. The American Heart Association (AHA) recommends that healthy adults eat fatty fish, such as salmon and tuna, at least twice per week to obtain the equivalent of 600 mg/day of omega-3 fatty acids. The AHA also recommends an even higher level of intake for those at risk for coronary heart disease (Nutrition & Your Health, 2005). However, in today's environment, both farmed and wild-caught fish contain contaminants such as methylmercury,

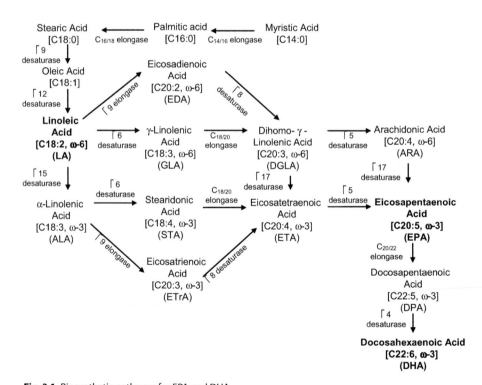

Fig. 3.1. Biosynthetic pathways for EPA and DHA.

polychorinated biphenyls, dioxins, and several other halogenated, persistent organic pollutants (Costa, 2007). Regulatory authorities recommend limited consumption of certain fish in children, pregnant women, and women of childbearing age.

Some plant oils (e.g., from flaxseed and canola) contain ALA, but humans do not efficiently convert this omega-3 fatty acid to either EPA or DHA. While fish oils are a common dietary source of both EPA and DHA, concerns over the sustainability of the fish oil supply in response to climate, disease, over-fishing, and the presence of contaminants have generated interest in alternative sources of omega-3 fatty acids (Qi et al., 2004; Barclay et al., 2005; Kiy et al., 2005; Wynn et al., 2005; Damude & Kinney, 2008).

An alternate source of DHA-enriched oil emerges from microalgal production (Barclay et al., 2005; Kiy et al., 2005; Wynn et al., 2005), and leads to a wide range of applications, including its use in dietary supplements, infant formulas, and functional foods and drinks. Some microalgae, notably *Schizochytrium*, use a biosynthetic pathway, based on a polyketide synthase, that converts acetyl-CoA directly to DHA, rather than the more conventional fatty acid synthase pathway (Metz et al., 2001, 2009). These organisms have not been shown to produce significant amounts of EPA.

Although "pure," concentrated EPA can be obtained from extraction of oils from fish, using rather difficult and expensive processes, no other commercial process exists for EPA production (Wen & Chen, 2005). EPA has unique, beneficial effects for human and animal health. For example, large-scale clinical trials, such as the Japan EPA Lipid Intervention Study (JELIS), have demonstrated that EPA reduces major coronary events by 19% in patients with a history of coronary artery disease (Yokoyama et al., 2007; Saito et al., 2008; Tanaka et al., 2008). Recent studies have also demonstrated the effectiveness of EPA in preventing and treating obesity, mental illness, metabolic syndrome, non-alcoholic steatohepatitis, and type 2 diabetes (Sinclair et al., 2005; Itoh et al., 2007; Mita et al., 2007; Ross et al., 2007; Satoh et al., 2007; Feart et al., 2008; Jazayeri et al., 2008; Tanaka et al., 2008). EPA also reduced arterial stiffness (Hall et al., 2008) and inhibited cancer cell adhesion, migration, and growth (Siddique et al., 2005; Slagsvold et al., 2009). EPA was more effective than DHA in the treatment of depression and bipolar disorder (Ross et al., 2007). The clinical uses of EPA in human and animal health will be greatly expanded once alternate sources are developed to provide "EPA-only" oil (containing no DHA).

We report here the development of a clean and sustainable source of both EPA and DHA through fermentation. Our approach was to introduce the genes encoding an omega-3 fatty acid biosynthesis pathway (Fig. 3.1) into an oleaginous yeast that synthesizes and stores triglycerides as an energy reserve when starved for nitrogen in the presence of an excess carbon source, such as glucose.

Selection of *Yarrowia lipolytica* as Production Host

Yarrowia lipolytica is one of the most intensively studied "non-conventional" yeast species and is not considered to be pathogenic (Wolzschu, 1979). Most strains of *Y. lipolytica* grow very efficiently with alkanes, fatty acids, ethanol, acetate, glucose, fructose, or glycerol as their sole carbon source. Several processes have been described that use *Y. lipolytica*, grown on either glucose or *n*-paraffins, for the commercial production of citric acid (Mattey, 1992; Ratledge, 2005). *Y. lipolytica* is Generally Recognized as Safe (GRAS) for the commercial production of food grade citric acid (see the U.S. Food and Drug Administration list of microbial-derived ingredients approved for use in food: Title 21, Part 173, Sec. 165). Wild-type strains produce a mixture of citric and isocitric acids at consumed substrate yields of up to 130%.

Y. lipolytica has an established history of robust fermentation performance on a commercial scale. With the emergence of single-cell protein (SCP) in the mid-1960s, industrial interest in this yeast arose from its ability to grow on cheap and abundant *n*-paraffins, its sole carbon source. *Y. lipolytica* was first produced commercially by British Petroleum, as a source of SCP for animal feeds, under the trade name of *Toprina*. There is no toxicity or carcinogenicity associated with this product (Ratledge, 2005; see Chapter 1). *Y. lipolytica* has also been used to produce a gamma-decalactone flavoring agent from the alkyl ricinoleate derivative of castor oil (Cardillo et al., 1991; Ercoli et al., 1992).

Some *Y. lipolitica* strains are oleaginous organisms that can accumulate up to 35–40% dry cell weight (dcw) as storage triglycerides in oil bodies when starved for nitrogen in the presence of excess glucose or alkanes. The composition of the lipid depends upon the carbon source (Athenstaedt et al., 2006; Beopoulos et al., 2008). The lipids from glucose-grown cells are comprised mainly of triglycerides and a small amount of sterols. Oleic acid is the most prominent fatty acid. Lipids from oleic acid-grown cells contain more sterols and oleic acid than glucose-grown cells.

A vast amount of data on the genetics and molecular biology of *Y. lipolytica* has been generated since the development of a transformation system (Davidow et al., 1985; Gaillardin et al., 1985; Chen et al., 1997; Vernis et al., 1997) and the determination of its genome sequence (Dujon et al., 2004). It has six chromosomes that encode about 6,500 genes. No plasmid DNA was detected in a systematic survey of wild-type *Y. lipolytica* isolates. Genetic transformation occurs when exogenous linear DNA integrates into the genome by homologous and non-homologous recombination (Davidow et al., 1985; Gaillardin et al., 1985; Weterings & Chen, 2008). *Y. lipolytica* has been used as a model system for studying the function of genes that encode peroxisome biogenesis factor proteins (Kie et al., 2006) and for expressing useful proteins (Madzak et al., 2004; Bordes et al., 2007).

It is easy to develop *Y. lipolytica* auxotrophic mutants. Transformants can be selected through the complementation of auxotrophic mutations, and they do not require the use of selectable marker genes to confer antibiotic resistance. The auxotrophic markers most commonly used are the *LYS5* gene, which encodes for saccharopine dehydrogenase (GenBank Accession No. M34929); the *LEU2* gene, which encodes for beta-isopropylmalate dehydrogenase (GenBank Accession No. AF260230); and *URA3* gene, which encodes for orotidine 5'-monophosphate decarboxylase (GenBank Accession No. AJ306421).

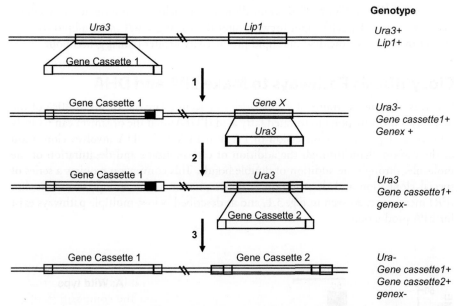

Fig. 3.2. Schematic diagram of *URA3* gene and FOA counter selection system. Gene cassettes integrated into the *Y. lipolytica* genome by homologous recombination through multiple rounds of transformation. Step 1: Integration of a first gene cassette into the *URA3* locus and selection for FOA resistant transformants (*Ura⁻*). Step 2: Integration of a wild-type *URA3* into the *gene-x* locus and selection for *Ura⁺* phenotype. Step 3: Integration of a second gene cassette into the *URA3* locus and selection for FOA resistant transformants (*Ura⁻*).

As shown in Fig. 3.2, the *URA3* gene can be used repeatedly in combination with 5-fluoroorotic acid (5-FOA) selection. 5-FOA is toxic to yeast cells that possess a functional *URA3* gene; thus, based on this toxicity, 5-FOA is especially useful for the selection and identification of *Ura⁻* mutant yeast strains (Bartel & Fields, 1997). More specifically, one can first knock out the native *URA3* gene to produce a strain having an *Ura⁻* phenotype, where selection occurs based on 5-FOA resistance. Then, a cluster of multiple chimeric genes (or a single chimeric gene) and a new *URA3* gene can be integrated into a different locus of the genome of *Y. lipolytica*, thereby producing a new strain with an *Ura⁺* phenotype. Subsequent integration by homologous recombination would produce a new *Ura3⁻* strain (again, identified using 5-FOA selection) when the introduced *URA3* gene is knocked out. Thus, the *URA3* gene (in combination with 5-FOA selection) can be used as a selection marker in multiple rounds of transformation to introduce a large number of genes into the genome, thereby readily permitting the addition of necessary genetic modifications.

Based on above information, we selected *Y. lipolytica* as a production host. We then screened over forty different strains from various public depositories and selected #20362 from the American Type Culture Collection (ATCC) for our pathway engineering. The ATCC #20362 strain allowed us to achieve our fermentation performance targets: a dry cell weight (dcw) greater than 100 g dcw/L, with a lipid content greater than 30% of the dcw, and lipid productivity greater than 1 g/liter.hour.

Biosynthesis Pathways to Make EPA and DHA

Whereas oleaginous strains of *Y. lipolytica* can accumulate oil for up to 40% of the dcw, LA is the only polyunsaturated fatty acid (PUFA) that the organism can natively synthesize (Fig. 3.3A). Conversion of LA into EPA and DHA involves elongation of the carbon chain through the addition of carbon atoms and desaturation of the molecule, through the addition of double bonds. This conversion requires a series of special elongation and desaturation enzymes present in the endoplasmic reticulim (ER) membrane. As seen in Fig. 3.1, and as described below, multiple pathways exist for EPA production.

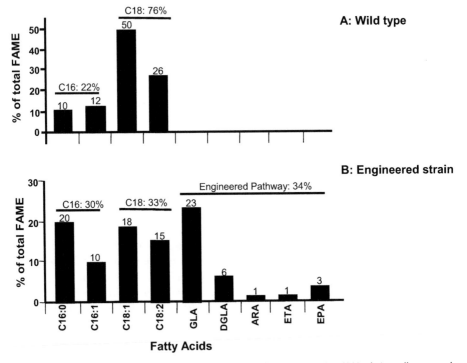

Fig. 3.3. Fatty acid profiles from wild-type and engineered *Yarrowia* strains. *Y. Lipolytica* cells were collected from oleaginous media by centrifugation and lipids were extracted. Fatty acid methyl esters (FAME) were prepared by transesterification of the lipid extract with sodium methoxide and analyzed with a Hewlett-Packard 6890 gas chromatograph equipped with a 30 m x 0.25 mm (i.d.) HP-INNOWAX (Hewlett-Packard) column.

Specifically, all pathways require the initial conversion of oleic acid to LA, the first of the omega-6 fatty acids, by a delta12-desaturase. Using the "delta6-desaturase/$C_{18/20}$-elongase pathway" (delta6 pathway) and LA as substrate, EPA is formed as follows: (1) delta6-desaturase converts LA to gamma-linolenic acid (GLA, 18:3n-6); (2) $C_{18/20}$-elongase converts GLA to dihomo-gamma-linoleic acid (DGLA, 20:3n-6); (3) delta5-desaturase converts DGLA to ARA [although most delta5-desaturases also convert eicosatetraenoic acid (ETA, 20:4n-3) to EPA with less efficiency]; and (4): delta17-desaturase converts ARA to EPA (although most delta17-desaturases also convert DGLA to ETA with less efficiency). The "delta6 pathway" can also use ALA as substrate to produce EPA: (1) delta15-desaturase converts LA to ALA; (2) delta6-desaturase converts ALA to steridonic acid (STA, 18:4n-3); (3) $C_{18/20}$ elongase converts STA to ETA; and (4) delta5-desaturase converts ETA to EPA. To date, no delta15-desaturase has been identified to convert LA to ALA with 100% efficiency. Transformed cells with a heterologous delta-15-desaturase gene will always contain both LA and ALA. Therefore, the "delta6 pathway" can simultaneously use both LA and ALA as primary substrates.

An alternate pathway for the biosynthesis of EPA is the "delta9-elongase/delta8-desaturase pathway" (delta9 pathway). It requires the following steps: (1) a delta9-elongase to convert LA to eicosadienoic acid (EDA, 20:2n-6); (2) a delta8-desaturase to convert EDA to DGLA; (3) a delta5-desaturase to convert DGLA to ARA; and (4) a delta17-desaturase to convert ARA to EPA. The "delta9 pathway" can also use ALA as a substrate to produce EPA in which (1) delta15-desaturase converts LA to ALA; (2) delta9-elongase converts ALA to eicosatrienoic acid (ETrA: 20:3n-3); (3) delta8-desaturase converts ETrA to ETA; and (4) delta5-desaturase converts ETA to EPA. If the transformed cells contain both LA and ALA, the "delta9 pathway" can simultaneously use both of them as primary substrates.

For the synthesis of DHA from EPA, two additional steps are required: (1) $C_{20/22}$-elongase converts EPA to docosapentaenoic acid (DPA, 22:5n-3) and (2) delta4-desaturase converts DPA to DHA. Some fish and seal oils contain omega-3 DPA. Incromega Trio oil, launched by Croda Ltd. in 2007, contains EPA (15%), DPA (7.5%), and DHA (30%). In some DHA oils originating from *Schizochytrium* (Barclay et al., 2005; see also Chapter 4), an omega-6 docosapentaenoic acid (DPA, 22:5n-6) is present that is not found in fish oil.

Feasibility Demonstration: Generation of *Y. lipolytica* Strains to Produce EPA via the "Delta6 Pathway"

The central research and development section of E.I. du Pont Nemous and Company initiated a program in 2002 to produce EPA in *Y. lipolytica*. Genes encoding for desaturases and $C_{18/20}$-elongases isolated from several different organisms (Knutzon et al., 1998; Huang et al., 1999; Sakuradani et al., 1999; Parker-Barnes et al., 2000) were licensed from the Ross Products Division of Abbott Laboratories to facilitate our effort to assemble a "delta6 pathway" for EPA biosynthesis in *Y. lipolytica*. We selected heterologous genes encoding a delta6-desaturase, a $C_{18/20}$-elongase, a delta5-desaturase (from *Morteriella alpina*), and a delta17-desaturase (from *Saprolegnia declina*). Since *Y. lipolytica* is physiologically and taxonomically very distant from *Saccharomyces cerevisiae*, the promoters from *S. cerevisiae* and other yeasts did not function in *Y. lipolytica* (data not shown). The heterologous genes introduced into *Y. lipolytica* needed *Y. lipolytica* promoters to ensure their expression. Therefore, we used the *Y. lipolytica* translation elongation factor (TEF) promoter (Muller et al., 1998) to individually drive the transcription of the four genes mentioned above.

As shown in Fig. 3.3B, integration of a single copy of these four genes into the genome of wild-type *Y. lipolytica* resulted in the synthesis of EPA, comprising about 3% of the total fatty acid methyl esters (FAME), with 34% of all fatty acids derived from the engineered pathway. This result conceptually proved that *Y. lipolytica* could be engineered to produce EPA and suggested that additional engineering improvements were needed to: 1) increase the carbon flux into the engineered pathway, 2) improve the efficiency of the $C_{18/20}$-elongases to convert more GLA into DGLA, and 3) enhance the expression of other pathway genes to increase the level of EPA. These needs were met via an integrated strategy, as we describe in the following sections.

Isolation and Use of Strong Promoters

In 2002, few promoters from *Y. lipolytica* had been isolated (Muller et al., 1998; Casaregola et al., 2000). Initially, we used the TEF promoter to drive expression of the genes encoding delta6-desaturase, C_{18-20}-elongase, delta5-desaturase, and delta17-desaturase. The TEF promoter is a constitutive one, but it is not a very strong promoter in other organisms.

Using genome-walking and PCR techniques, we isolated several *Y. lipolytica* promoters from the genes encoding for glyceraldehyde-3-phosphate dehydrogenase (GPD, E.C. 1.2.1.12), phosphoglycerate mutase (GPM, EC 5.4.2.1), and fructose-bisphosphate aldolase (FBA, E.C. 4.1.2.13). All of these promoters directed high-level gene expression. In order to compare their relative strengths, we made transcriptional fusions of these promoters with an *E. coli* β-glucuronidase (GUS) reporter gene (Jefferson, 1989), transformed *Y. lipolytica* strains with these constructs, and then analyzed the GUS activities individually (Fig. 3.4).

The results showed that the strength of the GPM promoter was approximately equivalent to that of the TEF promoter, while the GPD promoter was about one and on-half times as strong as the TEF promoter, and the FBA promoter was about three times as strong as the TEF promoter.

In the *N*-terminal coding region (169 bp) of the FBA gene, a 102 bp intron located between amino acids 20 and 21 was subsequently identified. Fusion of the FBA promoter and the *N*-terminal coding region which covers the first 23 amino acids and the intron (FBA$_{in}$), with the GUS reporter gene resulted in GUS activity about five times greater than the FBA promoter alone (Fig. 3.4). The 169 bp *N*-terminal coding region of FBA gene also enhanced the activity of other promoters when chimeric promoters were constructed (data not shown). These data demonstrate that there was a transcriptional enhancer located in the intron of the FBA gene.

A: GUS staining

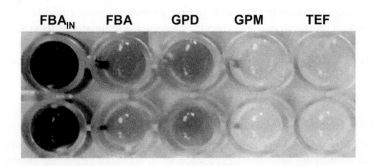

B: FluorometricGUS assays

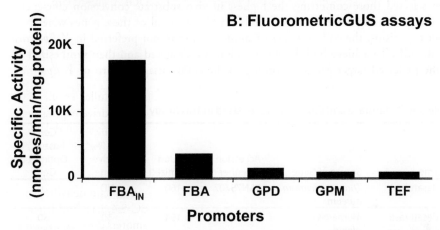

Fig. 3.4. Isolation of a set of promoters from *Y. lipolytica*. GUS activity was analyzed by fluorimetric determination of the production of 4-methylumbelliferone from corresponding beta-glucuronide. GUS staining was performed with the chromogenic substrate 5-bromo-4-chloro-3-indolyl beta-D-glucuronide (X-gluc). Transformed *Y. lipolytica* cells were kept in 50 mM sodium phosphate buffer (pH 7.0) containing X-gluc and incubated at 30 °C.

Promoters from a gene encoding glycerol-3-phosphate *O*-acyltransferase (GPAT, E.C. 2.3.1.15) and an ammonium transporter (YAT1, TC 2.A.49; GenBank Accession No. XM_504457) were also isolated. The YAT1 promoter was especially useful for directing the expression of omega-3 biosynthetic genes, since its activity increased approximately 35-fold under nitrogen-limiting conditions (data not shown). The relative strength of these promoters in oleaginous conditions was determined by quantitative GUS assays, and they are as follows: FBA$_{in}$ > YAT1 > FBA > GPD > GPAT > GPM = TEF.

Today, it is much easier to isolate promoters from *Y. lipolytica* based on the publication of the whole genome sequence (http://www.genolevures.org). Theoretically, there are about 6,500 promoters that could be used for different purposes (Goffeau et al., 1996; Dujon et al., 2004). Each promoter has unique property as a result of its use in different stages of cellular growth. A diverse promoter library could be generated through random mutagenesis using a specific promoter as template (Alper et al., 2005). Additional research on promoter characteristics is required before they can be utilized to maximize the expression of targeted genes in certain growth stages and conditions.

Synthesis of Codon-Optimized Genes for *Yarrowia lipolytica*

Initially, we screened wild-type genes that encode desaturases and elongases from a variety of sources by functional expression in *Y. lipolytica* (data not shown) and then selected those conferring the highest in vivo substrate conversion efficiencies for use in the engineered pathway (Table 3.A). Since all of these genes were from other organisms, the codon usages of most genes were not preferred in *Y. lipolytica* (Table 3.B). To achieve high-level expression, we designed and then synthesized all of the preferred target genes, according to the codon usage pattern of *Y. lipolytica*.

Table 3.A. Desaturase and Elongase Genes Used in This Study.

Enzyme	Source of Gene	Accession Number	Codon Optimized	Conv. Rate (%)* Native Gene	Conv. Rate (%)* Optimized Gene
Δ4-desaturase	*Thraustochytrium aureum*	AAN75707	166	ND	20
Δ5-desaturase	*Mortierella alpina*	AF067654	194	30	30
Δ5-desaturase	*Isochrysis galbana*	CCMP1323	193	7	32
Δ6-desaturase	*Mortierella alpina*	AF465281	144	37	51
Δ12-desaturase	*Fusarium moniliforme*	ABB8516	172	ND	74
Δ17-desaturase	*Saprolegia diclina*	AY373823	117	23	45
$C_{16/18}$ elongase	*Mortierella alpina*	AB071986	125	ND	43
$C_{18/20}$ elongase	*Mortierella alpina*	AX464731	85	30	47
$C_{18/20}$ elongase	*Thraustochytrium aureum*	AX464803	108	33	46
$C_{20/22}$ elongase	*Ostreococcus tauri*	AY591336	147	ND	67

Conversion rate: product / (substrate + product). ND: not done.

Table 3.B. Codon Usage in *Yarrowia lipolytica*.

Codon	AA	Frequency	Codon	AA	Frequency	Codon	AA	Frequency
GCA	A	0.142	GGG	G	0.068	AGC	S	0.120
GCC	A	0.418	GGT	G	0.263	AGT	S	0.082
GCG	A	0.113	CAC	H	0.603	TCA	S	0.094
GCT	A	0.336	CAT	H	0.397	TCC	S	0.251
AGA	R	0.164	ATA	I	0.043	TCG	S	0.187
AGG	R	0.047	ATC	I	0.497	TCT	S	0.266
CGA	R	0.431	ATT	I	0.459	TAA	*	0.420
CGC	R	0.086	CTA	L	0.061	TAG	*	0.388
CGG	R	0.152	CTC	L	0.261	TGA	*	0.192
CGT	R	0.119	CTG	L	0.386	ACA	T	0.172
AAC	N	0.780	CTT	L	0.152	ACC	T	0.421
AAT	N	0.220	TTA	L	0.020	ACG	T	0.139
GAC	D	0.640	TTG	L	0.120	ACT	T	0.268
GAT	D	0.360	AAA	K	0.210	TGG	W	1.000
TGC	C	0.503	AAG	K	0.790	TAC	Y	0.774
TGT	C	0.497	ATG	M	1.000	TAT	Y	0.226
CAA	Q	0.232	TTC	F	0.592	GTA	V	0.060
CAG	Q	0.768	TTT	F	0.408	GTC	V	0.322
GAA	E	0.287	CCA	P	0.126	GTG	V	0.382
GAG	E	0.713	CCC	P	0.430	GTT	V	0.237
GGA	G	0.327	CCG	P	0.124			
GGC	G	0.342	CCT	P	0.320			

We also incorporated the consensus sequence (5'-ACC<u>ATG</u>G-3') around the "ATG" translation initiation codon, manipulated the GC content to be about 54% (typical for genes from *Y. lipolytica*), and applied the general rules of RNA stability (Guhani-yogi & Brewer, 2001) into the synthetic genes. The data (Table 3.A) showed that the substrate conversion rate was increased in almost all of the codon-optimized genes, except for the delta5-desaturase gene derived from *Mortierella alpina*. In the case of the delta5-desaturase gene from *Isochrysis galbana*, *Y. lipolytica* strains transformed with the wild-type gene had a substrate conversion rate of 7%, while strains trans-formed with the codon-optimized gene had a substrate conversion rate of 32%, an increase in efficiency of about 4.6 times. Similarly, *Y. lipolytica* strains transformed with the wild type $C_{18/20}$-elongase gene from *M. alpina* had a substrate conversion rate of 30%, while the strains transformed with the codon-optimized gene had a substrate conversion rate of 47%, an increase in efficiency of about 1.6 times. The improved substrate conversion efficiency of these "codon-optimized" genes is hypothesized to result from more efficient translation of their encoded mRNAs in *Y. lipolytica*. All genes introduced into *Y. lipolytica* strains for the production of omega-3 and omega-6 fatty acids are now "codon-optimized."

"Pushing" and "Pulling" Carbons into the Engineered Pathway

The effect of gene copy number on the substrate conversion of selected steps in the engineered pathway was studied by generating two *Y. lipolytica* strains for GLA production. Strain 1 contained one copy each of a heterologous delta12-desaturase gene and a delta6-desaturase gene. Strain 2 contained two copies each of the heterologous delta12-desaturase gene and the delta6-desaturase gene. As shown in Fig. 3.5,

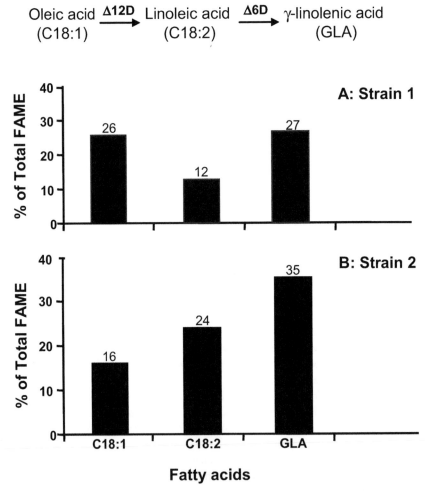

Fig. 3.5. Fatty acid profiles of transformed *Y. lipolytica* Strain 1 and Strain 2. Fatty acid analyses were performed, as stated in Figure 3.3. Strain 1 contained a single copy of a delta12 desaturase gene (*Δ12D*) and a delta6 desaturase gene (*Δ12D*). Strain 2 contained two copies each of *Δ12D* and *Δ6D*.

strain 1 produced about 27% GLA, 12% L A, and 26% oleic acid of total FAME. Strain 2 produced about 35% GLA, 24% LA, and 16% oleic acid of total FAME. Comparison of the fatty acid profiles revealed that strain 2 produced about 30% more GLA, about 38% less oleic acid, and about 50% more LA than strain 1. These data demonstrate that *Y. lipolytica* strains with two copies each of the heterologous delta12-desaturase gene and the delta6-desaturase gene pushed more carbon into the production of GLA.

Additional carbon was directed into the engineered pathway by amplifying the *M. alpina* $C_{16/18}$-elongase gene, which converts palmitic acid into stearic acid, and the *Fusarium moniliforme* delta12-desaturase gene, which converts oleic acid to LA (Fig. 3.1; Table 3.A). We "pulled" pathway intermediates to EPA by selectively amplifying the $C_{18/20}$-elongase, which converts GLA to DGLA,;the delta5-desaturase, which converts both DGLA to ARA and ETA to EPA; and the delta17-desaturase, which converts both DGLA to ETA and ARA to EPA.

Generation of *Y. lipolytica* Strain to Produce about 40% EPA of Total FAME via the "Delta6 Pathway"

A multi-step effort was applied to engineer strains of *Y. lipolytica* capable of producing high levels of EPA. The first step was to integrate a *F. moniliforme* delta12-desaturase gene, a *M. alpina* delta6-desaturase gene, and two $C_{18/20}$-elongase genes—one from *M. alpina* and another from *Thraustochytrium aureum* (Tables 3.1 and 3.3)—into the *URA3* locus of a wild-type *Y. lipolytica* ATCC #20362 strain to enable synthesis of DGLA. While there was no DGLA detected in the wild-type strain, gas chromatographic (GC) analyses of the lipid extracted from 5-FOA-resistant transformants identified a strain (M4) that produced about 8% of total FAME as DGLA (Table 3.C). Three copies of a *M. alpina* delta5-desaturase gene and a *Y. lipolytica* wild type *URA3* gene marker were then integrated into the *LEU2* locus of strain M4 to enable the synthesis of ARA. GC analyses of the lipid extracted from *Leu*⁻ transformants identified strain Y2047. While there was no ARA in the parental M4 strain, this new strain produced 10% of total FAME as ARA (Table 3.C). Three copies of a *Saprolegnia declina* delta17-desaturase gene and a *Y. lipolytica* wild-type *LEU2* gene marker were then designed to integrate into the *POX3* locus (GenBank Accession No. XP_503244) of strain Y2047, which would enable the synthesis of EPA. GC analyses of the lipid extracted from Leu+ transformants identified strain Y2048, producing 12.5% of total FAME as EPA, with 41.2% of total fatty acids produced via the engineered pathway (Table 3.C). Linear DNA fragment integration into the genome of *Y. lipolytica* could be achieved by both homologous and non-homologous recombination. We did not confirm that the DNA fragment containing three copies of the *S. declina* delta17-desaturase gene and the *Y. lipolytica* wild-type *LEU2* gene marker integrated at the *POX3* locus in the Y2048 strain.

Table 3.C. Genotypes and Fatty Acid Compositions of *Y. lipolytica* Strains.

Generation	0	1	2	3	4	5	7
Strain	WT	M4	Y2047	Y2048	Y2060	Y2072	Y2097
Gene							
Δ12-desaturase		1	1	1	1	2	3
Δ6-desaturase		1	1	1	1	1	2
$C_{18/20}$ elongase		2	2	2	2	3	4
Δ5-desaturase			3	3	3	4	5
Δ17-desaturase				3	3	3	3
$C_{16/18}$ elongase					1	1	2
Fatty Acid Composition (% Total FAME)	WT	M4	Y2047	Y2048	Y2060	Y2072	Y2097
C16:0	10	15	13	14	8	8	6
C16:1	12	4	9	10	4	4	1
C18:0	2	1	1	2	4	2	2
C18:1	39	5	9	17	20	17	9
C18:2	30	27	16	10	14	14	11
GLA		34	24	23	25	28	21
DGLA		8	8	3	4	4	4
ARA		10	1	1	2	1
ETA				2	2	2	3
EPA				13	13	15	40
Sum > C18:2	30	69	52	52	59	55	81

We next attempted to increase the amount of the C18:2 substrate for the delta6-desaturase by integrating a $C_{16/18}$-elongase gene into the *URA3* locus of strain Y2048. GC analyses of the lipid extracted from FOA-resistant transformants identified strain Y2060, containing about 11% less C16 fatty acids and 10% more C18 fatty acids than the parental strain (Table 3.3). Four more genes, encoding the *M. alpina* $C_{18/20}$-elongase, the *F. moniliforme* delta12-desaturase, the *I. galbana* delta5-desaturase, and the *Y. lipolytica* wild type *URA3* gene marker, were integrated into the native delta12-desaturase locus of strain Y2060 to generate strain Y2072. This new strain produced about 15% of total lipids as EPA (Table 3.C). We deliberately inactivated the native delta12-desaturase gene because the *F. moniliforme* homologue resulted in a more favorable distribution of the C18:2 substrate for the delta6-desaturase at the *sn*-2 position of phosphatidylcholine (results not shown). A second copy of the $C_{16/18}$-elongase gene was then integrated into the *URA3* locus of strain Y2072 to generate strain Y2072U3, which also produced 15% of the total lipids as EPA (Table 3.C). Although this second copy of the elongase gene did not increase the C18 content any further, the integration event restored the uracil auxotroph needed for subsequent transformation events. Finally, genes encoding the *M. alpina* $C_{18/20}$-elongase, a delta6-desaturase, a delta5-desaturase, a *F. moniliforme* delta12-desaturase gene, and the *Y. lipolytica* wild-type *URA3* gene marker were designed to integrate into the *LIP1* locus (Gen-Bank Accession No. Z50020) of strain Y2072U3 to generate strain Y2097. This new strain produced oil containing about 21% GLA and about 40% EPA of total FAME (Table 3.C). The Y2097 strain contained 19 copies of 10 different heterologous genes that integrated into its genome and produced about 70% of total fatty acids via the engineered pathway.

In certain clinical applications, EPA and GLA have a synergistic effect; therefore, it is beneficial to have oil containing both EPA and GLA. We also generated a series of strains that produced oil containing EPA and GLA with different ratios (EPA/GLA: 2/1, 3/2, 1/1, etc.) sufficient to suit various commercial market needs.

Comparison of the *Y. lipolytica* EPA oil with oils produced from microalgae and fish reveals that the yeast oil has several unique features (Table 3.D). The yeast oil contains less than 10% saturated fatty acids of total FAME, while fish and microalgae DHA oils contain more than 30% saturated fatty acids of total FAME. Additionally, the yeast oil also has about 80% PUFA of total FAME, while the fish and microalgae oils have only about 35% and 45% PUFA of total FAME, respectively.

Table 3.D. Comparison of Fatty Acid Composition among Different Oils.

Fatty Acid Composition (% Total FAME)	Yeast Oil	Krill Oil	Menhaden Oil	*C. cohnii* Oil	*Schizochytrium* Oil
C12:0				4	
C14:0		18	11	14–16	5–15
C16:0	6	19	19	10–14	24–28
C16:1	1	10		2–3	
C18:0	2	1	3		
C18:1	9	23	11	9–10	
C18:2	11	1	1		
GLA	21				
DGLA	4				
ARA	1				
ALA		1	1		
STA		3	3		
ETA	3				
EPA	40	6	18		
DPA (n6)					11–14
DPA (n3)			3		
DHA	0	2	13	50–60	35–40
Total Saturated	8	38	33	28–34	31–43
Total PUFA	81	13	39	50–60	46–54

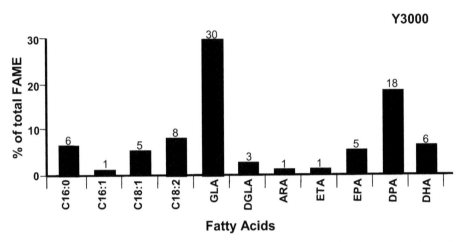

Fig. 3.6. Fatty acid profiles of transformed *Y. lipolytica* Strain Y3000. Fatty acid analyses were performed, as described in Fig. 3.3.

Generation of *Y. lipolytica* Strain to Produce about 6% DHA of Total FAME via the "Delta6 Pathway"

To demonstrate the synthesis of DHA (Fig. 3.1), we first generated strain Y2097U (*Ura3-*) in which the *URA3* gene was knocked out using a non-functional *Ura3* gene mutant. One copy each of the codon-optimized genes encoding an *Ostreococcus tauri* $C_{20/22}$-elongase (Meyer et al., 2004) and a *T. aureum* delta4-desaturase (Genbank accession No: AAN75707), as well as the *Y. lipolytica* wild type *URA3* gene, were integrated into the genome of strain Y2097U. Analyses of the lipid extracted from strain Y3000 revealed the presence of about 18% DPA and 6% DHA of total FAME (Fig. 3.6). There was no DPA or DHA detected in the Y2097U parental strain. The high substrate conversion efficiency (76%) of the *O. tauri* $C_{20/22}$-elongase is especially notable, compared to the lower conversion efficiencies of other elongases. The conversion efficiency of the delta4-desaturase was about 25% in strain Y3000. An increase in the copy number of the *T. aureum* delta4-desaturase gene or the use of more effective delta4-desaturase genes (Hashimoto et al., 2008; Hoffman et al., 2008) would improve the production of DHA.

Strategies to Improve EPA and DHA Production via the "Delta6 Pathway"

In order to engineer *Y. lipolytica* to produce EPA and DHA, we used a series of genes that encoded desaturases and elongases (Fig. 3.1; Table 3.A). Desaturases and elongases are multi-subunit enzyme complexes embedded in the ER membrane. Whereas desaturases introduce a double bond into an acyl-chain esterified to a phospholipid backbone, elongases catalyze the condensation of a malonyl-group to an acyl-chain esterified to coenzyme A (CoA). Thus, efficient omega-3 and omega-6 fatty acid biosynthesis requires active acyl-exchange between the phospholipid and CoA pools in the ER membrane.

One consistent feature of the fatty acid profile in the engineered *Y. lipolytica* strains (Table 3.C) was an accumulation of GLA in the triglyceride fraction (about 40% of total GLA; data not shown), suggesting that the $C_{18/20}$-elongation reaction caused a bottleneck in the pathway. The in vivo substrate conversion efficiencies, catalyzed by the delta6-, delta5- and delta17-desaturases, were about 86%, 90%, and 97%, respectively, indicating that these enzymes functioned well in strain Y2097. In contrast, the substrate conversion efficiency catalyzed by four copies of the $C_{18/20}$-elongase was only about 69% in that strain. We tried to increase the GLA substrate conversion efficiency by further amplifying the elongase gene, but with little effect. In contrast, when we fed GLA to the Y2097 cells, all of the substrate was rapidly and efficiently converted to DGLA (data not shown). These results suggested that, although strain Y2097 had sufficient elongase activity, the availability of substrate for elongation was limited by poor transfer of GLA to the CoA pool, coupled with effective competition from the acyltransferases catalyzing triglyceride biosynthesis. To increase GLA concentration in the CoA pool, we separately amplified genes encoding acyl-exchange reactions, including glycerol-3-phosphate acyltransferase (*GPAT*), lysophosphatidic acid acyltransferase (*LPAAT*), diacylglycerol acyltransferase (*DGAT1* and *DGAT2*), lysophosphatidylcholine acyltransferase (*LPCAT*), choline phosphotransferase (*CPT1*), acyl-CoA synthase (*ACS*), phospholipase C (*PCL1*), and phospholipase D (*SPO22*). These modifications each improved the GLA conversion efficiency up to 15% (data not shown). It is hypothesized that a combination of some of these genes could improve the efficiency of the $C_{18/20}$-elongase in engineered *Y. lipolytica* strains.

Growth conditions for cultivation of the recombinant strains were found to greatly affect the production of EPA. For example, a low percent EPA of the total FAME and minimal lipid content were produced in engineered *Y. lipolytica* strains when grown in regular minimal or YPD-rich media. In contrast, a high percent EPA of the total FAME and an increased lipid content were obtained in engineered strains cultured in nitrogen-limiting, oleaginous conditions. The growth temperature and pH also affected the production of EPA and the accumulation of oil. Engineered *Y. lipolytica* strains derived from ATCC #20362 produced the most oil with the highest amount of EPA at 32 °C and pH 7.0 in oleaginous conditions.

Production of EPA with No GLA via the "Delta9 Pathway"

In addition to the commercial market demand for oil containing both GLA and EPA, demand in alternate markets indicates they would prefer oils containing high amounts of EPA, with no GLA. As we mentioned above, oil from Y2097 contains about 21% GLA and about 40% EPA of total FAME. It would be difficult to convert all the intermediate GLA into subsequent products of the PUFA pathway using current genes encoding $C_{18/20}$-elongase and other supportive genes, such as those encoding various acyltransferases.

To produce oil that contains high levels of EPA, with no GLA, a "delta9 pathway" was introduced into *Y. lipolytica* (Fig. 3.1). Various genes that encode delta9-elongases and delta8-desaturases have been isolated (Wallis & Browse, 1999; Qi et al., 2002; data not shown). Upon the expression of codon-optimized delta9-elongase and delta8-desaturase genes in *Y. lipolytica*, yeast oil is now produced, which has a very high level of EPA, without the co-production of GLA (data not shown). Subsequent efforts have focused on the use of this pathway for the commercial production of EPA.

Summary

We have reported the development of a clean and sustainable source of omega-3 fatty acids by fermentation, which uses a metabolically engineered strain of the oleaginous yeast *Y. lipolytica*. While certain strains of *Y. lipolytica* can accumulate oil up to 40% of the dry cell weight, the only PUFA normally synthesized by the organism is LA (18:2). Coordinate expression of desaturase genes and elongase genes comprising a "delta6 pathway" was sufficient to demonstrate the synthesis of EPA. However, only an integrated strategy—based on the use of strong promoters, an increase in gene copy numbers, the push and pull of carbon into the engineered pathway, and the use of oleaginous condition—resulted in the generation of a high EPA production strain. This yeast oil has a unique fatty acid profile, with up to 40% EPA of total FAME, and a low saturated fatty acid content. It provides a clean and sustainable alternative to existing sources of omega-3 fatty acids. We also extended the pathway and demonstrated the synthesis of DHA. Versatile genetic tools provide a means to tailor the omega-3 fatty acid composition of the yeast oil for different applications. A commercial *Y. lipolytica* EPA production strain, based on expression of a "delta9 pathway," is currently being pursued.

Acknowledgments

We are very grateful to Ethel Jackson and John Pierce for their strong support. We thank Edgar Cahoon and Kevin Ripp for their technical input, Raymond Jackson and Catherine Byrne for their DNA-sequencing service, Jean-Francois Tomb and Shiping Zhang for bioinformatic analyses, and Neil Feltham and Kelley Norton for their legal guidance.

References

Alper, H.; C. Fischer; E. Nevoigt; G. Stephanopoulos. Turning Genetic Control through Promoter Engineering. *Proc Natl. Acad. Sci. USA* **2005**, *102*, 12678–12683.

Arita, M.; M. Yoshida; S. Hong; E. Tjonahen; J.N. Glickman; N.A. Petasis; R.S. Blumberg; C.N. Serhan. Resolvin E1, an Endogenous Lipid Mediator Derived from Omega-3 Eicosapentaenoic Acid, Protects against 2,4,6-Trinitrobenzene Sulfonic Acid-Induced Colitis. *Proc. Natl. Acad. Sci.* **2005,** *102,* 7671–7676.

Athenstaedt. K.; P. Jolivet; C. Boulard; M. Zivy; L. Negroni; J.M. Nicaud; T. Chardot. Lipid Particle Composition of the Yeast *Yarrowia lipolytica* Depends on the Carbon Source. *Proteomics* **2006,** *6,* 1450–1459.

Barclay, W.; C. Weaver; J. Metz. Development of a Docosahexaenoic Acid Production Technology Using *Schizochytrium*: A Historical Perspective. *Single Cell Oils,* 1ˢᵗ edn.; Z. Cohen, C. Ratledge, Eds.; AOCS Press: Champaign, IL, 2005; pp 36–52.

Bartel, P.L.; S. Fields. Yeast 2-Hybrid System; Oxford University: New York, 1997; 7, 109–147.

Belluzzi, A.; C. Brignola; M. Campieri; A. Pera; S. Boschi; M. Miglioli. Effect of an Enteric-Coated Fish-Oil Preparation on Relapses in Crohn's disease. *N. Engl. J. Med.* **1996,** *334,* 1557–1560.

Benatti. P.; G. Peluso; R. Nicolai; M. Calvani. Polyunsaturated Fatty Acids: Biochemical, Nutritional and Epigenetic Properties. *J. Am. Coll. Nutr.* **2003,** *23,* 281–302.

Beopoulos, A.; Z. Mrozova; F. Thevenieau; M.-T.L. Dall; I. Hapala; S. Papanikolaou; T. Chardot; J.-M. Nicaud. Control of Lipid Accumulation in the Yeast *Yarrowia lipolytica. Appl. Environ. Microbiol.* **2008,** *74,* 7779–7789.

Bordes, F.; F. Fudalej; V. Dossat; J.M. Nicaud; A. Marty. A New Recombinant Protein Expression System for High-Throughput Screening in the Yeast *Yarrowia lipolytica. J. Microbiol. Methods* **2007,** *70,* 493–502.

Breslow, J.L. N-3 Fatty Acids and Cardiovascular Disease. *Am. J. Clin. Nutr.* **2006,** *83,* 1477S–1482S.

Buydens-Branchery, L.; M. Branchey; J.R. Hibbeln. Associations between Increases in Plasma n-3 Polyunsaturated Fatty Acids Following Supplementation and Decreases in Anger and Anxiety in Substance Abusers. *Prog. Neuropsychopharmacol. Biol. Psychiatry* **2008,** *32,* 568–575.

Cardillo, R.; G. Fronza; C. Fuganti; P. Grasseli; A. Mele; D. Pizzi. Stereochemistry of the Microbial Generation of Delta-Decanolide, Gamma-Dodecanolide and Gamma-Nonanolide from C18 13-Hydroxy, C18 10-Hydroxy, and C19 14-Hydroxy Unsaturated Fatty Acids. *J. Org. Chem.* **1991,** *56,* 5237–5239.

Casaregola, S.; C. Neuvéglise; A. Lépingle; E. Bon; C. Feynerol; F. Artiguenave; P. Wincker; C. Gaillardin. Genomic Exploration of the Hemiascomycetous Yeasts: 17. *Yarrowia lipolytica. FEBS Lett.* **2000,** *487,* 95–100.

Chen, D.C.; J.M. Beckerich; C. Gaillardin; One-Step Transformation of the Dimorphic Yeast *Yarrowia lipolytica. Appl. Microbiol. Biotechnol.* **1997,** *48,* 232–235.

Costa, L.G. Contaminants in Fish: Risk-Benefit Considerations. *Arch. Ind. Hyg. Toxicol. (Croatia)* **2007,** *58,* 567–574.

Damude, H.G; A.J. Kinney. Engineering Oilseeds to Produce Nutritional Fatty Acids. *Physiol. Plant* **2008,** *132,* 1–10.

Davidow, L.S.; D. Apostolakos; M.M. O'Donnell; A.R. Proctor; D.M. Ogrydziak; R.A. Wing; I. Stasko; J.R. De Zeeuw. Integrative Transformation of the Yeast *Yarrowia lipolytica. Curr. Genet.* **1985,** *10,* 39–48.

Dujon, B.; D. Sherman; G. Fischer; P. Durrens; S. Casaregola; I. Lafontaine; J. De Montigny; C. Marck; C. Neuveglise; E. Talla; et al. Genome Evolution in Yeasts. *Nature* **2004,** *430,* 35–44.

Ercoli, B.; C. Fuganti; P. Grasselli; S. Servi; G. Allegrone; M. Barbeni. Stereochemistry of the Biogeneration of C-10 and C-12 Gamma Lactones in *Yarrowia lipolytica* and *Pichia ohmeri. Biotechnol. Lett.* **1992,** *14,* 665–668.

Féart. C; E. Peuchant; L. Letenneur; C. Samieri; D. Montagnier; A. Fourrier-Reglat; P. Barberger-Gateau. Plasma Eicosapentaenoic Acid Is Inversely Associated with Severity of Depressive Symptomatology in the Elderly: Data from the Bordeaux Sample of the Three-City Study. *Am. J. Clin. Nutr.* **2008,** *87,* 1156–1162.

Freeman. M.P.; M. Dacis; P. Sinha; K.L. Wisner; J.R. Hibbeln; A.J. Gelenberg. Omega-3 Fatty Acids and Supportive Psychotherapy for Perinatal Depression: A Randomized Placebo-controlled Study. J. Affect. Disord. **2008,** *110,* 142–148.

Freedman, S. D.; P.G. Blanco.; M.M. Zaman; J.C. Shea; M. Ollero; I.K. Hopper; D.A. Weed; A. Gelrud; M.M. Regan; M. Laposata; et al.. Association of Cystic Fibrosis with Abnormalities in Fatty Acid Metabolism. *N. Engl. J. Med.* **2004,** *350,* 560–569.

Funk, C.D. Prostaglandins and Leukotrienes: Advances in Eicosanoid Biology. *Science* **2001,** *294,* 1871–1875.

Gaillardin, C.; A.M. Ribet; H. Heslot. Integrative Transformation of the Yeast *Yarrowia lipolytica. Curr. Genet.* **1985,** *10,* 49–58.

Guhaniyoqi, J; G. Brewer. Regulation of mRNA Stability in Mammalian Cells. *Gene* **2001,** *265,* 11–23.

Goffeau, A.; B.G. Barrell; H. Bussey; R.W. Davis; B. Dujon; H. Feldmann; F. Galibert; J.D. Hoheisel; C. Jacq; M. Johnston; et al. Life with 6000 Genes. *Science* **1996,** *274(5287),* 546, 563–567.

Hall, W.L.; K.A. Sanders; T.A. Sanders; P.J. Chowienczyk. A High-Fat Meal Enriched with Eicosapentaenoic Acid Reduces Postprandial Arterial Stiffness Measured by Digital Volume Pulse Analysis in Healthy Men. *J. Nutr.* **2008,** *138,* 287–291.

Hashimoto, K.; A.C. Yoshizawa; S. Okuda; K. Kuma; D. Goto; M. Kanehisa. The Repertoire of Desaturases and Elongases Reveals Fatty Acid Variations in 56 Eukaryotic Genomes. *J. Lipid. Res.* **2008,** *49,* 183–191.

Hoffmann, M.; M. Wagner; A. Abbadi; M. Fulda; I. Feussner. Metabolic Engineering of ω3-Very Long Chain Polyunsaturated Fatty Acid Production by an Exclusively Acyl-CoA-dependent Pathway. *J. Biol. Chem.* **2008,** *283,* 22352–22362.

Hong, S.; K. Groner; P.R. Devchand; R.L. Moussignac; C.N. Serhan. Novel Docosatrienes and 17S-Resolvins Generated from Docosahexaenoic Acid in Murine Brain, Human Blood, and Glial Cells. Autacoids in Anti-inflammation. *J. Biol. Chem.* **2003,** *278,* 14677–14687.

Huang, Y.S.; S. Chaudhary; J.M. Thurmond; E.G. Bobik Jr; L. Yuan; G.M. Chan; S.J. Kirchner; P. Mukerji; D.S. Knutzon. Cloning of Delta12- and Delta6-desaturases from *Mortierella alpine* and Recombinant Production of Gamma-Linolenic Acid in *Saccharomyces cerevisiae. Lipids* **1999,** *34,* 649–659.

Itoh, M.; T. Suganami; N. Satoh; K. Tanimoto-Koyama; X. Yuan; M. Tanaka; H. Kawana; T. Yano; S. Aoe; M. Takeya; et al. Increased Adiponectin Secretion by Highly Purified

Eicosapentaenoic Acid in Rodent Models of Obesity and Human Obese Subjects. *Arterioscler. Thromb. Vasc. Biol.* **2007,** 918–1925.

Jacobson, T.A. Role of n-3 Fatty Acids in the Treatment of Hypertriglyceridemia and Cardiovascular Disease. *Am. J. Clin. Nutr.* **2008,** *87,* 1981S–1990S.

Jazayeri, S.; M. Tehrani-Doost; S.A. Keshavarz; M. Hosseini; A. Djazayery; H. Amini; M. Jalali; M. Peet. Comparison of Therapeutic Effects of Omega-3 Fatty Acid Eicosapentaenoic Acid and Fluoxetine, Separately and in Combination, in Major Depressive Disorder. *Aust. N. Z. J. Psychiatry* **2008,** *42,* 192–198.

Jefferson, R.A. The GUS Reporter Gene System. *Nature* **1989,** *342,* 837–838.

Kiel, J.A.; M. Veenhuis; I.J. van der Klei. Pex Genes in Fungal Genomes: Common, Rare or Redundant. *Traffic* **2006,** *7,* 1291–1303.

Kinsella, J.E.; B. Lokesh; S. Broughton; J. Whelan. Dietary Polyunsaturated Fatty Acids and Eicosanoids: Potential Effects on the Modulaton of Inflammatory and Immune Cells: An Overview. *Nutrition* **1990,** *6,* 24–44.

Kiy, T.; M. Rusing; D. Fabritius. Production of Docosahexaenoic Acid by the Marine Microalga, *Ulkenia* sp. *Single Cell Oils,* 1st edn.; Z. Cohen, C. Ratledge, Eds.; AOCS Press: Champaign, IL, 2005; pp 99–106.

Knutzon, D.S.; J.M. Thurmond; Y.S. Huang; S. Chaudhary; E.G. Bobik Jr; G.M. Chan; S.J. Kirchner; P. Mukerji; Identification of Delta5-desaturase from *Mortierella alpina* by Heterologous Expression in Bakers' Yeast and Canola. *J. Biol. Chem.* **1998,** *273,* 29360–29366.

Leaf, A. Historical Overview of n-3 Fatty Acids and Coronary Disease. *Am. J. Clin. Nutr.* **2008,** *87,* 1978S–19780S.

Lee, J.H.; J.H. O'Keefe; C.J. Lavie; R. Marchioli; W.S. Harris. Omega-3 Fatty Acids for Cardioprotection. *Mayo Clin. Proc.* **2008,** *83,* 324–332.

Lewis, S.J. Prevention and Treatment of Atherosclerosis: A Practitioner's Guide for 2008. *Am. J. Med.* **2009,** *122,* S38–S50.

Madzak, C.; C. Gaillardin; J.M. Beckerich. Heterologous Protein Expression and Secretion in the Non-conventional Yeast *Yarrowia lipolytica*: A Review. *J. Biotechnol.* **2004,** *109,* 63–81.

Mattey, M.; The Production of Organic Acids. *Crit. Rev. Biotechnol.* **1992,** *12,* 87–132

McColl, J.; Health Benefits of Omega-3 Fatty Acids. *Nutraceut.* **2003,** *4,* 35–40.

Metz, J.G.; P. Roessler; D. Facciotti; et al. Production of Polyunsaturated Fatty Acids by Polyketide Synthases in Both Prokaryotes and Eukaryotes. *Science* **2001,** *293,* 290–293.

Metz, J.G.; J. Kuner; B. Rosenzweig; J.C. Lippmeier; P. Roessler; R. Zirkle. Biochemical Characterization of Polyunsaturated Fatty Acid Synthesis in *Schizochytrium*: Release of the Products as Free Fatty Acids. *Plant Physiol. Biochem.* **2009,** *47,* 472–478.

Meyer, A; H. Kirsch; F. Domergue; A. Abbadi; P. Sperling; J. Bauer; P, Cirpus; T.K. Zank; H. Moreau; T.J. Roscoe; et al. Novel Fatty Acid Elongases and Their Use for the Reconstitution of Docosahexaenoic Acid Biosynthesis. *J. Lipid Res.* **2004,** *45,* 1899–1909.

Mita, T.; H. Watada; T. Ogihara; T. Nomiyama; O. Ogawa; J. Kinoshita; T. Shimizu; T. Hirose; Y. Tanaka; R. Kawamori. Eicosapentaenoic Acid Reduces the Progression of Carotid Intima-Media Thickness in Patients with Type 2 Diabetes. *Atherosclerosis* **2007,** *191,* 162–167.

Muller, S.; T. Sandal; P. Kamp-Hansen; H. Dalboge. Comparison of Expression Systems in the Yeasts *Sacromyces cerevisiae, Hansenula polymorpha, klyveromyces lactis, Schizosaccharomyces pompe* and *Yarrowia lipolytica*. Cloning of Two Novel Promoters from *Yarrowia lipolytica*. *Yeast* **1998**, *14*, 1267–1283.

Nutrition and Your Health: Dietary Guidelines for Americans. Dietary Guidelines Advisory Committee Report. U.S. Departments of Health and Human Services and Agriculture. Part D, Section 4: Fats. 2005.

Parker-Barnes, J.M.; T. Das; E. Bobik; A.E. Leonard; J.M. Thurmond; L.T. Chaung; Y.S. Huang; P. Mukerji. Identification and Characterization of an Enzyme Involved in the Elongation of n-6 and n-3 Polyunsaturated Fatty Acids. *Proc. Natl. Acad. Sci.* **2000**, *97*, 8284–8289.

Qi, B.; F. Beaudoin; T. Fraser; A.K. Stobart; J.A. Napier; C.M. Lazarus. Identification of a cDNA Encoding a Novel C18-Delta9-Polyunsaturated Fatty Acid-Specific Elongating Activity from the Docosahexaenoic Acid (DHA)-Producing Microalga, *Isochrysis galbana. FEBS Lett.* **2002**, *510*, 159–165.

Qi, B.; T. Fraser; S. Mugford; G. Dobson; O. Sayanova; J. Butler; J.A. Napier; A.K. Stobart; C.M. Lazarus. Production of Very Long Chain Polysaturated Omega-3 and Omega-6 Fatty Acids in Plants. *Nat. Biotechnol.* **2004**, *22*, 739–745.

Ratledge, C. Single Cell Oils for the 21st Century. *Single Cell Oils*, 1st edn.; Z. Cohen, C. Ratledge, Eds.; AOCS Press: Champaign, IL, 2005; pp 1–20.

Ross, B.M.; J. Seguin; L.E. Sieswerda. Omega-3 Fatty Acids as Treatments for Mental Illness: Which Disorder and Which Fatty Acid? *Lipids Health Dis.* **2007**, *6*, 21.

Saito, Y.; M. Yokoyama; H. Origasa; M. Matsuzaki; Y. Matsuzawa; Y. Ishikawa; S. Oikawa; J. Sasaki; H. Hishida; H. Itakura; et al. Effects of EPA on Coronary Artery Disease in Hypercholesterolemic Patients with Multiple Risk Factors: Sub-analysis of Primary Prevention Cases from the Japan EPA Lipid Intervention Study (JELIS). *Atherosclerosis* **2008**, *200*, 135–140.

Sakuradani, E.; M. Kobayashi; S. Shimizu. Delta6-Fatty Acid Desaturase from an Arachidonic Acid-Producing *Mortierella* Fungus. Gene Cloning and Its Heterologous Expression in a Fungus, *Aspergillus. Gene* **1999**, *238*, 445–453.

Salari, P.; A. Rezaie; B. Larijani; M. Abdollahi. A Systematic Review of the Impact of n-3 Fatty Acids in Bone Health and Osteoporosis. *Med. Sci. Monit.* **2008**, *14*, 37–44.

Satoh, N.; A. Shimatsu; K. Kotani; N. Sakane; K. Yamada; T. Suganami; H. Kuzuya; Y. Ogawa. Purified Eicosapentaenoic Acid Reduces Small Dense LDL, Remnant Lipoprotein Particles, and C-Reactive Protein in Metabolic Syndrome. *Diabetes Care* **2007**, *30*, 144–146.

Siddiqui, R.A.; M. Zerouga; M. Wu; A.M. Castillo; K. Harvey; G.P. Zaloga; W. Stillwell. Anticancer Properties of Propofol-Docosahexaenoate and Propofol-Ecosopentaenoate on Breast Cancer Cells. *Breast Cancer Res.* **2005**, *7*, R645–654.

Sinclair, A.; J. Wallace; M. Martin; N. Attar-Bashi; R. Weisinger; D. Li. The Effects of Eicosapentaenoic Acid in Various Clinical Conditions. *Health Lipids*; C.C. Akoh, Q.M. Lai, Eds.; AOCS Press; Champaign, IL, 2005; pp 361–394.

Slagsvold, J.E.; C.H. Prettersen; T. Follestad; H.E. Krokan; S.A. Schonberg. The Antiproliferation of EPA in HL60 Cells Is Mediated by Alterations in Calcium Homeostasis. *Lipids* **2009**, *44*, 103–113

Tanaka, K.; Y. Ishikawa; M. Yokoyama; H. Origasa; M. Matsuzaki; Y. Saito; Y. Matsuzawa; J. Sasaki; S. Oikawa; H. Hishida; et al. Reduction in the Recurrence of Stroke by Eicosapentaenoic

Acid for Hypercholesterolemic Patients: Subanalysis of the JELIS Trial. *Stroke* **2008**, *39*, 2052–2058

Teitelbaum, J.E.; W.A. Walker. Review: The Role of Omega 3 Fatty Acids in Intestinal Inflammation. *J. Nutr. Biochem.* **2001**, *12*, 21–32.

Vernis, L.; A. Abbas; M. Chasles; C.M. Gaillardin; C. Brun; J.A. Huberman; P. Fournier. An Origin of Replication and a Centromere Are Both Necessary to Establish a Replicative Plasmid in the Yeast *Yarrowia lipolytica*. *Mol. Cell. Biol.* **1997**, *17*, 1996–2004.

Wallis, J.G.; J. Browse. The Delta8-desaturase of *Euglena gracilis*: An Alternate Pathway for Synthesis of 20-Carbon Polyunsaturated Fatty Acids. *Arch. Biochem. Biophys.* **1999**, *365*, 307–316.

Wen, Z.; F. Chen. Prospects for Eicosapentaenoic Acid Production Using Microorganisms. *Single Cell Oils*, 1ˢᵗ edn.; Z. Cohen, C. Ratledge, Eds.; AOCS Press: Champaign, IL, 2005; pp 138–160.

Weterings. E.; D.J. Chen. The Endless Tale of Non-homologous End-Joining. *Cell. Res.* **2008**, *18*, 114–124.

Wolzschu, D.L.; F. W. Chandler; L. Ajello; D.G. Ahearn. Evaluation of Industrial Yeasts for Pathogenicity. *Sabouraudia* **1979**, *17*, 71–78.

Wynn, J.; P. Behrens; A. Sundarajan; J. Hansen; K. Apt. Production of Single Cell Oils by Dinoflagellates. *Single Cell Oils*, 1ˢᵗ edn.; Z. Cohen, C. Ratledge, Eds.; AOCS Press: Champaign, IL, 2005; pp 86–98.

Yokoyama, M.; H. Origasa; M. Matsuzaki; Y. Matsuzawa; Y. Saito; Y. Ishikawa; S. Oikawa; J. Sasaki; H. Hishida; H. Itakura; et al. Effects of Eicosapentaenoic Acid on Major Coronary Events in Hypercholesterolaemic Patients (JELIS): A Randomised Open-Label, Blinded Endpoint Analysis. *Lancet* **2007**, *369*, 1090–1098.

4

Development of a Docosahexaenoic Acid Production Technology Using *Schizochytrium*: Historical Perspective and Update

William Barclay, Craig Weaver, James Metz, and Jon Hansen

Martek Biosciences Boulder Corporation, 4909 Nautilus Ct. North, Suite 208, Boulder, Colorado, USA 80301

Introduction

Importance of Long Chain Omega-3 Fatty Acids in Human Health

Recognition of the nutritional importance of the long chain polyunsaturated omega-3 fatty acids (LC-PUFAs), eicosapentaenoic acid (EPA, 20:5n-3), and docosahexaenoic acid (DHA, 22:6n-3), began emerging in the mid-1980s. Research during the previous two decades has indicated that: (1) prior to the introduction of modern agricultural practices, human intake of these fatty acids was much higher; (2) populations whose intake of these fatty acids were higher exhibited a lower occurrence of chronic diseases, including cardiovascular diseases, arthritis, asthma, diabetes, etc.; and (3) unique compounds, called eicosanoids, were made from these fatty acids, and these eicosanoids had protective effects against the initiation and progressive development of these diseases (Nettleton, 1995; Simopoulos, 1991).

In the 1980s, the major source of LC-PUFAs in the human diet was fish or fish oil capsules. Consumption of fish in North America has historically been low, and fish oil capsules met with poor acceptance because of their inferior organoleptic (taste and odor) characteristics. Attempts to utilize fish oils as an ingredient in foods were also largely unsuccessful because of their strong fishy taste and odor problems. Attempts to improve the organoleptic characteristics of fish oil remained unsuccessful, and concerns emerged regarding the levels of environmental contaminants (PCBs, dioxins, and mercury) reported in fish (Ahmed, 1991), which continue to trouble us to this day (Jacobs et al., 2004).

It was recognized at the time that the primary source of omega-3 LC-PUFAs in fish was not from biosynthesis by the fish, but rather from their diet, a result of the plankton they consumed. We hypothesized that if one could develop a way to economically cultivate a microbial strain that was particularly rich in omega-3 LC-PUFAs, it would provide an acceptable source of these compounds that would avoid the organoleptic and environmental contaminant problems found in fish and fish oils. Although fish oils contain both EPA and DHA, and much of the research demonstrating the importance of omega-3 LC-PUFAs was conducted with fish oil, we made a decision to initially focus on the production of DHA. Emerging research indicated that, while EPA had significant anti-inflammatory effects (Yamashita et al., 2004), DHA was important in infant nutrition (British Nutrition Foundation, 1992), had positive effects on blood lipids and cardiovascular health in adults (Gaudette & Holub, 1991), was more effectively retained in the human body (10 weeks for DHA versus only about 2 weeks for EPA), and could be readily retroconverted to EPA (Von Schacky et al., 1985; Fischer et al., 1987). These and other factors suggested that DHA might be a better form of omega-3 LC-PUFAs to function as a food ingredient or nutritional supplement for consumers.

Sources of Omega-3 LC-PUFAs

In the 1980s, marine microalgae were recognized to be the primary microbial produc-ers of omega-3 LC-PUFAs. Bacteria and yeast were not known to produce omega-3 LC-PUFAs, although later, a few strains of bacteria were discovered in the intestines of deep sea fish that could make EPA, but only in phospholipid form (Yazawa et al., 1992). Since phospholipids comprise only a maximum of about 10–15% of the dry weight of microbes, a phospholipid producer of long chain omega-3 fatty acids would not prove to be an especially economical producer of DHA for food uses. However, a few microal-gae and fungi can accumulate up to about 80% of their dry weight as triacylglycerols. Cost effective production of DHA from microbes would require the use of microbial strains that could produce large amounts of triacylglycerols (Ratledge, 1984).

Although outdoor ponds and photobioreactors were the most commonly utilized production systems for microalgae at the time (Burlew, 1976; Barclay & McIntosh, 1986), the use of these systems was very expensive because of low productivities, low cell concentrations requiring large quantities of water to be centrifuged to recover the biomass, a lack of controls to maximize formation of the target product, and an inability to ensure food grade production conditions in outdoor production systems. Although heterotrophic production of algae in large scale fermentors had been proposed, research in this area has focused primarily on employing freshwa-ter strains of microalgae, which are not effective producers of omega-3 LC-PUFAs (Zajic et al., 1970). Additionally, since omega-3 LC-PUFAs were associated with the photosynthetic membranes of microalgae, there was initially a bias against using a non-photosynthetic production system for producing these fatty acids.

Two competitive processes for the production of long chain omega-3 fatty acids were under development at the time. Japanese researchers were attempting to modify a fermentation process for arachidonic acid (ARA, 20:4n-6), using the fungus *Mortierella* which can also produce EPA (Yamada et al., 1992). Process modifications included growing the fungus at low temperatures or chilling the harvested biomass for several days to induce the production of EPA. However, these modifications resulted in very low overall productivity. The second competitive process involved culturing the slow growing apochlorotic dinoflagellate *Crypthecodinium* via fermentation (Kyle et al., 1992). We believed that significant improvements in these two technologies would be necessary before they could enable the inexpensive production of DHA required to produce a cost-effective ingredient in foods.

Need for Alternative Technology Development

Recognizing the need for an alternative source of omega-3 LC-PUFAs, while reflecting on the state of the competitive technologies being developed at the time, we made a decision to pursue the development of an alternative technology using a different set of assumptions. First, we felt there were no known strains of microalgae that could be developed to effectively produce omega-3 LC-PUFAs at a cost that could compete with fish oil. Development of a microbial production technology would require isolation and identification of new, natural over-producers of omega-3 fatty acids. Second, these strains would need to produce long-chain omega-3 fatty acids at relatively high temperatures, a property which was counterintuitive to the belief at the time, that microalgae normally made these fatty acids under cold conditions in an attempt to keep their membranes flexible and functioning. Third, we recognized the fundamental need to address the significant corrosion problems associated with growing marine microorganisms in conventional stainless steel fermentors. Others had grown marine microalgae in glass-lined fermentors to avoid corrosion. However, on a large scale, these fermentors were very expensive, and only a few were commercially available around the world. Glass-lined fermentors only partially solved the problem, ignoring the associated issues with corrosion in down-stream processing equipment (which cannot be glass-lined) and with the disposal of waste high salt fermentation media at locations far from the ocean. Since the most likely production candidates (if they existed) would be marine strains, new strategies to cultivate them under conditions of very low salinity would have to be developed to minimize the corrosive effects of seawater-like fermentation media.

After identifying all of the problems with alternative technologies, we decided that the best way to rapidly, competitively, and comprehensively address all of these issues would be to utilize a bio-rational approach to strain collection/isolation/screening. This practice would best ensure the enrichment and identification of strains with, hopefully, all or most of the desirable production characteristics and greatly facilitate technology development. This paper describes the design and implementation of a bio-rational approach to microalgal technology development that resulted in the intense study and ultimate commercialization of products based on a unique group of microorganisms. Additionally, we outline the numerous benefits to microbial technology development that result from the use of this type of approach. Since we began developing this technology in 1987, many researchers have recognized the biotechnological potential of this group of microorganisms (thraustochytrids) (Lewis et al., 1999) and have contributed to the understanding of both their production potential (Bajpai et al., 1991a; Ward & Li, 1994; Singh & Ward, 1996; Singh et al., 1991; Nakahara et al., 1996), and their potential to produce compounds other than fatty acids (Jenkins et al., 1999; Aki et al., 2003).

Bio-rational Approach to Technology Development

A bio-rational approach first involves identifying the most desirable characteristics that an ideal production strain should possess, both in terms of the target production system and the target product. A strategy of isolating such a strain (if it exists in nature) is then developed, drawing on concepts derived from ecology, physiology, biochemistry, evolution, etc. The ultimate goal of our proposed technology was to inexpensively produce large quantities of DHA from heterotrophic microalgae in conventional stainless steel fermentors, utilizing glucose as the carbon source. With this in mind, a bio-rational collection/isolation/screening program was designed to isolate microorganisms with the combination of characteristics outlined in Table 4.A. These characteristics were deemed desirable in a microorganism that would be utilized for the economical production of omega-3 fatty acids. In a complementary fashion, a strategy was devised to collect microorganisms from habitats that experience a wide range of temperatures and salinities. Examples of these locations included micro-habitats in the following locations: marine tide pools, estuaries, and inland saline lakes, playas and springs. The working hypothesis was that, in addition to their ability

Table 4.A. Targeted Production and Product-Related Characteristics Selected for by the Bio-rational Collection/Isolation/Screening Program and the Bio-rationale behind Selection of These Characteristics.

Characteristics Important for Production in Fermentors:	Bio-rationale:
a) capable of heterotrophic growth	— produce on inexpensive carbon source (corn syrup)
b) unicellular (non-filamentous) and ≤ 25 μm in diameter	— minimize mixing energy needed in fermentor
c) thermotolerant (grow above 30 ºC)	— higher temperatures equate to faster production
d) euryhaline (grow especially at low salinity)	— minimize corrosion in the stainless steel fermentors
Characteristics Important for Utilization of Whole-Cell Algae or Extracted DHA Oil in Foods:	
e) high content of omega-3 highly unsaturated fatty acids	— high content of target product
f) preferably, a low content of saturated and omega-6 fatty acids	— undesirable, in some applications
g) preferably, non-pigmented, white or colorless cells	— target is to be invisible ingredient in foods

to grow at low salinities and high temperatures, some of these strains might be natural over-producers of DHA—an adaptation for their survival in these harsh environments—with DHA possibly playing a key role in helping to stabilize membrane function in rapidly fluctuating environments.

The implementation of this collection/isolation/screening program has been outlined in detail by Barclay (1992). Briefly, water samples from target aquatic niches were run through a sandwich filtration system that eliminated cells greater than 25 μm; polycarbonate filters containing biomaterial from 1–25 μm were placed on low and medium salinity nutrient agar plates and incubated at 30 °C; clear and white colonies, which were not yeast colonies, were picked, cultured, and then analyzed for their fatty acid profile and content by gas chromatography.

The initial results indicated that the bio-rational collection/isolation strategy had selected for two groups of microalgae or algae-like microbes: diatoms, and thraustochytrids. The newly isolated diatom strains, however, primarily produced EPA and were relatively slow-growing. Additionally, diatoms were known to have silica cell walls which might make them difficult to grow at high concentrations in fermentors because of problems with fragility and supplying silica in the fermentation medium.

We, therefore, decided to focus on the thraustochytrid strains. Thraustochytrids were relatively unknown at the time from a biotechnological perspective. Thraustochytrids are microalgae or microalgae-like microorganisms. The earliest research on thraustochytrids placed them in the fungi because of their heterotrophic nature and superficial resemblance to chytrids (Sparrow, 1936). However, later analyses using molecular biology techniques demonstrated that thraustochytrids were not fungi, but instead were related to the heterokont algae (Cavalier-Smith et al., 1996). Prior to the late 1980s, most studies on thraustochytrids had focused on their distribution, taxonomy, ultrastructure, and physiology. Because thraustochytrids are generally very lightly pigmented, they probably were under-reported in phytoplankton samples. Later analyses indicated that they comprise a significant portion of the phytoplankton community (Raghukumar & Shaumann, 1993).

Prior to our implementation of the biorational collection/isolation program, there had not been any reports in the scientific literature that thraustochytrids were good lipid producers. There were two reports that thraustochytrids had long-chain omega-3 fatty acids in their lipids (Ellenbogen et al., 1969; Findlay et al., 1986), but there were no reports of how much lipid these organisms could produce. In fact, as late as 1992, Kendrick and Ratledge (1992), reporting on their own data and that of Bajpai et al. (1991b), noted that *Thraustochytrium* contained oil contents of only 10–15% of the biomass and that the prospects for promoting a high amount of lipid in this microorganism appeared to be limited due to its lack of a key enzyme for lipid accumulation, ATP:citrate lyase. However, contrary to this belief, by using the biorational collection/isolation/screening procedures outlined above, we were able to isolate thraustochytrid strains that could produce large amounts of lipid and, at the same time, have all of the other targeted characteristics for fermentation production and product use. Most significantly, these strains also had three important properties: (1)

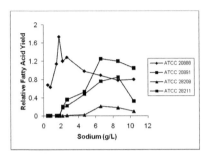

Fig. 4.1. Enhanced fatty acid yield of *Schizochytrium* sp. (ATCC 20888) (Isolated by the bio-rational approach) at low sodium concentrations, compared to fatty acid yield of previously-known strains of Thraustochytrids, which had been isolated by other methods. Experimental conditions are outlined in Barclay (1992). Fatty acid yields are are presented as relative to the average fatty acid yield of ATCC 20888 over the entire salinity range.

the isolated strains had very fast growth rates (6–9 doublings per day) compared with prior art strains of thraustochytrids (3–5 doublings per day); (2) high amounts of omega-3 fatty acids as a percentage of total fatty acids were produced even at elevated temperatures (30 °C); and (3) strains with enhanced lipid and DHA production could be grown at low salinities, and production at low salinities enhanced the production of DHA-rich lipids (Fig. 4.1).

Even though there was a reduction in omega-3 fatty acid production at the very lowest salinities (Table 4.B), we discovered that by varying both the types of sodium salt and nitrogen salt used in the fermentation media, the DHA concentration in the oil could be enhanced from 21 to 43% without a significant loss in biomass production at the very lowest salinities tested (Table 4.C). Since chloride is the component of marine media that contributes the most to metal corrosion, we worked to develop cultivation with a minimum content of both sodium and chloride (Barclay, 1994). Chloride levels for fermentation medium used in stainless steel fermentors are recommended to be less than 300 ppm in order to minimize stress cracking and corrosion. The results, shown in Table 4.C, indicated that sodium sulfate could be an effective substitute for sodium chloride in the cultivation media, and we were eventually able to develop media that resulted in excellent production at chloride levels less than

Table 4.B. Production of *Schizochytrium* sp. (ATCC 20888) in Low Salinity Medium. Experimental conditions are described in detail in Barclay (1994).

Na Conc. (g/L)	Cl Conc. (g/L)	Biomass (g/L)	Fatty Acids (% dwt)	DHA (% dwt)	Final Glucose (g/L)
4.88	7.1	1.8 ± 0.6	35.4 ± 1.0	9.2 ± 0.5	0.0
3.90	3.9	5.7 ± 0.7	37.0 ± 0.7	10.0 ± 0.3	0.2
2.93	4.3	1.7 ± 0.4	43.0 ± 0.2	10.9 ± 0.1	0.2
1.95	2.9	1.7 ± 0.6	29.8 ± 0.7	8.4 ± 0.1	1.6
0.98	1.4	0.4 ± 0.6	10.6 ± 2.4	3.6 ± 1.0	4.3

Table 4.C. Effect of Sodium Sulfate-Based Medium Compared with Sodium Chloride-Based Medium on Fatty Acid Content in *Schizochytrium* sp. (ATCC 20888). Experimental conditions are described in detail in Barclay (1994).

N Source (g/L)	DHA (% dwt)	TFA* (% dwt)	Biomass (g/L)
A) Na salt = sodium chloride; N source = sodium glutamate			
3.0	5.6	11.2	1.74
2.5	5.5	10.8	1.71
2.0	5.5	11.0	1.65
1.5	7.1	20.3	1.39
B) Na salt = sodium chloride; N source = peptone			
3.0	7.4	21.0	1.34
2.5	8.8	27.4	1.21
2.0	6.3	28.9	1.18
1.5	10.4	42.1	1.16
C) Na salt = sodium sulfate; N source = sodium glutamate			
3.0	8.7	31.9	1.34
2.5	8.7	31.9	1.34
2.0	9.5	41.4	1.30
1.5	8.9	43.6	1.26

*TFA = total fatty acid

300 ppm (Table 4.D). Thus, by experimenting to solve the significant corrosion problem presented by growing marine organism in stainless steel fermentors, we surprisingly found a way to also increase DHA production in strains of thraustochytrids. An additional surprise was that the sodium sulfate-based medium also appeared to further limit ectoplasmic net formation in *Schizochytrium*. One of the unique characteristics of thraustochytrids is that they produce ectoplasmic nets, which tend to link the cells

Table 4.D. Fermentation of *Schizochytrium* sp. (ATCC 20888) at Low Chloride Concentrations. Experimental conditions are described in detail in Barclay (1994).

Chloride Conc. (mg/L)	Na 2.37 g/L Biomass Yield (mg/L)	Na 4.0 g/L Biomass Yield (mg/L)
0.1	198 ± 21	158 ± 48
0.7	545 ± 120	394 ± 151
15.1	975 ± 21	758 ± 163
30.1	1140 ± 99	930 ± 64
59.1	1713 ± 18	1650 ± 14
119.1	1863 ± 53	1663 ± 46
238.1	1913 ± 11	1643 ± 39

together. These nets can lead to clumped growth in liquid cultures resulting in slower growing cells. The most rapidly growing strain isolated by the biorational method *Schizochytrium* sp. ATCC 20888, however, was uniquely characterized in part by very limited ectoplasmic net production (Porter, personal communication). Surprisingly, ectoplasmic net production appeared to be further reduced when we grew the cells in a sodium sulfate-based medium, an additional benefit resulting from our attempts to solve the corrosion problem.

Preliminary Toxicology Screen

In the late 1980s, little was known about the physiology and biochemistry of thraustochytrids as a group, and nothing had been published related to the presence or absence of toxins in this group of microalgae. As a result, we made a decision early in the technology development process to conduct a preliminary screen for toxins to ensure that this group of microalgae, which was becoming a focus in technology development and was novel from a biotechnological perspective, would be safe to use for the production of a DHA-rich oil that would function as a food ingredient. A literature survey on the phylum Heterokonta, to which the thraustochytrids belong, did not turn up any reports of toxins from thraustochytrids, and, in the Heterokonta, there were reports of only two very remotely related types of microalgae that produced toxins: (1) the diatom *Pseudonitzschia* sp. (and, possibly, one species of *Chrysochromulina* sp.), which produced a neurotoxin called domoic acid (Villac et al., 1993) and (2) two species of *Prymnesium* (*P. parvum* and *P. patelliferum*), which produced toxic phospho-proteolipids called prymnesium toxin (Silo, 1971). To be safe, we analyzed *Schizochytrium* sp. biomass for the presence of domoic acid using the standard HPLC method of Lawrence et al. (1991), and no trace of this compound was detected. Additionally, the presence of pyrmnesium toxin was evaluated with the sensitive bioassays of Vanhaecke et al. (1981) and Larsen et al. (1993). The results indicated an absence of prymnesium toxin in *Schizochytrium*. Results from additional simple toxicity screens (monitoring growth and reproduction)—which used laying hens and two aquaculture organisms, *Artemia nauplii* and *rotifers*, that were fed *Schizochytrium* and *Thraustochytrium* biomass produced in 1 liter fermentors—also suggested that there were no other toxins present in the thraustochytrid biomass. These preliminary safety results gave us the confidence to move ahead and begin to scale-up the production technology.

Fermentation Scale-up

Scale-up of the fermentation technology was conducted in partnership with Kelco Biopolymers, San Diego, a division of Merck Pharmaceutical and, later, a division of Monsanto. This effort was combined with a classical strain improvement program conducted in-house. The progress of scale-up work, in terms of several DHA production parameters, is summarized in Table 4.E. Initial productivity levels in our

Table 4.E. Major Process Changes that Occurred during Scale-up of the Fermentation Process and Their Positive Effect on Various Parameters, Related to DHA Production.

Major Scale-up Process Changes	Cell Conc. (g/L)	DHA Titer (g/L)	Lipid Conc. (% FAME)	DHA in Oil (% FAME)	DHA in Cells (% dry wt.)	DHA Productivity (g/L.h)
Initial Lab Scale [low salinity, monosodium glutamate as N source (Barclay, 1992)]	21	2	39	26	10	0.05
Scale-up of Lab Fermentation to 10,000 L Scale	40	4	35–40	25–28	8–12	0.07
Low Chloride + Ammonium sulfate as N source	65–70	8	35–40	32	10–13	0.10
Fed Batch + Low DO + +direct harvest on drum dryers + ammonium as N source (Bailey et al., 2003)	170–210	40–50	50–73	35–45	22–25	0.45–0.55

DO = dissolved oxygen

lab at the 14 liter stage were about 22 g/L biomass in 48 hours using 40 g/L glucose (Barclay, 1992). Several approaches were utilized to accomplish these objectives, including a variety of statistically designed multivariate experiments, limiting nutrient chemostat studies, initial rate investigations, and a variety of nutrient feeding and other process control strategies. Kelco engineers quickly doubled both DHA titer and productivity at the 14 liter level. Using their data and a media optimization program—which exploited alternative nitrogen sources and enhanced lipid production at low salinities, especially at low chloride levels—the engineers were able to achieve cell concentrations of 65–70 g/L with a DHA titer of about 8 g/L, doubling DHA productivity to 0.1 g/L/hr.

Scale-up of the fermentation technology continued with a focus on the use of fed-batch approaches to supply both the carbon and nitrogen sources. The fed-batch carbon feed allowed for optimum control of glucose concentration throughout the fermentation. This control was particularly important for the very high cell densities that were demonstrated through further process improvements, since much higher levels of carbon source were required. A variety of sterilization and feeding system designs were also employed in this rapid and successful scale-up. A simple, but effective and operationally simple, approach to nitrogen feeding was employed in which the nitrogen source was used to automatically regulate the pH of the medium. Overall, scale-up to large scale was very successful. Furthermore, the rapid improvements and scale-up made in this program were incorporated into the safety assessment discussed below.

During this time, we also developed data indicating that low dissolved oxygen levels greatly facilitated DHA production. This finding ran contrary to the conventional biochemical wisdom that high dissolved oxygen levels facilitate DHA production, as molecular oxygen was considered necessary for the enzymatic process that involves the formation of double (unsaturated) bonds in fatty acids (Hadley, 1985). Until recently, it was assumed that all PUFAs were synthesized via variations of a single, basic biochemical pathway (i.e., a soluble fatty acid synthase system produced medium chain saturated fatty acids, which were then modified by a series of membrane-associated elongation and oxygen-dependent desaturation reactions). The ability to achieve high biomass densities with enhanced DHA levels, while running at very low levels of dissolved oxygen, had the potential to allow for greater flexibility in the fermentor configuration and operations at scale, thereby enabling options for reducing capital and energy requirements. Evidence for an alternative pathway for PUFA synthesis came initially from studies of marine bacteria. Although many bacteria, such as *E. coli*, were known to produce monounsaturated fatty acids (primarily *cis*-vaccenic acid, 18:1 Δ-11), it was generally thought that bacteria were incapable of PUFA synthesis. This opinion changed when DeLong and Yayanos (1986) reported detecting EPA and DHA in several strains of psychrophilic marine bacteria. A second major advance occurred when Yazawa (1996) identified a segment of genomic DNA from an EPA-producing marine bacterium (*Shewanella* strain SCRC-2738) that, when transferred to *E. coli*, conferred EPA synthesis on those cells. Further analysis led to identification of the genes necessary and sufficient for EPA accumulation in the transformed *E. coli* (Yazawa, 1996). Although the proteins encoded by introduced *Shewanella* genes contained regions with homology to enzymes associated with fatty acid synthesis, it was not clear how these specific activities contributed to EPA synthesis. Based on biochemical assays of extracts from *Shewanella*, Watanabe et al. (1997) suggested that EPA was the product of an aerobic desaturase and elongase pathway, presumably encoded by five genes.

Metz et al. (2001) provided a different interpretation. They noted that all of the enzymatic activities required for *de novo* synthesis of unsaturated fatty acids could be identified in the *Shewanella* proteins. Some of the individual domains closely resembled those found in polyketide synthases, while others more closely resembled those of fatty acid synthesis systems. Polyketide synthase systems use the same basic reactions of FAS, but often do not complete the cycles resulting in highly derivatized end products that typically contain keto and/or hydroxyl groups, as well as carbon-carbon double bonds in the *trans* configuration (Hopwood & Sherman, 1990). Additionally, the linear products of polyketide synthase systems can be cyclized to produce very complex molecules such as antibiotics, toxins, and many other secondary products (Hopwood & Sherman 1990; Keating & Walsh, 1999). Domains with homology to known fatty acid desaturases were notably lacking in the *Shewanella* proteins. Culturing *E. coli* that expressed the *Shewanella* genes under anaerobic conditions provided additional evidence, eliminating the potential involvement of desaturases; EPA accumulation was not inhibited by the lack of oxygen. A rationale for the anaerobic incorporation of *cis* double bonds is provided by the presence of two domains with homology to the *E. coli* FabA protein. FabA is a dual function enzyme responsible for the formation of *cis* vaccenic acid via a dehydration reaction (forming a *trans* double bond in the carbon chain) in combination with a reversible *trans-cis* isomerization.

In combination, the data indicated that four of the *Shewanella* genes encoded subunits of an enzyme (a PUFA synthase) capable of *de novo* synthesis of EPA from the same small precursor molecule (malonyl-CoA) utilized in normal fatty acid synthesis (Yu et al., 2000). The other gene, encoding a single domain protein, was identified as encoding an accessory enzyme (a phosphopantetheinyl transferase) that activates several domains of the complex (acyl carrier protein domains) by the addition of an essential co-factor. Genes encoding proteins with homology to the *Shewanella* EPA synthesis proteins have been found in several other PUFA containing marine bacteria, including ones that accumulate DHA instead of EPA (Facciotti et al., 2000; Allen & Bartlett, 2002; Tanaka et al., 1999).

When first recognized as such, PUFA synthase gene sequences were available from only a few bacteria. The sequencing of bacterial genomes has greatly increased in recent years, and now a search of public databases will reveal dozens of new homologs to these genes. All of the bacteria that contain the full synthase gene set are marine species, and most of those come from only a few genera (e.g., *Shewanella*, *Photobacterium*, and *Moritella*). Although direct links between the presence of the PUFA synthase gene cluster and the accumulation of PUFA have only been established in a few cases, it now seems likely that all of the PUFA found in marine bacteria are produced via the PUFA synthase pathway. Although some diversity has been found, the basic domain architecture can be recognized in all of the predicted translation products of these gene sets (Okuyama et al., 2007).

We, with the assistance of our collaborators at the Calgene Campus of Monsanto (Davis, California), discovered that *Schizochytrium* contained genes encoding proteins homologous to those associated with EPA synthesis in the marine bacterium *Shewanella* sp. strain SCRC-2738. The possibility that *Schizochytrium* possessed a similar pathway for DHA synthesis provided a rationale for the seemingly contradictory oxygen effects on DHA production. A key aspect of this novel PUFA biosynthetic system is that the *cis* double bonds are incorporated into the growing carbon chain via a dehydrase/ isomerization mechanism rather than being inserted by oxygen-dependent desaturation reactions (Metz et al., 2001). *Schizochytrium* sp (ATCC 20888) represented the first eukaryotic microorganism ever identified to contain a PUFA synthase system.

Since the hypothetical scheme for the PUFA synthase reactions was presented in 2001 by Metz et al. (2001), some progress has been made by directly testing the roles of individual domains of these complex enzymes. One remarkable feature of all of the PUFA synthases examined to date is the presence of multiple (4–10) tandem acyl-carrier protein (ACP) domains. Jiang et al. (2008) examined the ACP repeats found in *Shewanella japonica* using directed mutagenesis and expression of those synthase genes in *E. coli*. They demonstrated that each of the individual ACP domains could support PUFA (specifically EPA) production by itself. The amount of PUFA that accumulated in the heterologous host depended on the number of functional ACP domains. An essential step of some cycles in the proposed PUFA synthase reaction scheme is the generation of fully reduced carbons via the action of an enoyl-reductase (ER) activity. Based on its homology to a triclosan-resistant ER from *Streptococcus pneumonia*, this activity was proposed to be associated with the protein (domain) encoded by the bacterial *pfaD* gene (Metz et al., 2001). That hypothesis was recently confirmed by detection of ER activity during in vitro assays of the PfaD protein from *Shewanella oneidensis* MR-1, which is overexpressed in *E. coli* (Bumpus et al., 2008). The findings related to the ACP and ER domains obtained with the bacterial systems may be directly applicable to the PUFA synthase found in *Schizochytrium*. Although far fewer gene sequences are available for the thraustochytrid PUFA synthases, progress has also been made in that area. Hauvermale et al. (2006) expressed the *Schizochytrium* PUFA synthase genes in *E. coli* and found production of both DHA and DPA n-6 in those cells. This result demonstrated that the single synthase present in *Schizochytrium* produces the two PUFAs that accumulate in that organism. The independence of PUFA synthases from the standard fatty acid synthesis pathways and the relatively pure end-product profiles has made them attractive targets for heterologous expression. In particular, the productivity and temperature characteristics of the thraustochytrid PUFA synthases may be of value for particular applications. Additionally, it was recently demonstrated that the DHA and DPA n-6 products of the *Schizochytrium* synthase are released as free fatty acids and that this thioesterase activity is an integral component of the enzyme (Metz et al., 2009). Knowledge of the release mechanism has implications for attempts to use the system for the production of PUFA in alternative hosts, such as oil seed crop plants.

The discovery that *Schizochytrium* sp. did not require oxygen for long chain unsaturated fatty acid production synthesis helped us begin to grow this organism at very high densities. The final scale-up of fermentation involved the incorporation of four key components: (1) low chloride media; (2) fed-batch supplies of carbon and nitrogen sources; (3) use of low dissolved oxygen levels to induce DHA formation; and (4) harvest of the resulting biomass by recovering the fermentation broth directly on drum driers (avoiding the use of centrifugation). Using this process, outlined in Bailey et al. (2003), biomass yields of greater than 200 g/L could be achieved with DHA concentrations greater than 20% in dried biomass. This process resulted in DHA productivities of greater than 0.55 g/L.h.

During the fermentation scale-up process from laboratory scale 14-liter fermentors to 150,000-liter commercial fermentors, the following key improvements in DHA production parameters were achieved: (1) final cell dry weight concentrations improved from 21 to 170–210 g/L; (2) DHA concentrations in the resulting oil increased from 26% to 35–45% of total fatty acids; (3) DHA concentration, as a percentage of the dry weight of the cells, increased from 10% to 22–25%; and (4) DHA productivity increased from 0.05 to 0.45–0.55 g/L.h. a 10-fold increase. These cell concentrations and DHA productivities are the highest ever reported from a microbial technology for the production of highly unsaturated fatty acids.

Electron micrographic analysis of the oil bodies of *Schizochytrium* sp., using high pressure freeze substitution, has suggested the presence of secondary and tertiary

Table 4.F. Example Fatty Acid Profile of the Refined Oil Produced from the *Schizochytrium* Fermentation Process.

Fatty Acid	Fatty Acid Content	
	(FAME, mg/g)	(% TFA)
12:0	2	0.2
14:0	71	7.6
14:1	1	0.1
15:0	4	0.4
16:0	205	22.1
16:1	4	0.4
18:0	5	0.5
18:1	7	0.8
18:2	5	0.5
18:3(n-6)	3	0.3
18:4	3	0.3
20:0	1	0.1
20:4(n-6)	9	1.0
20:4(n-3)	9	1.0
20:5(n-3)	21	2.3
22:0	1	0.1
22:5(n-6)	154	16.6
22:6(n-3)	380	40.9
24:0	2	0.2
24:1	2	0.2
Others	40	4.3

semicrystalline structures of triacylglycerols within the oil bodies (Ashford et al., 2000). A similar observation has also been reported in oil bodies of *Schizochytrium limacinum* (Morita et al., 2006). The conventional commercial methods used to extract and refine DHA-rich oil, resulting from the *Schizochytrium* biomass, have been outlined by Zeller et al. (2001), and the fatty acid profile of the refined DHA-rich oil is illustrated in Table 4.F.

Safety of the Biomass and Extracted DHA-Rich Oil

After the scale-up of a representative and reproducible fermentation process was completed, a series of standard toxicology studies were conducted to demonstrate the safety of this whole cell, DHA-rich biomass and extracted oil. The strategy behind the selection of these studies and the results is discussed in greater detail in Chapter 15, Safety Evaluations of Single Cell Oils, of this collection. These studies included: (1) mutagenicity studies (in vitro Ames and mammalian cell line, in vitro human peripheral blood lymphocytes, and in vivo mouse micronucleus) (Hammond et al., 2002); (2) a 13-week sub-chronic rat feeding study (Hammond et al., 2001b); (3) developmental toxicity evaluation in rats and rabbits (Hammond et al., 2001a); and (4) a single-generation rat reproduction study (Hammond et al., 2001c). An acute gavage trial (extracted oil) in mice was also conducted along with target animal safety studies in laying hens (Abril et al., 2000), broiler chickens, and swine (Abril et al., 2003). The NOAEL (No Observable Adverse Effect Level) for all safety tests was at the highest dietary level of the oil tested in each study. Based on the amount of DHA-rich oil fed to the test animals, these levels were: 1) one generation rat reproduction study: 8800 mg oil/kg.day; 2) rat teratology study: 8400 mg oil/kg.day; 3) rabbit teratology study: 720 mg oil/kg.day; 4) 13-week rat feeding study: 1600 mg oil/kg.day; 4) laying hen target animal safety test (16 wks): 1600 mg oil/kg.day; 5) broiler target animal safety test (7 wks): 950 mg oil/kg.day.

The results of these safety studies allowed a GRAS (Generally Recognized as Safe) determination to be made by a committee of food safety experts, qualified by scientific training and experience, for use of the whole cell biomass in poultry feed and for use of the extracted DHA-rich oil in foods. Subsequently, in 2004, the US FDA did not object to the GRAS determination for the safe use of DHA oil from *Schizochytrium* in foods. Novels Foods approval was also awarded for use of the oil in foods in Australia and New Zealand in 2002 and in Europe in 2003.

Products

The DHA-rich biomass resulting from this fermentation process has been evaluated for use in enrichment applications in aquaculture (Barclay & Zeller, 1996) and is now used throughout the world in aquaculture feeds for enrichment applications and as a nutritional ingredient in feed for larval fish and shrimp (e.g., see www.aquafauna.com/Diets&Feeds.htm). It has also been used in poultry feed in numerous countries to produce DHA-enriched eggs (e.g., see www.goldcirclefarms.com) and poultry meat, and it has been used in Europe to produce DHA-enriched milk from dairy cows. Research has indicated that the DHA-rich biomass can also be fed to swine to enrich the resulting meat with DHA without any compromise in meat characteristics, sensory attributes or tenderness (Abril & Barclay, 1998; Marriott et al., 2002a,b). The extracted DHA-rich oil is sold in capsules as a nutritional supplement and has been used commercially as an ingredient in nutritional bars, soy milk drinks, bakery products, including breads, cookies, and muffins, and in dairy products, including milk, yogurt, spreads, margarines, and cheeses.

New Research Related to *Schizochytrium* Production

The success of the technology described in this paper has stimulated further research into the commercial potential of *Schizochytrium* and other thraustochytrids. This research has focused primarily on four key areas: (1) the isolation of new strains;

(2) the identification of culture conditions for improving DHA production; (3) different fermentation strategies for producing *Schizochytrium*; and (4) the use of thraustochytrid cultures, including *Schizochytrium*, for producing compounds other than PUFAs and in various waste recycling processes. Research on the isolation of new strains has centered on finding new strains with different PUFA profiles (Burja et al., 2006), higher contents of specific PUFAs or lipid types (Fan et al., 2007; Patell & Rajyashri, 2007), and/or improved productivities for PUFAs (Burja et al., 2006; Fan et al., 2007; Komazaura et al., 2004).

With regard to increasing the LCPUFA content of lipids extracted from thraustochytrids, research has centered both on controlling culture conditions and on post-harvest processing of the cells prior to extraction of the oils. Raghukumar et al. (2002) and Raghukumar & Jain (2005) described a method for increasing the EPA and DHA content of selected thraustochytrid strains, including *Schizochytrium*, by cultivating them in high viscosity culture media. Other investigators have enhanced the growth, lipids, and/or fatty acid profile of thraustochytrids, including *Schizochytrium*, by adjusting cultivation time, initial pH, and culture media composition (Wu et al., 2005); by optimizing temperature and salinity (Unagul et al., 2006b); by modifying the ionic composition of the medium (Unagul et al., 2006a); or by incorporating specific additives (i.e. cyanocoblamin and *p*-toluic acid) in the fermentation media (Shirasaka et al., 2005). Song et al. (2007) used response surface methodology to optimize several fermentation parameters and improve biomass and DHA production in *Schizochytrium limacinum*. The four most important variables they identified were the temperature, aeration rate, agitation, and growth stage of the fermentation inoculum. Additionally, other investigators have explored continuous, or fed-batch culture strategies for improving the production of *Schizochytrium* (Wumpelmann, 2007; Ganuza & Izquiredo, 2007; Ganuza et al., 2008).

The use of *Schizochyrium* cultures in several waste recycling systems has also been suggested. It has long been known that *Schizochytrium* can grow very well on glycerol. Exploiting this fact, Chi et al. (2007b) and Pyle et al. (2008) recently proposed the production of DHA from *Schizochytrium* grown on waste glycerol, which is a by-product of biodiesel production processes. Chi et al. (2007a) also proposed producing DHA from *Schizochytrium* grown on hydrolyzed waste cull potatoes, while Yamasaki et al. (2006) described using barley-based distillery wastewaters for cultivating *Schizochytrium* to produce polyunsaturated fatty acids.

The creation of products other than PUFAs in *Schizochytrium* and other thraustochytrids is actively being investigated, and many of these concepts have been reviewed in depth by Raghukumar (2008). Long (2002) and others (Aki et al., 2003; Burja et al., 2006; Yamasaki et al., 2006; Yamaoka, 2004) have suggested the cultivation of *Thraustochytrium* and *Schizochytrium* strains for the production of xanthophylls and carotenoids including lutein, astaxanthin, β-carotene, and canthaxanthin. Jain et al. (2005) and Kollenmareth et al. (2007) have proposed *Schizochytrium*-based processes for the simultaneous production of DHA-rich lipids and exocellular polysaccharides. Additionally, Fan and Chen (2007) have described thraustochytrids as potentially rich sources of squalene and sterols under certain culture conditions.

Conclusion

A bio-rational approach was used to develop a production technology for long-chain omega-3 fatty acids. This innovative collection/isolation/screening program resulted in the isolation of strains of a unique group of microorganisms, thraustochytrids, with significant biotechnological potential. The best strains contained all of the targeted production characteristics that enabled the inexpensive production of long-chain omega-3 fatty acids. These strains were unicellular and small (<25 μm), exhibited fast growth, excellent omega-3 fatty acid rich lipid production at low salinities, and high omega-3 fatty acid content, even at high temperatures. The strains also exhibited two completely unexpected characteristics: (1) enhanced long-chain omega-3 fatty acid production in low salinity, especially, low chloride production media and (2) enhanced production of long-chain omega-3 fatty acids under very low dissolved oxygen concentrations. Scale-up of this technology, using a strain of *Schizochytrium* sp., resulted in a fermentation technology with the highest cell concentrations (>200 g/L), highest unsaturated lipid productivities (>0.55 g/L.h), and highest cellular concentrations of DHA (>22 % of dry weight) ever reported for a microbial production technology. In addition to the unique combination of fermentation characteristics, another key to these exceptional results was the unique genetic system that these strains possessed for the production of LC-PUFAs. These strains represented the first eukaryotic microbes ever discovered to utilize a polyketide-like PUFA synthase system for production of LC-PUFAs. The success of this technology continues to stimulate new research into *Schizochytrium*, including isolating and identifying new strains of this organism, making improvements in the production technology, expanding the range of products produced from *Schizochytrium*, and finding new uses for its novel PUFA synthase gene set.

Acknowledgments

From a technical perspective, we owe a special debt of gratitude to those who participated on the Project Alpha Team and other scale-up teams at Kelco Biopolymers, San Diego, including Patrick Adu-Peasah, Mark Applegate, Richard Bailey, Vince Bevers, Brewster Brock, Don Crawford, Sandra Diltz, Don DiMasi, Brian Englehart, Olivia Faison, Jim Flatt, Eunice Flores, Chris Guske, Jon Hansen, Ian Hodgson, Mike Hughs, Tony Javier, Bojolane Kan, Tatsuo Kaneko, Richard Langston, Jerry Lucas, Mark Macias, Dave Matthews, Darlene McGhee, Megan McMahon, Brian Mueller, Pete Mirrasoul, Heather Nenow, Jay Peard, Jerry Peik, Tom Ramseier, Paul Roessler, Craig Ruecker, Wayne Sander, John Stankowski, Vladimir Sluzky, Robert Speights, George Veeder, Eugene Vivino, Melanie Writer, Sam Zeller, and Ruben Abril, Patricia Abril, Amy Ashford, Frank Overton, and Kent Meager, formerly of

OmegaTech, Inc. This project would never have succeeded without the passion and leadership of Craig Ruecker and Wayne Sander of Kelco Biopolymers who kept the scale-up and regulatory approval projects successfully moving forward despite a series of significant management changes at Kelco.

References

Abril, R.; W. Barclay. Production of Docosahexaenoic Acid-Enriched Poultry Eggs and Meat Using an Algae-Based Feed Ingredient. *World Rev. Nutr. Diet* **1998,** *83,* 77–88.

Abril, J.R.; W. Barclay; P.G. Abril. Safe Use of Microalgae (DHA GOLD) in Laying Hen Feed for the Production of DHA-Enriched Eggs. *Egg Nutrition and Biotechnology;* J.S. Sim, S. Nakai, W. Guenter, Eds.; CAB International: Wallingford UK, 2000; pp 197–202.

Abril, R.; J. Garrett; S.G. Zeller; W.J. Sander; R.W. Mast. Safety Assessment of DHA-Rich Microalgae from *Schizochytrium* sp. V. Target Animal Safety/Toxicity Study in Growing Swine. *Regul. Toxicol. Pharm.* **2003,** *37,* 73–82.

Ahmed, F.E. *Seafood Safety.* Food and Nutrition Board, Institute of Medicine; National Academy Press: Washington, DC, 1991.

Aki, T.; K. Hachida; M. Yoshinaga; Y. Kata; T. Yamasaki; S. Kawamoto; T. Kakizono; T. Maoka; S. Shigeta; O. Suzuki; K. Ono. Thraustochytrid as a Potential Source of Carotenoids. *J. Am. Oil Chem. Soc.* **2003,** *80,* 789–794.

Allen, E.E.; D.H. Bartlett. Structure and Regulation of the Omega-3 Polyunsaturated Fatty Acid Synthase Genes from the Deep-Sea Bacterium *Photobacterium profundum* Strain SS9. *Microbiology* **2002,** *148,* 1903–1913.

Ashford, A.; W.R. Barclay; C.A. Weaver; T.H. Giddings; S. Zeller. Electron Microscopy May Reveal Structure of Docosahexaenoic Acid-Rich Oil within *Schizochytrium* sp. *Lipids* **2000,** *35,* 1377–1386.

Bailey, R.B.; D. DiMasi; J.M. Hansen; P.J. Mirrasoul; C.M. Ruecker; G.M. Veeder; T. Kaneko; W.R. Barclay. U.S. Patent 6,607,900, 2003.

Bajpai, P.K.; P. Bajpai; O.P. Ward. Optimization of Production of Docosahexaenoic Acid (DHA) by *Thraustochytrium aureum* ATCC 34304. *J. Am. Oil Chem. Soc.* **1991a,** *68,* 509–514.

Bajpai, P.; P.K. Bajpai; O.P. Ward. Production of Docosahexaenoic Acid by *Thraustochytrium aureum. Appl. Microbiol. Biotechnol.* **1991b,** *35,* 706–710.

Barclay, W.R. U.S. Patent 5,130,242, 1992.

Barclay, W.R. U.S. Patent 5,340,742, 1994.

Barclay, W.R.; R.P. McIntosh, Eds. Algal Biomass Technologies: An Interdisclipinary Perspective. *Beihefte zur Nova Hedwigia* **1986,** *83,* 1–273.

Barclay, W.R.; S. Zeller. Nutritional Enhancement of n-3 and n-6 Fatty Acid in Rotifers and *Artemia nauplii* by Feeding Spray-Dried *Schizochytrium* sp. *J. World Aquacul. Soc.* **1996,** *27,* 314–322.

Bowles, R.D.; A.E. Hunt; G.B. Bremer; M.G. Duchars; R.A. Eaton. Long-Chain n-3 Polyunsaturated Fatty Acid Production by Members of the Marine Protistan Group the Thraustochytrids: Screening of Isolated and Optimization of Docosahexaenoic Acid Production. *J. Biotechnol.* **1990,** *70,* 193–202.

British Nutrition Foundation. *Unsaturated Fatty Acids: Nutritional and Physiological Significance*; The British Nutrition Task Force, Eds.; Chapman & Hall: London, 1992.

Bumpus, S.B.; N.A. Magarvey; N.L. Kelleher; C.T. Walsh; C.T. Calderone. Polyunsaturated Fatty Acid-like *Trans*-Enoyl Reductases Utilized in Polyketide Biosynthesis. *J. Am. Chem. Soc.* **2008,** *130,* 11614–11616.

Burja, A.M.; H. Radianingtyas; A. Windust; C.J. Barrow. Isolation and Characterization of Polyunsaturated Fatty Acid Producing *Thraustochytrium* Species: Screening of Strains and Optimization of Omega-3 Production. *Appl. Microbiol. Biotechnol.* **2006,** *72,* 1161–1169.

Burlew, J.S., Ed. *Algal Culture From Laboratory to Pilot Plant.* Carnegie Institution of Washington: Washington, DC, 1976.

Cavalier-Smith, T.; M.T.E.P. Allsopp; E.E. Chao. Thraustochytrids Are Chromists Not Fungi: 18sRNA Signatures of Heterokonta. *Philos. Trans. R. Soc. London B: Biol. Sci.* **1994,** *346,* 387–397.

Chi, Z.; B. Hu; Y. Liu; C. Frear; Z. Wen; S. Chen. Production of ω-3 Polyunsaturated Fatty Acids from Cull Potato Using an Algae Culture Process. *Appl. Biochem. Biotech.* **2007a,** *136–140,* 805–816.

Chi, A.; D. Pyle; Z. Wen; C. Frear; S. Chen. A Laboratory Study of Producing Docosahexaenoic Acid from Biodiesel-Waste Glycerol by Microalgal Fermentation. *Process Biochem.* **2007b,** *42,* 1537–1545.

DeLong, E.F.; A.A. Yayanos. Biochemical Function and Ecological Significance of Novel Bacterial Lipids in Deep-Sea Prokaryotes. *Appl. Environ. Microbiol.* **1986,** *51,* 730–737.

Ellenbogen, B.B.; S. Aaronson; S. Goldstein; M. Belsky. Polyunsaturated Fatty Acids of Aquatic Fungi: Possible Phylogenetic Significance. *Comp. Biochem. Physiol.* **1969,** *29,* 805–811.

Facciotti, D.; J. Metz; M. Lassner. U.S. Patent 6,140,486, 2000.

Fan, K.W.; F. Chen. Production of High-Value Products by Marine Microalgae Thraustochytrids. *Bioprocessing for Value-Added Products From Renewable Resources: New Technologies and Applications*; S.T. Yang, Ed.; Elsevier: Amsterdam, 2007; pp 293–324.

Fan, K.W.; Y. Jiang; Y.W. Faan; F. Chen. Lipid Characterization of Mangrove Thraustochytrid— *Schizochytrium mangrovei. J. Agric. Food Chem.* **2007,** *55,* 2906–2910.

Findlay, R.H.; J.W. Fell; N.K. Coleman. Biochemical Indicators of the Role of Fungi and Thraustochytrids in Mangrove Detrital Systems. *Biology of Marine Fungi*; S.T. Moss, Ed.; Cambridge University Press: London, 1980; pp 91–103.

Fischer, S.; A. Vischer; V. Preac-Mursic; P.C. Weber. Dietary Docosahexaenoic Acid is Retroconverted in Man to Eicosapentaenoic Acid, which Can Be Quickly Transformed to Prostaglandin I$_3$. *Prostaglandins* **1987,** *34,* 367–375.

Ganuza, E.; A.J. Anderson; C. Ratledge. High-Cell-Density Cultivation of *Schizochytrium* sp. in an Ammonium/pH-Auxotat Fed-Batch System. *Biotechnol. Lett.* **2008,** *30,* 1559–1564.

Ganuza, E.; M.S. Izquiredo. Lipid Accumulation in *Schizochytrium* G13/2S Produced in Continuous Culture. *Appl. Microbiol. Biotechnol.* **2007,** *76,* 985–990.

Gaudette, D.C.; B.J. Holub. Docosahexaenoic Acid (DHA) and Human Platelet Reactivity. *J. Nutr. Biochem.* **1991,** *2,* 116–121.

Hadley, N.F. *The Adaptive Role of Lipids in Biological Systems*; John Wiley & Sons: New York, 1985; pp 29–31.

Hammond, B.G.; D.A. Mayhew; J.F. Holson; M.D. Nemec; R.W. Mast; W.J. Sander. Safety Assessment of DHA-Rich Microalgae from *Schizochytrium* sp. II. Developmental Toxicity Evaluation in Rats and Rrabbits. *Regul. Toxicol. Pharmacol.* **2001a,** *33,* 205–217.

Hammond, B.G.; D.A. Mayhew; M.W. Naylor; C.M. Ruecker, R.W. Mast; W.J. Sander. Safety Assessment of DHA-Rich Microalgae from *Schizochytrium* sp. I. Subchronic Rat Feeding Study. *Regul. Toxicol. Pharmacol.* **2001b,** *33,* 192–204.

Hammond, B.G.; D.A. Mayhew; K. Robinson; R.W. Mast; W.J. Sander. Safety Assessment of DHA-Rich Microalgae from *Schizochytrium.* III. Single-Generation Rat Reproduction Study. *Regul. Toxicol. Pharmacol.* **2001c,** *33,* 356–362.

Hammond, B.G.; D.A. Mayhew; L.D. Kier; R.W. Mast; W.J. Sander. Safety Assessment of DHA-Rich Microalgae from *Schizochytrium* sp. IV. Mutagenicity Studies. *Regul. Toxicol. Pharmacol.* **2002,** *35,* 255–265.

Hauvermale, A.; J. Kuner; B. Rosenzweig; D. Guerra; S. Diltz; J.G. Metz. Fatty Acid Production in *Schizochytrium* sp.: Involvement of a Polyunsaturated Fatty Acid Synthase and a Type I Fatty Acid Synthase. *Lipids* **2006,** *41,* 739–747.

Hopwood, D.A.; D.H. Sherman. Molecular Genetics of Polyketides and Its Comparison to Fatty Acid Biosynthesis. *Annu. Rev. Microbiol.* **1990,** *24,* 7–66.

Jacobs, M.N.; A. Covaci; A. Gheorghe; P. Schepens. Time Trend Investigation of PCBs, PBDEs, and Organochlorine Pesticides in Selected n-3 Polyunsaturated Fatty Acid Rich Dietary Fish Oil and Vegetable Oil Supplements: Nutritional Relevance for Human Essential n-3 Fatty Acid Requirements. *J. Agric. Food Chem.* **2004,** *52,* 1780–1788.

Jain, R.; S. Raghukumar; R. Tharanathan; N.B. Bhosle. Extracellular Polysaccharide Production by Thraustochytrid Protists. *Mar. Biotechnol.* **2005,** *7,* 184–192.

Jenkins, K.M.; P.R. Jensen; W. Fenical. Thraustochytrosides A-C: New Glycosphingolipids from a Unique Marine Protist, *Thraustochytrium globosum.* *Tet. Lett.* **1999,** *40,* 7637–7640.

Jiang, H.; R. Zirkle; J.G. Metz; L. Braun; L. Richter; S.G. Van Lanen; B. Shen. The Role of Tandem Acyl Carrier Protein Domains in Polyunsaturated Fatty Acid Biosynthesis. *J. Am. Chem. Soc.* **2008,** *130,* 6936–6937.

Keating, T.A.; C.T. Walsh. Initiation, Elongation, and Termination Strategies in Polyketide and Polypeptide Antibiotic Biosynthesis. *Curr. Opin. Chem. Biol.* **1999,** *3,* 598–606.

Kendrick, A.J.; C. Ratledge. Microbial Polyunsaturated Fatty Acids of Potential Commercial Interest. *SIM News* **1992,** *42,* 59–65.

Kollenmarenth, O.I.; O.P. Lollenmareth; S.R. Vedamuthu; L.K. Arlagadda; S. Raghukumar. International Patent Application WO2007/074479, 2007.

Komazawa, H.; M. Kojima; T. Aki; K. Ono; M. Kawakami. U.S. Patent Application US2004/0161831, 2004.

Kyle, D.J.; V.J. Sicotte; J.J. Singer; S. Reeb. Bioproduction of Docosahexaenoic Acid (DHA) by Microalgae. *Industrial Applications of Single Cell Oil*; C. Ratledge, D. Kyle, Eds.; American Oil Chemists' Society: Champaign, IL, 1992; pp 287–300.

Larsen, A.; W. Eikrem; E. Paasche. Growth and Toxicity in *Prymnesium patelliferum* (Prymnesiophyceae) Isolated from Norwegian Waters. *Can. J. Bot.* **1993,** *71,* 1357–1362.

Lawrence, C.F.; C.F. Charbonneau; C. Menard. Liquid Chromatographic Determination of Domoic Acid in Mussels Using AOAC Paralytic Shellfish Poison Extraction Procedure: Collaborative Study. *J. Assoc. Off. Anal. Chem.* **1991,** *74,* 68–72.

Lewis, T.E.; P.D. Nichols; T.A. McMeekin. The Biotechnological Potential of Thraustochytrids. *Mar. Biotechnol.* **1999,** *1,* 580–587.

Long, T.V. U.S. Patent Application US2002/0015978, 2002.

Marriott, N.G.; J.E. Garrett; M.D. Sims; R. Abril. Performance Characteristics and Fatty Acid Composition of Pigs Fed a Diet with Docosahexaenoic Acid. *J. Muscle Foods* **2002a,** *13,* 265–277.

Marriott, N.G.; J.E. Garrett; M.D. Sims; H. Wang; R. Abril. Characteristics of Pork with Docosahexaenoic Acid Supplemented in the Diet. *J. Muscle Foods* **2002b,** *13,* 253–263.

Metz, J.G.; P. Roessler; D. Facciotti: C. Levering; F. Dittrich; M. Lassner; R. Valentine; K. Lardizabal; F. Domergue; A. Yamada; et al. Production of Polyunsaturated Fatty Acids by Polyketide Synthases in Both Prokaryotes and Eukaryotes. *Science* **2001,** *293,* 290–293.

Metz, J.G.; J. Kuner; B. Rosenzweig; J.C. Lippmeier; P. Roessler; R. Zirkle. Biochemical Characterization of Polyunsaturated Fatty Acid Synthesis in *Schizochytrium*: Release of the Products as Free Fatty Acids. *Plant Physiol. Biochem.* **2009,** *47,* 472–478.

Morita, E.; Y. Kumon; T. Nakahara; S. Kagiwada; T. Noguchi. Docosahexaenoic Acid Production and Lipid-Body Formation in *Schizochytrium limacinum* SR21. *Mar. Biotechnol.* **2006,** *8,* 319–327.

Nakahara, T.; T. Yokochi; S. Higashihara; S. Tanaka; T.Yaguchi; D. Honda. Production of Docosahexaenoic Acid and Docosapentaenoic Acid by *Schizochytrium* sp. Isolated from Yap Islands. *J. Am. Oil Chem, Soc.* **1996,** *73,* 1421–1426.

Nettleton, J.A. *Omega-3 Fatty Acids and Health*; Chapman & Hall: New York, 1995.

Okuyama, H.; Y. Orikasa; T. Nishida; K. Watanabe; N. Morita. Bacterial Genes Responsible for the Biosynthesis of Eicosapentaenoic and Docosapentaenoic Acids and Their Heterologous Expression. *Appl. Environ. Microbiol.* **2007,** *73,* 665–670.

Patell, V.M.; K.R. Rajyashri. International Patent Publication Number WO07/068997, 2007.

Pyle, D.J.; R.A. Garcia; Z. Wen. Producing Docosahexaenoic Acid (DHA)-Rich Algae from Biodiesel-Derived Crude Glycerol: Effects of Impurities on DHA Production and Algal Biomass Composition. *J. Agric. Food Chem.* **2008,** *56,* 3933–3939.

Raghukumar, S.; K. Schaumann. An Epifluorescence Microscopy Method for Direct Enumeration of the Fungi-like Marine Protists, the Thraustochytrids. *Limnol. Oceanogr.* **1993,** *38,* 182–187.

Raghukumar, S.; D.R. Cahndramohan; E. Desa. U.S. Patent 6,410,282, 2002.

Raghukumar, S.; R. Jain. U.S. Patent Application US2005/0019880, 2005.

Raghukumar, S. Thraustochytrid Marine Protists: Production of PUFAs and Other Emerging Technologies. *Mar. Biotechnol.* **2008,** *10,* 631–640.

Ratledge, C. Microbial Oils and Fats—An Overview. *Biotechnology for the Oils and Fats Industry*; P. Dawson, C. Ratledge, J. Rattray, Eds.; American Oil Chemists' Society: Champaign, IL, 1984; pp 119–127.

Shilo, M. Toxins of Chrysophyceae. *Microbial Toxins*; S. Kadis, A. Ciegler, A.J. Ajl, Eds.; Academic Press: New York, 1971; pp 67–103.

Shirasaka, N.; Y. Hirai; H. Najabayashi; H. Yoshuzumi. Effect of Cyanocobalamin and *p*-Toluic Acid on the Fatty Acid Composition of *Schizochytrium limacinum* (Thraustochytriaceae, Labrinthulomycota). *Mycoscience* **2005**, *46,* 358–363.

Simopoulos, A.P. Omega-3 Fatty Acids in Health and Disease and in Growth and Development. *Am. J. Clin. Nutr.* **1991**, *54,* 438–463.

Singh, A.; O.P. Ward. Production of High Yields of Docosahexaenoic Acid by *Thraustochytrium roseum* ATCC 28210. *J. Ind. Microbiol.* **1996**, *16,* 370–373.

Singh, A.; S. Wilson; O.P. Ward. Docosahexaenoic Acid (DHA) Production by *Thraustochytrium* sp. ATCC 20892. *World. J. Microbiol. Biotechnol.* **1996**, *12,* 76–81.

Song, X.; X. Zhang; C. Kuang; L. Zhu; N. Guo. Optimization of Fermentation Parameters for the Biomass and DHA Production of *Schizochytrium limacinum* OUC88 Using Response Surface Methodology. *Process Biochem.* **2007**, *42,* 1391–1397.

Sparrow, F.K. Biological Observations on the Marine Fungi of Woods Hole Waters. *Biol. Bull. Mar. Biol. Lab.* **1936**, *70,* 236–263.

Tanaka, M.; A., Ueno; K. Kawasaki; I. Yumoto; S. Ohgiya; T. Hoshino; K. Ishizaki; H. Okuyama; N. Morita. Isolation of Clustered Genes that Are Notably Homologous to the Eicosapentaenoic Acid Biosynthesis Gene Cluster from the Docosahexaenoic Acid-Producing Bacterium *Vibrio marinus* Strain MP-1. *Biotechnol. Lett.* **1999**, *21,* 939–945.

Unagul, P.; C. Assantachai; S. Phadungruengluij; T. Pongsuteeragul; M. Suphantharika; C. Verduyn. Biomasss and Docosahexaenoic Acid Formation by *Schizochytrium mangrovei* Sk-02 at Low Salt Concentrations. *Bot. Mar.* **2006a**, *49,* 182–190.

Unagul, P.; C. Assantachai; S. Phadungruengluij; M. Suphantharika; C. Verduyn. Properties of the Docosahexaenoic Acid-Producer *Schizochytrium mangrovei* Sk-02: Effects of Glucose, Temperature and Salinity and Their Interaction. *Bot. Mar.* **2006b**, *48,* 387–394.

Vanhaecke, P.; G. Persoone; C. Claus; P. Sorgeloos. Proposal for a Short-Term Toxicity Test with *Artemia nauplii. Ecotoxicol. Environ Saf.* **1981**, *5,* 382–387.

Villac, M.C.; D.L. Roelke; T.A. Villareal; G.A. Fryxell. Comparison of Two Domoic Acid-Producing Diatoms: A Review. *Hydrobiologia* **1993**, *269–270,* 213–224.

Von Schacky, C.; S. Fischer; P.C. Weber. Long Term Effects of Dietary Marine Omega-3 Fatty Acids upon Plasma and Cellular Lipids, Platelet Function, and Eicosanoid Formation in Humans. *J. Clin. Invest.* **1985**, *76,* 1626–1631.

Ward, O.P.; Z. Li. Production of Docosahexaenoic Acid by *Thraustochytrium roseum. J. Ind. Microbiol.* **1994**, *13,* 234–241.

Watanabe, K.; K. Yazawa; K. Kondo; A. Kawaguchi. Fatty Acid Synthesis of an Eicosapentaenoic Acid-Producing Bacterium: *De novo* Synthesis, Chain Elongation, and Desaturation Systems. *J. Biochem.* **1997**, *122,* 467–473.

Wu, S.T.; S.T. Yu; L.P. Lin. Effect of Culture Conditions on Docosahexaenoic Acid Production by *Schizochytrium* sp. S31. *Process Biochem.* **2005**, *40,* 3103–3108.

Wumpelmann, M. U.S. Patent Application US20070015263, 2007.

Yamada, H.; A. Shimizu; Y. Shinmen; K. Akimoto; H. Kawashima; S. Jareonkitmongkol. Production of Dihomo-γ-Linolenic Acid, Arachidonic Acid, and Eicosapentaenoic Acid by

Filamentous Fungi. *Industrial Applications of Single Cell Oil*; C. Ratledge, D. Kyle, Eds.; American Oil Chemists' Society: Champaign, IL, 1992; pp 118–138.

Yamaoka, Y. U.S. Patent Application US2004/0253724, 2004.

Yamasaki, T.; T. Aki; M. Shinosaki; M. Taguchi; S. Kawamoto; K. Ono. Utilization of *Shochu* Distillery Wastewater for Production of Polyunsaturated Fatty Acids and Xanthophylls Using Thraustochytrid. *J. Biosci. Bioeng.* **2006,** *4,* 323–327.

Yamashita, N.; A. Yokoyama; T. Hamazaki; S. Yano. Inhibition of Natural Killer Cell Activity of Human Lymphocytes by Eicosapentaenoic Acid. *Biochem. Biophys. Res. Commun.* **1986,** *138,* 1058–1067.

Yazawa, K. Production of Eicosapentaenoic Acid from Marine Bacteria. *Lipids* **1996,** *31,* S297–S300.

Yazawa, K.; K. Watanabe; C. Ishikawa; K. Kondo; S. Kimura. Production of Eicosapentaenoic Acid from Marine Bacteria. *Industrial Applications of Single Cell Oil*; American Oil Chemists' Society: Champaign, IL, 1992; pp 29–51.

Yu, R.; A. Yamada; K. Watanabe; K. Yazawa; H. Takeyama; T. Matsunaga; R. Kurane. Production of Eicosapentaenoic Acid by a Recombinant Marine Cyanobacterium, *Synechococcus* sp. *Lipids* **2000,** *35,* 1061–1064.

Zajic, J.E.; Y.S. Chiu. Heterotrophic Culture of Algae. *Properties and Products of Algae*; J.E. Zajic, Ed.; Plenum Press: New York, 1970; pp 1–47.

Zeller, S.; W. Barclay; R. Abril. Production of Docosahexaenoic Acid by Microalgae. *Omega-3 Fatty Acids: Chemistry, Nutrition, and Health Effects*; F. Shahidi, J.W. Finley, Eds.; American Chemical Society: Washington, DC, 2001; pp 108–124.

5

Arachidonic Acid: Fermentative Production by *Mortierella* Fungi

Hugo Streekstra
DSM Food Specialties, PO Box 1, 2600 MA Delft, the Netherlands

Arachidonic Acid

Arachidonic acid (ARA; 20:4n-6) is a long-chain polyunsaturated fatty acid (LCPUFA) with twenty carbon atoms and four double bonds (Fig. 5.1) . Its systematic name is (all-*cis*)-5,8,11,14-eicosatetraenoic acid, and it is an important structural component of the lipids in the central neural system, including those in the brain. ARA is also a biosynthetic precursor to several classes of biologically active metabolites, such as eicosanoids.

At present, the commercial demand for ARA is dominated by its application in infant formula. Human milk contains significant amounts of two LCPUFAs: ARA and docosahexaenoic acid (DHA; 22:6n-3). There is a significant body of evidence showing that neural development in growing infants benefits from the provision of these fatty acids, either through breast-feeding or their inclusion in infant formula (Fleith & Clandinin, 2005; Birch et al., 2007), and that infants may also benefit in other ways from LCPUFA consumption (Pastor et al., 2006). These benefits are most apparent in underprivileged infants, such as prematurely born ones (Henriksen et al., 2008).

The inclusion of ARA in infant formula is a well-established application. More applications are being considered—for instance, the use of ARA in dietary supplements for pregnant women and nursing mothers. LCPUFA supplementation may also be beneficial for individuals affected by a range of disorders associated with disturbed fatty acid profiles (Horrobin et al., 2002; Pantaleo et al., 2004; Kotani et al., 2006) or impaired membrane function (Pazirandeh et al., 2007; Oe et al., 2008), but specific applications have not yet been developed. There are also applications in animal feed (aquaculture) and pet food. For fish (Koven et al., 2003; Lund et al., 2007) and shellfish (Seguineau et al., 2005; Nghia et al., 2007), ARA is an important nutrient, essential for development after hatching and for the growth and metamorphosis of larvae and fry. For carnivorous mammals, such as cats, ARA is also an essential nutrient (Pawlosky et al., 1997; Morris, 2004). Finally, applications in plant cultivation have been considered (Rozhnova et al., 2001; Groenewald & van der Westhuizen, 2004).

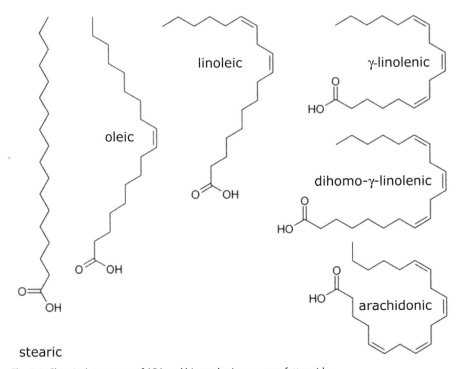

oleic

linoleic

γ-linolenic

dihomo-γ-linolenic

arachidonic

stearic

Fig. 5.1. Chemical structures of ARA and biosynthetic precursor fatty acids.

Sources of Arachidonic Acid

Plants do not contain significant levels of LCPUFA. In the case of ARA, well-accepted and concentrated animal sources are not available, even though most of the ARA in our diet does come from animal-derived foods. Therefore, microbial sources have been sought, and found.

Haskins et al. (1964) reported that ARA accounted for more than 25% of the fatty acids in *Mortierella renispora* lipids. This early report is easily missed because the data are "hidden" in a paper on a different subject. Perhaps for this reason *Mortierella* was not mentioned in an early overview of the PUFAs in fungi (Shaw, 1966), although ARA production by other fungi, algae, and Oomycetes (microorganisms that are no longer classified as fungi) was reported there. In 1974, the data from Haskins et al. were included in a massive review on fungal lipids (Wassef, 1974), making this information easily accessible. From that moment on, the genus *Mortierella* was an obvious place to search for ARA-producing strains, because it was there that the highest levels of ARA had been found.

When researchers started looking for sources of ARA in the 1980s—notable exploration was done in Japan by Lion Corp. (Totani & Oba, 1987) and Suntory (Shinmen et al., 1989)—they investigated a number of groups, including many species of

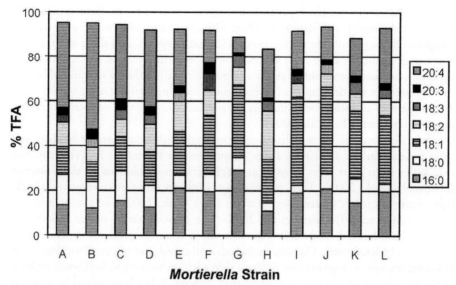

Fig. 5.2. Major fatty acids of various *Mortierella* strains.

Mortierella, along with other Zygomycetes (such as *Entomophthora*) and Oomycetes (such as *Pythium*). It became apparent that the vast majority of strains currently considered to belong to the genus *Mortierella* produce ARA (Eroshin et al., 1996) and that *M. alpina* stands out as a high producer (Higashiyama et al., 2002; Ho et al., 2007; Hou, 2008; Fig. 5.2), with it often exceeding 50% of the total fatty acids found in this microorganism. Another property characteristic of the lipid fraction in this organism is its sterol composition (Nes & Nichols, 2006), the major sterol species being desmosterol, in contrast to the ergosterol that is often considered to be typical for fungi. *M. alpina* has been shown to be a safe organism for the production of food ingredients (Streekstra, 1997).

Finally, this organism shows a high ability to accumulate intracellular lipids, mainly triglycerides (Certik & Shimizu, 2003; Ho & Chen, 2008b), under appropriate conditions (Wynn et al., 1999; Zhang & Ratledge, 2008). ARA is found both in polar and apolar (triglyceride) lipids, and the triacylglycerol fraction is used for the commercial product. More recently, algae have also been explored as potential producers of ARA (Khozin-Goldberg et al., 2002; Seguineau et al., 2005) (see also Chapter 10).

Some Properties of *Mortierella alpina* and *M. alpina* Lipids

Mortierella is a genus of filamentous fungi within the Zygomycetes (Tanabe et al., 2004; White et al., 2006; Ho & Chen, 2008a). This group of fungi has a characteristic sexual cycle that incorporates zygospores and specific pheromones (Schimek et al., 2003), as well as an a vegetative (asexual) sporulation cycle (Lounds et al., 2007). It is also characteristic that the mycelia do not contain septated cells. Rather, the mycelium is a tube comprised of sections filled with cytoplasm and empty sections. The cytoplasm harbors multiple nuclei, even in vegetative spores, and there is no uninucleate stage in the life cycle. This absence complicates the selection and maintenance of strains, because even a single cell is a population, rather than a true individual. It also complicates the development of high-producing industrial strains, for instance, by mutagenesis and selection, although the successful application of such techniques has been reported (Zhu et al., 2004).

As with many fungi, mycelial growth in liquid culture can either be dispersed or in pellets (Park et al., 2006). Pellet cultures have a (desirably) low viscosity, but may suffer from poor mass transfer and degeneration of biomass in their interior (Hamanaka et al., 2001). Their morphology is influenced by environmental conditions, such as carbon source concentration (Park et al., 2002) and the presence of specific amino acids (Koizumi et al., 2006). The aseptate nature of the zygomycetous mycelium has led to the fear that these fungi could be particularly sensitive to shear stress in fermenters. Damage to the cell wall could lead to cytoplasm loss in an extended section, a situation not unlike a ship without watertight bulkheads sinking faster than one with separate compartments. Indeed, it has been found in a mixed pellet/dispersed culture of *M. alpina* that most of the ARA accumulation occurred in the pellet fraction (Higashiyama et al., 1999). This result was interpreted as the mycelial fragments having been "shaved" off the pellets by the shear force and being damaged in the process.

The morphological properties of *M. alpina* are quite variable among different isolates. This divergence applies both to the tendency to sporulate on solid media—a desirable property for strain selection and maintenance—and the tendency to grow dispersed or as pellets in a liquid culture. The dispersed morphology is much less common in standard media.

The fatty acid spectrum of *M. alpina* lipids is dominated by intermediates of the biosynthetic pathway leading to ARA. This profile seems to be strongly influenced by the intrinsic properties of the biosynthetic pathway, since different N-sources gave surprisingly similar fatty acid percentages in shake flasks (Fig. 5.3); such congruence occurred even though the production of biomass and the absolute production level of ARA differed by a factor of ten between the highest- and the lowest-producing conditions. Nevertheless, the fatty acid composition does show a pronounced kinetic effect; in older, non-growing cultures, the ARA percentage increased when compared with younger cultures (Fig. 5.4). This increase was most pronounced in the triacylglycerol (storage lipid) fraction (Fig. 5.5; Certik & Shimizu, 2000; Eroshin et al., 2002). It takes place during active sugar consumption, a phase where no net lipid-free biomass is being produced due to exhaustion of the nitrogen source (see below). In this phase, the rate of triacylglycerol (TAG) accumulation is very high. It may be that the desaturases and elongases cannot completely match this rate and need some time to produce ARA as the endpoint of the pathway, which would cause a temporary accumulation of its biosynthetic precursors.

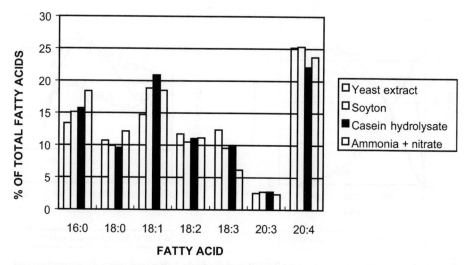

Fig. 5.3. Fatty acid profile of *M. alpina* PUF101: Various N-sources.

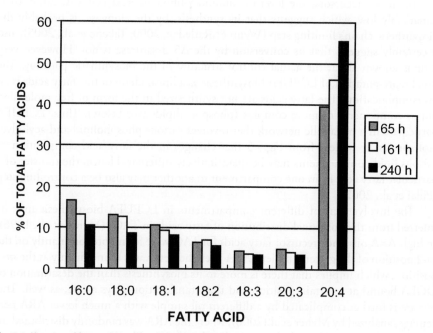

Fig. 5.4. Kinetics of fatty acid profile of *M. alpina* PUF101.

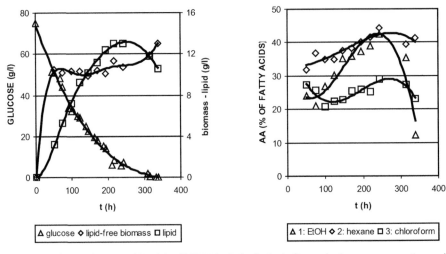

Fig. 5.5. Production kinetics of *M. alpina* PUF101 in shake-flasks. **Left panel:** glucose consumption and biomass and lipid production; lipid-free biomass is calculated as the difference between culture dry weight and intracellular lipids. **Right panel:** ARA percentage in successive solvent extracts.

In these precursors, the level of dihomo-γ-linolenic acid (DGLA; 20:3n-6) is remarkably low, which suggests that its synthesis, by the elongase, is probably the biosynthetic chain's limiting step (Wynn & Ratledge, 2000; Takeno et al., 2005), and it certainly suggests that its conversion by the Δ5 desaturase is not. However, very little is known about the actual control structure of the biosynthetic pathway. The usual representation of LCPUFA biosynthesis as a linear chain of free fatty acids is an oversimplification. The fatty acids are not synthesized in the form of free molecules, but as components of more complex (phospho-)lipids (see below). Thus, LCPUFA participate in a metabolic network that involves various phospholipid and acyl-glycerol species. It is clear from Fig.5.5. that changes in the fatty acid composition of various lipid compartments may be quantitatively different. Hence, the transfer of a particular fatty acid from one compartment to another may also be a controlling step (Pillai et al., 2002).

The involvement of different compartments in LCPUFA biosynthesis may be inferred from the positional distribution of fatty acids in TAGs (Myher et al., 1996). In high-ARA oils, the precursor fatty acids for TAG were found predominantly on the *sn*-2 position of the glycerol moiety. ARA itself, however, was found mainly at the *sn*-1 position, which implies that there is more to its biosynthesis than the desaturation of DGLA bound at a certain position and that translocation plays a role, as well. This picture is further complicated by a different oil sample with a much lower ARA percentage, analyzed by Myher et al.(2006), where the ARA was randomly distributed on the glycerol moiety. Nevertheless, this distribution would also require translocation at some stage because the *sn*-3 position is not available for ARA biosynthesis; such biosynthesis is believed to take place on a phospholipid scaffold (see below).

TAG oils with high levels of LCPUFAs may have some special properties. An ARA content of 40% of total fatty acids implies that about 20% of the TAG species must contain two ARA residues (if all of the remaining 80% contain only one residue). At an ARA content of 50%, this fraction has increased to 50% of all TAGs. Such species are considered to be poor substrates for most lipases (Myher et al., 1996). This finding may explain the extremely high accumulation of ARA under some conditions, if multiple ARA-substituted species represent metabolic "dead ends" in the fungus.

Older work by the Japanese groups has shown that very high ARA levels are found when fermentation is extended beyond the exhaustion of the carbon source. In the 1980s, a resting phase of several days was inserted between fermentation and harvesting of the oil to increase the percentage of ARA. This period of maturation usually does not give a higher productivity of ARA, since it is associated with lipid breakdown, but it can be useful when oil with the highest possible ARA content is desired. Indeed, extremely high percentages of ARA are possible (Yuan et al., 2002).

Shimizu's group at Kyoto University—in collaboration with Suntory—has shown that intermediates of the biosynthetic pathway to ARA, and a range of other LCPUFAs, can be produced in this organism by changing the culture conditions, using metabolic inhibitors, and/or inducing specific mutations (Certik & Shimizu, 1999; Kawashima et al., 2000; Ogawa et al., 2002). For instance, the use of a Δ5 desaturase-defective mutant leads to the production of DGLA, whereas low temperatures promote the formation of eicosapentaenoic acid (EPA; 20:5n-3), which involves a Δ3 desaturase that does not participate in the ARA biosynthesis pathway (Sakuradani et al., 2005; Sakuradani et al., 2009b). Mutants that are defective in "early" enzymes of the pathway, in combination with having a high activity of "late" enzymes, may show accumulation of LCPUFAs with unsaturation patterns not commonly encountered in nature (Takeno et al., 2005; Abe et al., 2006; Zhang et al., 2006). Such fatty acids are normally only found as minor components of oils, formed because the specificity of biosynthetic enzymes is not absolute.

LCPUFA Biosynthesis in *Mortierella alpina*

In fungi, unsaturated fatty acids are formed from stearic acid (18:0) by an elongase and integral, membrane-bound, fatty-acid desaturases, which sequentially insert double bonds (Domergue et al., 2003). The desaturases (Sperling et al., 2003) require oxygen and cytochrome *b*5 as co-factors, similar to other microsomal desaturases from plants and yeast. Electrons are transferred from NADH-dependent cytochrome *b*5 reductase, via the cytochrome *b*5, to the desaturase.

The desaturases have been difficult to purify in an active form, mainly due to their hydrophobicity. However, some of their properties are known. They contain eight highly conserved histidine residues, which are catalytically essential. The consensus sequence is sufficiently robust to use degenerate primers for picking up new desaturase genes.

The first desaturation step is catalyzed by the Δ9-desaturase, which uses stearoyl-CoA as substrate. Subsequent desaturation steps are believed to take place on the fatty acyl chains of phospholipids. The enzyme acyl-CoA:1-acyl-lysophosphatidylcholine acyltransferase (LCPAT) is responsible for channeling fatty acids to the *sn*-2 position of phosphatidyl-choline for desaturation and PUFA production.

Classical mutants have been isolated, which were defective (Certik et al., 1998; Jareonkitmongkol et al., 2002; Sakuradani et al., 2004) or enhanced (Sakuradani et al., 2002) in one or more of the desaturase activities. Thanks to the efforts of various research groups, the structural genes of the desaturases and the elongase, along with some auxiliary genes, have been cloned from *M. alpina*. Their functionality has generally been assessed by expression in a heterologous, non-LCPUFA-producing host (Table 5.A).

Table 5.A. Enzymes and Genes from *Mortierella alpina* Involved in ARA Biosynthesis.

Enzyme	Substrate	Product	Name	Remarks	Ref.
Δ9 desaturase	18:0	18:1n9	ole1p ole2p Δ9-1,2,3	Expressed in *S. cerevisiae* and *A. oryzae* Mutation analysis	(Wongwathanarat et al., 1999) (Sakuradani et al., 1999) (MacKenzie et al., 2002) (Abe et al., 2006)
Δ12 desaturase	18:1 n9	18:2n6	MaΔ12	Expressed in *S. cerevisiae* and *A.oryzae* Mutation analysis	(Sakuradani et al., 1999) (Huang et al., 1999) (Sakuradani et al., 2009a)
Δ6 desaturase	18:2n6	18:3n6	Δ6I Δ6II	Expressed in *A. oryzae* Muation analysis	(Huang et al., 1999) (Sakuradani et al., 1999) (Sakuradani & Shimizu, 2003) (Abe et al., 2005a)
Elongase	18:3n6	20:3n6	GLELO MAELO	Expressed in *S. cerevisiae*	(Parker-Barnes et al., 2000) (Sakuradani et al., 2008)
Δ5 desaturase	20:3n6	20:4n6	M.A5	Expressed in *S. cerevisiae* and Canola Mutation analysis	(Michaelson et al., 1998) (Knutzon et al., 1998) (Abe et al., 2005b)
Cytochrome b_5				Expressed in *E. coli*	(Kobayashi et al., 1999)
NADH-cytochrome b_5 reductase			Cb5R-I Cb5R-II	Expressed in *A. oryzae*	(Sakuradani et al., 1999) (Certik et al., 1999)

This heterologous expression is equivalent to the reconstitution of a section of the ARA biosynthetic pathway. Using genes from different sources (including *M. alpina* genes), it has proved possible to detect ARA production in *S. cerevisiae* from exogenously supplied linolenic acid (18:2n-6) (Beaudoin et al., 2000). Genes from *M. alpina* have also allowed the production of ARA in plants, such as soybeans (Chen et al., 2006). An interesting competing concept is to leave the genes and the enzymes in the fungus and use the *M. alpina* biomass to enrich plant oil with LCPUFAs (Dong & Walker, 2008).

To achieve a high productivity of LCPUFAs from cheap carbon sources, it would be advantageous to start with an oleaginous (lipid-accumulating) microorganism as a host for genetic engineering. Unfortunately, genetic systems for such organisms have not been developed to the same extent that they have for the more conventional yeasts and filamentous fungi. Recently, progress has been made towards the genetic transformation of *M. alpina* (MacKenzie et al., 2000; Lounds et al., 2003; Takeno et al., 2004a; Takeno et al., 2004b; Takeno et al., 2005), which could be an important tool to influence the flux of carbon through the biosynthetic pathway towards ARA and other LCPUFAs. Another tool in the study of the biosynthetic pathway is a cell-free lipid assembly in microsomal membrane preparations, obtained from *M. alpina* (Chatrattanakunchai et al., 2000).

Fermentation of *Mortierella alpina*

Higashiyama (2003) published a review of the fermentation conditions for ARA production with *M. alpina*, which continues to be relevant. The reader is referred to this text for a more in-depth treatment of this subject and for references to additional literature.

Generally, microorganisms show the highest accumulation of TAGs under nitrogen limitation. The production of ARA oil by *M. alpina* is no exception. The dosage of the nitrogen source determines the amount of productive (lipid-free) biomass that can be formed (Yu et al., 2003). This level should be tuned to the technological limits of the equipment. The carbon dosage then determines how much oil can be accumulated by this biomass (Jin et al., 2008a, 2008b). Since very high levels of carbon dosage may cause inhibitory effects—although *M. alpina* is able to tolerate fairly high glucose concentrations (Zhu et al., 2003)—there is usually an optimal C/N ratio (Koike et al., 2001; Jang et al., 2005). The best way to avoid inhibitory effects is to dose the carbon as a non-limiting feed. Another concept that has been applied is the use of solid-state fermentation, reported in the older literature (not reviewed here), but also revisited recently (Jang & Yang, 2008).

The concept of the C/N ratio has been a persistent one in relevant literature. However, rarely has it been applied beyond the operational optimization to which I have just referred. The reason for its persistence may well be that lipid production is one of the few biotechnological processes that proceeds quite well in a simple batch culture, and, in such a system, the C/N ratio is an important parameter.

However, in physiological terms, the situation is not so clear. The mechanism that nitrogen limitation induces oil accumulation is well-supported. It follows that a certain, minimal C/N ratio is required to achieve this physiological state. Of course, much more carbon will be needed in practice to allow the subsequent high accumulation of lipid, but it remains open to question whether the absolute concentration of the carbon source has an effect, as well. The effects may be different for each microorganism, because they would depend on the affinity of uptake systems, catabolic enzymes, and sensors for the carbon substrate or its metabolites. If there is a low-affinity step involved, the magnitude of the carbon excess could play a role in determining the rate of oil accumulation. However, in *M. alpina*, at least, oil accumulation proceeds quite well at a modest external concentration of the carbon source.

In a dual feed system, the nitrogen and the carbon source(s) can be fed independently (Hwang et al., 2005) or, at least, in different relative proportions in different stages of the fermentation (Zhu et al., 2006). The N-feed should be controlling the process, whereas the C-feed is regulated to ensure a continuous, but moderate, excess of the C-source. Once the desired maximal biomass concentration has been reached, the N-feed can be (all but) stopped (Jin et al., 2007), and the C-feed can be used to extend the lipid accumulation. This process can be kept going until overall productivity decreases or until the desired percentage of ARA in the oil is reached.

Fig. 5.6 shows two laboratory-scale fermentations, executed according to such a regime, at two different temperatures. The Respiratory Quotient (RQ; the quotient of

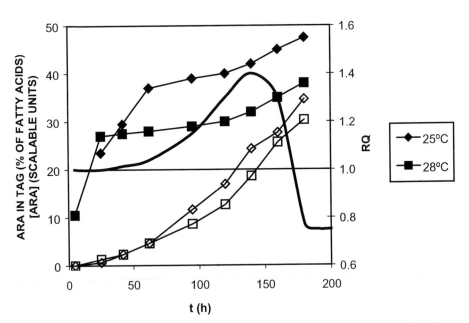

Fig. 5.6. Lab-scale fermentation and the influence of the growth temperature. Open symbols indicate ARA production (in scale-invariant arbitrary units; left axis). Closed symbols indicate ARA percentage of total fatty acids (left axis). The bold line indicates the Respiratory Quotient (RQ; right axis).

CO_2 produced and O_2 consumed) is a convenient on-line indicator of the metabolic state; growth on glucose leads to a RQ of about 1, and lipid production increases the RQ, whereas lipid consumption decreases it. All these conditions are encountered during the process. Biomass production is not indicated in the figure, but it was mostly complete within 80 hours.

It is clear that the lower temperature led to a higher proportion of ARA in the lipid fraction. However, the actual productivity of ARA was not much higher due to the lower metabolic rate, both intrinsic and imposed by limitation of the cooling capacity. The observation that growth temperature influences the fatty acid spectrum has also been made for other strains of *M. alpina* (Shinmen et al., 1989).

As the drop in RQ shows, the organism is capable of metabolizing the lipid that has been formed. This process can be employed as a method to increase the ARA-content of the oil—as mentioned previously—but it is also a potential problem, breaking down the lipid after the fermentation has been stopped. It is, therefore, important to inactivate the biomass as quickly as possible after the production phase. As this step may be considered to be the first of down-stream processing, this subject is dealt with in Chapter 9.

At present, the ARA production process with *M. alpina* is the only current example of a commercial fungal lipid being produced by full-scale fermentation, extraction, and refining. This process takes place in 200 m³ fermenters in the U.S. (see Chapter 20). An increasingly filled molecular biological toolbox would allow efficient production of a whole range of relevant LCPUFAs in this organism and in other fungi (see Chapter 2). It is to be hoped that commercial demand will drive the development of such processes to further exploit this fascinating and impressive biochemical pathway.

References

Abe, T.; E. Sakuradani; T. Asano; H. Kanamaru; Y. Ioka; S. Shimizu. Identification of Mutation sites on Δ6 Desaturase Genes from *Mortierella alpina* 1S-4 Mutants. *Biosci. Biotechnol. Biochem.* **2005a**, *69*, 1021–1024.

Abe, T.; E. Sakuradani; T. Ueda; S. Shimizu. Identification of Mutation Sites on Δ5 Desaturase Genes from *Mortierella alpina* 1S-4 Mutants. *J. Biosci. Bioeng.* **2005b**, *99*, 296–299.

Abe, T.; E. Sakuradani; T. Asano; H. Kanamaru; S. Shimizu. Functional Characterization of Δ9 and ω9 Desaturase Genes in *Mortierella alpina* 1S-4 and Its Derivative Mutants. *Appl. Microbiol. Biotechnol.* **2006**, *70*, 711–719.

Beaudoin, F.; L.V. Michaelson; S.J. Hey; M.J. Lewis; P.R. Shewry; O. Sayanova; J.A. Napier. Heterologous Reconstitution in Yeast of the Polyunsaturated Fatty Acid Biosynthetic Pathway. *Proc. Natl. Acad. Sci. USA* **2000**, *97*, 6421–6426.

Birch, E.E.; S. Garfield; Y. Castaneda; D. Hughbanks-Wheaton; R. Uauy; D. Hoffman. Visual Acuity and Cognitive Outcomes at 4 Years of Age in a Double-blind, Randomized Trial of Long-Chain Polyunsaturated Fatty Acid-Supplemented Infant Formula. *Early Hum. Dev.* **2007**, *63*, 279–284.

Certik, M.; S. Shimizu. Biosynthesis and Regulation of Microbial Polyunsaturated Fatty Acid Production. *J. Biosci. Bioeng.* **1999,** *87,* 1–14.

Certik, M.; S. Shimizu. Kinetic Analysis of Oil Biosynthesis by an Arachidonic Acid-Producing Fungus, *Mortierella alpina* 1S-4. *Appl. Microbiol. Biotechnol.* **2000,** *54,* 224–230.

Certik, M.; S. Shimizu. Isolation and Lipid Analyses of Subcellular Fractions from the Arachidonic Acid-Producing Fungus *Mortierella alpina* 1S-4. *Biologia* **2003,** *58,* 1101–1110.

Certik, M.; E. Sakuradani; S. Shimizu. Desaturase-Defective Fungal Mutants: Useful Tools for the Regulation and Overproduction of Polyunsaturated Fatty Acids. *Trends Biotechnol.* **1998,** *16,* 500–505.

Certik, M.; E. Sakuradani; M. Kobayashi; S. Shimizu. Characterization of the Second Form of NADH-Cytochrome *b*5 Reductase Gene from Arachidonic Acid-Producing Fungus *Mortierella alpina* 1S-4. *J. Biosci. Bioeng.* **1999,** *88,* 667–671.

Chatrattanakunchai, S.; T. Fraser; K. Stobart. Oil Biosynthesis in Microsomal Membrane Preparations from *Mortierella alpina*. *Biochem. Soc. Trans.* **2000,** *28,* 707–709.

Chen, R.; K. Matsui; M. Ogawa; M. Oe; M. Ochiai; H. Kawashima; E. Sakuradani; S. Shimizu; M. Ishimoto; M. Hayashi; Y. Murooka; Y. Tanaka. Expression of Δ6, Δ5 Desaturase and gl-elo Elongase Genes from *Mortierella alpina* for Production of Arachidonic Acid in Soybean [*Glycine max* (L.) Merrill] Seeds. *Plant Sci.* **2006,** *170,* 399–406.

Domergue, F.; A. Abbadi; C. Ott; T.K. Zank; U. Zahringer; E. Heinz. Acyl Carriers Used as Substrates by the Desaturases and Elongases Involved in Very Long-Chain Polyunsaturated Fatty Acids Biosynthesis Reconstituted in Yeast. *J. Biol. Chem.* **2003,** *278,* 35115–35126.

Dong, M.; T.H. Walker. Addition of Polyunsaturated Fatty Acids to Canola Oil by Fungal Conversion. *Enzyme Microb. Technol.* **2008,** *42,* 514–520.

Eroshin, V.K.; E.G. Dedyukhina; T.I. Chistyakova; V.P. Zhelifonova; C.P. Kurtzman; R.J. Bothast. Arachidonic-Acid Production by Species of *Mortierella*. *World J. Microbiol. Biotechnol.* **1996,** *12,* 91–96.

Eroshin, V.K.; E.G. Dedyukhina; A.D. Satroutdinov; T.I. Chistyakova. Growth-Coupled Lipid Synthesis in *Mortierella alpina* LPM 301, a Producer of Arachidonic Acid. *Microbiology (Moscow)* **2002,** *71,* 169–172.

Fleith, M.; M.T. Clandinin. Dietary PUFA for Preterm and Term Infants: Review of Clinical Studies. *Crit. Rev. Food Sci. Nutr.* **2005,** *45,* 205–229.

Groenewald, E.G.; A.J. van der Westhuizen. The Effect of Applied Arachidonic Acid on the Formation of Prostaglandins in Plantlets from Excised Apices of the Short-Day Plant, *Pharbitis nil. S. Afr. J. Bot.* **2004,** *70,* 206–209.

Hamanaka, T.; K. Higashiyama; S. Fujikawa; E.Y. Park. Mycelial Pellet Intrastructure and Visualization of Mycelia and Intracellular Lipid in a Culture of *Mortierella alpina*. *Appl. Microbiol. Biotechnol.* **2001,** *56,* 233–238.

Haskins, R.H.; A.P. Tulloch; R.G. Micetich. Steroids and the Stimulation of Sexual Reproduction in a Species of *Pythium*. *Can. J. Microbiol.* **1964,** *10,* 187–194

Henriksen, C.; K. Haugholt; M. Lindgren; A.K. Aurvag; A. Ronnestad; M. Gronn; R. Solberg; A. Moen; B. Nakstad; R.K. Berge; et al. Improved Cognitive Development among Preterm Infants Attributable to Early Supplementation of Human Milk with Docosahexaenoic Acid and Arachidonic Acid. *Pediatrics* **2008,** *121,* 1137–1145.

Higashiyama, K. Industrial Production of Arachidonic Acid by Filamentous Fungi, *Mortierella*. *Recent Research Developments in Biotechnology & Bioengineering* **2003,** *5,* 79–95.

Higashiyama, K.; S. Fujikawa; E.Y. Park; M. Okabe. Image Analysis of Morphological Change during Arachidonic Acid Production by *Mortierella alpina* 1S-4. *J. Biosci. Bioeng.* **1999,** *87,* 489–494.

Higashiyama, K.; S. Fujikawa; E. Y. Park; S. Shimizu. Production of Arachidonic Acid by *Mortierella* Fungi. *Biotechnol. Bioproc. Eng.* **2002,** *7,* 252–262.

Ho, S.Y.; Y. Jiang; F. Chen. Polyunsaturated Fatty Acids (PUFAs) Content of the Fungus *Mortierella alpina* Isolated from Soil. *J. Agric. Food Chem.* **2007,** *55,* 3960–3966.

Ho, S.Y.; F. Chen. Genetic Characterization of *Mortierella alpina* by Sequencing the 18S-28S Ribosomal Gene Internal Transcribed Spacer Region. *Lett. Appl. Microbiol.* **2008a,** *47,* 250–255.

Ho, S.Y.; F. Chen. Lipid Characterization of *Mortierella alpina* Grown at Different NaCl Concentrations. *J. Agric. Food Chem.* **2008b,** *56,* 7903–7909.

Horrobin, D.F.; K. Jenkins; C.N. Bennett; W.W. Christie. Eicosapentaenoic Acid and Arachidonic Acid: Collaboration and Not Antagonism Is the Key to Biological Understanding. *Prostaglandins Leukotrienes Essent. Fatty Acids* **2002,** *66,* 83–90.

Hou, C.T. Production of Arachidonic Acid and Dihomo-gamma-linolenic Acid from Glycerol by Oil-Producing Filamentous Fungi, *Mortierella* in the ARS Culture Collection. *J. Indust. Microbiol. Biotechnol.* **2008,** *35,* 501–506.

Huang, Y.S.; S. Chaudhary; J.M. Thurmond; E.G. Bobik; L. Yuan; G.M. Chan; S.J. Kirchner; P. Mukerji; D.S. Knutzon. Cloning of Δ12- and Δ6-Desaturases from *Mortierella alpina* and Recombinant Production of Gamma-Linolenic Acid in *Saccharomyces cerevisiae*. *Lipids* **1999,** *34,* 649–659.

Hwang, B.H.; J.W. Kim; C.Y. Park; C.S. Park; Y.S. Kim; Y.W. Ryu. High-Level Production of Arachidonic Acid by Fed-Batch Culture of *Mortierella alpina* Using NH_4OH as a Nitrogen Source and pH Control. *Biotechnol. Lett.* **2005,** *27,* 731–735.

Jang, H.D.; Y.Y. Lin; S.S. Yang. Effect of Culture Media and Conditions on Polyunsaturated Fatty Acids Production by *Mortierella alpina*. *Bioresour. Technol.* **2005,** *96,* 1633–1644.

Jang, H.D.; S.S. Yang. Polyunsaturated Fatty Acids Production with a Solid-State Column Reactor. *Bioresour. Technol.* **2008,** *99,* 6181–6189.

Jareonkitmongkol, S.; E. Sakuradani; S. Shimizu. Isolation and Characterization of a Δ9-Desaturation-Defective Mutant of an Arachidonic Acid-Producing Fungus, *Mortierella alpina* 1S-4. *JAOCS* **2002,** *79,* 1021–1026.

Jin, M.J.; H. Huang; K. Zhang; J. Yan; Z. Gao. Metabolic Flux Analysis on Arachidonic Acid Fermentation (in Chinese). *J. Chem. Eng. Chin. Univ.* **2007,** *21,* 316–321.

Jin, M.J.; H. Huang; A.H. Xiao; Z. Gao; X. Liu; C. Peng. Enhancing Arachidonic Acid Production by *Mortierella alpina* ME-1 Using Improved Mycelium Aging Technology. *Bioproc. Biosyst. Eng.* **2008a,** *32,* 117–122.

Jin, M.J.; H. Huang; A.H. Xiao; K. Zhang; X. Liu; S. Li; C. Peng. A Novel Two-Step Fermentation Process for Improved Arachidonic Acid Production by *Mortierella alpina*. *Biotechnol. Lett.* **2008b,** *30,* 1087–1091.

Kawashima, H.; K. Akimoto; K. Higashiyama; S. Fujikawa; S. Shimizu. Industrial Production of Dihomo-Gamma-Linolenic Acid by a Δ5 Desaturase-Defective Mutant of *Mortierella alpina* 1S-4 Fungus. *JAOCS* **2000,** *77,* 1135–1138.

Khozin-Goldberg, I.; C. Bigogno; P. Shrestha; Z. Cohen. Nitrogen Starvation Induces the Accumulation of Arachidonic Acid in the Freshwater Green Alga *Parietochloris incisa* (*Trebuxiophyceae*). *J. Phycol.* **2002**, *38*, 991–994.

Knutzon, D.S.; J.M. Thurmond; Y.S. Huang; S. Chaudhary; E.G. Bobik; G.M. Chan; S.J. Kirchner; P. Mukerji. Identification of Δ5-Desaturase from *Mortierella alpina* by Heterologous Expression in Bakers' Yeast and Canola. *J. Biol. Chem.* **1998**, *273*, 29360–29366.

Kobayashi, M.; E. Sakuradani; S. Shimizu. Genetic Analysis of Cytochrome *b*(5) from Arachidonic Acid-Producing Fungus, *Mortierella alpina* 1S-4: Cloning, RNA Editing and Expression of the Gene in *Escherichia coli*, and Purification and Chararacterization of the Gene Product. *J. Biochem. (Tokyo)* **1999**, *125*, 1094–1103.

Koike, Y.; H.J. Cai; K. Higashiyama; S. Fujikawa; E.Y. Park. Effect of Consumed Carbon to Nitrogen Ratio on Mycelial Morphology and Arachidonic Acid Production in Cultures of *Mortierella alpina*. *J. Biosci. Bioeng.* **2001**, *91*, 382–389.

Koizumi, K.; K. Higashiyama; E.Y. Park. Effects of Amino Acid on Morphological Development and Nucleus Formation of Arachidonic Acid-Producing Filamentous Micro-organism, *Mortierella alpina*. *J. Appl. Microbiol.* **2006**, *100*, 885–892.

Kotani, S.; E. Sakaguchi; S. Warashina; N. Matsukawa; Y. Ishikura; Y. Kiso; M. Sakakibara; T. Yoshimoto; J. Z. Guo; T. Yamashima. Dietary Supplementation of Arachidonic and Docosahexaenoic Acids Improves Cognitive Dysfunction. *Neurosci. Res.* **2006**, *56*, 159–164.

Koven, W.; R. van Anholt; S. Lutzky; I. Ben Atia; O. Nixon; B. Ron; A. Tandler. The Effect of Dietary Arachidonic Acid on Growth, Survival, and Cortisol Levels in Different-Age Gilthead Seabream Larvae (*Sparus auratus*) Exposed to Handling or Daily Salinity Change. *Aquaculture* **2003**, *228*, 307–320.

Lounds, C.; A. Watson; M. Alcocer; A. Carter; D. MacKenzie; D. Archer. Pathways for Synthesis of Polyunsaturated Fatty Acids in the Oleaginous Zygomycete *Mortierella alpina*. In Proceedings of the 22nd Fungal Genetics Conference, Pacific Grove, CA, 2003.

Lounds, C.; J. Eagles; A.T. Carter; D.A. MacKenzie; D.B. Archer. Spore Germination in *Mortierella alpina* is Associated with a Transient Depletion of Arachidonic Acid and Induction of Fatty Acid Desaturase Gene Expression. *Arch. Microbiol.* **2007**, *188*, 299–305.

Lund, I.; S.J. Steenfeldt; B.W. Hansen. Effect of Dietary Arachidonic Acid, Eicosapentaenoic Acid and Docosahexaenoic Acid on Survival, Growth and Pigmentation in Larvae of Common Sole (*Solea solea* L.). *Aquaculture* **2007**, *273*, 532–544.

Mackenzie, D.A.; P. Wongwathanarat; A.T. Carter; D.B. Archer. Isolation and Use of a Homologous Histone H4 Promoter and a Ribosomal DNA Region in a Transformation Vector for the Oil-Producing Fungus *Mortierella alpina*. *Appl. Environ. Microbiol.* **2000**, *66*, 4655–4661.

MacKenzie, D.A.; A.T. Carter; P. Wongwathanarat; J. Eagles; J. Salt; D.B. Archer. A Third Fatty Acid Δ9-Desaturase from *Mortierella alpina* with a Different Substrate Specificity to *ole*1p and *ole*2p. *Microbiology UK* **2002**, *148*, 1725–1735.

Michaelson, L.V.; C.M. Lazarus; G. Griffiths; J.A. Napier; A.K. Stobart. Isolation of a Δ5-Fatty Acid Desaturase Gene from *Mortierella alpina*. *J. Biol. Chem.* **1998**, *273*, 19055–19059.

Morris, J.G. Do Cats Need Arachidonic Acid in the Diet for Reproduction? *J. Anim. Physiol. Anim. Nutr.* **2004**, *88*, 131–137.

Myher, J.J.; A. Kiksis; K. Geher; P.W. Park; D.A. Diersen-Schade. Stereospecific nalysis of Triacylglycerols Rich in Long-Chain Polyunsaturated Fatty Acids. *Lipids* **1996**, *31,* 207–215.

Nes, W.D.; S.D. Nichols. Phytosterol Biosynthesis Pathway in *Mortierella alpina. Phytochemistry* **2006**, *67,* 1716–1721.

Nghia, T.T.; M. Wille; S. Vandendriessche; Q. Vinh; P. Sorgeloos. Influence of Highly Unsaturated Fatty Acids in Live Food on Larviculture of Mud Crab *Scylla paramamosain* (Estampador 1949). *Aquacult. Res.* **2007**, *38,* 1512–1528.

Oe, H.; T. Hozumi; E. Murata; H. Matsuura; K. Negishi; Y. Matsumura; S. Iwata; K. Ogawa; K. Sugioka; Y. Takemoto; et al. Arachidonic Acid and Docosahexaenoic Acid Supplementation Increases Coronary Flow Velocity Reserve in Japanese Elderly Individuals. *Heart* **2008**, *94,* 316–321.

Ogawa, J.; E. Sakuradani; S. Shimizu. Production of C-20 Polyunsaturated Fatty Acids by an Arachidonic Acid-Producing Fungus *Mortierella alpina* 1S-4 and Related Strains. *Lipid Biotechnol.* **2002**, , 563–574.

Pantaleo, P.; F. Marra; F. Vizzutti; S. Spadoni; G. Ciabattoni; C. Galli; G. La Villa; P. Gentilini; G. Laffi. Effects of Dietary Supplementation with Arachidonic Acid on Platelet and Renal Function in Patients with Cirrhosis. *Clin. Sci.* **2004**, *106,* 27–34.

Park, E.Y.; T. Hamanaka; K. Higashiyama; S. Fujikawa. Monitoring of Morphological Development of the Arachidonic-Acid-Producing Filamentous Microorganism *Mortierella alpina. Appl. Microbiol. Biotechnol.* **2002**, *59,* 706–712.

Park, E.Y.; K. Koizumi; K. Higashiyama. Analysis of Morphological Relationship between Micro- and Macromorphology of *Mortierella* Species Using a Flow-through Chamber Coupled with Image Analysis. *J. Eukaryot. Microbiol.* **2006**, *53,* 199–203.

Parker-Barnes, J.M.; T. Das; E. Bobik; A.E. Leonard; J.M. Thurmond; L.T. Chaung; Y.S. Huang; P. Mukerji. Identification and Characterization of an Enzyme Involved in the Elongation of n-6 and n-3 Polyunsaturated Fatty Acids. *Proc. Natl. Acad. Sci. U.S.A.* **2000**, *97,* 8284–8289.

Pastor, N.; B. Soler; S.H. Mitmesser; P. Ferguson; C. Lifschitz. Infants Fed Docosahexaenoic Acid- and Arachidonic Acid-Supplemented Formula Have Decreased Incidence of Bronchiolitis/ Bronchitis the First Year of Life. *Clin. Pediatr.* **2006**, *45,* 850–855.

Pawlosky, R.J.; Y. Denkins; G. Ward; N. Salem. Retinal and Brain Accretion of Long-Chain Polyunsaturated Fatty Acids in Developing Felines: The Effects of Corn Oil-based Maternal Diets. *Am. J. Clin. Nutr.* **1997**, *65,* 465–472.

Pazirandeh, S.; P.R. Ling; M. Ollero; F. Gordon; D.L. Burns; B.R. Bistrian. Supplementation of Arachidonic Acid Plus Docosahexaenoic Acid in Cirrhotic Patients Awaiting Liver Transplantation: A Preliminary Study. *J. Parenter. Enteral Nutr.* **2007**, *31,* 511–516.

Pillai, M.; A. Ahmad; T. Yokochi; T. Nakahara; Y. Kamisaka. Biosynthesis of Triacylglycerol Molecular Species in an Oleaginous Fungus, *Mortierella ramanniana* Var. *angulispora. J. Biochem.* **2002**, *132,* 121–126.

Rozhnova, N.A.; G.A. Gerashchenkov; T.L. Odintsova; S.M. Musin; V.A. Pukhal'skii. Protective Effect of Arachidonic Acid during Viral Infections Synthesis of New Proteins by in vitro Plants. *Russ. J. Plant Physiol.* **2001**, *48,* 780–787.

Sakuradani, E.; S. Shimizu. Gene Cloning and Functional Analysis of a Second Δ6-Fatty Acid Desaturase from an Arachidonic Acid-Producing *Mortierella* Fungus. *Biosci. Biotechnol. Biochem.* **2003**, *67,* 704–711.

Sakuradani, E.; M. Kobayashi; S. Shimizu. Δ6-Fatty Acid Desaturase from an Arachidonic Acid-Producing *Mortierella* Fungus—Gene Cloning and Its Heterologous Expression in a Fungus, *Aspergillus*. *Gene* **1999a,** *238,* 445–453.

Sakuradani, E.; M. Kobayashi; S. Shimizu. Identification of an NADH-Cytochrome *b*(5) Reductase Gene from an Arachidonic Acid-Producing Fungus, *Mortierella alpina* 1S-4, by Sequencing of the Encoding cDNA and Heterologous Expression in a Fungus, *Aspergillus oryzae*. *Appl. Environ. Microbiol.* **1999b,** *65,* 3873–3879.

Sakuradani, E.; M. Kobayashi; T. Ashikari; S. Shimizu. Identification of Δ12-Fatty Acid Desaturase from Arachidonic Acid-Producing *Mortierella* Fungus by Heterologous Expression in the Yeast *Saccharomyces cerevisiae* and the Fungus *Aspergillus oryzae*. *Eur. J. Biochem.* **1999c,** *261,* 812–820.

Sakuradani, E.; N. Kamada; Y. Hirano; M. Nishihara; H. Kawashima; K. Akimoto; K. Higashiyama; J. Ogawa; S. Shimizu. Production of 5,8,11-Eicosatrienoic Acid by Δ5 and Δ6 desaturation activity-enhanced mutant derived from a Δ12 desaturation Activity-Defective Mutant of *Mortierella alpina* 1S-4. *Appl. Microbiol. Biotechnol.* **2002,** *60,* 281–287.

Sakuradani, E.; Y. Hirano; N. Kamada; M. Nojiri; J. Ogawa; S. Shimizu. Improvement of Arachidonic Acid Production by Mutants with Lower n-3 Desaturation Activity Derived from *Mortierella alpina* 1S-4. *Appl. Microbiol. Biotechnol.* **2004,** *66,* 243–248.

Sakuradani, E.; T. Abe; K. Iguchi; S. Shimizu. A Novel Fungal ω3-Desaturase with Wide Substrate Specificity from Arachidonic Acid-Producing *Mortierella alpina* 1S-4. *Appl. Microbiol. Biotechnol.* **2005,** *66,* 648–654.

Sakuradani, E.; S. Murata; H. Kanamaru; S. Shimizu. Functional Analysis of a Fatty Acid Elongase from Arachidonic Acid-Producing *Mortierella alpina* 1S-4. *Appl. Microbiol. Biotechnol.* **2008,** 1–7.

Sakuradani, E.; T. Abe; K. Matsumura; A. Tomi; S. Shimizu. Identification of Mutation Sites on Δ12 Desaturase Genes from *Mortierella alpina* 1S-4 Mutants. *J. Biosci. Bioeng.* **2009a,** *107,* 99–101.

Sakuradani, E.; T. Abe; S. Shimizu. Identification of Mutation Sites on ω3 Desaturase Genes from *Mortierella alpina* 1S-4 Mutants. *J. Biosci. Bioeng.* **2009b,** *107,* 7–9.

Schimek, C.; K. Kleppe; A.R. Saleem; K. Voigt; A. Burmester; J. Wostemeyer. Sexual Reactions in *Mortierellales* Are Mediated by the Trisporic Acid System. *Mycol. Res.* **2003,** *107,* 736–747.

Seguineau, C.; P. Soudant; J. Moal; M. Delaporte; P. Miner; C. Quere; J.E. Samain. Techniques for Delivery of Arachidonic Acid to Pacific Oyster, *Crassostrea gigas,* Spat. *Lipids* **2005,** *40,* 931–939.

Shaw, R. The Polyunsaturated Fatty Acids of Microorganisms. *Adv. Lipid Res.* **1966,** *4,* 107–74.

Shinmen, Y.; S. Shimizu; K. Akimoto; H. Kawashima; H. Yamada. Production of Arachidonic Acid by *Mortierella* Fungi: Selection of a Potent Producer and Optimization of Culture Conditions for Large-Scale Production. *Appl. Microbiol. Biotechnol.* **1989,** *31,* 11–16.

Sperling, P.; P. Ternes; T.K. Zank; E. Heinz. The Evolution of Desaturases. *Prostaglandins Leukotrienes Essent. Fatty Acids* **2003,** *68,* 73–95.

Streekstra, H. On the Safety of *Mortierella alpina* for the Production of Food Ingredients, Such as Arachidonic Acid. *J. Biotechnol.* **1997,** *56,* 153–165.

Tanabe, Y.; M. Saikawa; M.M. Watanabe; J. Sugiyama. Molecular Phylogeny of *Zygomycota* Based on EF-1 Alpha and RPB1 Sequences: Limitations and Utility of Alternative Markers to rDNA. *Molec. Phylogen. Evol.* **2004,** *30,* 438–449.

Takeno, S.; E. Sakuradani; S. Murata; M. Inohara-Ochiai; H. Kawashima; T. Ashikari; S. Shimizu. Cloning and Sequencing of the *ura*3 and *ura*5 Genes, and Isolation and Characterization of Uracil Auxotrophs of the Fungus *Mortierella alpina* 1S-4. *Biosci. Biotechnol. Biochem.* **2004a,** *68,* 277–285.

Takeno, S.; E. Sakuradani; S. Murata; M. Inohara-Ochiai; H. Kawashima; T. Ashikari; S. Shimizu. Establishment of an Overall Transformation System for an Oil-Producing Filamentous Fungus, *Mortierella alpina* 1S-4. *Appl. Microbiol. Biotechnol.* **2004b,** *65,* 419–425.

Takeno, S.; E. Sakuradani; S. Murata; M. Inohara-Ochiai; H. Kawashima; T. Ashikari; S. Shimizu. Molecular Evidence that the Rate-Limiting Step for the Biosynthesis of Arachidonic Acid in *Mortierella alpina* is at the Level of an Elongase. *Lipids* **2005a,** *40,* 25–30.

Takeno, S.; E. Sakuradani; A. Tomi; M. Inohara-Ochiai; H. Kawashima; T. Ashikari; S. Shimizu. Improvement of the Fatty Acid Composition of an Oil-Producing Filamentous Fungus, *Mortierella alpina* 1S-4, through RNA Interference with Δ12-Desaturase Gene Expression. *Appl. Environ. Microbiol.* **2005b,** *71,* 5124–5128.

Takeno, S.; E. Sakuradani; A. Tomi; M. Inohara-Ochiai; H. Kawashima; S. Shimizu. Transformation of Oil-Producing Fungus, *Mortierella alpina* 1S-4, Using Zeocin, and Application to Arachidonic Acid Production. *J. Biosci. Bioeng.* **2005c,** *100,* 617–622.

Totani, N.; K. Oba. The Filamentous Fungus *Mortierella alpina,* High in Arachidonic Acid. *Lipids* **1987,** *22,* 1060–1062.

Wassef, M.K. Fungal Lipids. *Adv. Lipid Res.* **1974,** *15,* 159–232.

White, M.M.; T.Y. James; K. O'Donnell; M.J. Cafaro; Y. Tanabe; J. Sugiyama. Phylogeny of the *Zygomycota* Based on Nuclear Ribosomal Sequence Data. *Mycologia* **2006,** *98,* 872–884.

Wongwathanarat, P.; L.V. Michaelson; A.T. Carter; C.M. Lazarus; G. Griffiths; A.K. Stobart; D.B. Archer; D.A. MacKenzie. Two Fatty Acid Δ9-Desaturase Genes, *ole*1 and *ole*2, from *Mortierella alpina* Complement the Yeast *ole*1 Mutation. *Microbiology UK* **1999,** *145,* 2939–2946.

Wynn, J.P.; A.A. Hamid; C. Ratledge. The Role of Malic Enzyme in the Regulation of Lipid Accumulation in Filamentous Fungi. *Microbiology UK* **1999,** *145,* 1911–1917.

Wynn, J.P.; C. Ratledge. Evidence that the Rate-Limiting Step for the Biosynthesis of Arachidonic Acid in *Mortierella alpina* Is at the Level of the 18:3 to 20:3 Elongase. *Microbiology UK* **2000,** *146,* 2325–2331.

Yu, L.J.; W.M. Qin; W.Z. Lan; P.P. Zhou; M. Zhu. Improved Arachidonic Acids Production from the Fungus *Mortierella alpina* by Glutamate Supplementation. *Bioresour. Technol.* **2003,** *88,* 265–268.

Yuan, C.L.; J. Wang; Y. Shang; G.H. Gong; J.M. Yao; Z.L. Yu. Production of Arachidonic Acid by *Mortierella alpina* I-49-N-18. *Food Technol. Biotechnol.* **2002,** *40,* 311–315.

Zhang, Y.; C. Ratledge. Multiple Isoforms of Malic Enzyme in the Oleaginous Fungus, *Mortierella alpina. Mycol. Res.* **2008,** *112,* 725–730.

Zhang, S.; E. Sakuradani; S. Shimizu. Identification and Production of n-8 Odd-Numbered Polyunsaturated Fatty Acids by a Δ12 Desaturation-Defective Mutant of *Mortierella alpina* 1S-4. *Lipids* **2006,** *41,* 623–626.

Zhu, M.; L.J. Yu; Y.X. Wu. An Inexpensive Medium for Production of Arachidonic Acid by *Mortierella alpina. J. Ind. Microbiol. Biotechnol.* **2003,** *30,* 75–79.

Zhu, M.; L.J. Yu; Z. Liu; H.B. Xu. Isolating Strains of High Yield of Arachidonic Acid. *Lett. Appl. Microbiol.* **2004,** *39,* 332–335.

Zhu, M.; L. J. Yu; W. Li; P.P. Zhou; C.Y. Li. Optimization of Arachidonic Acid Production by Fed-Batch Culture of *Mortierella alpina* Based on Dynamic Analysis. *Enzyme Microb. Technol.* **2006,** *38,* 735–740.

Production of Single Cell Oils by Dinoflagellates

**James Wynn,[a] Paul Behrens,[b] Anand Sundararajan,[c]
Jon Hansen,[c] and Kirk Apt[b]**

[a]MBI International 3900 Collins Road, Lansing, MI 48910, USA; b Martek Biosciences Corporation, 6480 Dobbin Road, Columbia, MD 21045, USA; c Martek Biosciences Winchester Corporation, 555 Rolling Hills Lane, Winchester, KY 40391, USA

Introduction

Dinoflagellates are a group of over 2000 species of eukaryotic algae that, alongside diatoms, play an important ecological role as primary producers at the base of aquatic ecosystems (Taylor & Pollingher, 1987). The dinoflagellates are distinctive morphologically and in terms of their genetic organization.

This group of microalgae derives their name from the pair of "whirling" flagella that provide the cells with locomotion. One flagellum (the posterior flagellum) extends outward from the cell, whilst the second (transverse) flagellum runs in a lateral groove around the middle of the cell, between the cellulosic plates (theca) that make up part of the rigid dinoflagellate cell wall. The combined action of these twin flagella result in a tumbling spiral swimming action of the motile stages in the dinoflagellate life-cycle. The cellulosic plates are themselves somewhat unusual in that they are situated between the inner and outer cell membrane and are, therefore, bounded by a membranous sheath (Kwok et al., 2007).

The genetic organization of the dinoflagellates is unique. Despite the organisms being eukaryotic, their chromosomes lack histones and remain condensed throughout the interphase (Rizzo, 2002). The genomes are also characterized by extensive redundancy and, therefore, tend to be extremely large (Rizzo, 2002)—up to and over fifty-fold larger than the human genome!

Although the dinoflagellates are algae, only half of this group is photosynthetic, while the rest is heterotrophic, either free-living or endosymbionts (e.g., in coral, etc.). Despite their widespread occurrence, impressive productivity in nature, and ecological importance, the dinoflagellates are of limited biotechnological importance, due mainly to the relative difficulty of cultivating this group of organisms in fermentors and in synthetic culture media.

One dinoflagellate that is of significant biotechnological importance, however, is *Crypthecodinium cohnii*. This marine heterotrophic species is the production organism used by Martek Biosciences for the production of a docosahexaenoic acid (DHA)-rich single cell oil (SCO), DHASCO™ or life'sDHA™.

Crypthecodinium cohnii was identified as early as the late 1960s or early 1970s as a primary producer of DHA, the longest, most unsaturated fatty acid commonly found in nature (Beach & Holz, 1973; Harrington & Holz, 1968; Tuttle & Loeblich, 1975). However, issues with growing this organism to a high cell density and the lack of demand for a high potency (and non-fish) source of DHA were sufficient to preclude the commercial production of a DHA-rich oil from this organism until the mid-1990s.

Significance of Docosahexaenoic Acid

DHA is the longest, most unsaturated fatty acid commonly found in nature. It possesses a carbon chain of 22 carbon atoms and has six methylene interrupted double bonds, leading to its designation as 22:6 n-3. DHA is a fatty acid commonly associated with fish oils and, therefore, with the health benefits that they provide. It is not as commonly recognized that fish do not actually produce this "healthy" fatty acid de novo (Sargent & Tacon, 1999) but, instead, obtain it in their diets from primary producers (including the dinoflagellates) at the base of aquatic food chains.

Like fish, humans and other animals cannot synthesize DHA de novo, as they lack the key fatty acid desaturases required for its production (specifically, a $\Delta12$ and a $\Delta15/n$-3 desaturase). Although, in theory, humans can synthesize DHA from a precursor fatty acid, α-linolenic acid (18:3 n-3, ALA), the actual conversion of this fatty acid to DHA is very limited. Recent studies agree that the conversion of dietary ALA to DHA is very low (as little as 0.1%), though conversion to EPA is far greater (Plourde & Cunnane, 2007; Williams & Burdge, 2006). One study noted an apparent discrepancy between the rate of conversion in males and females (female conversion being higher) and suggested this divergence was an evolutionary adaptation that reflects the importance of DHA in neonatal nutrition (Williams & Burdge, 2006). The limited ability humans have to elongate and desaturate ALA to DHA, together with high consumption of 18:2 n-6 in the typical Western diet, leads to a favoring of the n-6 pathway (to ARA and other n-6 fatty acids), rather than the n-3 pathway, which leads to DHA. This preference then creates a significant dietary requirement for preformed DHA in a healthy human diet. Unfortunately, levels of preformed DHA in Western diets tend to be low (0.1–0.5 g/day) because of a low acceptance of oily fish—the most abundant sources of DHA (Williams & Burdge, 2006).

Although DHA is found in appreciable amounts only in a restricted number of foods (predominantly, in oily fish) and its intake in the Western diet is low, the role of DHA in human health is highly significant. DHA, along with another long chain polyunsaturated fatty acid (LC-PUFA), arachidonic acid (20:4 n-6), is the major polyunsaturated fatty acid in human neural and eye tissues. The grey matter of the

Table 6.A. Fatty Acids Profile of Human Grey Matter (derived from Lalovic et al., 2007).

Fatty Acid	Percentage of Total Fatty Acids
16:0 (palmitic acid	21.3
16:1 (palmitoleic acid)	2.0
18:0 (stearic acid)	18.9
18:1 n-7	4.3
18:1 n-9 (oleic acid)	15.9
20:4 n-6 (arachidonic acid)	9.2
22:4 n-6 (adrenic acid)	6.3
22:6 n-3 (docosahexaenoic acid)	14.4

All other fatty acids are present at < 1% total fatty acids.

human brain is particularly rich in this fatty acid (see Table 6.A), with DHA making up 15% of all the fatty acids found in this fat-rich tissue (Lalovic et al., 2007). Indeed, it has been suggested that ready access to fish and shellfish (foods rich in DHA) and a high dietary intake of DHA was instrumental in the ability of humans to develop a large forebrain (Broadhurst et al., 2002).

It is well accepted that DHA is a key nutrient for infant development, both in utero and during the postnatal period (Agostoni et al., 1995; Birch et al., 2000; Carlson et al., 1993; see also Chapter 16, Nutritional aspects of SCOs: Applications of ARA and DHA Oils). Developing infants obtain supplies of DHA from their mothers in utero—via the placenta (particularly, during the final trimester of pregnancy)—and then through human milk while nursing. However, DHA is found in neither plant oils nor cow's milk, and so it is not present in infant formula unless specifically added. Due to its critical importance in the development and health of very young children, the World Health Organization (WHO), the British Nutritional Foundation (BNF), the European Society of Pediatric Gastroenterology and Nutrition (ESPGAN), and the International Society for the Study of Fats and Lipids (ISSFAL) have recommended that DHA should be included (along with ARA) in all infant formula.

As well as the critical role that DHA plays in neural/eye development, there is significant (and accumulating) evidence that increased dietary DHA plays a role in the maintenance of human health and helps ward off chronic health issues, such as heart disease, depression, and even some forms of dementia later in life.

Commercial Success of Dinoflagellate-derived SCO

Even though *C. cohnii* was recognized as a potential source of DHA-rich oil in the early 1970s, it took over 20 years for this oil to become a commercial reality. The reasons for this delay were both technical (the inability to cultivate this organism in a commercially viable fashion) and commercial (the lack of a market need for a non-fish source of DHA). By the mid 1990s, however, both the technology and market conditions were correct, and so DHASCO entered the market both as an adult supplement (in direct competition with fish oil supplements) and as an additive for infant formula (in combination with an ARA as Formulaid™).

As the science behind the benefits of preformed DHA in the diets of neonates, in general, and premature neonates, in particular (as premature neonates are deprived of the "DHA-loading" that occurs during the final trimester of gestation), became more apparent, suitable sources of DHA were investigated for addition to infant formula. For the reasons outlined below, the "traditional" source of dietary, preformed DHA, fish oils, were contra-indicated for use in infant formula, although the use of fish oil for this application still persists in some countries. Use of *C. cohnii* (dinoflagellate) DHA-rich oil as a supplement for infant formula was begun in the mid-1990s outside of the United States after the United Kingdom's Committee on Toxicology of the Advisory Committee for Novel Foods and Processes and the Ministry of Health of The Netherlands both found there were no safety concerns with the use of DHA from Martek's dinoflagellate-derived SCO. In the United States, Martek filed a Generally Recognized as Safe (GRAS) notification for the Formulaid combination of DHASCO and ARASCO with the FDA in December 1999. After over a year of scrutiny, in May 2001 the FDA completed a "favorable" review of the document, which effectively allowed the inclusion of *C. cohnii* DHA-rich oil (in combination with microbial ARA-rich oil) in infant formula in the U.S.. The following year the first Formulaid-supplemented infant formula appeared on retail shelves, and such was the success of this launch that by the time this chapter was written (2009), over 95% of infant formula sold in the U.S. (and, approximately, 30% of the infant formula sold globally) contains dinoflagellate-derived, DHA-rich oil.

Dinoflagellate Oils versus Oils from Traditional Sources

Although DHA-rich oil from the dinoflagellate *Crypthecodinium cohnii* has been an undoubted commercial success, this success was not easily achieved. The technical challenges encountered when cultivating this organism at a commercial scale were not inconsequential. However, as for any microbial (single cell) oil, it was the competition with "traditionally" sourced oils, from plants or animals (including fish), that was the real hurdle to commercial success. It is a paradigm that has yet to be disproved (though, current work in the area of biodiesel may yet achieve this success) that, due to the cost of microbial cultivation, a microbial oil cannot compete with an equivalent oil from a plant or animal source. Currently, oils from plants and animals cost in the range of $ 0.60–$ 2.00/kg, depending on the source (Anonymous, 2009), whereas microbial oils are sold at closer to $100/kg.

Fortunately for Martek, there are several distinctive features about the dinoflagellate DHA-rich oil produced using *C. cohnii* that provide clear and sufficient differentiation from any oil that can be obtained commercially from a plant or animal. Plants, although a commercially important source of oils and fats, do not synthesize

LC-PUFAs (i.e., fatty acids with more than 2 double bonds and more than 18 carbons). Whilst plant sources of 18-carbon PUFA are commercially exploited, borage and evening primrose for 18:3 n-6, flax and several nuts for 18:3 n-3, and, more recently, *Echium* for an oil containing stearidonic acid (18:4 n-3), no "non-GM" oil contains docosahexaenoic acid (or its precursor, eicospentaenoic acid—EPA, 20:5 n-3). As a result, although high quality and cheap plant oils are available (in vast quantities and from multiple sources), none of these oils can be used as a source of dietary, preformed DHA.

Animal fats and oils are more diverse than plant oils in their fatty acid profiles (see Table 6.A), and animal (particularly, fish) oils are well known to contain significant amounts of LC-PUFAs, including DHA. However, there are confounding factors that make these fish oils unsuitable for their inclusion in infant formula, in particular, and can limit their use for human dietary supplementation, in general. Fish and animals lack certain fatty acid desaturases that are required for the de novo synthesis of LC-PUFAs. As a result, the fatty acid profiles of animal/fish oils reflect their own dietary inputs more than the metabolic capabilities of the organism itself (as is the case with plant and microbial oils). The influence of dietary intake on the composition of animal oils is a major factor that limits the use of these oils for certain applications.

As an animal's fatty acid profile is dependent on the fatty acids within the animal's diet, the composition of animal fats/oils are generally more complex than those of plants/microbes. Due to this complexity, the absolute amount of any given fatty acid tends to be lower in animal oils than the level that can be obtained through the judicious selection of a plant or microbial oil. Furthermore, as the fatty acid composition of the oil is a reflection of the animal's diet, the composition of the oil can be impacted by changes in diet and, therefore, can vary more widely due to geography, climate, and time of harvest. Contamination of fish and fish oils with lipophilic environmental pollutants, PCBs, dioxins, and methyl-mercury have also become more of a widespread concern over the past few years. Although thorough processing of fish oils is effective in removing these undesirable contaminants, this processing is complicated by variations in the level of contamination found in fish organs at the start of the process. Depending on the fish species and the location of the catch, the level and type of environmental contamination of the crude fish oil can vary to a considerable degree. Therefore, the processing of fish oil to preclude the presence of PCBs, dioxins, and heavy metals is not a "one size fits all" process, and diligent testing of the starting material and the finished oil is necessary to ensure a product devoid of contamination. Recent studies in both the European Union and North America have highlighted the issue of pesticide residues in fish oil supplements. The EU study, published in 2006, analyzed commercial fish oil supplements purchased between 2000 and 2002. Although the samples contained lower levels of contaminants than had been detected in similar samples obtained 4 years earlier, all samples contained detectable dioxins and PCBs (Fernandes et al., 2006). Of these samples, 12 of the 33 would have exceeded levels permitted by the EU (2 ng/kg oil), although this limit was not introduced until after the oil samples were collected. Similarly, a North American study found detectable levels of PCBs in thirty commercial fish oil samples sourced between 2005 and 2007. This study emphasized that the levels of these contaminants in fish oils were generally low a sentiment echoed by a fish industry response to the study (Heller, 2009) but varied greatly between different oils from different fish species and even between different oils from the same fish species (Rawn et al., 2008).

Along with the potential issues that follow the environmental contamination of fish oils, there are increasing fears concerning the sustainability of fish oils as a source of LC-PUFAs (Miller et al., 2008). Overfishing has caused a significant decline in the populations of many "food-fish," leading to an increase in price and to the imposition of fishing quotas by many governments. In light of declining "wild" stocks, fish farming appears to be a sustainable alternative for the supply of DHA for human consumption (in either fish or oil form). However, the apparent sustainability of aquaculture in comparison with wild-caught fish is somewhat illusory. Fish, although they are good sources of DHA, cannot synthesize DHA de novo and, therefore, must rely on their diet to supply pre-formed DHA (only small quantities being produced from dietary precursors). Therefore, in order to successfully rear fish commercially, a diet containing DHA is required (Miller et al., 2008). As a result, and paradoxically, fish farming is a major user of "wild-caught" fish DHA, either in the form of fish oil or fish meal (Sargent, 1997). As the source of DHA in farmed fish is generally wild-caught fish, this activity does not really add any sustainable DHA to the human food chain. The exception here would be when the farmed fish are fed a diet containing microbial DHA. One such product that is commercially available is Aquagrow DHA, which is formulated using dinoflagellate (*C. cohnii*) oil-rich biomass.

Microbial oils (including those from the dinoflagellate *C. cohnii*) are not subject to the limitations outlined for plant and animal oils and, as a consequence, exhibit advantages over "traditional" oils that can help offset their high cost. As microbes (grown on glucose or some other simple sugar) synthesize all their cell lipid fatty acids de novo, the profile of these lipids is relatively simple and can be very rich in specific fatty acids. Unlike plants, however, microbes can synthesize significant amounts of LC-PUFAs. One such species is the dinoflagellate *C. cohnii*, which is the subject of this chapter. *C. cohnii* is exceptional not only because it is capable of accumulating large stores of storage lipid (triacylglycerol), but also because the fatty acid profile of this lipid contains high levels of DHA, without the significant buildup of PUFA intermediates. As a result, the oil extracted from this organism contains DHA but no other PUFA. This result is a crucial factor for the inclusion of *C. cohnii* oil in infant formula (the most economically significant application) as the presence of another LC-PUFA (eicosapentaenoic acid, EPA, 20:5 n-3), a precursor in the synthesis of DHA and a common fatty acid in fish oils, is contra-indicated for infant formula inclusion due to a link with neonate growth retardation (Carlson et al., 1993; see also Chapter 16).

Microbial oils, due to their origin in fermentation, are also far more predictable in both supply and quality than either plant or animal oils. Climate (year to year variation), disease, and the geographical source of fish and plant oils can impact supply, quality, and price of "traditional" oils. Microbial oils, such as the dinoflagellate oil DHASCO, produced in closed fermentation vessels are manufactured under closely controlled conditions. Therefore, the supply and quality of these oils can be guaranteed from batch to batch, season to season, and year to year. Cultivation in a closed tank system with well-defined inputs also allows dinoflagellate-derived oils to be devoid of any potential environmental contaminants.

Commercial Production of Dinoflagellate Oil

Strain Selection and Optimization

Although the production of DHA as a major constituent of the triacylglycerol (TAG) accumulated by *C. cohnii* was reported in the late 1960s and early 1970s, the strains used in these early studies were not amenable to large-scale cultivation. These problems were associated with the organism's sensitivity to both the shear and high dissolved O_2 in stirred tank fermentors. Despite these problems, Martek Biosciences Corp. embarked on a program to identify a strain of *C. cohnii* capable of growing in fermentors and then to develop that strain further, allowing large-scale production of DHA-rich SCO from this organism.

Many strains of *C. cohnii* were obtained from public culture collections and screened for growth and DHA production in fermentors. From this screen, a single strain, obtained from the UTEX culture collection, was identified as the most promising on the basis of growth rate and DHA production. This strain became the parent from which all Martek production strains were developed via classical strain selection—although the strains in use today bear little or no resemblance to this original strain. Strain selection and cultivation from single colonies on agar plates, as well as yielding potentially improved strains, ensures genetic homogeneity of cultures used for production and guards against phenotypic "drift."

Using classical strain-screening techniques, many strains were isolated that had potentially attractive attributes. Screening for increased lipid production, decreased byproduct formation, and improved growth under conditions suitable for large-scale production have all been successfully accomplished—although by no means does every strain identified in a laboratory-based screen make it as far as production! As each potential strain is isolated, a dedicated group is involved in optimization of culture media and conditions to assess and realize the full potential of each strain before the Fermentation Group takes the strain forward and carries out scale-up trials.

Strains and culture conditions have been developed that not only allow high cell density cultivation in stirred tank fermentors but also permit growth under the decreased Cl^- concentration required to avoid excessive corrosion of steel fermentation tanks (see below) (Behrens et al., 2005 & 2006). At the same time as producing strains amenable to large-scale cultivation, the strain selection/optimization program has been very efficient at increasing the productivity of the production strain.

Industrial Production of Dinoflagellate Oil (DHASCO)

The basic fermentation process used to cultivate *C. cohnii* for the subsequent extraction of cell lipid resembles other commercial fermentation processes in many respects (Fig. 6.1). A cryovial of a certified proprietary strain (developed as described above) is thawed and used to inoculate a seed train, including shake-flasks and fermentors, so as to maintain an inoculum volume of 5–10% (v/v) for each successive step, up to a final vessel volume ≈ 200 m³. Martek currently has many hundreds of thousands of liters of fermentation capacity dedicated to DHASCO production at plants in Winchester, KY and Kingstree, SC (Fig. 6.2). As *C. cohnii* is a relatively slow-growing organism, in comparison to commonly fermented bacteria, the seed train takes a considerable period and so must be carefully coordinated to ensure inocula are available for each main tank as they are turned around. The slow growth rate of *C. cohnii*, in comparison to bacteria, also has implications in terms of susceptibility to contamination. As a result, plant hygiene is paramount and "clean-in-place" (CIP) and "sterilize-in-place" (SIP) regimes have to be closely monitored and maintained in order to avoid contamination and, subsequently, the loss of fermentation batches.

The culture medium throughout the seed train is kept constant so as to avoid stressing the organism and slowing the growth rate. One significant feature of the medium used at commercial scale is the Cl⁻ level employed. As a marine microbe, *C. cohnii* requires a saline environment for growth. The Cl⁻ of

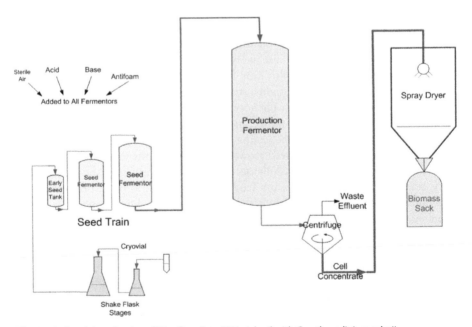

Fig. 6.1. Industrial production of Dinoflagellate DHA-rich oil with *Crypthecodinium cohnii*.

Fig. 6.2. Production Facilities for Dinoflagellate Oils (A: Winchester, Kentucky; B: Kingstree, South Carolina).

seawater, and seawater-based media commonly employed in academic research (Tuttle & Loeblich, 1975), is ≈ 19000 ppm and is not compatible with stainless steel cultivation tanks. As a result, media and adapted strains have been developed so that cultivation at only a fraction of the Cl⁻ concentration of seawater is possible.

To further minimize the risks associated with the high Cl⁻ requirement of *C. cohnii* (even at the lower levels possible with the current production strain), many of the tanks used by Martek for the cultivation of *C. cohnii* are constructed of high-grade stainless steel (types 317L, 2205, or AL6XN).

Significant lipid (TAG) accumulation does not occur during active growth in nutrient-replete medium but occurs during the idiophase after a culture nutrient other

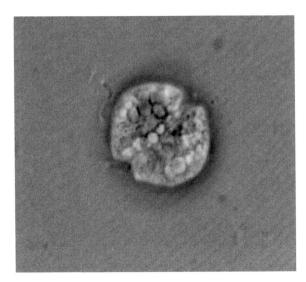

Fig. 6.3. Cell of *Crypthecodinium cohnii*, packed with DHA-rich oil bodies (cell is approximately 10μ).

than the carbon (C)-source is depleted (usually, nitrogen, N; see also Chapter 1). Therefore, the *C. cohnii* fermentation is a C-fed batch and progresses in two stages. The first is the active growth phase, during which the lipid content of the biomass is modest (approximately, 20% w/w dry wt) and the cells are motile. Once the N-source is depleted, C is continuously supplied to the fermentor and, as cell growth and division is halted by the lack of N for de novo protein and nucleotide synthesis, the supplied C is converted into storage lipid (TAG), rich in DHA. During this lipid-accumulating phase, *C. cohnii* cells lose their flagella, becoming "cyst-like" cells packed with DHA-rich lipid bodies (Fig. 6.3). Cell lipid at this stage constitutes in excess of half the cell dry weight. Maintaining the C-concentration in the cultivation vessel is important to optimize lipid accumulation not only by promoting synthesis but also to avoid the utilization of internal storage lipid. Induction of β-oxidation by cells experiencing C-limitation causes an increase in the levels of free fatty acids and di/monoacylglycerols in the final hexane extract, which complicates processing and decreases both final yield and final oil stability. The possible induction of β-oxidation at the end of the fermentation, during storage and harvest/drying, is another potential source of losses of oil quality and quantity. To decrease the impact of this step, the harvesting time must be kept to a minimum, and operating conditions must be closely controlled.

Although the strain improvement program has identified a number of improved strains of *C. cohnii*, production of DHASCO at a commercial scale using these strains is not always straightforward and can suffer from many of the scale-up challenges associated with other production organisms and processes. The very large fermentors employed by Martek have the advantage of being more cost-effective

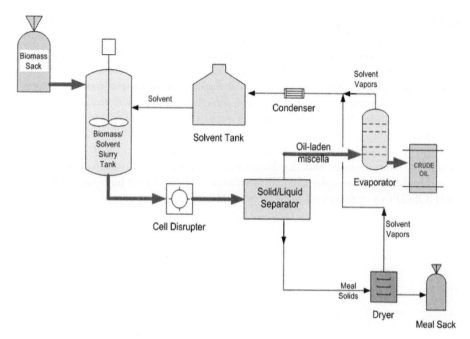

Fig. 6.4. Outline of extraction process used for production of DHASCO.

than smaller tanks, but they have some trade-offs and operational complications. The most obvious change in using large culture vessels is that they exhibit a higher absolute pressure at the bottom of the tank than smaller vessels do. Depending on mixing time and mass transfer coefficients, this pressure can result in increased levels of dissolved gases, which can impact the growth and productivity of *C. cohnii*. Therefore, sufficient mixing is required to avoid the depletion of O_2 and build-up of CO_2 without exposing the organism to excessive shear.

Once the culture fermentation vessel has produced sufficient biomass and lipid, it is harvested by continuous centrifugation, which includes a washing step to remove culture components. The harvested, wet biomass is then spray dried to <10% moisture (Fig. 6.1). Removing the water stabilizes the lipid-rich biomass and avoids lipid degradation due to biological processes. The temperature during spray drying is another factor that must be optimized to minimize drying time while maximizing oil quality. In general, the lowest temperatures possible are used throughout the processing of DHA-rich oil in order to avoid compromising oil quality. Oxidation of the intracellular oil by exposure to O_2 is kept to a minimum by freezing the spray-dried biomass and storing it under N_2 until extraction.

Extraction of DHA-rich lipid from the cells of *C. cohnii* is carried out by hexane extraction, an procedure essentially identical to that used routinely to extract vegetable and animal oils (see also Chapter 9, Downstream Processing, Extraction

And Purification of SCOs). Efficient extraction of the biomass is achieved by mixing the dried biomass with hexane to form slurry, which is then passed through a disruptor. This disruption breaks open the *C. cohnii* cells and makes the intracellular lipid bodies more available for extraction. After contact with the solvent, the hexane/oil mixture (miscella) is separated from the oil depleted biomass and passed on to an evaporator to remove the hexane and yield the crude oil that is stored at a low temperature and under N_2 (to avoid oxidation of the PUFA-rich oil) prior to processing (Fig. 6.4).

The crude oil is processed essentially as vegetable oil (see Fig. 6.5). Refining, bleaching, winterization, and deodorization yields an orange, translucent oil that, after the addition of tocopherols as antioxidants, is blended with high-oleic sunflower oil (HOSO) to a standardized 40% (w/w) DHA.

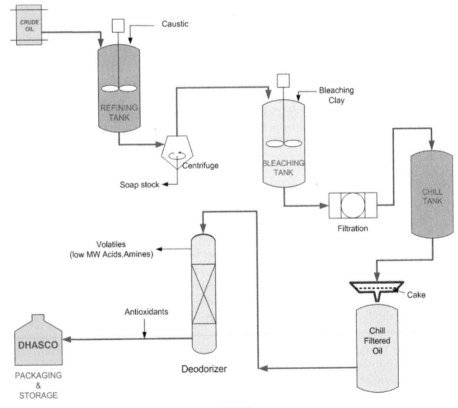

Fig. 6.5. Outline of processing used to produce DHASCO.

Characteristics of DHASCO

DHASCO is a free-flowing oil that contains a standardized 40% by weight of docosahexaenoic acid (DHA, 22:6n-3), as is displayed in Table 6.B. The processed *C. cohnii* oil contains a higher percentage of DHA but is blended to 40% w/w DHA with HOSO to reach the specified potency. The oil is >95% w/w triacylglycerol, with the remaining <5% w/w being mostly non-saponifiable material. In this regard, DHASCO is typical of food-grade vegetable oils.

The oil has a low aromatic impact (as defined by Sensory Spectrum, NY, USA), the most prevalent attribute being a "green/beany" note. Although not as bland as canola or sunflower oil, the flavor qualities of DHASCO lack the strong, undesirable "fish" or "paint" notes associated with even the most refined fish oils.

Analysis has confirmed that DHASCO is devoid of heavy metals (e.g., arsenic, mercury, and lead). Likewise, pesticide residues are not found in DHASCO; researchers have tested for seventy-four pesticides, and none of these pesticides were present at detectable levels.

Oil quality is carefully preserved during production and processing and maintained in the final product by adding 250 ppm each of ascorbyl palmitate and tocopherols (as antioxidants) and by storing the oil in N_2-purged containers at low temperatures. The stability of the oil is such that it has a shelf life of 2 years from the date of shipping, if stored frozen. Regardless of the method of storage (either frozen or at room temperature), DHASCO will maintain a peroxide value (PV) of <5 (routinely, <1) meq/kg oil for 2 years, if maintained in N_2-purged vacuum bags.

Table 6.B. Fatty Acid Profiles of Some Commercially Available Oils Containing PUFA.

Source	C. cohnii Oil (DHASCO™)	Soybean Oil	Sunflower Seed Oil	Borage Oil	Evening Primrose Oil	Cod Liver Oil	Lard
Type of Organism	microalga	plant	plant	plant	plant	fish	animal
FATTY ACID PROFILE (RELATIVE % OF TOTAL FATTY ACIDS)							
16:0	18	12	6	10	6	13	26
16:1	2	-	-	-	-	6	4
18:0	1	3	4	4	2	2	13
18:1	15	23	25	16	8	27	45
18:2	1	56	61	40	75	10	10
18:3n-3	-	6	-	Tr.	Tr.	3	1
18:3n-6	-	-	-	22	8	-	2
20:4n-6	-	-	-	-	-	1	-
20:5n-3	-	-	-	-	-	10	-
22:5n-6	-	-	-	-	-	-	-
22:6n-3	39	-	-	-	-	5	-

Tr. denotes < 0.1% of total fatty acids.

Safety of DHASCO

Due partly to its novel (i.e., microbial) origin and partly to its intended application (as part of a supplement to be added to infant formula), the safety of DHASCO has been thoroughly examined. It is true to say that DHASCO is one of the most (if not *the* most) carefully tested oils, in terms of safety and efficacy (see Chapters 15–18).

References

Agostoni, C; E. Riva; S. Trojan; R. Bellu; M. Giovannini. Docosahexaenoic Acid Status and Developmental Quotient of Healthy Term Infants. *Lancet* **1995,** *346,* 638.

Anonymous. Global Agriculture and Rural America in Transition. Agricultural Outlook Forum 2009. [Online] Feb 26–27, 2009. http://www.usda.gov/oce/forum (accessed).

Beach, D.H.; J.J. Holz. Environmental Influences on the Docosahexaenoate Content of the Triacylglycerols and Phosphatidylcholine of a Heterotrophic, Marine Dinoflageallate *Cryptheco-dinium cohnii. Biochim. Biophys. Acta* **1973,** *36,* 56–65.

Behrens, P.W.; D.J. Kyle. Microalgae as a Source of Fatty Acids. *J. Food Lipids* **1996,** *3,* 259–272.

Behrens, P.W.; J.M. Thompson; K. Apt; J.W Pfeifer; J.P. Wynn; J.C. Lippmeier; J. Fichtali; J. Hansen. Production of High Levels of DHA in Microalgae Using Modified Amounts of Chloride and Potassium. W.O. Patent 2005/035,775, 2005.

Behrens, P.W.; J.M. Thompson; K. Apt; J.W. Pfeifer; J.P. Wynn; J.C. Lippmeier; J. Fichtali; J. Hansen. Production of DHA in Microalgae in Low pH Medium. U.S. Patent 2006/100,279, 2006.

Birch, E.E.; S. Garfield; D.R. Hoffman; R. Uauy; D.G. Birch. A Randomized Controlled Trial of Early Dietary Supply of Long Chain Polyunsaturated Fatty Acids and Mental Development in Term Infants. *Dev. Med. Child Neurol.* **2000,** *42,* 174–181.

Broadhurst, C.L.; Y. Wang; M.A. Crawford; S.C. Cunnane; J.E. Parkington; W.F. Schmidt. Brain-Specific Lipids from Marine, Lacustrine or Terrestrial Food Resources: Potential Impact on Early African Homo Sapiens. *Comp. Biochem. Physiol. B Biochem. Mol. Biol.* **2002,** *131,* 653–673.

Carlson, S.E.; S.H. Werkman; P.G. Rhodes; E.A. Tolley. Visual-Acuity Development in Healthy Preterm Infants: Effect of Marine-Oil Supplementation. *Am. J. Clin. Nutr.* **1993,** *58,* 35–42.

Fernandes, A.R.; M. Rose; S. White; D.N. Mortimer; M. Gem. Dioxins and Polychlorinated Biphenyls (PCBs) in Fish Oil Dietary Supplements: Occurrence and Human Exposure in the UK. *Food Addit. Contam.* **2006,** *23,* 939–947.

Harrington G.W.; J.J. Holz. The Monoenoic and Docosahexaenoic Fatty Acids of a Heterotrophic Dinoflagellate. *Biochim. Biophys. Acta* **1968,** *164,* 137–139.

Heller L. Omega-3 Contamination Study is Misleading, Says Industry. [Online] http://www .nutringedients.com (accessed Jan 29, 2009).

Kwok, A.C.; C.C.M. Mak; F.T.W. Wong; J.T.Y. Wong. Novel Method for Preparing Spheroplasts from Cells with an Internal Cellulosic Cell Wall. *Eukaryot. Cell* **2007,** *6,* 563–567.

Lalovic, A.; É. Levy; L. Canetti; A. Sequeira; A. Montoudis; G. Turecki. Fatty Acid Composition in Postmortem Brains of People Who Committed Suicide. *J. Psychiatry Neurosci.* **2007,** *32,* 363–370.

Miller, M.R.; P.D. Nichols; C.G. Carter. N-3 Oil Sources for Use in Aquaculture—Alternatives to the Unsustainable Harvest of Wild Fish. *Nutr. Research Revs.* **2008,** *21,* 85–96.

Plourde, M.; S.C. Cunnane. Extremely Limited Synthesis of Long Chain Polyunsaturates in Adults: Implications for Their Dietary Essentiality and Use as Supplements. *Appl. Physiol. Nutr. Metab.* **2007,** *32,* 619–634.

Rawn, D.F.K.; K. Breakell; V. Verigin; H. Nicolidakis; D. Sit; M. Feeley. Persistent Organic Pollutants in Fish Oil Supplements on the Canadian Market: Polychlorinated Biphenyls and Organochlorine Insecticides. *J. Food Sci.* **2008,** *74,* 14–19.

Rizzo, P.J. Those Amazing Dinoflagellate Chromosomes. *Cell Res.* **2002,** *13,* 215–217.

Sandanger, T.M.; M. Brustad; E. Lund; I.C. Burkow. Change in Levels of Persistent Organic Pollutants in Human Plasma after Consumption of a Traditional Northern Norwegian Fish Dish— Molje (Cod, Cod Liver, Cod Liver Oil and Hard Roe). *J. Environ. Monit.* **2003,** *5,* 160–165.

Sargent, J.R. Fish Oils and Human Diet. *Brit. J. Nutr.* **1997,** *78(1),* S5–S13.

Sargent, J.R.; A.G. Tacon. Development of Farmed Fish: A Nutritionally Necessary Alternative to Meat. *Proc. Nutr. Soc.* **1999,** *58,* 377–383.

Taylor, F.J.R.; U. Pollingher. Ecology of Dinoflagellates. In Taylor, F.J.R. *The Biology of Dinoflagellates*; Blackwell Science Publications: Oxford, UK, 1987; pp 398–529.

Tuttle, R.C.; A.R. Loeblich. An Optimal Growth Medium for the Dinoflagellate *Crypthecodinium cohnii. Phycologia* **1975,** *14,* 1–8.

Williams, C.M.; G. Burdge. Long-Chain n-3 PUFA: Plant v. Marine Sources. *Proc. Nutr. Soc.* **2006,** *65,* 42–50.

Wynn, J.P.; A.J. Anderson. Microbial Polysaccharides and Single Cell Oils. *Basic Biotechnology,* 3rd edn.; Cambridge Press: UK, 2006; pp 381–401.

Alternative Carbon Sources for Heterotrophic Production of Docosahexaenoic Acid by the Marine Alga *Crypthecodinium cohnii*

Lolke Sijtsma,[a] Alistair J. Anderson,[b] and Colin Ratledge[b]
[a]*Wageningen University and Research Centre, Agrotechnology and Food Innovations b.v., P.O. Box 17, 6700 AA Wageningen, The Netherlands; [b]Department of Biological Sciences, University of Hull, Hull, HU6 7RX, UK*

Introduction

In recent years interest in polyunsaturated fatty acids (PUFA), especially the n-3 PUFA, has increased considerably due to their various physiological functions in the human body and their beneficial effects on human health (Kromhout et al., 1985; Albert et al., 1998: Horrocks & Yeo, 1999: Nordøy et al., 2001). Until recently, fish oils have been the only major sources of these fatty acids (FA); now we have the advent of single cell oils (SCO) with the identification of several marine microorganisms that contain substantial quantities of PUFA. These oils are now considered to be the major sources of the important FA that are covered in the various chapters in this book.

Of key importance is the heterotrophic marine dinoflagellate, *Crypthecodinium cohnii*, that has been studied intensively (Harrington & Holz, 1968; Beach & Holz, 1973; de Swaaf, 2003; de Swaaf et al., 1999, 2001, 2003a, 2003b, 2003c; Mendes et al., 2007) and that, together with *Schizochytrium* and related genera (see Chapters 1, 4, & 6), represents the major commercial source of docosahexaenoic acid (DHA, 22:6n-3). This is the only SCO in which DHA is the sole PUFA (Kyle 1994, 1996). *C. cohnii* can accumulate a high percentage of DHA [25–60% of the total fatty acids (TFA)] in its triacylglycerols (TAG) with only trivial amounts of other PUFA (see Chapter 6). Important parameters for optimal DHA productivity include growth rate, final biomass concentration, the total lipid content, and the DHA proportion of the lipid (Kyle 1994, 1996). In most of the documented commercial cultivation processes, glucose is used as the carbon and energy source. Glucose, of course, represents an easily accessible feedstock for many industrial fermentation processes and is usually obtained, in the form of glucose syrups, from the hydrolysis (chemical, as well as enzymological) of corn starch. It is inexpensive, readily available, and can be stored as a concentrated, self-sterilizing solution, which is, thus, stable. It is also water-soluble so that fermentation media can be made and sterilized without difficulty and then transferred into the final fermentor as a single solution. However, it is not the only possible substrate that could meet these criteria. This chapter discusses the use of alternative carbon sources—acetic acid and ethanol—on growth, lipid accumulation, and DHA productivity of *C. cohnii* in fed-batch cultures. A comment is also made regarding the possible uses of glycerol as a feedstock material.

Acetic acid is produced by the petrochemical industry on a very large scale. It costs about three times more than glucose at about $600/t (*Chemical Market Reporter* September 2008). Its drawback for large-scale fermentation use is that, in concentrated form, it requires careful handling; any spillage or contact with the skin must be dealt with promptly. It is not, however, classed as a "strong" acid and is not as aggressive as an inorganic acid, such as HCl or HNO_3. It, therefore, does not need the stringent precautions necessary when handling large quantities of these acids. In a fermentor, its concentration will be very diluted; subsequently, hazards at this level of the operation will be minimal.

Ethanol can be produced by fermentation; this process is performed extensively in countries such as Brazil, New Zealand, and the USA through extensive "gasohol" programs. However, since it is produced by glucose fermentation (originating from corn starch or from the sucrose in sugar cane), its cost will be higher than that of glucose. "Food-grade" ethanol, which would be needed for producing "food-grade" DHA, sells at $760-800/t and is currently more expensive than acetic acid. Its main disadvantage is its flammability when stored and transferred around a production site in its undiluted form. Also, and not of insignificant consequence, processes using ethanol may need to be continuously scrutinized by regulatory authorities to prevent the use of the ethanol for purposes other than for which it was intended. This necessity may place unwanted restrictions on its suitability as a fermentation feedstock.

Both substrates, like glucose, are water-soluble and, therefore, present no problems in mixing in the fermentation medium. Both substrates are also "pure" and appropriate for the production of products destined for human consumption. They have no residual, nonfermentable components, so the final waste, spent fermentation liquor, can be disposed of with a minimal amount of treatment. For fed-batch cultivations, in which both of these substrates are used, the feed supply of both substrates has the advantage of being self-sterilizing and reduces the risk of contamination.

Uses of Alternative Carbon Sources

The advantages, if any, of using either acetic acid or ethanol as a fermentation feedstock, instead of glucose, should be seen as either offering improved productivity or overall reduction of costs. However, there is a another practical, commercial consideration; the use of these, and perhaps other, substrates may circumvent any patents on processes that have specified the use of glucose as a feedstock. If a process using an alternative feedstock also yields improvements in cellular composition, gives additional advantages for product recovery, or provides improved disposal methods for the waste processing streams from the fermentation, then these will be extra incentives to switch substrates. Thus, the advantages or disadvantages of using an alternative feedstock are not exclusively associated with the material cost of producing the final product.

In dealing with both acetic acid and ethanol as substrates, one is immediately aware that, unlike glucose, these materials are toxic to most cells when used at concentrations higher than 1–2%. Since it is impractical to run a fermentation process with so little carbon substrate (the final cell biomass yields could not be more than 5 g/L and would, in all probability, be less than this estimation), it is necessary in both cases to operate the fermentations as fed-batch cultures. In such processes, additional substrate is supplied to the fermentors at a rate designed to keep the prevailing concentration of the substrate below the level at which it would begin to inhibit growth. The question then becomes how to attain a rate of substrate addition that, on the one hand, does not exceed the critical inhibitory concentration but, on the other hand, maintains a sufficient concentration of substrate that does not limit the growth rate. However, as we shall demonstrate in the following sections, this is not an insuperable task. Indeed, most industrial fermentation processes use some type of fed-batch process mode to achieve optimum productivity; this indicates that there are many precedents that could be followed in using acetic acid or ethanol as substrates.

It is usually a prerequisite that, in order to run a successful fed-batch culture, it is necessary to monitor (directly or indirectly) the prevailing concentration of the key nutrient being supplied to the fermentor. Although devices for monitoring ethanol (or methanol) are now available, acetic acid is not easy to monitor directly on-line in a fermentor; its concentration has to be measured by some suitable indirect means. With this substrate, the authors were fortunate to be able to identify prior publications in which this problem had already been tackled and, importantly, resolved.

Use of Acetic Acid

Acetic acid is toxic to many microorganisms when the concentration is over about 5–10 g/L; few organisms, if any, will grow on it at over 20 g/L, even at neutral pH values. Some initial and preliminary indication of acetic acid toxicity towards *C. cohnii* and other microalgae is given by Vazhappily and Chen (1998), who reported that *C. cohnii* might be able to grow on acetate if its concentration was no more than 1 g/L; since the cultivation medium contained other utilizable carbon materials, this result is not entirely unambiguous. Some initial work in the authors' laboratories (de Swaaf, unpublished work) indicated that *C. cohnii* could grow on sodium acetate at 3 g/L, but this growth was accompanied by a sharp rise in pH that prevented good growth. This problem turns out to be most crucial when using acetate as feedstock.

When added to a fermentation medium, acetic acid has to be neutralized or least brought to the pH level used in the growth medium meaning the addition of a counterion, usually Na^+ or K^+. When the cells start to grow, the acetate (CH_3COO^-) is consumed and replaced by hydroxyl ions (OH^-) that, as a consequence, cause the pH to rise. In effect, NaOH is being produced by the consumption of the acetate ion. It is this rapid increase in pH that stops cells from growing. Consequently, if the rise in pH is to be avoided, it has to be titrated with an acid; what better acid to use than acetic acid itself? Thus, a reservoir of acetic acid can be attached to the fermentor and linked to the pH meter and dosing pump; this equipment will then add acetic acid to the fermenatation medium on demand from the culture itself.

If a low concentration of acetic acid is used in the initial culture medium, it will cause the cells to grow and consume the acetate; further acetic acid will then be pumped into the fermentor to offset this rise in pH. The pH is controlled at its optimum value for growth. Consequently, the faster the organism grows, the faster will be the rate of acetic acid addition. Thus, the organism is always growing at its fastest possible rate. This concept was first set forth by Martin and Hempfling (1976), who gave the fermentation system the name "phauxostat." The name derives from a device that can be used to maintain the pH of a culture (in this case, *Escherichia coli*) by controlling the rate of substrate (glucose, glycerol, or lactic acid) addition into a fermentor based on changes in the culture pH.

The first use of acetic acid in a phauxostat (or pH auxostat, as it is now called) was, however, reported by Sowers et al. (1984) who grew various acetotrophic, methane-producing bacteria on acetic acid and achieved an increase in cell yield that was 18 times larger than any that had been previously achieved. An application of this technique for SCO production was achieved by du Preez et al. (1995) who showed that acetic acid was an excellent feedstock to produce an oil rich in γ-linolenic acid using *Mucor circinelloides* or *Mucor rouxii*. The interest of this latter group in using acetic acid as a fermentation substrate arose because of the availability of inexpensive acetic acid as a by-product of the Fischer-Tropsch synthesis process of manufacturing gasoline from coal, as carried out by Sasol Industries Ltd. of South Africa, and for which no satisfactory large-scale use had been identified. Patents on this process were granted (Kock & Botha, 1993; Kock & Botha, 1995).

Thus, in spite of the uncertain growth of *C. cohnii* on acetate achieved in preliminary studies (Vazhappily & Chen, 1998), it seemed that acetate could be used as a carbon source and that the application of a pH auxostat would then be an appropriate system to evaluate the possible use of this substrate to produce a SCO rich in DHA. *C. cohnii* was initially grown in a sea salt/yeast extract medium with sodium acetate between 4 and 16 g/L; acetic acid (at 50%) was pumped in to maintain the pH at 6.5 (Ratledge et al., 2001). The best growth rate was achieved with an initial concentration of 8 g acetate/L (Fig. 7.1). There was only a slight variation in the lipid content of the cells (45–50% of the biomass) with initial acetate concentrations of up to 12 g/L and no real variation in the content of DHA in the oil (40–50%) at these concentrations. Since the cells needed to induce the synthesis of new enzymes

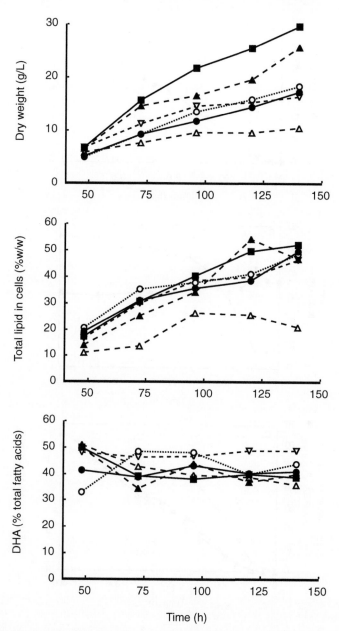

Fig. 7.1. (A) growth, (B) lipid content of cells, and (C) docosahexaenoic acid (DHA) content of total fatty acids for *Crypthecodinium cohnii* grown at pH 6.5 in a pH auxostat culture with various concentrations of sodium acetate in the original medium and with additional acetic acid then supplied into the culture on demand. Sodium acetate at (solid circle) 4, (open circle) 6, (solid square) 8, (solid triangle) 10, (inverted open triangle) 12, and (open triangle) 16 g/L. *Source:* Ratledge et al., 2001.

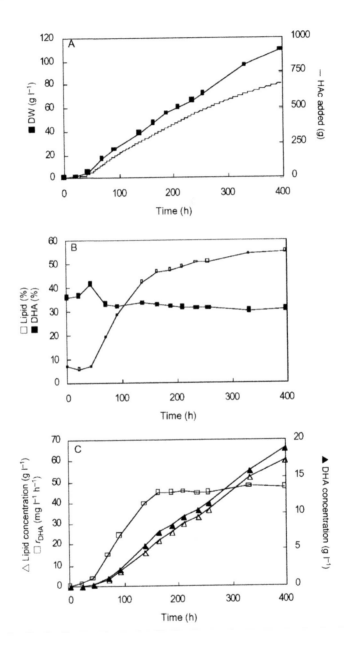

Fig. 7.2. Fed-batch cultivation of *C. cohnii* with a feed consisting of pure acetic acid. Initial medium: 10 g yeast extract/L, 25 g sea salt/L, 8 g NaAc/L; a 10% (v/v) inoculum was used. (A) Biomass dry weight (DW) and acetic acid added (HAc added); (B) Lipid content of dry biomass (Lipid), DHA content of lipid (DHA); (C) Lipid concentration, overall volumetric productivity of DHA (r_{DHA}) and DHA concentration. *Source:* de Swaaf et al., 2003a.

Table 7.A. Fatty Acid Profiles (as relative % w/w) of the Total Lipid and Neutral Lipids (e.g., Triacylglycerol) Fraction from *Crypthecodinium cohnii* ATCC 30772 Grown on Glucose in Batch Culture and Acetic Acid in a Ph-Auxostat Culture[a].

Fatty Acid	Glucose-Grown		Acetic Acid-Grown	
	Total Lipid	Neutral Lipids[a]	Total Lipid	Neutral Lipids[a]
10:0	7	7	tr.[b]	tr.
12:0	24	19	7	8
14:0	26	22	21	16
16:0	8	9	18	16
16:1	1	2	2	2
18:0	1	1	2	1
18:1	7	9	11	10
22:6 (DHA)[c]	26	31	39	47

[a]Represents 74–75% of the total lipid in each case.
[b]tr = <0.5%
[c]DHA, Docosahexaenoic Acid
(Source: Ratledge et al., 2001)

in order to be able to use acetic acid (see Fig. 7.3 below), it was not surprising that more rapid growth was achieved using an inoculum grown on acetate rather than one grown on glucose (Ratledge et al., 2001).

Overall, this system gives results superior to those obtained when glucose is used as a feedstock (de Swaaf et al., 1999, 2001; Ratledge et al., 2001), producing as much as a 60% increase in the growth rate, a 70% increase in the lipid content of the cells (Ratledge et al., 2001), and a 50% increase in the level of DHA in TFA (Table 7.A). This increase is also reflected in the much higher level of DHA in the TAG fraction (i.e., edible oil fraction) of these cells. In other respects, the composition of the other lipid components of extracted lipids is approximately equivalent between the two types of cells (Ratledge et al., 2001). The authors interpret this improved performance of *C. cohnii* on acetate compared to glucose as being at least partially due to it always growing at the maximum possible rate.

These findings have been confirmed by de Swaaf et al. (2003) who, using extended cultivation (up to 400 h), achieved final cell biomass yields of 109 g/L (Fig. 7.2). However, such high cell density cultures required vigorous mixing to sustain sufficient aerobic conditions; this transfer of oxygen into the system was complicated by increases in culture viscosity. The increase in viscosity was the result of the production of viscous extracellular polysaccharides (de Swaaf et al., 2001). In large-scale industrial production of DHA, production of polysaccharides may cause problems, though the addition of a commercial polysaccharide-hydrolase preparation to the cultures can decrease the viscosity. The stirring rate needed to maintain a fixed dissolved oxygen tension can then be decreased (de Swaaf et al., 2003).

The overall conversion of acetic acid to biomass in the various pH auxostat cultures is calculated to be about 0.13 g biomass/g acetic acid used (de Swaaf et al., 2003). This, though, is not a high conversion ratio. In part, it occurs because the energy content of acetic acid is less than that of glucose (Linton & Rey, 1989). Improvements in this biomass yield seem possible, however, because previous work with both yeasts and bacteria has achieved growth yields over 0.35 g biomass/g acetic acid (Ratledge, 2006). Biomass yields for microorganisms grown on glucose, though, are normally between 0.45 and 0.5 g/g (Ratledge, 2006).

Another reason for this rather low growth yield by *C. cohnii* when using acetate may be attributed to the secretion of succinic acid in these cultures (Hopkins, unpublished work). The authors have also noted in the acetate pH auxostat cultures that

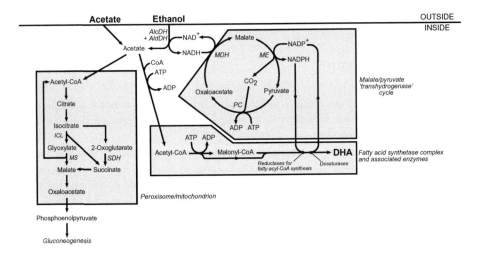

Fig. 7.3. Possible pathway for DHA biosynthesis from acetic acid and ethanol in *C. cohnii*. Ethanol in its conversion to acetate (acetyl-CoA) generates 2 mol of reduced nicotinamide adenince dinucleotide (NADH): one from the alcohol dehydrogenase (*AlcDH*) reaction and the other from the aldehyde dehydrogenase (*AldDH*) reaction. Further metabolism of acetate metabolism (from both ethanol and acetic acid) generates NADH in the mitochondrion and/or peroxisome via the reactions of the tricarboxylic acid. The conversion of acetate into biomass precursors (i.e., through gluconeogenesis) requires the synthesis of two enzymes to be induced: isocitrate lyase (*ICL*) and malate synthase (*MS*). Succinate is now generated by *ICL* and succinate dehydrogenase (*SDH*) implying a metabolic imbalance in this organism and, as it accumulates in the culture medium, this implies that the rate of succinate utilization, by conversion to fumarate (not shown, but this is then converted to malate), is less than the combined rates of its synthesis. NADH is converted to reduced nicotinamide adenince dinucleotide phosphate (NADPH), which is needed for fatty acid synthesis and fatty acid desaturases, *via* a presumptive malate/pyruvate transhydrogenase cycle involving malic enzyme (*ME*), pyruvate carboxylase (*PC*), and malate dehydrogenase (*MDH*). This cycle has been demonstrated in other oleaginous microorganisms (Ratledge & Wynn, 2004) but not yet in *C. cohnii*.

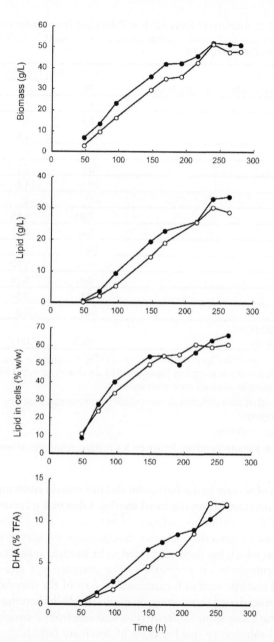

Fig. 7.4. Growth of *C. cohnii* ATCC 30772 in a pH auxostat culture using acetic acid as feedstock with the inclusion of 10 g propionic acid/L in the culture medium. The two curves given are from duplicate cultures run simultaneously in 5 L fermentors; (Kanagachandran, Anderson, & Ratledge, unpublished work).

Table 7.B. Analysis of Fatty Acids in Total Lipid from an Acetic Acid pH-Stat Culture of C. cohnii Grown in a Medium Containing 1% (v/v) Propionic Acid[a].

Fatty Acid	Fementation Time (h)	
(Relative % w/w Total Fatty Acids)	96	240
10:0	1	2.8
12:0	6.7	8.1
14:0	19.4	13.8
14:1	—	~0.7[b]
15:0	tr.[c]	0.1
16:0	20.5	14.2
16:1	~1[b]	~2[b]
17:0	0.03	0.05
18:0	1.4	0.05
18:1	14.9	15.4
19:0	—	0.02
20:0	0.4	0.25
22:1	0.1	0.3
22:6	34.1	39.9
Total Odd Chain Fatty Acids	0.03	0.17

[a]The course of the fermentation is shown in Fig. 7.4; only the results from two times of sampling (96 and 240 h) are shown here, but all other samples showed consistent values.

[b]Values that are estimated as exact values were not calculated by GC program.

[c]tr. Trace = <0.01%

(Source: Kanagachandran, Anderson,& Ratledge, unpublished work).

the concentration of acetate in the fermentor did not remain constant with time but, in fact, gradually decreased from the usual starting value of 8 g acetate/L to less than 1 g/L, in spite of the continuous addition of fresh acetic acid (Wynn, unpublished work). This decrease suggests that there is a divergence of the acetate into some extracellular metabolite, which has been confirmed to be succinic acid accumulating at up to 5 g/L. This accumulation is attributed to the existence of a rate-limiting step in the conversion of succinic acid to fumaric acid as part of the tricarboxylic acid cycle (Fig. 7.3); it was exacerbated in acetate-grown cells in which there are now two routes to produce succinate: one from isocitrate lyase and one from 2-oxoglutarate.

An empirical solution to this build up of succinate (which is clearly a "waste" of acetate and accounts, at least in part, for the low biomass yield from acetate) was to add propionic acid at 1% (v/v) to the medium. Fig. 7.4 shows the increased performance of the organisms under these conditions. There is a sustained increase in biomass to about 50 g/L, and the total lipid now reaches 60% of the cell biomass by the end of growth. The amount of DHA in the lipid remained at approximately 35–40% of the TFA (Table 7.B).

The reason why propionate improves cellular growth when using acetate is not entirely clear, but it may be connected to the fact that propionate provides additional oxaloacetate to the cells. This removes a metabolic bottleneck, as indicated by the accumulation of succinate for the reasons given previously. The provision of oxaloacetate could come about via a transcarboxylase reaction that converts propionate into pyruvate and is well-known in other microorganisms (Doelle, 1969). The pyruvate, by the action of pyruvate carboxylase, would then provide additional oxaloacetate and other precursors needed to synthesize numerous cell components (Ratledge, 2006). Evidently, cells grown on acetate are not in metabolic balance; the addition of propionate appears to improve this balance and, consequently, increases the efficiency of biomass production.

Contrary to expectations, the added propionate does not appear to be used to produce odd chain length FA in the lipid (Table 7.B). The overall FA profile for the acetate-grown cells in which propionate is included in the medium does not show any trace of odd chain length FA, the characteristic product from this substrate, until the very end of growth. But even here, they do not constitute more than 0.2% of the TFA. Normally, one might have expected up to 10% of the TFA to be of odd chain lengths as they had been derived from propionate being used by the FA-synthesizing enzymes. Thus, all of the added propionate is used in other aspects of the cells' metabolism, which clearly improves the performance where acetate is the sole carbon source.

Ethanol as Carbon Source

Ethanol was considered as a potential feedstock to produce SCOs for some time in work carried out in Czechoslovakia, the former USSR, and Japan in the 1970s and 1980s, which used several species of oleaginous yeast (Ratledge, 1988). During this research, a microcomputer-controlled, fed-batch fermentation was carried out by Yamauchi et al. (1983) in which *Lipomyces starkeyi* was grown on ethanol maintained at its optimal concentration of 2.5 g/L. This technique allowed extremely high cell densities to be attained: 153 g/L over 140 h, with a cellular oil content in excess of 50%. There was an overall conversion coefficient of 0.23 g lipid/g ethanol used, which is in keeping with the very high theoretical conversion ratio of 0.54 g lipid/g ethanol that was calculated (Ratledge, 1988). For glucose, the theoretical conversion ratio would be about 0.32 g (Ratledge, 1988).

With respect to DHA production in *C. cohnii*, the authors previously identified both the cultivation scale and volumetric productivity (r_{DHA}) as major factors in determining the economic feasibility of this process (Sijtsma et al., 1998). Factors that determine r_{DHA} include biomass concentration, lipid content of the cells, DHA content of the lipid, and cultivation time. Obviously, a high DHA content of the biomass is also desirable from the viewpoint of product recovery.

In laboratory-scale batch cultivations, the r_{DHA} for *C. cohnii* ATCC 30772 reported on glucose is 19 mg L^{-1} h^{-1} (de Swaaf et al., 1999). Similar productivities were observed in fed-batch cultivations grown with a concentrated (50% w/v) glucose feed (de Swaaf et al., 2003). In pH-controlled fed-batch cultivations with acetic acid as the carbon source, productivities of up to 48 mg DHA L^{-1} h^{-1} were achieved (Figs. 7.2, and 7.4) (de Swaaf, 2003; de Swaaf et al., 2003a, 2003b). This finding clearly indicates the strong impact that the carbon source can have on DHA productivity by *C. cohnii*.

Like acetic acid, ethanol could be a potential carbon source for *C. cohnii*, provided that it is not toxic to the cells, can enter the cells, and can be metabolized by them. The ability to utilize ethanol suggests the presence of an alcohol dehydrogenase, to convert ethanol to acetaldehyde, and an acetaldehyde dehydrogenase, to convert acetaldehyde to acetate (Fig. 7.3). As a carbon source for large-scale cultivation, ethanol may have advantages over acetic acid due to its ability to generate higher biomass yields per mol of carbon substrate (Linton & Rye, 1989; Ratledge 1988). For practical reasons—particularly, ethanol's flammable nature—specific requirements will be necessary to store and add it to large-scale applications. Furthermore, the application of ethanol as a fermentation feedstock requires equipment to control the ethanol concentration in a bioreactor at appropriate levels, as well as a controlled addition system.

The possible utilization of ethanol by cultures of *C. cohnii* was studied in initial shake-flask cultures. The alga was grown on a complex medium containing yeast extract, sea salt, and variable concentrations of ethanol (de Swaaf et al., 2003b).

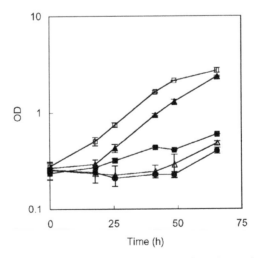

Fig. 7.5. Influence of initial ethanol concentration on growth of *C. cohnii* in shake-flask cultures. The medium contained: 2 g yeast extract/L; 25 g sea salt/L; a 10% (v/v) inoculum was used, and the medium was supplemented with ethanol at 0 (■), 5 ( ), 10 (▲), 15 (Δ) or 25 (•) g /L. An optical density (OD) value of 1 corresponds to a cell dry wt value of approximately 1 g/L. *Source:* de Swaaf et al., 2003b.

Growth when yeast extract was used as the sole carbon source was negligible. With ethanol at 5 or 10 g L⁻¹, cells were able to grow well (Fig. 7.5). The specific growth rate at these ethanol concentrations, calculated from the exponential part of the growth curves, was 0.05 h⁻¹. In contrast to cultures grown on 5 g ethanol L⁻¹, cultures grown on 10 g ethanol L⁻¹ exhibited a significant lag phase. As ethanol concentrations reached 15 g L⁻¹ and higher, growth of *C. cohnii* substantially lessened. These data

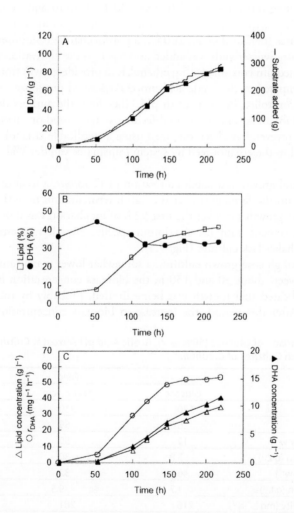

Fig. 7.6. Fed-batch cultivation of *C. cohnii* with a feed consisting of pure ethanol. *A.* Biomass dry weight (DW) and ethanol addition. *B.* Lipid content of dry biomass (Lipid), DHA content of lipid (DHA). *C.* Lipid concentration, overall volumetric productivity of DHA (rDHA) and DHA concentration. The second, slightly higher, line indicates the total amount of ethanol added. *Source:* de Swaaf et al., 2003b.

clearly demonstrate that ethanol, like acetic acid, cannot be used directly in batch cultures to achieve high biomass concentrations.

In order to avoid toxicity and provide sufficient ethanol for energy, growth, and metabolic activities, de Swaaf et al. (2003b) developed a process for ethanol fed-batch cultivations in a 2 L, computer-controlled laboratory bioreactor. As growth and lipid production require a sufficient supply of O_2, the dissolved O_2 tension (DOT) should preferably be kept over 30% air saturation; this goal can be accomplished by automatically controlling the stirrer speed (range: 200-1250 rpm) and by flushing with filter-sterilized air. Furthermore, since viscosity increases considerably in very high cell density cultures (de Swaaf et al., 2003a), a polysaccharase (Glucanex from Novo Nordisk, Neumatt, Switzerland) was added at 0.5 g/L to the cultivation broth.

The initial medium contained 5.5 g ethanol/L to provide growth from the start of inoculation. Compared to the acetic acid process described previously, ethanol addition cannot be controlled by changes in pH. Therefore, the authors developed (de Swaaf et al., 2003b) an automatic ethanol-feeding system based on changes in DOT. Similar results, however, have been obtained using a sterilizable ethanol sensor in the medium coupled to the ethanol-feeding system (Sijtsma & van der Wal, unpublished work).

In the ethanol-grown, fed-batch cultivation of *C. cohnii*, a total of 300 g ethanol was added into the fermentor over the 220 h fermentation time (Fig. 7.6). The estimated specific growth rate over the first 52 h of incubation was 0.047 h^{-1}, which was in good agreement with the maximum specific growth rate estimated from ethanol-grown shake-flask cultures (Fig. 7.5).

Compared to glucose-grown cultures, a somewhat lower maximum growth rate was found. Between about 50 and 150 h, the biomass concentration increased linearly, which indicated that growth was being limited, probably by the supply and uptake of O_2. After this period, the increase in biomass concentration leveled off

Table 7.C. Comparison of Glucose (50% w/v), Acetic Acid pH Auxostat Culture, and Ethanol-Fed-Batch Cultures of C. cohnii.

	Feed		
	Glucose	Acetic Acid	Ethanol
Time (h)	120	210	200
Biomass (g·L^{-1})	26	59	77
Lipid Content (% w·w^{-1})	15	50	41
Lipid Concentration (g·L^{-1})	3.8	30	31
DHA in Lipid (% w·w^{-1})	46	32	33
DHA Concentration (g·L^{-1})	1.7	9.5	10.4
Biomass Productivity (mg·L^{-1}·h^{-1})	216	281	385
Lipid Productivity (mg·L^{-1}·h^{-1})	31.6	143	155
DHA Productivity (mg·L^{-1}·h^{-1})	14	45	52

Selected parameters are shown for time point 120 h (glucose), 200 h (ethanol), and 210 h (acetic acid). (Source: Modified from de Swaaf et al., 2003b)

(Fig. 7.6). Over this time, the amount of lipid within the biomass increased from 9 to 35%; it reached a final value of 42% at the end of the fermentation process (Fig. 7.6). During the first 120 h of the process, the DHA content of the lipid varied between 44 and 32%. It remained constant at 33% during the final 100 h (Fig. 7.6). The final dry weights of cells, lipid, and DHA concentrations were 83, 35, and 11.7 g L^{-1}, respectively (Fig. 7.6). The latter value resulted in a volumetric DHA production rate (r_{DHA}) of 53 mg L^{-1} h^{-1}. The calculated biomass yield on ethanol was 0.31 g biomass/g ethanol used (de Swaaf et al., 2003c), about 2.4 times higher than the value calculated for acetic acid.

Lipid Production from C$_2$ Carbon Sources

Compared with glucose, the use of either acetic acid or ethanol as a carbon source resulted in a remarkable increase in DHA productivity by *C. cohnii* (Table 7.C). The major reasons for this improved productivity appear to originate from both acetic acid and ethanol feeding directly into the pool of acetyl-CoA needed for lipogenesis (Fig. 7.3). The ability to maintain a high concentration of this key metabolite must, therefore, be crucial for lipid biosynthesis in *C. cohnii* and has been commented on elsewhere (Sijtsma et al., 1998; Verduyn et al., 1991; Sijtsma & de Swaaf, 2003).

In glucose-grown cells, the main flux of carbon involves glucose uptake, glycolysis, transport of pyruvate into the mitochondrion, conversion of pyruvate into citrate, transport of citrate into the cytosol, and cleavage of citrate by adenosine triphosphate (ATP):citrate lyase to yield acetyl-CoA. This has been well explored in both yeasts and fungi (Ratledge & Wynn, 2002) and serves as a base for understanding the possible processes for lipogenesis in other microorganisms.

If one assumes that a similar metabolic system operates in *C. cohnii* for DHA biosynthesis (and this may be a very big assumption), then one can see from Fig. 7.3 that ethanol will produce higher yields of either cells or DHA (or both) than it will acetic acid. This occurs because ethanol, being more reduced than acetate, is able to generate more reducing power [in the form of reduced nicotinamide adenine dinucleotide (NADH)] than acetate, and large amounts of acetate are needed to convert acetyl-CoA into FA. However, NADH is not used per se, but needs to be converted by a "transhydrogenase" cycle involving malic enzyme, pyruvate carboxylase, and malate dehydrogenase (Fig. 7.3) into reduced nicotinamide adenine dinucleotide phosphate (NADPH). The same system is probably involved in providing NADPH for the various desaturases needed to produce DHA. In all, 26 mol of NADPH is needed for each mol of DHA that is synthesized by the conventional fatty acid synthase route: 2 for each of the ten condensation reactions (used to reduce the beta-ketoacyl and alpha/beta-unsaturated fatty acyl intermediates), plus 1 for each of the six desaturases (see also Chapter 1). Since each mol of ethanol produces 50% more NADH than each mol of acetate (Heijnen, 2006), we can easily appreciate why ethanol can generate a higher biomass yield than acetate. NADH, besides being used for the biosynthesis of

FA, is the principal source of energy (ATP) in the cells through its re-oxidation via the oxidative phosphorylation sequence (Ratledge, 2006). Thus, ethanol generates not only acetyl-CoA (as does acetate), but, more importantly, it produces more energy than acetate and, thus, gives higher yields of cells per unit weight.

Whether ethanol is used as a substrate for the large-scale production of DHA—rather than acetic acid or, indeed, in preference to, glucose—we will clearly have to await development of appropriate commercial interests. It may be of relevance to point out that the conversion efficiency of ethanol to lipid is considerably higher than for both glucose and acetic acid (Ratledge, 1988, Yamauchi et al., 1983) and thus may represent the most economical feedstock for SCO production.

Glycerol as Carbon Source

Because of the current plentiful and cheap supply of crude glycerol from biodiesel manufacture, there has been considerable interest in using it as a possible fermentation feedstock for various processes, including its use as an alternative carbon source for the production of SCO. Both Meesters et al. (1996) and Yokochi et al. (1998) used a purified source of glycerol for the production of SCO using, respectively, *Cryptococcus curvatus* and *Schizochytrium limacinum*. More recently, it has been used as a biodiesel waste product for the cultivation of *Schizochytrium* spp. to produce DHA-rich SCO (Yokochi et al., 1998; Chi et al., 2007; Pyle et al., 2008). Papanikolaou and Aggelis (2009) have also used this material for the production of an SCO, using *Yarrowia lipolytica* with the simultaneous production of citric acid. Although we have no data for growth of *C. cohnii* and DHA production with biodiesel-derived glycerol as carbon source, we are also not aware of any publication covering this topic. We believe glycerol would be as efficient as glucose for the production of biomass and lipid (de Swaaf et al., 1999; Mendes et al., 2007). The yield coefficients for the conversion of glucose to biomass are usually between 0.4 and 0.5 g cell/g glucose, and yields with glycerol would be expected to be approximately the same (Ratledge, 2006). This is because glycerol ($C_3H_8O_3$) is effectively half of glucose ($C_6H_{12}O_6$); its metabolism is, therefore, like that of glucose as it uses the main glycolytic pathway, and it will generate pyruvate directly after its initial conversion to 3-phosphoglycerol (see Ratledge and Wynn, 2002 and, also, Papanikolaou & Aggelis, 2009 for diagrams). Being a C3 substrate, there is now no need for the induction of the glyoxlate bypass enzymes (see Fig. 7.3). The pyruvate produced from glycerol is converted into acetyl-CoA for fatty acid biosynthesis and, also, into oxaloacetate and, finally, into malate (Fig. 7.3) to generate NADPH via the action of the malic enzyme—the essential reductant for the fatty acid synthetic pathway. Yields of FA, including DHA, would be similar to those achieved with glucose as substrate (see Table 7.C).

There is, however, one major problem in particular with the use of biodiesel-derived glycerol for the production of SCO and for DHA-rich oils. It is not of food-grade quality. There are many impurities in the crude glycerol (see Pyle et al., 2008), including residual methanol and various fatty acids soaps. Although these materials may not affect the productivities of DHA (Pyle et al., 2008), the very presence of methanol obviates the use of material for the production of a product destined for human (and probably animal) consumption. In addition, the fats used for biodiesel production may include waste cooking oils in which various carcinogens may be present. Thus, crude glycerol cannot be considered a viable feedstock for DHA production. The same considerations would apply to the production of any food-destined fermentation product, including citric acid. Of course, crude glycerol would be fine for the production of SCO to be used as a biofuel (see Chapter 15) or indeed any product outside of the food chain.

Acknowledgments

This work was financially supported by the European Community (Q5RS-2000-30271) and the Dutch Ministry of Agriculture, Nature Management, and Fisheries. M. E. de Swaaf is acknowledged for his contribution to the work carried out in The Netherlands.

References

Albert, C.M.; C.H. Hennekens; C.J. O'Donnell; U.A. Ajani; V.C. Carey; W.C. Willett; J.N. Ruskin; J.E. Manson. Fish consumption and risk of sudden cardiac death. *J. Am. Med. Ass.* **1998,** *279,* 23–28.

Beach, D.H.; G.G. Holz. Environmental influences on the docosahexaenoate content of the triacylglycerols and phosphatidylcholine of a heterotrophic, marine dinoflagellate, *Crythecodinium cohnii. Biochim. Biophys. Acta,* **1973,** *316,* 56–65.

Chi, Z.; D. Pyle; Z. Wen; C. Frear; S. Chen. A laboratory study of producing docosahexaenoic acid from biodiesel-waste glycerol by microalgal fermentation. *Proc. Biochem.* **2007,** *42,* 1537–1545.

de Swaaf, M.E. Docosahexaenoic acid production by the marine alga *Crypthecodinium cohnii.* Ph.D. Thesis, TU Delft, The Netherlands, 2003, pp 61–98.

de Swaaf, M.E; T.C. de Rijk; G. Eggink; L. Sijtsma. Optimisation of docosahexaenoic acid production in batch cultivations by *Crypthecodimium cohnii. J. Biotechnol.* **1999,** *70,* 185–192.

de Swaaf, M.E.; G.J. Grobben; G. Eggink; T.C. de Rijk; P. van der Meer; L. Sijtsma. Characterisation of extracellular polysaccharides produced by *Crypthecodinium cohnii. Appl. Microbiol. Biotechnol.* **2001,** *57,* 395–400.

de Swaaf, M.E.; L. Sijtsma; J.T. Pronk. High-cell-density fed-batch cultivation of the docosahexaenoic-acid producing marine alga *Crypthecodinium cohnii. Biotechnol. Bioeng.* **2003a,** *81,* 666–672.

de Swaaf, M.E.; J.T. Pronk; L. Sijtsma. Fed-batch cultivation of the docosahexaenoic acid producing marine alga Crypthecodinium cohnii on ethanol. *Appl. Microbiol. Biotechnol.* **2003b,** *61,* 40–43.

de Swaaf, M.E.; T.C. de Rijk; P. van der Meer; G. Eggink; L. Sijtsma. Analysis of docosahexaenoic acid biosynthesis in Crypthecodinium cohnii by ^{13}C labelling and desaturase inhibitor experiments. *J. Biotechnol.* **2003c,** *103,* 21–29.

Doelle, H.W. Bacterial Metabolism. Academic Press: New York and London, 1969; pp 266–268.

du Preez, J.C.; M. Immelman; J.L.K. Kock; S.G. Kilian. Production of gamma-linolenic acid by Mucor circinelloides and Mucor rouxii with acetic acid as carbon substrate. *Biotech. Lett.* **1995,** *17,* 933–938.

Harrington, G.W.; G.G. Holz. The monoenoic and docosahexaenoic fatty acids of a heterotrophic dinoflagellate. *Biochim. Biophys. Acta* **1968,** *164,* 137–139.

Heijnen, J.J. Stoichiometry and kinetics of microbial growth from a thermodynamic perspective. *Basic Biotechnology,* 3rd edn.; Ratledge, C.; B. Kristiansen, Eds.; Cambridge University Press: Cambridge, UK, 2006; pp 55–71.

Horrocks, L.A.; Y.K Yeo. Health benefits of docosahexaenoic acid (DHA). *Pharmacol. Res.* **1999,** *40,* 211–225.

Kock, J.L.F.; A. Botha. South African Patent 91/9749, **1993.**

Kock, J.L.F.; A. Botha. U.S. Patent 5,429,942, **1995.**

Kromhout, D.; E.B. Bosschieter; C. de Lezenne Coulander. The inverse relation between fish consumption and 20-year mortality from coronary heart disease. *N. Engl. J. Med.* **1985,** *312,* 1205–1209.

Kyle, D.J. U.S. Patent 5,374,657, **1994.**

Kyle, D.J. Production and use of a single cell oil which is highly enriched in docosahexaenoic acid. *Lipid Technol.* **1996,** *8,* 107–110.

Linton, J.D.; A.J. Rye. The relationship between the energetic efficiency in different micro-organisms and the rate and type of metabolite overproduced. *J. Ind. Microbiol.* **1989,** *4,* 85–96.

Martin, G.A.; W.P. Hempfling. A method for the regulation of microbial population density during continuous culture at high growth rates. *Arch. Microbiol.* **1976,** *107,* 41–47.

Meesters, P.A.E.P.: G.N.M. Huijberts; G. Eggink. High-cell-density cultivation of the lipid accumulating yeast *Cryptococcus curvatus* using glycerol as a carbon source. *Appl. Microbiol. Biotech.* **1996,** *45,* 575–579.

Mendes, A.; P. Guerra; V. Madeira; F. Ruano; T. Lopes da Silva; A. Reis. Study of docosahexaenoic acid production by the heterotropjic microalgae *Crypthecodinium* cohnii CCMP 316 using carob pulp as a promising carbon source. *World J. Microbiol. Biotechnol.* **2007,** 1209–1215.

Nordøy, A.; R. Marchioli; H. Arnesen; J. Videbæk. n-3 Polyunsaturated fatty acids and cardiovascular diseases. *Lipids* **2001,** *36,* S127–S129.

Papanikolaou, S.; G. Aggelis. Biotechnological valorization of biodiesel dereived glycerol waste through production of single cell oil and citric acid by *Yarrowia lipolytica. Lipid Technol.* **2009,** *21,* 83–87.

Pyle, D.J.; R.A. Garcia; Z. Wen. Producing docosahexaenoic acid (DHA)-rich algae from biodiesel-derived crude glucerol: effects of impurities on dha production and algal biomass composition. *J. Agric. Food Chem.* **2008,** *56,* 3933–3939.

Ratledge, C. Biochemistry, stoichiometry, substrates and economics. *Single Cell Oil;* Moreton, R.S., Ed.; Longman Scientific and Technical: Harlow, UK, 1988; pp 33–70.

Ratledge, C. Biochemistry and physiology of growth and metabolism. *Basic Biotechnology,* 3rd edn,; Ratledge, C.; B. Kristiansen, Eds.; Cambridge University Press: Cambridge, UK, 2006; pp 25–71.

Ratledge, C.; J.P. Wynn. The biochemistry and molecular biology of lipid accumulation in oleaginous microorganisms. *Adv. App. Microbiol.* **2002,** *51,* 1–51.

Ratledge, C.; K. Kanagachandran; A.J. Anderson; D.J. Grantham; J.M. Stephenson. Production of docosahexaenoic acid by *Crypthecodinium cohnii* grown in a ph-auxostat culture with acetic acid as principal carbon source. *Lipids* **2001,** *36,* 1241–1246.

Sijtsma, L.; M.E. de Swaaf. Biotechnological production and applications of the omega-3 polyunsaturated fatty acid docosahexaenoic acid. *Appl. Microbiol. Biotechnol.* **2004,** *64,* 146–153.

Sowers, K.R.; M.J. Nelson; J.G. Ferry. Growth of acetotrophic, methane-producing bacteria in a pH auxostat. *Curr. Microbiol.* **1984,** *11,* 227–230.

Sijtsma, L.; J. Springer; P.A.E.P. Meesters; M.E. de Swaaf; G. Eggink. Recent advances in fatty acid synthesis in oleaginous yeasts and microalgae. *Rec. Res. Dev. Microbiol.* **1998,** *2,* 219–232.

Vazhappily, R.; F. Chen. Eicosapentaenoic acid and docosahexaenoic acid production potential of microalgae and their heterotrophic growth. *J. Am. Oil Chem. Soc.* **1998,** *75,* 393–397.

Verduyn, C.; A.H. Stouthamer; W.A. Scheffers; J.P. van Dijken. A theoretical evaluation of growth yields of yeasts. *Ant. van Leeuwenhoek Int. J. Gen. Mol. Microbiol.* **1991,** *59,* 49–63.

Yamauchi, H.; H. Mori; T. Kobayashi; S. Shimizu. Mass production of lipids by *Lipomyces starkeyi* in microcomputer-aided fed-batch culture, *J. Ferment. Technol.* **1983,** *61,* 275–280.

Yokochi, Y.; D. Honda; T. Higashihara; T. Nakahara. Optimization of docosahexaenoic acid production by *Schizochytrium limacinum* SR21. *Appl. Microbiol. Biotechnol.* **1998,** *49,* 72–76.

Production of Eicosapentaenoic Acid Using Heterotrophically Grown Microalgae

Zhiyou Wen[a] and Feng Chen[b]

[a] Department of Biological Systems Engineering, Virginia Polytechnic Institute and State University, Blacksburg, VA 24061, USA; [b] School of Biological Sciences, The University of Hong Kong, Pokfulam Road, Hong Kong, P.R. China

Introduction

The therapeutic significance of n-3 polyunsaturated fatty acids (PUFAs), such as eicosapentaenoic acid (20:5n-3, EPA) and docosahexaenoic acid (22:6n-3, DHA), has been clearly indicated by recent clinical and epidemiological studies (Nettleton, 1995). EPA performs many vital functions in biological membranes and serves as a precursor of a variety of lipid regulators in cellular metabolism; as a result, it plays an important role in the treatment of various human diseases, such as rheumatoid arthritis, heart disease, cancers, schizophrenia, and bipolar disorder (Gill & Valivety, 1997; Emsley et al., 2003; Ponizovsky et al., 2003; Su et al., 2003; Peet, 2004). These findings have led to considerable interest in developing commercial processes for EPA production (Belarbi et al., 2000; Grima et al., 2003).

Marine fish oil is presently the richest source for EPA and is currently used for its commercial production; however, the recovery and purification of EPA from fish oils are expensive (Belarbi et al., 2000). Microorganisms, including microalgae, lower fungi, and marine bacteria, are the primary producers of EPA; fish usually obtain EPA via bioaccumulation in the food chain (Barclay et al., 1994; Yazawa, 1996). Much effort is being devoted to developing a commercially feasible technology to produce EPA directly from microorganisms. The aim of the present paper is to review the recent advances in EPA production by microorganisms, in particular, by microalgae.

EPA

Structure, Occurrence, and Significance

EPA is an important n-3 PUFA. Another important and related n-3 PUFA is DHA. The chemical structures of EPA and DHA are shown in Fig. 8.1. In living cells, EPA is

EPA (20:5n-3)

DHA (22:6n-3)

Fig. 8.1. Chemical structure of eicosapentaenoic acid (EPA) and docosahexaenoic acid (DHA). Source: Wen & Chen (2003), with permission from Elsevier.

widely distributed among neutral lipids, glycolipids, and phospholipids (Chen et al., 2007). The EPA distribution among different types of lipids is also dependent on the growth conditions present in living cells, such as temperature and salinity (Chen et al., 2008a, 2008b).

EPA plays an important role in higher animals and humans, as a precursor of a group of eicosanoids that are crucial in regulating developmental and regulatory physiology. The eicosanoids are hormone-like substances, including prostaglandins (PG), thromboxanes (TX), and leukotrienes (LT). Arachidonic acid (AA, 20:4n-6) and EPA are precursors of eicosanoid compounds (Fig. 8.2). However, the eicosanoids from these two fatty acids (FAs) are different, both structurally and functionally, and are sometimes even antagonistic in their effects. A balanced uptake of EPA/AA can prevent eicosanoid dysfunctions and may be effective in treating a number of illnesses and metabolic disorders (Nettleton, 1995; Gill & Valivety, 1997).

EPA is a potential anti-inflammatory agent (Calder, 1997; Babcock et al., 2000). One possible mechanism is that EPA-derived eicosanoids compete with AA-derived eicosanoids by forming leukotrienes 5 (LT_5), and, hence, reduce the level of LT_4 in rheumatoid arthritis patients (Fig. 8.2). The n-6 eicosanoids have a strong effect on body tissues, whereas n-3 eicosanoids possess a different or weaker potency, with respect to various cellular responses (Gurr, 1999). EPA also possesses therapeutic activity against cardiovascular diseases. For example, EPA can prevent atherosclerosis by decreasing the level of low-density lipoproteins (Bonaa et al., 1992). EPA appears to affect the electrical behavior, rhythms, and chemical responses of the heart; therefore, it reduces the likelihood of heart attack and arrhythmias (abnormalities of the heartbeat). EPA is capable of reducing the tendency toward thrombosis by reducing the level of fibrinogen, an activation factor in the occurrence of thrombosis (Hostmark et al., 1988). Recently, clinical evidence has shown that EPA plays a significant role in the treatment of schizophrenia and bipolar disorder by

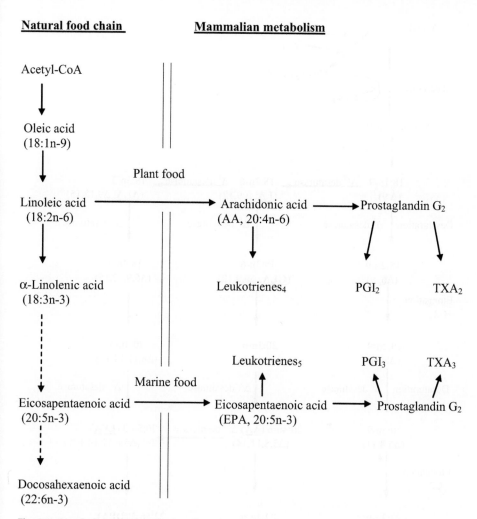

Fig. 8.2. Metabolic pathways of n-3 and n-6 eicosanoids from arachidonic acid (AA) and eicosapentaenoic (EPA). Source: Braden & Carroll (1986)

increasing the omega-3 FA content of erythrocyte membrane and plasma (Emsley et al., 2003; Ponizovsky et al., 2003; Su et al., 2003; Peet, 2004).

Biosynthesis

The biosynthesis of PUFA occurs through a series of reactions (Fig. 8.3) that can be divided into two distinct steps. The first step is the de novo synthesis of oleic acid (OA, 18:1n-9) from acetate. Almost all biological systems, including those found in

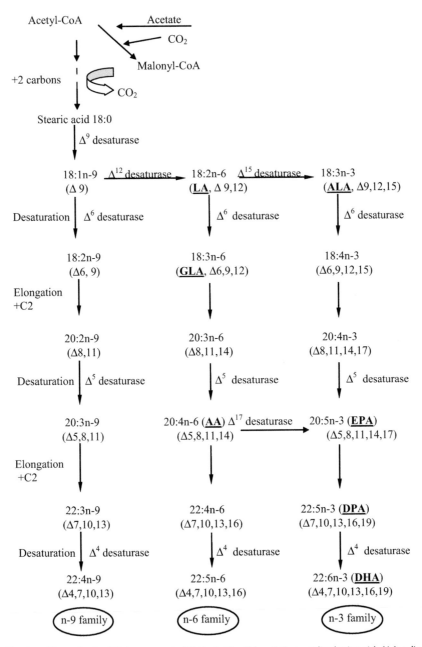

Fig. 8.3. Biosynthesis of Polyunsaturated Fatty Acids. Abbreviations: α-linolenic acid, ALA; γ-linolenic acid, GLA; arachidonic acid, AA; docosapentaenoic acid, DPA; and linoleic acid, LA. Source: Wen & Chen (2003), with permission from Elsevier.

microorganisms, insects, higher plants, and animals, are capable of performing this synthesis, with OA as the major product. The biosynthesis starts with the carboxylation of acetyl-CoA. Acetyl-CoA is synthesized from acetate or pyruvate by the action of glycolytic enzymes, and then converted into malonyl-CoA, which is used to drive a condensation reaction to extend the acyl group to stearic acid (18:0).

In the second step of PUFA synthesis, OA is desaturated by a Δ12 desaturase to form linoleic acid (LA, 18:2n-6) and then by a Δ15 desaturase to form α-linolenic acid (ALA, 18:3n-3). The n-9, n-6, and n-3 fatty acid families are formed from these precursors by a series of desaturation and elongation reactions. The biosynthesis of the three families of FAs is shown in Fig. 8.3. The three parent FAs—OA, LA, and ALA—compete with each other for the Δ6 desaturase. The affinity of this enzyme to the substrate and the amount of substrate available determine which metabolic pathway is predominant (Gurr, 1985). Generally, the first Δ6 desaturation is the limiting step, and ALA has the highest affinity for Δ6 desaturase, followed by LA and OA. Most algae, fungi, bacteria, mosses, insects, and some invertebrates possess the desaturase and elongase required for the synthesis of various PUFAs. They are the primary producers of these FAs in nature (Blomquist et al., 1991; Ratledge, 1993; Apt & Behrens, 1999; Chiou et al., 2001). In contrast, higher plants and animals lack the requisite enzymes and, thus, rarely produce PUFAs above C_{18} (Gill & Valivety, 1997).

In general, EPA can be synthesized through either the n-6 route (i.e., from LA to arachidonic acid and, subsequently, to EPA) or the n-3 route (i.e., from ALA to EPA) (Shimizu et al., 1989). Some fungi species, such as *Mortierella alpina* and *Pythium irregulare*, synthesize EPA mainly through n-6 route, rather than n-3 route, because these fungi exhibit high activity of Δ12, and Δ17 desaturase, which converts arachidonic acid into EPA (Fig. 8.3) (Shimizu et al., 1989; Jareonkitmongkol et al., 1993). When ALA-rich flaxseed oil was added to the culture, the fungal cells could use these external precursors (ALA) to synthesize EPA (Shimizu et al., 1989; Athalye et al., 2009).

Sources

Fish Oils

Fish oil is the conventional source of EPA, but there are number of disadvantages to using it for the production of a pharmaceutical/clinical grade material.

The quality of fish oil depends on the fish species, seasons, and geographical locations of the catching sites. Marine fish oil is a complex mixture of FAs with varying chain lengths and degrees of unsaturation. For clinical applications, however, it is necessary to produce an oil containing EPA as the sole PUFA. Even an oil containing a mixture of only EPA/DHA or EPA/AA is not good enough for clinical applications because of their different pharmaceutical functions, not to mention the complexity of the lipid and FAs in fish oil. Moreover, the purification of EPA from low-grade fish oil requires a number of physical steps, including the use of large-scale, preparative, high-performance liquid chromatography (HPLC), which are prohibitively expensive (Belarbi et al., 2000). In addition, fish oil has its own peculiar taste and odour (Barclay et al., 1994). Marine fish stocks are also subject to seasonal and climatic variations (Gill & Valivety, 1997) and, in some locations, are physically dwindling due to chronic over-fishing. Last, but by no means least, is the content of deleterious synthetic chemicals and that of organo-heavy metals, including mercury compounds, in fish oils. These materials have been ingested by the fish and then concentrated in their liver, which is then the principal organ used for obtaining oils. All these factors argue against the continued use of fish oils as a supply of EPA (and of DHA) and encourage the use of alternative sources, which, at the present time, have to be microbial.

Microorganisms
A variety of phylogenetically and physiologically diverse bacteria, isolated from cold marine environments, produce EPA as a part of their normal metabolism (Yazawa et al., 1988; Ringo et al., 1992; Hamamoto et al., 1994; Yazawa, 1996; Yano et al., 1997; Bowman et al., 1998; Kato & Nogi, 2001; Gentile et al., 2003; Satomi et al., 2003). The marine bacterium *Shewanella* has been widely studied for its EPA content (Table 8.A). Other EPA-producing bacterial genera include *Alteromonas, Flexibacter, Psychroflexus*, and *Vibrio* (Yazawa et al., 1988; Shimizu et al., 1989; Ringo et al., 1992; Hamamoto et al., 1994; Bowman et al., 1998). The bacterial synthesis of FAs, like EPA, can be affected by varying culture conditions, such as temperature and carbon and energy sources, as well as bacterial growth phase (Hamamoto et al., 1994). One apparent disadvantage for bacterial EPA production is that EPA predominates in total fatty acids (TFA) only when bacteria grow at very low temperature and high pressure (Yazawa et al., 1988; Shimizu et al., 1989; Hamamoto et al., 1994; Gentile et al., 2003). Consequently, it is difficult to practically exploit these bacteria for the commercial production of EPA.

Many lower fungi belonging to the genera *Mortierella, Pythium*, and *Saprolegnia* are capable of producing considerable amounts of EPA (Shimizu et al., 1988a, 1988b; Bajpai et al., 1992; Shirasaka & Shimizu, 1995; Cheng et al., 1999; Stredansky et al., 2000). The fungus *Mortierella alpina* has been widely investigated because of its potential for EPA production (Table 8.A). Temperature is crucial in regulating the formation of different fatty acids in this fungus. It produces exclusively very high levels of AA at temperatures over 20 °C. At low temperatures (12 °C), however, this fungus accumulates large amounts of EPA (Shimizu et al., 1988b). The conversion of AA to EPA at lower temperatures is considered to be catalysed by the Δ17 desaturase (Fig. 8.3) (Shimizu et al., 1989). Exogenous addition of α-linolenic acid to cultures of *M. alpina* can also lead to the enhancement of EPA yield (Shirasaka & Shimizu, 1995). Similar results have been reported with *M. elongata* NRRL 5513 (Bajpai et al., 1992).

Among the various microorganisms, microalgae provide the most abundant source of EPA. EPA has been found in a wide variety of marine microalgae, including in the classes Bacillariophyceae (diatoms), Chlorophyceae, Chrysophyceae,

Table 8.A. Proportions of PUFAs in Bacteria and Fungi.

Organisms	PUFA (% TFA) 20:4 (AA)	20:5 (EPA)	Cultivation/Isolation Method	Ref.
BACTERIA				
Shewanella benthica	-	16	Intestine of holothurian, South Atlantic Ocean, depth 4575 m, 4 °C	Kato & Nogi, 2001
Shewanella woodyi	-	16	Detritue, Alboran Sea, depth 370 m, 25 °C	Kato & Nogi, 2001
Shewanella sp.	1.9	6.5	Antarctic seawater, 4 °C, 30 days	Gentile et al., 2003
Shewanella marinintestina	1.1	17.5	Squid body, Yokohama, Japan, 20 °C	Satomi et al., 2003
Shewanella schlegeliana	2.2	18.6	Black porgy intestine, 20 °C	Satomi et al., 2003
Shewanella sairae	0.7	15.2	Saury intestine, Pacific Ocean, 20 °C	Satomi et al., 2003
Shewanella pealeana	1.6	19.1	Marine broth, 20 °C, 2 days	Satomi et al., 2003
Shewanella hanedai	1.4	23.1	Marine broth, 15 °C, 2 days	Satomi et al., 2003
Shewanella gelidimarina	0.7	14.0	Marine broth, 15 °C, 2 days	Satomi et al., 2003
FUNGI				
Mortierella alpina 20–17	22.9	4.9	GY medium, 28 °C, 6 days	Shimizu et al., 1989
Mortierella alpina 1–83	24.2	19.8	GY medium, 12 °C, 7 days	Shimizu et al., 1988a
Mortierella alpina 2O–17	38.7	17.1	GY medium, 12 °C, 7 days	Shimizu et al., 1988a
Mortierella alpina 1S–4	28.4	13.9	GY medium, 12 °C, 7 days	Shimizu et al., 1988a
Mortierella parvispora	14.4	10.9	YM medium, 12 °C, 7 days	Shimizu et al., 1988a
Mortierella hygrophila	13.6	10.4	GY medium, 12 °C, 7 days	Shimizu et al., 1988a
Saprolegnia sp. 28YTF-1	1.0	3.6	PYM medium, 28 °C, 7 days	Shirasaka & Shimizu, 1995
Pythium ultimum	6.1	8.2	Solid substrate (linseed oil, barley), 21 °C	Stredansky et al., 2000
Pythium irregulare	24	27	1% glucose (basal medium), 24 °C, 8 days	Cheng et al., 1999

Abbreviations: PUFA, polyunsaturated fatty acids; EPA, eicosapentaenoic acid; AA, arachidonic acid; TFA, total fatty acids.

Cryptophyceae, Eustigamatophyceae, and Prasinophyceae (Tables 8.B and 8.C). A study of *Porphyridium cruentum* has established the production potential of EPA by this microalga (Cohen, 1990). Ohta et al. (1993) also reported a high proportion of EPA in *Porphyridium propureum*. Most diatoms contain high EPA content (Table 8.C). The biotechnological potential of diatoms has been reviewed recently by Lebeau and Robert (2003b). The diatoms *Phaeodactylum tricornutum* and *Nitzschia laevis* have been intensively investigated for their EPA production potentials (Ohta et al., 1993; Garci et al., 2000; Wen & Chen, 2000b). The marine

Table 8.B. Proportions of PUFA in Marine Microalgae[a].

Organisms	PUFA (% TFA)			References
	20:4 (AA)	20:5 (EPA)	22:6 (DHA)	
CHRYSOPHYCEAE				
Monochrysis lutheri	1	19	-	Yongmanitchai & Ward, 1989
Pseudopedinella sp.	1	27	-	Yongmanitchai & Ward, 1989
Coccolithus huxleyi	1	17	-	Yongmanitchai & Ward, 1989
Cricosphaera carterae	3	20	-	Yongmanitchai & Ward, 1989
Cricosphaera elongata	2	28	-	Yongmanitchai & Ward, 1989
Isochrysis galbana	-	15	7.5	Grima et al., 1992
EUSTIGMATOPHYCEAE				
Monodus subterraneus	4.7	33	-	Hu et al., 1997
Nannochloropsis sp.	-	35	-	Sukenik, 1991
Nannochloris sp.	-	27	-	Yongmanitchai & Ward, 1989
Nannochloris salina	1	15	-	Yongmanitchai & Ward, 1989
CHLOROPHYCEAE				
Chlorella minutissima	5.7	45	-	Seto et al., 1984
Prasinophyceae				
Hetermastrix rotundra	1	28	7	Yongmanitchai & Ward, 1989
CRYPTOPHYCEAE				
Chromonas sp.	-	12.0	6.6	Renaud et al., 1999
Cryptomonas maculata	2	17	-	Yongmanitchai & Ward, 1989
Cryptomonas sp.	-	16	10	Yongmanitchai & Ward, 1989
Rhodomonas sp.	-	8.7	4.6	Renaud et al., 1999

[a]All algae cultures were grown photoautotrophically.

Abbreviations: DHA, docosahexaenoic acid. For other abbreviations, see Table 8.A.

Source: Wen & Chen (2003), with permission from Elsevier.

diatom *Odontella aurita* is also reported to be rich in EPA, as well as in essential trace elements, and can be grown in simple seawater or salt ground-water in outdoor tanks. In December 2002, the marine diatom *Odontella aurita* was filed to the Advisory Committee on Novel Foods and Processes of European Union (EU) to market the alga as novel food (Committee Paper for Discussion, 2009). The cultivation process of *O. aurita* has been favorably commented upon by the French Competent Authority and is under review by other Member States of the EU. In contrast to the large number of EPA-containing microalgae that are known, only a few microalgal species have been clearly demonstrated to have industrial production potential (Tables 8.B and 8.C). This result is mainly due to the low specific growth rate and low cell density of the microalgae when grown under conventional photoautotrophic conditions. In open pond systems, for example, a typical cell density was just 0.5 g L^{-1} (Chen, 1996).

Table 8.C. PUFA Compositions of Diatoms.

Organisms	PUFA (% TFA)			Growth Mode	References
	20:4 (AA)	20:5 (EPA)	22:6 (DHA)		
Asterionella japonica	11	20	-	Photoautotrophic	Yongmanitchai & Ward, 1989
Amphora coffeaformis	4.9	1.4	0.3	Photoautotrophic	Renaud et al., 1999
Biddulphia sinensis	-	24	1	Photoautotrophic	Yongmanitchai & Ward, 1989
Chaetoceros sp.	3.0	16.7	0.8	Photoautotrophic	Renaud et al., 1999
Cylindrotheca fusiformis	-	18.8	-	Heterotrophic	Tan & Johns, 1996
Fragilaria pinnata	8.7	6.8	1	Photoautotrophic	Renaud et al., 1999
Navicula incerta	-	25.2	-	Heterotrophic	Tan & Johns, 1996
Navicula pelliculosa	-	9.4	-	Heterotrophic	Tan & Johns, 1996
Navicula saprophila	2.7	16.0	-	Mixotrophic[a]	Kitano et al., 1997
Nitzschia closterium	-	15.2	-	Photoautotrophic	Renaud et al., 1994
Nitzschia frustulum	-	23.1	-	Photoautotrophic	Renaud et al., 1994
Nitzschia laevis	6.2	19.1	-	Heterotrophic	Wen & Chen, 2000b
Phaeodactylum tricornutum	-	34.5	-	Photoautotrophic	Yongmanitchai & Ward, 1991
Skeletonema costatum	-	29.2	-	Photoautotrophic	Blanchemain & Grizeau, 1999

[a]The algal cells were grown in light, with acetate as the carbon source.

For abbreviations, see Tables 8.A and 8.B.

Source: Wen & Chen (2003), with permission from Elsevier.

EPA Production by Microalgae

To date, microalgae have been seen as promising candidates for the commercial production of EPA because the following traits: (1) most microorganisms have a high cellular EPA content; (2) microalgae require relatively simple nutrients and mild environmental factors; (3) their cultivation conditions are easy to control; and (4) some microalgae are capable of heterotrophic growth, which offers the possibility of greatly increasing cell density and productivity by using high cell-density culture techniques.

Factors Influencing EPA Production

Culture Age

Oleaginous microalgae tend to store their energy source in lipid form as the culture ages. In contrast, the cellular content of PUFAs (including EPA) tends to follow a sigmoid curve: the PUFA content increases until the culture approaches the late growth or early stationary phase of growth and then decreases gradually at the late stationary and death phases (Yongmanitchai & Ward, 1989). In heterotrophic cultures of the marine diatom *Nitzschia laevis* the cellular content of EPA increases as the culture ages, but the proportion of EPA (as % TFA) remains relatively stable (Wen & Chen, 2001b).

Nutritional Factors

Several major nutrients are necessary for microalgal growth, including carbon, nitrogen, and phosphorus sources. Diatoms also require silicon. In addition, all microalgae require a variety of salts, such as sodium, potassium, calcium, and trace elements (Fe^{3+}, Cu^{2+}, Co^{2+}, and Zn^{2+}), to maintain a natural marine environment. Vitamin B_{12} and biotin are also required, in many cases.

Carbon sources are necessary to provide the energy and carbon skeletons for cell growth. Under photoautotrophic conditions, algae use CO_2 as the carbon source. Heterotrophic microalgae, however, are capable of growing in darkness, and, therefore, they must derive their energy from at least one organic carbon source, which is often provided in the form of acetate or glucose (Springer et al., 1994; Vazhappilly & Chen, 1999a, 1998b). Other carbon sources - including mono-, di-, and polysaccharides, such as fructose, sucrose, lactose, and starch - may also be usable by some species. Some obligate phototrophic algae can be converted into heterotrophs by genetic engineering and are capable of utilizing organic carbon sources (Zaslavskaia et al., 2001). Vegetable oils, such as linseed, corn, and canola oils, may promote growth and/or EPA production, depending on the microalgal species used.

A recent study has explored the potential of using crude glycerol for producing EPA by *P. irregulare*. As a major byproduct of biodiesel manufacturing processes, crude glycerol has posed a challenge to the biodiesel industry, particularly due to the difficulty in disposing of it. Preliminary results have indicated that crude glycerol is a good carbon for cell growth and EPA accumulation. The major impurities contained in crude glycerol, soap, and methanol were inhibitory to cell growth. Soap can be precipitated from the liquid medium through pH adjustment, while methanol can be evaporated from the medium during autoclaving (Athalye et al., 2009). Furthermore, a chemical characterization of the crude, glycerol-derived *P. irregulare* revealed that the biomass has a balanced nutritional value (protein, lipid, and carbohydrate); and does not contain any heavy metal contaminations, such as mercury or lead (Athalye et al., 2009). In the future, depending on whether the product can pass FDA scrutiny, the whole biomass might be usable as a food-grade product, or, perhaps, EPA can be purified from it.

Simple nitrogen sources, such as nitrate and urea, can support growth and EPA production in the microalgae *P. tricornutum* (Yongmanitchai & Ward, 1991) and *N. laevis* (Wen & Chen, 2001b). When ammonium was used as the nitrogen source, it was found that the pH of the medium tends to fluctuate markedly, which ultimately inhibits the growth and EPA production of the diatom (Wen & Chen, 2001b). For example, if an ammonium salt, such as sulphate or chloride, is used, assimilation of the ammonium ion leaves either sulphuric or hydrochloric acid behind. These are then the reasons for pH change and loss of cell viability. Complex nitrogen sources, such as yeast extract, tryptone, and corn steep liquor, can promote growth of most microalgal species by providing amino acids, vitamins, and growth factors (Aasen et al., 2000; Wen & Chen, 2001b). However, the contributions to cost of these nitrogen sources are not significant when compared with the carbon source used because their concentrations are usually very low.

Generally, the C/N ratio of the medium may influence the final cellular lipid content by controlling the switch between protein and lipid syntheses (Gordillo et al., 1998). A high C/N ratio favors lipid accumulation, which is triggered by nitrogen depletion in the culture (Ratledge, 1989). However, a high C/N ratio led to a lower proportion of unsaturated fatty acids in the microalga *Chlorella sorokiniana* (Chen & Johns, 1991) The reason might be that, under shortage of nitrogen, triacylglycerols (which are mostly PUFA-poor) are accumulated. Subsequently, the share of PUFA-rich polar lipids decreases.

Phosphate plays an important role in cellular energy transfer. Syntheses of phospholipids and nucleic acids also involve phosphate, and, thus, the phosphate level may significantly affect the cellular content of n-3 PUFAs. Yongmanitchai and Ward (1991) investigated the effect of phosphate concentrations on the EPA production of *P. tricornutum* and found that the EPA yield was higher at a higher phosphate concentration. However, in a culture of *Pythium irregulare*, a high initial phosphate concentration lowered the EPA yield (Stinson et al., 1991). The optimal phosphate concentration for EPA production in *Porphyridium purpureum* was 3 mM (in a 0.3 to 30 mM range) (Ohta et al., 1993)

Many diatoms contain a considerable amount of EPA (Table 8.C). Diatoms need silicate to form their frustules (cell walls composed of amorphous silica). Silicon metabolism in diatoms has been reviewed (Martin-Jezequel et al., 2000). The effect of silicate on the growth of diatoms usually follows the Monod equation (Wen & Chen, 2000a). Under silicate-deficient conditions, the diatom cell uses its intracellular silicon pool to support its physiological activities (Werner, 1977). The lipid contents of several marine diatoms increase as the supply of silicate to the medium decreases (Enright et al., 1986; Taguchi et al., 1987). Similarly, the EPA content of *N. laevis* increased when silicate became the limiting factor (Wen & Chen, 2000a). The reason for this phenomenon is that, in silicate-limited cultures, the cell tends to alter its metabolism and divert energy, which had been previously allocated for silicate uptake, into lipid storage (Coombs et al., 1967).

Environmental Factors

Photosynthesis requires light; efficient photosynthesis requires an abundance of light for all cells. The effects of light intensity and light/dark cycle on microalgal growth and n-3 PUFA production have been extensively investigated (Seto et al., 1984; Ohta et al., 1993; Cohen, 1994; Akimoto et al., 1998; Gordillo et al., 1998). The EPA contents of many microalgae were lower in heterotrophic cultures than in photoautotrophic cultures (Vazhappilly & Chen, 1998a, 1998b); the biosynthesis of n-3 PUFAs was enhanced under light, but faster growth was obtained under heterotrophic conditions (Chen, 1996). It appears that cultivation of mixtrophic cultures under a light/dark regime could yield optimal EPA productivities. This claim, however, requires further investigation.

Temperature is a very important factor affecting cell growth, lipid composition, and formation of microalgae by n-3 PUFAs. Low-temperature stress leads to a relatively high n-3 PUFAs content, which is often believed to result from the higher level of intracellular O_2, as the enzymes responsible for the desaturation and elongation of PUFAs depend on the availability of O_2 (Chen & Johns, 1991; Singh & Ward, 1997). However, an accurate interpretation of this phenomenon is very difficult, because a drop in temperature will decrease the growth rate and, simultaneously, increase the solubility of O_2 and CO_2 in the medium. Therefore, a chemostat cell culture is needed in the future to explain the temperature effect as all the other parameters can be kept constant in such a system.

A temperature shift strategy has been employed to enhance the overall production of n-3 PUFAs (including EPA), because the optimal temperature for microalgal growth is often higher than that for n-3 PUFAs formation. Such a phenomenon has been observed in many different algal species, including *Porphyridium cruentum* (Springer et al., 1994), *Nannochloropsis* sp. (Sukenik, 1991), and *Pythium irregulare* (Stinson et al., 1991). However, the optimal temperature for the growth of *Porphyridium purpureum* also yields the highest EPA content (Ohta et al., 1993). These results suggest that the effect of temperature on cell growth and EPA production should be studied for individual microalgal species.

pH is another physicochemical factor that affects the production of n-3 PUFAs by microalgae. Stinson et al. (1991) studied the effect of initial pH on the EPA yield of *Pythium irregulare*. Although initial pH values were quite different, the pH values were similar at the end of cultivation. However, in autotrophic cultures of *Phaeodactylum tricornutum*, the final pH values were quite different when the initial pH values ranged from 6.0 to 8.8. Although the cell dry wt concentrations were almost the same at different pH values, the EPA yield reached a maximum when the initial pH was 7.6 (Yongmanitchai & Ward, 1991).

High levels of EPA have been detected in many species of marine microalgae. In contrast, very few species of freshwater algae contain high levels of EPA. Medium salinity may influence the physiological properties of marine microalgae. A green microalga, *Dunaliella* sp., has been investigated in detail to determine how medium salinity affects its FA composition. The content of n-3 PUFAs decreased as medium salinity increased (Xu & Beardall, 1997). In contrast, Seto et al. (1984) investigated the effect of salt on the growth rate and FA composition of *Chlorella minutissima* by adding various amounts of seawater concentrate (SWC) or NaCl to an initially salt-free medium. Their results showed that cells grown in SWC or a NaCl-enriched medium contained higher percentages of EPA. In cultures of the marine diatom *N. laevis*, EPA yield was the highest at half of the salinity of artificial seawater (Wen & Chen, 2001a).

Cultivation Systems for Microalgae

An efficient large-scale cultivation system is necessary to explore the development of a process for the commercial production of EPA (Lebeau & Robert, 2003a). Although most microalgal species are obligate photoautotrophs, which require light for growth, a number of microalgae are capable of heterotrophic growth in the dark with one or more organic substrates as their sole carbon and energy source. For this type of microalgae, fermentation technology can be adopted and modified for large-scale production of microalgal products.

Photoautotrophic Cultivation Systems

Mass cultivation of microalgae originates from the development of open ponds, the oldest and simplest systems in which algae are cultured under conditions identical to their natural environment. Generally, commercial scale-up of open ponds is difficult because of problems with contamination by unwanted algae, bacteria, and predatory protozoa (Chen, 1996). In addition, optimal culture conditions are difficult to maintain in open ponds, and recovery of the biomass from dilute culture is expensive (Grima et al., 2003). As a result, only a few algal genera that grow in a selective and specialized environment have been successfully cultivated in open ponds. The EPA productivity of *P. cruentum* in open ponds was only 0.5 mg $L^{-1}d^{-1}$ in winter and 1 mg $L^{-1}d^{-1}$ in summer (Cohen & Heimer, 1992). It has been estimated that EPA productivity in open ponds can only reach 4–8 mg $L^{-1}d^{-1}$, at best (Ratledge, 1997).

Closed algal photobioreactors have been employed to overcome the problems encountered in open ponds (Grima et al., 1999; Tredici, 1999). These systems are made of transparent plastics and are generally placed outdoors for illumination by natural light; the cultivation vessels have a large surface-to-volume ratio (Fig. 8.4). Although enclosed systems can reduce the chances of contamination, the growth of microalgae is still suboptimal due to variations in temperature and light intensity.

In principle, enclosed photobioreactors with artificial light and a separated CO_2 supply are similar to conventional fermentors. Some photobioreactors also have O_2 removal devices to reduce the toxic effect of high O_2 concentrations on algal growth (Tredici, 1999; Fernandez et al., 2001). Photobioreactors have been employed to produce EPA from certain microalgae, such as *Nannochloropsis* sp. (Zittelli et al., 2000), *Monodus subterraneus* (Hu et al., 1997), and *P. tricornutum* (Yongmanitchai & Ward, 1991; Molina et al., 2001). A high EPA productivity of almost 60 mg $L^{-1}d^{-1}$ was obtained in the culture of *M. subterraneus* (Hu et al., 1997). Enclosed photobioreactors have several disadvantages, though, such as difficult scale-up, complexity in configuration and construction, and very high capital costs. Moreover, light limitation cannot be entirely overcome, since light penetration into the culture is inversely proportional to the cell concentration (Chen & Johns, 1995). Possible film buildup on the plastic surface of the photobioreactors may further decrease light uptake by the algal cells.

Fig. 8.4. A typical photobioreactor for outdoor cultivation of *Nannochloropsis* sp. for EPA production. Source: Zittelli et al. (1999), used with permission.

Heterotrophic Cultivation Systems

In microalgal culture, heterotrophic growth can be a cost-effective alternative to photoautotrophic growth (Chen, 1996). In heterotrophic culture, organic carbon sources, such as sugars or organic acids, can be used as the sole carbon and energy source. This mode of culture eliminates the requirement for light and, therefore, offers the possibility of greatly increasing cell density and productivity in batch culture. A heterotrophic batch culture may be further modified to become a high cell-density culture, such as fed-batch, chemostat, or perfusion culture. The development of high cell-density cultures for EPA production would also lead to a lower cost for EPA recovery and purification.

Some microalgae can grow rapidly heterotrophically (Vazhappilly & Chen, 1998a, 1998b). Generally, an organism used for heterotrophic production should possess the following abilities: to divide and metabolize in the dark, to grow on inexpensive and easily sterilized media, to adapt rapidly to a new environment (short or no lag-phase when inoculated to fresh media), and to withstand hydrodynamic stresses in fermentors and peripheral equipment. Some diatoms can also produce EPA heterotrophically

(Tan & Johns, 1996; Wen & Chen, 2000b), which indicates this type of culture may provide an effective and feasible means for the large-scale production of EPA.

Mixotrophic Cultivation Systems

Although heterotrophic growth is a cost-effective way to produce a high level of EPA, not all algal species are capable of heterotrophic growth, probably due to the fact that some photosynthetic pigments (e.g., phycocyanin, carotenoid, and chlorophyll) are essential for algae but cannot be synthesized heterotrophically. To overcome this limitation, the mixotrophic growth mode, in which algae perform both photosynthetic and oxidative metabolisms, simultaneously provides great potential for enhancing EPA production from not-heterotrophic algae. The diatom *N. leavis* has exhibited the best EPA production performance in mixotrophic growth on glucose, as compared to phototrophic and heterotrophic conditions (Wen & Chen, 2000b). In the presence of acetate, *Navicula saprophila* also displayed the greatest EPA content and productivity at mixotrophic conditions (Kitano et al., 1997). The mixotrophic culture of *P. tricornutum* was also studied extensively with various organic carbon sources (acetate, starch, lactic acid, glycine, glucose, and glycerol) (Garcia et al., 2005). The algal growth performance was highly dependent on the carbon sources and their dosage used. It has been reported that the biomass, total lipid content, and EPA production level in mixotrophic culture of *P. tricornutum* was greatly enhanced when glycerol and fructose were used as carbon sources (Ceron Garcia et al., 2006).

Cultivation Strategies for EPA Production by Microalgae

The competitiveness of microalgae-derived EPA when compared with fish oil EPA depends largely on the high EPA yield and productivity attained by microalgal cultures. The commonly used cultivation mode is batch culture. To achieve high cell-density of microalgae and, thus, high yield and productivity of EPA, strategies such as fed-batch, continuous, and perfusion cultures may be employed.

Fed-Batch Culture

Fed-batch is a commonly used culture technique in the fermentation industry. This mode of culture can attain a high cell-density by avoiding the limitation of, or inhibition by, substrates. A fed-batch process has been developed for the high cell-density culture of *N. laevis* for enhanced production of EPA (Wen et al., 2002). Fed-batch culture led to a cell dry weight concentration of 22.1 g L^{-1} and an EPA yield of 695 mg L^{-1}, both of which are much higher than equivalent levels achieved in batch cultures (Wen et al., 2002).

Although fed-batch culture can eliminate substrate limitation, it cannot overcome the inhibition caused by the toxic metabolites produced by the cells. When cell density reaches a high level, the accumulation of toxic metabolites becomes significant and, as a result, limits further growth of the cells. Therefore, other efficient culture techniques should be developed in order to enhance EPA yield and productivity.

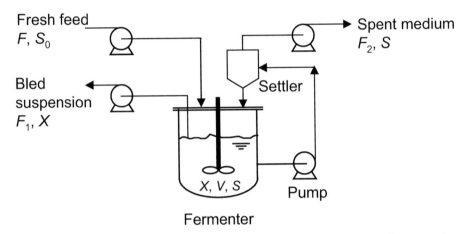

Fig. 8.5. Schematic diagram of the perfusion culture with cell bleeding system. X, cell concentration; V, culture volume; S, glucose concentration in medium; S_0, glucose concentration, in feed; F, flow rate of feed; F_1, flow rate of bleeding; F_2, flow rate of perfusion. Source: Wen & Chen (2003), with permission from Elsevier.

Continuous Culture

Continuous culture allows a higher EPA productivity than batch or fed-batch culture. It is also used as an important tool to study the basic physiological behavior of algal cells, as kinetic parameters, such as specific growth rate, cell density, and productivity, can be kept at a steady state. Continuous culture of *Isochrysis galbana* was previously investigated for achieving high EPA productivity (Grima et al., 1993, 1994; Otero et al., 1997b). Similarly, continuous cultures of *P. tricornutum* and *N. laevis* have been reported to produce EPA (Miron et al., 1999, 2002; Molina et al., 2001; Wen & Chen, 2002a).

A large number of extracellular metabolic products have been identified in microalgae, in general (Shimizu, 1996; Srivastava et al., 1999), and in diatoms, in particular (Eppley, 1977). To avoid the inhibition of toxic by-products and the limitation of substrate to growth, a continuous culture with cell-recycle (called perfusion culture) has been developed in heterotrophic culture of *N. laevis* (Wen & Chen, 2002b). The perfusion culture is further modified to permit cell bleeding during the perfusion operation (Fig. 8.5). This strategy allows continuous harvesting of the algal cells, while removing inhibitory compounds during cultivation. Thus, high EPA productivity can be achieved by harvesting the EPA-containing algal biomass (Wen & Chen, 2001c). The perfusion-cell bleeding culture allows a much higher EPA productivity than the simple perfusion culture does. At a bleeding rate of 0.67 d^{-1} and a perfusion rate of 0.6 d^{-1}, the EPA productivity achieved is 175 mg L^{-1} d^{-1} (Table 8.D). This EPA productivity is the highest ever reported in microalgal cultures.

Table 8.D. Comparison of EPA Productivity of Microalgae Under Various Culture Conditions[a].

Organisms	Culture Vessels	Culture Modes	EPA productivity $(mg \cdot L^{-1} \cdot d^{-1})$	References
Phaeodactylum tricornutum	Glass tubes	Batch	19.0	Yongmanitchai & Ward, 1991
Phaeodactylum tricornutum	Glass tanks	Continuous	25.1	Yongmanitchai & Ward, 1992
Isochrysis galbana	Fermenters	Continuous	7.2	Grima et al., 1993
Isochrysis galbana	Fermenters	Continuous	15.3	Grima et al., 1994
Monodus subterraneus	Erlenmeyer flasks	Continuous	25.7	Cohen, 1994
Monodus subterraneus	Flat plate reactors	Semi-continuous	58.9	Hu et al., 1997
Isochrysis galbana	Glass tubes	Semi-continuous	4.6	Otero et al., 1997b
Phaeodactylum tricornutum	Glass tubes	Semi-continuous	5.2	Otero et al., 1997a
Porphyridium cruentum	Flasks	Batch	3.6	Akimoto et al., 1998
Isochrysis galbana	Cylindrical fermenters	Continuous	23.8	Sevilla et al., 1998
Nannochloropsis sp.	Tubular photobioreactors	Continuous	32.0	Zittelli et al., 1999
Phaeodactylum tricornutum[b]	Glass vessels	Batch	33.5	Garci et al., 2000
Nitzschia laevis[c]	Fermenters	Perfusion-cell bleeding	174.6	Wen & Chen, 2001c

[a]Unless specified, the microalgae were grown photoautotrophically.

[b]The microalga was grown in mixotrophic growth conditions(i.e., grown under light, with glycerol as the carbon source).

[c]The microalga was grown heterotrophically.

For abbreviations, see Table 8.A.

Source: Wen & Chen (2003), with permission from Elsevier.

Prospects of EPA Production by Microorganisms

EPA and DHA products have now been well recognized by consumers. To date, microalgae-derived DHA products have already been introduced into the infant formula and adult food markets. DHA-containing microalgae have also been made available as aquafeed. In contrast, fish oil is still the sole commercial source of EPA. The main reason for impeding the commercialization of microbial EPA production is that the yield and productivity of EPA are still very low, which results in high operation cost and high recovery/purification cost. For example, the current DHA production process for the dinoflagellate *Crypthecodinium cohnii* has achieved a productivity of over 50 mg L^{-1} h^{-1} by simple fed-batch cultivation (see Chapter 6). In contrast, the highest EPA productivity is only 7.25 mg L^{-1} h^{-1}, using a relatively complicated, perfusion-cell bleeding culture technique. A comparison of DHA and EPA production from microalgae is given in Table 8.E. Low EPA productivity is attributed to both the low cellular EPA content and the relatively low growth rate of algae (Table 8.E). Thus, the microalgae derived-EPA product is not yet commercially feasible. Another reason microbial EPA production is rarely commercialized is that EPA-producing algae have relative complex FA profiles, compared with DHA-containing non-photosynthetic algae (Wen & Chen, 2001a; De Swaaf et al., 2003b). Low EPA productivity with relatively complex FA composition means high operational costs and high costs for EPA recovery, which limit the applications of microalgae for EPA production.

Several areas of research need to be developed to continue the commercialization of EPA production by microalgae. First, research should focus on heterotrophic growth of microalgae. Under photoautotrophic conditions, the growth rate of microalgae is very low, and light limitation further limits cell growth. EPA productivity could be significantly enhanced if the selected algae could grow heterotrophically. In addition, high cell-density culture techniques may be applied using industrial-scale fermentors. Although fermentors used in heterotrophic culture are more expensive than photobioreactors, the enhancement of EPA productivity could pay for the high cost of fermentors.

Second, EPA-producing strains can be improved by mutation and genetic engineering. For example, Cohen et al. (1992) selected cell lines of *Spirulina platensis* and *Porphyridium cruentum* with the herbicide Sandoz 9785. The herbicide-resistant mutant of *P. cruentum* was able to over-produce EPA. Lopez-Alonso et al. (1996) also selected mutant strains of *Phaeodactylum tricornutum* for EPA production, and one mutant (II242) contained 44% more EPA than the wild type.

The first cloning and expression of the large heterologous EPA synthesis gene cluster was accomplished in a marine cyanobacterium, *Synechococcus* sp. (Takeyama et al., 1997). The EPA synthesis gene cluster (approx. 38 kb, GenBank accession number U73935) was isolated from a marine bacterium, *Shewanella putrefaciens* (Yazawa, 1996). It contained eight open-reading frames, three of which included the genes that encode carbon chain elongation enzymes, while the rest included desaturase genes (Yazawa, 1996). Another transconjugant, *Synechococcus* sp., could produce a maximum EPA yield of 3.9 mg L^{-1} (Yu et al., 2000). Although the EPA yields obtained in these

Table 8.E. Comparison of EPA and DHA Production by Microalgae.

Alga	Product	Culture Mode[a]	Substrate	Biomass (g·L^{-1})	EPA/DHA Content (% DW)	Yield (g·L^{-1})	Productivity (mg·L^{-1}·h^{-1})	References
Crypthecodinium cohnii	DHA	Fed-batch	Glucose	26	6.9	1.7	14	de Swaaf et al., 2003b
C. cohnii	DHA	Fed-batch	Acetate	61	15.7	9.5	45	de Swaaf et al., 2003b
C. cohnii	DHA	Fed-batch	Ethanol	83	14.1	11.7	53	de Swaaf et al., 2003a
Nitzschia laevis	EPA	Fed-batch	Glucose	40	3.0	1.1	3.1	Wen et al., 2002
N. laevis	EPA	Perfusion-bleeding	Glucose	10	2.6	0.3	7.3	Wen & Chen, 2001c

[a]Heterotrophic culture mode was employed for all cultures.

For abbreviations, see Tables 8.A and 8.B.

studies remained low (Takeyama et al., 1997; Yu et al., 2000), they demonstrated the possibility of employing this approach for modifying the lipid composition, which might lead to a further improvement in EPA productivity. Some remarkable findings in making transgenic microalgae for enhanced EPA production have been recently reported (Zaslavskaia et al., 2001). These researchers introduced the gene encoding a glucose transporter (glut1 or hup1) into an obligate photoautotrophic diatom, *Phaeodactylum tricornutum*, to enable the alga to thrive on exogenous glucose and to produce EPA in the absence of light. Researchers at DuPont Co., also engineered the yeast *Yarrowia lipolytica* by cloning heterologous genes encoding a delta6 desaturase, an 18–20 elongase, a delta5 desaturase, and a delta17 desaturase into the cells,; they successfully produced EPA at a level of 3% of TFA (see Chapter 3).

Conclusion

EPA is a precursor of a large variety of bioactive metabolites and performs diverse physiological functions in the human body. Evidence of the beneficial effects of EPA has brought this FA to the attention of food and pharmaceutical markets, worldwide. The increasing applications for EPA and its inadequate conventional sources (i.e., fish oils) have led to an extensive search for alternative sources of production, including microalgae, lower fungi, and marine bacteria.

The EPA production potential of microalgae depends on the characteristics of the specific algal species and the cultivation strategies developed. Heterotrophic cultivation is a cost-effective means for producing EPA on a large scale. However, investigations of heterotrophic EPA production from microalgae are still in their infancy. An in-depth understanding of the factors that affect EPA production is thus needed. In the future, the application of genetically modified microorganisms may be the most efficient means to attain improved production of EPA.

References

Aasen, I.M.; T. Moretro; T. Katla; L. Axelsson; I. Storro. Influence of Complex Nutrients, Temperature and pH on Bacteriocin Production by *lactobacillus sakei* CCUG 42687. *Appl. Microbiol. Biotechnol.* **2000**, *53*, 159–166.

Akimoto, M.; A. Shirai; K. Ohtaguchi; K. Koide. Carbon Dioxide Fixation and Polyunsaturated Fatty Acid Production by the Red Alga Porphyridium Cruentum. *Appl. Biochem. Biotechnol.* **1998**, *73*, 269–278.

Alonso, D.L.; C.I.S. del Castillo; E.M. Grima; Z. Cohen. First Insights into Improvement of Eicosapentaenoic Acid Content in *Phaeodactylum tricornutum* (bacillariophyceae) by Induced Mutagenesis. *J. Phycol.* **1996**, *32*, 339–345.

Apt, K.E.; P. W. Behrens. Commercial developments in microalgal biotechnology. *J. Phycol.* **1999**, *35*, 215–226.

Athalye, S. K.; R.A. Garcia; Z.Y. Wen. Use of Biodiesel-Derived Crude Glycerol for Producing Eicosapentaenoic Acid (EPA) by the Fungus *pythium irregulare*. *J. Agri. Food Chem.* **2009**, *57*, 2793–2744.

Babcock, T.; W.S. Helton; N.J. Espat. Eicosapentaenoic Acid (EPA): An Anti-inflammatory Omega-3 Fat with Potential Clinical Applications. *Nutrition* **2000,** *16,* 1116–1118.

Bajpai, P.K.; P. Bajpai; O.P. Ward. Optimization of Culture Conditions for Production of Eicosapentaenoic Acid by Mortierella-elongata NRRL-5513. *J. Ind. Microbiol.* **1992,** *9,* 11–17.

Barclay, W.R.; K.M. Meager; J.R. Abril. Heterotrophic Production of Long-Chain Omega-3-Fatty-Acids Utilizing Algae and Algae-like Microorganisms. *J. Appl. Phycol.* **1994,** *6,* 123–129.

Belarbi, E.H.; E. Molina; Y. Chisti. A Process for High Yield and Scaleable Recovery of High Purity Eicosapentaenoic Acid Esters from Microalgae and Fish Oil. *Enzyme Micro. Technol.* **2000,** *26,* 516–529.

Blanchemain, A.; D. Grizeau. Increased Production of Eicosapentaenoic Acid by Skeletonema Costatum Cells after Decantation at Low Temperature. *Biotechnol. Tech.* **1999,** *13,* 497–501.

Blomquist, G.J.; C.E. Borgeson; M. Vundla. Polyunsaturated Fatty-Acids and Eicosanoids in Insects. *Insect Biochem.* **1991,** *21,* 99–106.

Bonaa, K.H.; K.S. Bjerve; A. Nordoy. Habitual Fish Consumption, Plasma Phospholipid Fatty-Acids, and Serum-Lipids—The Tromso Study. *Am. J. Clinical Nutri.* **1992,** *55,* 1126–1134.

Bowman, J.P.; S.A. McCammon; T. Lewis; J.H. Skerratt; J.L. Brown; D.S. Nichols; T.A. McMeekin. Psychroflexus Torquis Gen. Nov., sp. Nov., a Psychrophilic Species from Antarctic Sea Ice, and Reclassification of Flavobacterium Gondwanense (Dobson et al., 1993) as Psychroflexus Gondwanense Gen. Nov., Comb. Nov. *Microbiology UK* **1998,** *144,* 1601–1609.

Braden, L.M.; K.K. Carroll. Dietary Polyunsaturated Fat in Relation to Mammary Carcinogenesis in Rats. *Lipids* **1986,** *21,* 285–288.

Calder, P.C. N-3 Polyunsaturated Fatty Acids and Cytokine Production in Health and Disease. *Annals Nutri. Metabol.* **1997,** *41,* 203–234.

Ceron Garcia, M.C.; F.G. Camacho; A.S. Miron; J.M.F. Sevilla; Y. Chisti; E.M. Grima. Mixotrophic Production of Marine Microalga *Phaeodactylum tricornutum* on Various Carbon Sources. *J. Microbiol. Biotechnol.* **2006,** *16,* 689–694.

Chen, F. High Cell Density Culture of Microalgae in Heterotrophic Growth. *Trends Biotechnol.* **1996,** *14,* 421–426.

Chen, F.; M.R. Johns. Effect of C/N Ratio and Aeration on the Fatty-Acid Composition of Heterotrophic Chlorella-Sorokiniana. *J. Appl. Phycol.* **1991,** *3,* 203–209.

Chen, F.; M.R. Johns. A Strategy for High Cell-Density Culture of Heterotrophic Microalgae with Inhibitory Substrates. *J. Appl. Phycol.* **1995,** *7,* 43–46.

Chen, G.Q.; Y. Jiang; F. Chen. Fatty Acid and Lipid Class Composition of the Eicosapentaenoic Acid-Producing Microalga, Nitzschia laevis. *Food Chem.* **2007,** *104,* 1580–1585.

Chen, G.Q.; Y. Jiang; F. Chen. Salt-Induced Alterations in Lipid Composition of Diatom *Nitzschia laevis* (bacillariophyceae) under Heterotrophic Culture Condition. *J. Phycol.* **2008a,** *44,* 1309–1314.

Chen, G.Q.; Y. Jiang; F. Chen. Variation of Lipid Class Composition in *Nitzschia laevis* as a Response to Growth Temperature Change. *Food Chem.* **2008b,** *109,* 88–94.

Cheng, M.H.; T.H. Walker; G.J. Hulbert; D.R. Raman. Fungal Production of Eicosapentaenoic and Arachidonic Acids from Industrial Waste Streams and Crude Soybean Oil. *Biores. Technol.* **1999,** *67,* 101–110.

Chiou, S.Y.; W.W. Su; Y.C. Su. Optimizing Production of Polyunsaturated Fatty Acids in *Marchantia polymorpha* Cell Suspension Culture. *J. Biotechnol.* **2001**, *85*, 247–257.

Cohen, Z. The Production Potential of Eicosapentaenoic and Arachidonic Acids by the Red Alga *Porphyridium cruentum*. *J. Am. Oil Chem. Soc.* **1990**, *67*, 916–920.

Cohen, Z. Production Potential of Eicosapentaenoic Acid by *Monodus subterraneus*. *J. Am. Oil Chem. Soc.* **1994**, *71*, 941–945.

Cohen, Z.; S. Didi; Y.M. Heimer. Overproduction of Gamma-Linolenic and Eicosapentaenoic Acids by Algae. *Plant Physiol.* **1992**, *98*, 569–572.

Cohen, Z.; Y.M. Heimer. Production of Polyunsaturated Fatty Acids (EPA, ARA, and GLA) by the Microalgae *Porphyridium* and *Spirulina*. *Industrial Applications of Single Cell Oils*; D.J. Kyle, C. Ratledge, Ed.; AOCS Press: Champaign, IL, 1992; pp 243–273.

Committee Paper for Discussion. Microalga (*odontella aurita*): A Notification under Article 5 of the Novel Food Regulation (ec) 258/97. Published online: http://www.foodstandards.gov.uk/multimedia/pdfs/acnfp593.pdf (accessed Jul 2009).

Coombs, J.; P.J. Halicki; O. Holm-Hansen; B.E. Volcani. Studies on the Biochemistry and Fine Structure of Silicate Shell Formation in Diatoms. II. Changes in Concentration of Nucleoside Triphosphates in Silicon-Starvation Synchrony of *Navicula pelliculosa* (breb.) hilse. *Exp. Cell Res.* **1967**, *47*, 315–328.

de Swaaf, M.E.; J.T. Pronk; L. Sijtsma. Fed-Batch Cultivation of the Docosahexaenoic-Acid-Producing Marine Alga *Crypthecodinium cohnii* on Ethanol. *Appl. Microbiol. Biotechnol.* **2003a**, *61*, 40–43.

de Swaaf, M.E.; L. Sijtsma; J.T. Pronk. High-Cell-Density Fed-Batch Cultivation of the Docosahexaenoic Acid-Producing Marine Alga *Crypthecodinium cohnii*. *Biotechnol. Bioeng.* **2003b**, *81*, 666–672.

Emsley, R.; P. Oosthuizen; S.J. van Rensburg. Clinical Potential of Omega-3 Fatty Acids in the Treatment of *Schizophrenia*. *CNS Drugs* **2003**, *17*, 1081–1091.

Enright, C.T.; G.F. Newkirk; J.S. Craigie; J.D. Castell. Growth of Juvenile Ostrea-Edulis l Fed *Chaetoceros gracilis* Schutt of Varied Chemical-Composition. *J. Exp. Marine Bio. Eco.* **1986**, *96*, 15–26.

Eppley, R.W. The Growth and Culture of Diatom. *The Biology of Diatoms*; D. Werner, Ed.; Blackwell Scientific: Oxford, 1977; pp 24–64.

Fernandez, F.G.A.; J.M.F. Sevilla; J.A.S. Perez; E.M. Grima; Y. Chisti. Airlift-Driven External-Loop Tubular Photobioreactors for Outdoor Production of Microalgae: Assessment of Design and Performance. *Chem. Eng. Sci.* **2001**, *56*, 2721–2732.

Garcia, M.C.C.; J.M.F. Sevilla; F.G.A. Fernandez; E.M. Grima; F.G. Camacho. Mixotrophic Growth of Phaeodactylum tricornutum on Glycerol: Growth Rate and Fatty Acid Profile. *J. Appl. Phycol.* **2000**, *12*, 239–248.

Garcia, M.C.C.; A.S. Miron; J.M.F. Sevilla; E.M. Grima; F.G. Camacho. Mixotrophic Growth of the Microalga Phaeodactylum tricornutum—Influence of Different Nitrogen and Organic Carbon Sources on Productivity and Biomass Composition. *Process Biochem.* **2005**, *40*, 297–305.

Gentile, G.; V. Bonasera; C. Amico; L. Giuliano; M.M. Yakimov. *Shewanella* sp ga-22, a Psychrophilic Hydrocarbonoclastic Antarctic Bacterium Producing Polyunsaturated Fatty Acids. *J. Appl. Microbiol.* **2003**, *95*, 1124–1133.

Gill, I.; R. Valivety. Polyunsaturated Fatty Acids. 1. Occurrence, Biological Activities and Applications. *Trends Biotechnol.* **1997,** *15,* 401–409.

Gordillo, F.J.L.; M. Goutx; F.L. Figueroa; F.X. Niell. Effects of Light Intensity, CO_2 and Nitrogen Supply on Lipid Class Composition of *Dunaliella viridis. J. Appl. Phycol.* **1998,** *10,* 135–144.

Grima, E.M.; J.A.S. Perez; J.L.G. Sanchez; F.G. Camacho; D.L. Alonso. EPA from *Isochrysis galbana*—Growth-Conditions and Productivity. *Process Biochem.* **1992,** *27,* 299–305.

Grima, E.M.; J.A.S. Perez; F.G. Camacho; J.L.G. Sanchez; D.L. Alonso. N-3 PUFA Productivity in Chemostat Cultures of Microalgae. *Appl. Microbiol. Biotechnol.* **1993,** *38,* 599–605.

Grima, E.M.; J.A.S. Perez; F.G. Camacho; J.M.F. Sevilla; F.G.A. Fernandez. Effect of Growth-Rate on the Eicosapentaenoic Acid and Docosahexaenoic Acid Content of *Isochrysis galbana* in Chemostat Culture. *Appl. Microbiol. Biotechnol.* **1994,** *41,* 23–27.

Grima, E.M.; F.G.A. Fernandez; F.G. Camacho; Y. Chisti. Photobioreactors: Light Regime, Mass Transfer, and Scaleup. *J. Biotechnol.* **1999,** *70,* 231–247.

Grima, E.M.; E.H. Belarbi; F.G.A. Fernandez; A.R. Medina; Y. Chisti. Recovery of Microalgal Biomass and Metabolites: Process Options and Economics. *Biotechnol. Adv.* **2003,** *20,* 491–515.

Gurr, M.I. Biosynthesis of Fats. *The Role of Fats in Human Nutrition;* F. B. Padley, J. Podmore, Ed.; Ellis Horwood Ltd.: Chichester, 1985; pp 24–28.

Gurr, M.I. The Nutritional and Biological Properties of the Polyunsaturated Fatty Acids. *Lipids in Nutrition and Health: A Reappraisal;* M.I. Gurr, Ed.; The Oily Press: Bridgewater, 1999; pp 119–160.

Hamamoto, T.; N. Takata; T. Kudo; K. Horikoshi. Effect of Temperature and Growth-Phase on Fatty-Acid Composition of the Psychrophilic *Vibrio* sp Strain No-5710. *FEMS Microbiol. Lett.* **1994,** *119,* 77–81.

Hostmark, A.T.; T. Bjerkedal; P. Kierulf; H. Flaten; K. Ulshagen. Fish Oil and Plasma-Fibrinogen. *Bri. Med. J.* **1988,** *297,* 180–181.

Hu, Q.; Z.Y. Hu; Z. Cohen; A. Richmond. Enhancement of Eicosapentaenoic Acid (EPA) and Gamma-Linolenic Acid (GLA) Production by Manipulating Algal Density of Outdoor Cultures of *Monodus subterraneus* (Eustigmatophyta) and *Spirulina platensis* (cyanobacteria). *Eur. J. Phycol.* **1997,** *32,* 81–86.

Jareonkitmongkol, S.; S. Shimizu; H. Yamada. Production of an Eicosapentaenoic Acid-Containing Oil by a Delta-12 Desaturase-Defective Mutant of *Mortierella alpina* 1S-4. *J. Am. Oil Chem. Soc.* **1993,** *70,* 119–123.

Kato, C.; Y. Nogi. Correlation between Phylogenetic Structure and Function: Examples from Deep-Sea Shewanella. *FEMS Microbiol. Eco.* **2001,** *35,* 223–230.

Kitano, M.; R. Matsukawa; I. Karube. Changes in Eicosapentaenoic Acid Content of Navicula saprophila, Rhodomonas salina and Nitzschia sp under Mixotrophic Conditions. *J. Appl. Phycol.* **1997,** *9,* 559–563.

Lebeau, T.; J.M. Robert. Diatom Cultivation and Biotechnologically Relevant Products. Part I: Cultivation at Various Scales. *Appl. Microbiol. Biotechnol.* **2003a,** *60,* 612–623.

Lebeau, T.; J.M. Robert. Diatom Cultivation and Biotechnologically Relevant Products. Part II: Current and Putative Products. *Appl. Microbiol. Biotechnol.* **2003b,** *60,* 624–632.

Martin-Jezequel, V.; M. Hildebrand; M.A. Brzezinski. Silicon Metabolism in Diatoms: Implications for Growth. *J. Phycol.* **2000**, *36*, 821–840.

Miron, A.S.; A.C. Gomez; F.G. Camacho; E.M. Grima; Y. Chisti. Comparative Evaluation of Compact Photobioreactors for Large-Scale Monoculture of Microalgae. *J. Biotechnol.* **1999**, *70*, 249–270.

Miron, A.S.; M.C.C. Garcia; F.G. Camacho; E.M. Grima; Y. Chisti. Growth and Biochemical Characterization of Microalgal Biomass Produced in Bubble Column and Airlift Photobioreactors: Studies in Fed-Batch Culture. *Enzyme Micro. Technol.* **2002**, *31*, 1015–1023.

Molina, E.; J. Fernandez; F.G. Acien; Y. Chisti. Tubular Photobioreactor Design for Algal Cultures. *J. Biotechnol.* **2001**, *92*, 113–131.

Nettleton, J.A. *Omega-3 Fatty Acids and Health.* Chapman & Hall: New York, 1995.

Ohta, S.; T. Chang; O. Aozasa; N. Ikegami; H. Miyata. Alterations in Fatty-Acid Composition of Marine Red Alga *Porphyridium purpureum* by Environmental-Factors. *Botanica Marina* **1993**, *36*, 103–107.

Otero, A.; D. Garcia; J. Fabregas. Factors Controlling Eicosapentaenoic Acid Production in Semicontinuous Cultures of Marine Microalgae. *J. Appl. Phycol.* **1997a**, *9*, 465–469.

Otero, A.; D. Garcia; E.D. Morales; J. Aran; J. Fabregas. Manipulation of the Biochemical Composition of the Eicosapentaenoic Acid-Rich Microalga *Isochrysis galbana* in Semicontinuous Cultures. *Biotechnol. Appl. Biochem.* **1997b**, *26*, 171–177.

Peet, M. Nutrition and Schizophrenia: Beyond Omega-3 Fatty Acids. *Prostaglandins Leukotrienes Essen. Fatty Acids* **2004**, *70*, 417–422.

Ponizovsky, A.M.; G. Barshtein; L.D. Bergelson. Biochemical Alterations of Erythrocytes as an Indicator of Mental Disorders: An Overview. *Harvard Review Psychiatry* **2003**, *11*, 317–332.

Ratledge, C. Biotechnology of Oils and Fats. *Microbial Lipids;* C. Ratledge, S.G. Wilkinson, Ed.; Academic Press: London, 1989; Vol. 2, pp 567–668.

Ratledge, C. Single-Cell Oils—Have They a Biotechnological Future. *Trends Biotechnol.* **1993**, *11*, 278–284.

Ratledge, C. Microbial Lipid. *Biotechnology: Products of Second Metabolism*, 2nd edn.; H. Kleinkauf, H. Von Dohre. Eds.; VCH Press: Weinheim, 1997; Vol. 7, pp 133–197.

Renaud, S.M.; D.L. Parry; L.V. Thinh. Microalgae for Use in Tropical Aquaculture .1. Gross Chemical and Fatty-Acid Composition of 12 Species of Microalgae from the Northern-Territory, Australia. *J. Appl. Phycol.* **1994**, *6*, 337–345.

Renaud, S.M.; L.V. Thinh; D.L. Parry. The Gross Chemical Composition and Fatty Acid Composition of 18 Species of Tropical Australian Microalgae for Possible Use in Mariculture. *Aquaculture* **1999**, *170*, 147–159.

Ringo, E.; P.D. Sinclair; H. Birkbeck; A. Barbour. Production of Eicosapentaenoic Acid (20-5 n-3) by Vibrio pelagius Isolated from Turbot (Scophthalmus maximus) Larvae. *Appl. Environ. Microbiol.* **1992**, *58*, 3777–3778.

Satomi, M.; H. Oikawa; Y. Yano. *Shewanella marinintestina* sp nov., *Shewanella schlegeliana* sp nov and *Shewanella sairae* sp nov., Novel Eicosapentaenoic-Acid-Producing Marine Bacteria Isolated from Sea-Animal Intestines. *Inter. J. Sys. Evol. Microbiol.* **2003**, *53*, 491–499.

Seto, A.; H.L. Wang; C.W. Hesseltine. Culture Conditions Affect Eicosapentaenoic Acid Content of *Chlorella minutissima*. *J. Am. Oil Chem. Soc.* **1984**, *61*, 892–894.

Sevilla, J.M.F.; E.M. Grima; F.G. Camacho; F.G.A. Fernandez; J.A.S. Perez. Photolimitation and Photoinhibition as Factors Determining Optimal Dilution Rate to Produce Eicosapentaenoic Acid from Cultures of the Microalga *Isochrysis galbana*. *Appl. Microbiol. Biotechnol.* **1998**, *50*, 199–205.

Shimizu, Y. Microalgal Metabolites: A New Perspective. *Ann. Rev. Microbiol.* **1996**, *50*, 431–465.

Shimizu, S.; H. Kawashima; Y. Shinmen; K. Akimoto; H. Yamada. Production of Eicosapentaenoic Acid by *Mortierella* Fungi. *J. Am. Oil Chem. Soc.* **1988a**, *65*, 1455–1459.

Shimizu, S.; Y. Shinmen; H. Kawashima; K. Akimoto; H. Yamada. Fungal Mycelia as a Novel Source of Eicosapentaenoic Acid—Activation of Enzyme(s) Involved in Eicosapentaenoic Acid Production at Low-Temperature. *Biochem. Biophy. Res. Com.* **1988b**, *150*, 335–341.

Shimizu, S.; H. Kawashima; K. Akimoto; Y. Shinmen; H. Yamada. Microbial Conversion of an Oil Containing Alpha-Linolenic Acid to an Oil Containing Eicosapentaenoic Acid. *J. Am. Oil Chem. Soc.* **1989**, *66*, 342–347.

Shirasaka, N.; S. Shimizu. Production of Eicosapentaenoic Acid by *Saprolegnia* sp 28ytf-1. *J. Am. Oil Chem. Soc.* **1995**, *72*, 1545–1549.

Singh, A.; O.P. Ward. Microbial Production of Docosahexaenoic Acid (DHA, C22:6). *Advances in Appl. Microbiol.*; **1997**, *45*, 21–312.

Springer, M.; H. Franke; O. Pulz. Increase of the content of Polyunsaturated Fatty-Acids in Porphyridium-cruentum by Low-Temperature Stress and Acetate Supply. *J. Plant Physiol.* **1994**, *143*, 534–537.

Srivastava, V.C.; G.J. Manderson; R. Bhamidimarri. Inhibitory Metabolites Production by the Cyanobacterium *Fischerella muscicola*. *Microbiol. Res.* **1999**, *153*, 309–317.

Stinson, E. E.; R. Kwoczak; M. J. Kurantz. Effect of cultural conditions on production of eicosapentaenoic acid by *Pythium irregulare*. *J. Ind. Microbiol.* **1991**, *8*, 171–178.

Stredansky, M.; E. Conti; A. Salaris. Production of Polyunsaturated Fatty Acids by *Pythium ultimum* in Solid-State Cultivation. *Enzyme Micro. Technol.* **2000**, *26*, 304–307.

Su, K.P.; S.Y. Huang; C.C. Chiu; W.W. Shen. Omega-3 Fatty Acids in Major Depressive Disorder—A Preliminary Double-blind, Placebo-controlled Trial. *Eur. Neuropsychopharmacol.* **2003**, *13*, 267–271.

Sukenik, A. Ecophysiological Considerations in the Optimization of Eicosapentaenoic Acid Production by *Nannochloropsis* sp (Eustigmatophyceae). *Biores. Technol.* **1991**, *35*, 263–269.

Taguchi, S.; J.A. Hirata; E.A. Laws. Silicate Deficiency and Lipid-Synthesis of Marine Diatoms. *J. Phycol.* **1987**, *23*, 260–267.

Takeyama, H.; D. Takeda; K. Yazawa; A. Yamada; T. Matsunaga. Expression of the Eicosapentaenoic Acid Synthesis Gene Cluster from *Shewanella* sp. in a Transgenic Marine Cyanobacterium, *Synechococcus* sp. *Microbiol. UK* **1997**, *143*, 2725–2731.

Tan, C.K.; M.R. Johns. Screening of Diatoms for Heterotrophic Eicosapentaenoic Acid Production. *J. Appl. Phycol.* **1996**, *8*, 59–64.

Tredici, M.R. Bioreactors, Photo. *Encyclopedia of Bioprocess Technology: Fermentation, Biocatalysis, and Bioseparation*; M.C. Flickinger, S.W. Drew, Eds.; Wiley: New York, 1999; Vol. 1, pp 395–419.

Vazhappilly, R.; F. Chen. Eicosapentaenoic Acid and Docosahexaenoic Acid Production Potential of Microalgae and Their Heterotrophic Growth. *J. Am. Oil Chem. Soc.* **1998a,** *75,* 393–397.

Vazhappilly, R.; F. Chen. Heterotrophic Production Potential of Omega-3 Polyunsaturated Fatty Acids by Microalgae and Algae-like Microorganisms. *Botanica Marina* **1998b,** *41,* 553–558.

Wen, Z.Y.; F. Chen. Heterotrophic Production of Eicosapentaenoid Acid by the Diatom *Nitzschia laevis*: Effects of Silicate and Glucose. *J. Ind. Microbiol. & Biotechnol.* **2000a,** *25,* 218–224.

Wen, Z.Y.; F. Chen. Production Potential of Eicosapentaenoic Acid by the Diatom *Nitzschia laevis. Biotechnol. Lett.* **2000b,** *22,* 727–733.

Wen, Z.Y.; F. Chen. Application of Statistically-Based Experimental Designs for the Optimization of Eicosapentaenoic Acid Production by the Diatom *Nitzschia laevis. Biotechnol. Bioeng* **2001a,** *75,* 159–169.

Wen, Z.Y.; F. Chen. Optimization of Nitrogen Sources for Heterotrophic Production of Eicosapentaenoic Acid by the Diatom *Nitzschia laevis. Enzyme Micr. Technol.* **2001b,** *29,* 341–347.

Wen, Z.Y.; F. Chen. A Perfusion-Cell Bleeding Culture Strategy for Enhancing the Productivity of Eicosapentaenoic Acid by *Nitzschia laevis. Appl. Microbiol. Biotechnol.* **2001c,** *57,* 316–322.

Wen, Z.Y.; F. Chen. Continuous Cultivation of the Diatom *Nitzschia laevis* for Eicosapentaenoic Acid Production: Physiological Study and Process Optimization. *Biotechnol. Progress* **2002a,** *18,* 21–28.

Wen, Z.Y.; F. Chen. Perfusion Culture of the Diatom *Nitzschia laevis* for Ultra-high Yield of Eicosapentaenoic Acid. *Process Biochem.* **2002b,** *38,* 523–529.

Wen, Z.Y.; F. Chen. Heterotrophic Production of Eicosapentaenoic Acid by Microalgae. *Biotechnol. Adv.* **2003,** *21,* 273–294.

Wen, Z.Y.; Y. Jiang; F. Chen. High Cell Density Culture of the Diatom *Nitzschia laevis* for Eicosapentaenoic Acid Production: Fed-Batch Development. *Process Biochem.* **2002,** *37,* 1447–1453.

Werner, D. Silicate Metabolism. *The Biology of Diatom*; D. Werner, Ed.; Blackwell Scientific: Oxford, 1977; pp 111–149.

Xu, X.Q.; J. Beardall. Effect of Salinity on Fatty Acid Composition of a Green Microalga from an Antarctic Hypersaline Lake. *Phytochem.* **1997,** *45,* 655–658.

Yano, Y.; A. Nakayama; K. Yoshida. Distribution of Polyunsaturated Fatty Acids in Bacteria Present in Intestines of Deep-sea Fish and Shallow-sea Poikilothermic Animals. *Appl. Env. Microbiol.* **1997,** *63,* 2572–2577.

Yazawa, K. Production of Eicosapentaenoic Acid from Marine Bacteria. *Lipids* **1996,** *31,* S297–S300.

Yazawa, K.; K. Araki; N. Okazaki; K. Watanabe; C. Ishikawa; A. Inoue; N. Numao; K. Kondo. Production of Eicosapentaenoic Acid by Marine-Bacteria. *J. Biochem.* **1988,** *103,* 5–7.

Yongmanitchai, W.C.; O.P. Ward. Omega-3 Fatty-Acids—Alternative Sources of Production. *Process Biochem.* **1989,** *24,* 117–125.

Yongmanitchai, W.; O.P. Ward. Growth of and Omega-3-Fatty-Acid Production by *Phaeodactylum tricornutum* under Different Culture Conditions. *Appl. Env. Microbiol.* **1991,** *57,* 419–425.

Yongmanitchai, W.; O.P. Ward. Growth and Eicosapentaenoic Acid Production by *Phaeodactylum tricornutum* in Batch and Continuous Culture Systems. *J. Am. Oil Chem. Soc.* **1992,** *69,* 584–590.

Yu, R.; A. Yamada; K. Watanabe; K. Yazawa; H. Takeyama; T. Matsunaga; R. Kurane. Production of Eicosapentaenoic Acid by a Recombinant Marine Cyanobacterium, *Synechococcus* sp. *Lipids* **2000,** *35,* 1061–1064.

Zaslavskaia, L.A.; J.C. Lippmeier; C. Shih; D. Ehrhardt; A.R. Grossman; K.E. Apt. Trophic Obligate Conversion of an Photoautotrophic Organism through Metabolic Engineering. *Science* **2001,** *292,* 2073–2075.

Zittelli, G.C.; F. Lavista; A. Bastianini; L. Rodolfi; M. Vincenzini; M.R. Tredici. Production of Eicosapentaenoic Acid by *Nannochloropsis* sp. Cultures in Outdoor Tubular Photobioreactors. *J. Biotechnol.* **1999,** *70,* 299–312.

Zittelli, G.C.; R. Pastorelli; M.R. Tredici. A Modular Flat Panel Photobioreactor (MFPP) for Indoor Mass Cultivation of *Nannochloropsis* sp. under Artificial Illumination. *J. Appl. Phycol.* **2000,** *12,* 521–526.

Downstream Processing, Extraction, and Purification of Single Cell Oils

Colin Ratledge,[a] Hugo Streekstra,[b] Zvi Cohen,[c] and Jaouad Fichtali[d]

[a] Department of Biological Sciences, University of Hull, Hull, HU6 7RX, UK; [b] DSM Food Specialties, PO Box 1, 2600 MA Delft, The Netherlands; [c] The Microalgal Biotechnology Laboratory, Albert Katz Department for Drylands Biotechnologies, Jacob Blaustein Institute for Desert Research, Ben Gurion University of the Negev, Sde-Boker Campus 84990, Israel; [d] Martek Biosciences Corp., 555 Rolling Hills Lane, Winchester, Kentucky, KY 40391, USA.

General Considerations for SCO Extraction

In all cases, whether we are dealing with yeasts, molds (fungi), or algae, there is the common problem of biomass recovery from the fermenter and the subsequent extraction of the oil. Given that all of the single cell oils (SCOs) that will be discussed in this chapter are ones rich in a particular polyunsaturated fatty acid (PUFA), then there is an obvious need to avoid sustained high temperatures or conditions that could lead to the oxidation and rancidity of the oils during their extraction. Also, as SCOs are of very recent origin, their extraction initially caused some concern that novel processes would have to be developed that might add significantly to the overall manufacturing costs. In reality, however, extraction of SCOs proved relatively simple, and no new extraction machinery had to be built to accommodate the microscopic cells of microorganisms, nor was there a need to develop new solvent extraction methods. In practice, individuals skilled in the art of oil extraction found that extracting oil from microorganisms was no more difficult than extracting oil from other sources of valuable oils.

In a typical SCO process (Ratledge, 2004) using a nonphotosynthetic organism, the average fermenter would have a capacity of about 100,000 liters (100 m^3), which could be expected to yield between 50 and 100 kg biomass per m^3. Thus, at the end of a fermentation run, which might typically last for 4 to 6 days, 100,000 liters of culture broth must be processed; in this broth, the biomass must be separated from the liquor, sometimes by centrifugation, though more often by rotary vacuum filtration or by direct filtration. The spent liquor is of no value, though it will probably require some processing before it can be safely discharged into a river or water system and satisfy the local environmental regulations. The still-wet biomass—which contains the oil, but which, at this stage, may typically still contain 80% water—may, therefore, be of the order of 25 tons (~25 m^3), 5 tons of dry cells with 20 tons water. This biomass has to be quickly dried to a point where the oil is stabilized. The biomass, unless it has been pre-treated in some way, contains many enzymes that continue to be active, and some, such as lipases and esterases, are activated by the very process of removing the cells from their source of nutrients. Under certain conditions, therefore, the cells may start to consume the very oils that they have accumulated within themselves.

Once the biomass has been stabilized and the enzymes inactivated, usually by heating, water can be removed by further pressure filtration; this process may be followed, if necessary, by drying with a spray drier. The final preparation of dry cells in a powder form is then available for oil extraction. However, because all the enzymes that could degrade the oil have been heat-inactivated, the cells can be stored for some weeks before they need be extracted.

Solvent extraction, using hexane, is the preferred means of obtaining the oil. Extraction equipment employed by the industry to obtain oils from small amounts of plant material (e.g., some exotic oils and the essential oils) can be used without adaptation, as the microbial biomass at this stage approximates to dried and powdered plant material. The large-scale equipment used by oil extractors for the recovery of oils from commodity plant sources (e.g., soybeans, sunflower seeds, rapeseed, corn, etc.) are too large to be used for the extraction of relatively small quantities of microbial biomass. However, smaller-scale equipment that can handle 10–50 tons of material per day is appropriate. Operators of such equipment have not found any

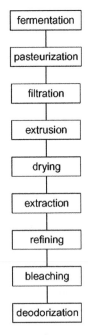

Fig. 9.1. Overview of industrial process operations for extraction of SCOs.

intrinsic difficulties in handling microbial biomass for oil recovery. Further refinement of the oil (see Fig. 9.1), to clean it and make it free of phospholipids and various nonsaponifiable materials, is carried out to produce the final, usually clear, bright oil. These final processes are carried out for all oils destined for direct consumption, and, consequently, the same procedures, as well as equipment, used for plant oils are also used for SCOs.

One final point should be stressed: many microbial oils contain relatively large quantities of their own antioxidants, so SCOs have been found to be much more stable than corresponding oils from plants or marine animals. This attribute is particularly significant during the initial steps of oil extraction. The standard commercial practice is, though, to add further antioxidants to the oils to satisfy the customer that all the correct measures have been taken to ensure the complete stability of the oil.

Oil Extraction Processes

Extraction of SCO-GLA from Mucor circinelloides

The first commercially viable SCO process was the production of an oil rich in gamma-linolenic acid (GLA) from *Mucor circinelloides* (Ratledge, 2005; Sinden, 1987). This procedure was developed by J. & E. Sturge, Selby, North Yorkshire, UK, whose core business was citric acid production using *Aspergillus niger*. The oil was sold under the name Oil of Javanicus and was available from 1985 to 1990; the process was eventually halted because of the advent of cheaper alternative sources of GLA oils, namely, borage oil (also known as starflower oil) and evening primrose oil.

The GLA-SCO process was run in one of the existing 220 m³ fermenters, which were normally used for citric acid production (Sinden, 1987), giving a yield of cells with about 50 kg dry biomass/m³ and an oil content of 25% within 72–90 hours of cultivation. The physical removal of biomass from the fermenter and its separation from the culture medium took more than 6 hours.

Examination of the first batch of oil extracted from *M. circinelloides* indicated the presence of free fatty acids at about 3%–5% of the total oil, which then contributed certain undesirable characteristics to the oil. These free fatty acids were quickly realized to be artifacts of the downstream processing system, as they were not present if the oil was quickly extracted from small samples of biomass taken from laboratory level fermenters. Clearly, the microbial cells were metabolically active: at the time of harvesting, all of the glucose in the culture medium had been consumed—indeed, this exhaustion was the desired endpoint of each fermentation run—but the lack of glucose then caused the cells to activate their own lipases and phospholipases to mobilize the oil that they had been accumulating during their time in the fermenter. The cells, after all, were physically starving and considered it time for them to consume their oil reserves to ensure their own continued viability.

The solution to this unwanted lipase activity, which was degrading the oil and resulting in the formation of free fatty acids, was to heat the broth in the fermenter in the final stages of the run. In practice, because fermentation processes are exothermic, the simple expedient was to switch off the cooling system and allow the fermenter to heat up to 55–60 °C, holding it at this temperature for at least 30 minutes prior to commencing the harvest. When this procedure was put into practice, the presence of free fatty acids in the final oil was negligible, and the oil met all subsequent tests for its stability, safety, and suitability for use as a dietary supplement.

The biomass, once heat-stabilized, was de-watered using filtration and drying, and then passed to a company (Bush Boake & Allen, Long Melford, UK) that specialized in the extraction of essential (terpenoid-type) oils from various plant materials. The equipment that was used was unmodified from used daily by the company and had the advantage of being able to handle the relatively small volumes of biomass that were available. Conventional commercial oil extraction from plant seeds uses huge extraction units built to deal with tens, if not hundreds, of tons of material per hour; these units are entirely inappropriate for handling the small amount of fungal biomass available at any one time. The volume of oil that is "held up" in such systems is often of the order of 100 tons. so that 5 tons of GLA oil would be simply "lost" in one of these units—or, at the very least, heavily contaminated with oils that were already in the system.

Over 98% of the oil within the cells of *M. circinelloides* could be extracted directly, using hexane, in one of the speciality extractors. Crude oil, however, required further processing to remove the nonsaponifiable materials—mainly phospholipids, some sterols, and non-lipid components that had co-extracted into the oil itself. Standard procedures of oil refinement, including deodorization, as practiced by the oil industry, were successfully applied to the crude oil extract, thereby generating a bright, pale yellow oil. Again, however, small pilot-scale equipment had to be found that could handle the reduced volumes of oil involved. For this part of the work, pilot-scale equipment, used by Simon Rosedown Ltd. (now DeSmets), Hull, UK for trial runs on small amounts of unusual plant oils, was commissioned. Once more, the oil was purified to a high level using existing equipment, without need for any modification.

For all intents and purposes, oil extraction from dry fungal biomass posed no significant problem, provided it was carried out in equipment of an appropriate size. Once the oil had been extracted, its further refinement and purification followed conventional procedures and, again, posed no problems.

The specifications of the oil are given in Table 9.A. Although an antioxidant was added to the final oil, this supplement was merely precautionary because the oil itself was highly stable to oxidation as, evidently, the natural antioxidants within the fungal cells had been co-extracted with the oil.

The production organism had a long historical association with traditionally fermented foods, such as tempeh and tapé—the organism was originally known as *Mucor javanicus*, which indicates its origins in Java—and this was an important factor in the oil being given GRAS status (Generally Recognised as Safe), although additional feeding trials to animals were, of course, carried out.

The *Mucor* process ceased production in 1990 because of competition with borage oil, which contained slightly higher levels of GLA (22%), but could be produced at

Table 9.A. Specifications of the GLA-SCO from *Mucor circinelloides* as produced by J. & E. Sturge, Selby, North Yorkshire, UK. Trade Name: Oil of Javanicus.

Oil	
Appearance:	Pale yellow, clean and bright
Specific Gravity:	0.92 at 20 °C
Peroxide Value:	3 maximum
Melting Point:	12–14°C
Added Antioxidant:	Vitamin E
Free Fatty Acids:	<< 1%
Triacylglycerol Content of Oil:	>97%
Fatty Acyl Composition (Rel. % w/w) of Oil	
14:0	1–1.5%
16:0	22–25%
16:1(n-9)	0.5–1.5%
18:0	5–8%
18:1(n-9)	38–41%
18:2(n-6)	10–12%
18:3(n-6)	15–19%
18:3(n-3)	0.2%

a cheaper (though subsidized) price. Over the 6 years of production, about 50–60 tons of GLA-SCO were produced. The oil had excellent long-term stability and, even without the addition of antioxidants, showed little or no deterioration in its GLA content over at least 10 years of storage at room temperature, in air, and in sunlight.

Although the commercial viability of this oil was short-lived, it nevertheless demonstrated for the first time that SCO production was achievable and that the oil itself was equal or better than the best plant oils in its safety and lack of toxicity. A more complete account of this process has been given elsewhere (Ratledge, 2006).

Extraction of SCO-ARA from Mortierella alpina

The application of ARA-rich oil for infant nutrition has led to the development of several commercial production processes over the past 15 years, all using *Mortierella alpina* (Higashiyama et al., 2002; Kyle, 1997; see also Chapter 5). *M. alpina* was selected because of its high level of ARA, which can exceed 50% of total fatty acids. Moreover, it is an oleaginous fungus that is able to accumulate high levels of triacylglycerol (TAG) lipids, and it is considered safe (Streekstra, 1997). ARA is found both in the polar lipids and in the TAGs. The TAG fraction is currently used as the commercial product.

For a number of years, this oil has been produced in full-scale fermentation and downstream processing facilities. Companies active in commercial production include Suntory (Japan), Martek (USA), and DSM (Delft, The Netherlands), and relevant literature also suggests developments in China (Wuhan Alking Bioengineering Co. Ltd.) (Yuan et al., 2002) and, possibly, also in South Korea. The DSM process, which is outlined here, produces an oil that is incorporated into infant formula, in conjunction with a compatible source of DHA. This incorporation has now been approved by many regulatory authorities, including the FDA (Food and Drug Administration) of the United States (see also Chapter 15).

For details concerning the properties of the production organism, strain selection, and the fatty acid composition of the oils, the reader is referred to Chapter 5.

Process Design

A schematic overview of the steps of this process, which are common to most oleaginous microorganisms used for SCO production, is given in Fig. 9.1.

Fermentation

Starting from a working cell bank (either spores or vegetative mycelia), shake-flasks and inoculum fermenters are used for the initial phases of biomass production. Lipid production is not yet a goal in these stages; rather, the goal is the generation of sufficient biomass with the desired macroscopic morphology (Park et al., 2002).

As for the main fermentation, TAGs—which include ARA—are, generally, best produced under nitrogen limitation (Ratledge, 2001, 2004). Specific conditions for the fermentation process are discussed in Chapter 5. In summary, efficient production conditions can be maintained by providing a limited feed of the N source and a non-limiting, but controlled, feed of the C source.

It is usually advisable to let the C source concentration drop to a low level by the end of the fermentation, since it is an expensive raw material. Any excess would be wasted during downstream processing and may even have undesirable effects, such as color formation, as well as increasing the BOD of the final effluent leaving the production plant.

The organism is capable of consuming the lipid that it has laid down during the production phase. This phenomenon has sometimes been exploited because it is associated with an increase in the ARA percentage of the lipid fraction (Zhu et al., 2002), though, of course, the amount of lipid is now diminished.

Biomass Stabilization

Self-consumption of the lipid, however, becomes a problem when it occurs after fermentation has been stopped, potentially affecting the quantity and quality of the product. Two relevant metabolic processes may occur here: (1) lipase activity, which lowers the TAG content of the oil, and (2) the subsequent catabolism of the fatty acids. To avoid the occurrence of these processes, the biomass should be inactivated as quickly as possible once the production phase has ended. To investigate this, the wet biomass from a pilot plant fermentation was stored for 24 hours at either -50, 4, 25 or 63 °C. The samples stored at 4 and 25 °C were tested with or without prior

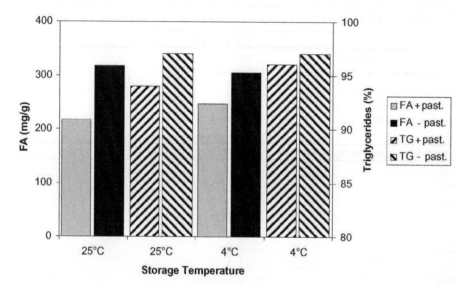

Fig. 9.2. Pasteurization of wet biomass of *Mort. alpina* for oil stabilization. **FA:** Total fatty acid content of dry biomass (mg/g)—solid bars. **TG:** Triacylglycerols in hexane extract (% w/w)—striped bars.

pasteurization at 63 °C. No changes were observed at -50 °C and at 63 °C, showing that biomass is stable at temperatures that are too low or too high for metabolic activity. The data for the samples kept at 4 and 25 °C are shown in Fig. 9.2. It appears that prior pasteurization at 63 °C was required and sufficient to stabilize the oil and total fatty acid levels of the biomass when kept at these intermediate temperatures.

In the DSM process, the biomass is dried by extrusion drying. This action yields low-dusting granules with a narrow size distribution, which are quite suitable for subsequent extraction. The dry, inactivated biomass is a rather stable production intermediate, but cool storage under N_2 is necessary for optimal quality, because the biomass is susceptible to oxidative damage. When kept for some weeks in air at room temperature, the levels of unsaturated fatty acids decrease. This finding also applies to samples stored for analysis.

Extraction, Refining, and Final Product

The dry biomass particles are extracted with hexane (Kyle, 1997; Zhu et al., 2002), which produces a crude oil that is quite stable. At this stage, many hexane-soluble cell components are still present, and it is commonly held that they include endogenous antioxidants. These components are greatly reduced in the subsequent steps of refining, bleaching, and deodorization. Therefore, the refined oil is protected by an antioxidant system, usually composed of mixed natural tocopherols and ascorbyl palmitate. These steps are standard for producing high-quality edible oils; the relatively small

Table 9.B. Specifications of the Commercial ARA-SCO Obtained from *Mortierella alpina* by the DSM Process.

General	Composition
Arachidonic Acid (g/kg oil)	min. 350
Arachidonic Acid (%)	min. 38
Peroxide Value (meq/kg oil)	< 5
Unsaponifiable Matter (% m/m)	max. 3
Free Fatty Acids (%)	< 0.4
Appearance	clear yellow liquid
Fatty Acid Composition (Ranges of Data Given as Rel. % of Total Fatty Acids)	
14:0	1–3
16:0	12–18
18:0	10–14
18:1	10–14
18:2	5–8
18:3	2–5
20:0	0.5–2
20:3	2–5
20:4	35–43
22:0	0.2–4
24:0	1–4
Stabilization	
Mixed Natural Tocopherols	250–500 ppm
Ascorbyl Palmitate	250 ppm

production scale of ARA is the only aspect of this process that differs, along with the oxidative susceptibility of the LC-PUFAs in this specialty oil.

The commercial product is a clear, yellow TAG oil with specified limits for unsaponifiables and free fatty acids (see Table 9.B). The final ARA content is adjusted to about 40% of total fatty acids by adding vegetable oil. This oil is used in conjunction with a compatible source of DHA for providing LC-PUFA in infant formulas.

Extraction of DHA-rich Oils from *Crypthecodinium cohnii* and *Schizochytrium* sp.

Fermentation and Harvesting

Crypthecodinium cohnii and *Schizochytrium* sp. are two microorganisms used for the commercial production of PUFAs by Martek Biosciences. Details of the two processes are provided, respectively, in Chapters 6 and 3.

Schizochytrium sp. is a heterotrophic microalgae belonging to the order Thraustochytriales within the phylum Heterokonta, which can yield about 40% (w/w) of

Table 9.C. Fatty Acid Composition of Oils Produced by *Cryptheco-dinium cohnii* and *Schizochytrium* sp. Prior to Final Processing and Blending (See Table 9.D for Final Specification of the Former Oil).

	C. cohnii	*Schizochytrium* sp.
10:0	0–0.2	–
12:0	3–5	0–0.5
14:0	14–16	9–15
16:0	10–14	24–28
16:1	2–3	0.2–0.5
18:0	0–0.3	0.5–0.7
18:1	9–10	–
18:3(n-3)	–	–
20:3(n-6)	–	0–0.5
20:4(n-3)	–	0.5–1
22:5(n-6)	–	11–14
22:6(n-3)	50–60	35–40

DHA from its total fatty acid production. *Crypthecodinium cohnii* is a unique, heterotrophic, marine dinoflagellate in that DHA is almost exclusively the only PUFA present in its lipid and can be as high as 65% of the total fatty acids (Kyle, 1996). Table 9.C shows a typical fatty acid composition of both microorganisms as produced by fermentation (i.e., prior to processing and blending).

DHA is contained entirely within the cells and is distributed in both structural lipids (e.g., phospholipids) and storage lipids. The latter consists of TAGs, is heat labile, and, consequently, requires due diligence during the downstream harvesting of cells and extraction of oil. The viscosity of the broth may increase drastically due to the production of extracellular polysaccharides (De Swaaf et al., 2001). In addition, the pH tends to change rapidly during the final holding time. All of these factors tend to increase the difficulties of product recovery. To ensure good recovery of the oil at optimum quality, speed of operation is an overriding factor, because the product is sensitive to contamination, cell lysis, and oxidation of the PUFAs. The processing equipment must, therefore, be of the correct type and size to ensure that the harvest broth can be processed within a satisfactory time limit. Nevertheless, this time limit may be extended by stabilizing the broth, using heat or preservatives. Generally, the broth should be concentrated using centrifugation or ultrafiltration in order to make it compatible with the subsequent drying operation and reduce the drying energy cost.

Drying is required to produce a stable form of biomass that can be stored for an extended time without any microbial, chemical, or sensory deterioration. Biomass normally has a limit to the time and temperature to which it can be exposed without inviting unacceptable decomposition of the oil. The choice is, thus, between long drying times at lower temperatures or a brief exposure to more severe conditions. The normal choice is to use spray or flash dryers with short exposure times. Spray dryers can handle large volumes and are suited to the size of the fermentors used for microalgae. It is important to minimize cell lysis prior to drying in order to preserve the quality of the oil and minimize its stickiness to the dryer. Drying conditions should be tailored for a heat-sensitive material, ensuring minimum loss of product, potency, nutritional value, or flavor deterioration. This should take into consideration the concentration of solids in the broth being harvested, temperature, air humidity, final moisture, nozzle design, and feed pressure. Both dewatering and drying can have a significant impact on extraction performance and product quality. Target moisture for dried biomass is usually between 4 and 6%.

It must be remembered that fermentation and product recovery are integral parts of the overall process. Because of the interaction between these two stages, neither one should be developed or modified independently, as this act might result in problems that cause unnecessary expenses.

Pre-Treatment and Cell Disruption

Both microorganisms under consideration in this section are protected by extremely tough cell walls. In order to release the contents of these cells, a number of methods for disintegration have been developed. These methods fall into three major categories, including chemical, biological, and physical. Some methods have severe limitations with regard to cost, large-scale application, or compatibility with the product.

Knowledge of cell wall structure and composition is, therefore, important for optimizing chemical methods and achieving cell lysis without damage to the DHA oils. For mechanical methods, the size, shape and degree of cross-linking found in structural polymers are important factors to consider when determining the ease of disruption. Nevertheless, mechanical methods—especially wet milling in high-speed agitator bead mills and high pressure homogenizers—have performed well for the large-scale disruption of microorganisms with tough cell walls, including microalgae. It is desirable to achieve cell disruption that is as complete as possible through the optimization of processing variables, including flow-rate, pressure, temperature, and disruption chamber design and operation. Some of the variables involved in cell disruption have been reviewed (Sobus & Holmlund, 1976). The disintegration process, therefore, will strongly influence the solid-liquid separation in downstream processing and overall extraction yield.

The ease of cell disruption is also related to fermentation growth conditions. Fast growth rates, in general, produce cells with weaker cell walls, because they do not have time to produce materials for reinforcing the walls. *Schizochytrium* sp., a faster-growing algae than *Crypthecodinium cohnii*, possesses an intrinsically weaker cell wall, and, as a consequence, the energy required for its disruption is significantly lower.

The findings from these two marine organisms are in keeping with work done with other microorganisms. For example, yeast cells that are in the exponential phase of growth are more susceptible to disruption than cells in the stationary phase (Sobus & Holmlund, 1976); fast-growing bacteria possess weaker cell walls than ones that grow more slowly and are easier to disintegrate by impingement (Kula & Schütt, 1987). During starvation or limited growth conditions—used to promote lipid accumulation in oleaginous microorganisms—physiological signals may trigger the cells to reinforce the cell wall to prepare for survival (Engler & Robinson, 1981). Similar phenomena have also been observed with microalgae. The mechanical stability of microalgae is, therefore, not a constant but depends on the strain being used, growth conditions, and, importantly, the history of the biomass. In conclusion, biology and upstream variables have a major impact on downstream processing cost, speed, and efficiency.

Pre-treatment of broth or dried biomass is not necessary, but could be useful in special situations, such as long-term storage or shipping for further processing. Although the main objective for pre-treatment is usually stabilization, pre-treatment methods and conditions could also improve extractability by weakening the cell wall, for instance. The most common pre-treatment is subjecting the broth to heat at pasteurization conditions. Great care should be taken, however, in order to minimize the potential degradation of PUFAs or other chemical reactions that would impart an off-flavor and make downstream processing more problematic.

Extraction and Refining

Historically, the three most common processes for recovering oil from plant seeds are hydraulic pressing, expeller pressing, and solvent extraction. Solvent extraction originated as a batch process in Europe in 1870. The modern solvent-based process usually consists of the extraction, by successive counter-current washes with hexane, of previously cracked, flaked, ground, or pressed oleaginous material. The extracted meal is conveyed to a solvent recovery system, usually a desolventizer or toaster. The hexane is removed from the oil using evaporators and reused.

The same process could be applied to SCO microorganisms, with some modifications (e.g., using hexane as a solvent). The main differences are in the pre-treatment of the cells and the disruption method being used, as described above.

Cell disruption generates rather a wide distribution, according to particle size, of cell debris that needs to be removed. Adjustments in equipment design and operating conditions are necessary in order to process streams that are different in physical and chemical characteristics than those generated by oilseeds.

Fig. 9.3 illustrates the main unit operations used in extracting SCOs. The many advantages of this approach include cost, efficiency, and quality.

Once the oil has been obtained, a micella winterization process is required to remove high melting point triacyglycerols (i.e., those with the greatest saturated fatty acyl residue) and other impurities. This step is necessary when a clear oil at room temperature is required. It is also preferable, if the oil is being used in capsules, for it to remain clear when stored in the refrigerator; a cloudy oil can be seen in the capsules and may be off-putting to a customer who does not understand the physical chemistry of oils.

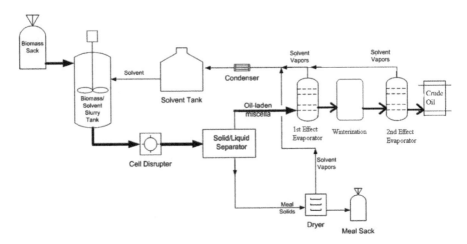

Fig. 9.3. Extraction flow diagram for recovery of SCO from *Crypthecodinium cohnii*.

Lipid extraction from intact cells, using solvents, has had limited success and is generally limited to laboratory practice. Different solvents were tested for their ability to extract lipids and other products from intact microorganisms. For instance, methanol/benzene mixtures have been used to extract lipids from yeast (Save et al, 1997), but such procedures have very limited application for the large-scale recovery of SCOs. Supercritical extraction is another alternative but needs further developments to make it attractive in terms of processing cost and extraction yield.

DHA crude oil is unfit for consumption because of impurities, odor, and taste, as well as a lack of clarity. It, therefore, needs to be refined, which can be achieved using standard vegetable oil refining steps, including degumming, caustic refining, bleaching, and deodorization. Impurities and minor components that are removed or reduced in the refining step include free fatty acids, water, phospholipids, minerals, carotenoids, sterols, tocopherols/tocotrienols, waxes, and residual cell debris. As the oil is sensitive to oxidation (up to 60% DHA prior to blending—see Table 9.C), process conditions and speed of operation are more critical than when vegetable oils are processed. These conditions have been optimized at Martek Biosciences Corp. for odor, taste, and oxidative stability. The deodorized oil is blended with high oleic sunflower oil to standardize the DHA concentration and then is stabilized by adding antioxidants, mainly, ascorbyl palmitate and tocopherols.

The cost of extracting and refining SCOs is, however, much higher than for vegetable oils, which benefit from economy of scale and decades of optimizing and fine-tuning the operation. In addition, SCOs require a higher degree of good manufacturing practice (GMP) (closer to pharmaceutical standards than to those used by the food industry) and a very stringent control of oxidation (which limits the performance of equipment being optimized).

Table 9.D. Typical Analysis of DHASCO™ Produced by
Crypthecodinium cohnii.

Analysis	Typical Result
Free Fatty Acids (%)	0.03–0.1
Peroxide Value (meq/kg)	0–0.5
Anisidine Value	2–8
Unsaponifiable Matters (%)	1–2
Moisture and Volatiles (%)	0–0.02
Insoluble Impurities	Below detection
Trans Fatty Acids	Below detection
Heavy Metals	Below detection
Tocotrienols (ppm)	400–500
Fatty Acid Composition (Rel. % w/w)	
10:0	0–0.5
12:0	2–5
14:0	10–15
16:0	10–14
16:1	1–3
18:0	0–2
18:1	10–30
22:6(n-3)	40–45

Quality Aspects

Unlike crude oils from oilseeds and fish, the crude oils from *Crypthecodinium cohnii* and *Schizochytrium* sp., produced by fermentation, are free from pesticides, aflatoxins, organophospho-insecticides, organochlorio-insecticides, heavy metals, and other pollutants often found in fish oils, such as polychlorinated biphenyls (Shim et al., 2003). This simplifies the final refining procedures aimed at removing impurities that have been co-extracted from the cells, along with the TAGs, and, therefore, does not compromise the quality of the final oil.

Typical analysis of refined, bleached, and deodorized oil, after blending these oils with high oleic sunflower oil, is given in Table 9.D.

Rigorous quality control is implemented at Martek Biosciences Corp. to ensure consistency and quality. At each step of the operation, exposure to heat, air, light, and heavy metals is minimized. State-of-the-art analytical work and a trained sensory panel are used to improve and maintain the highest quality standards for food and therapeutic applications.

The SCOs from microalgae possesses a remarkable oxidative and flavor stability. Fig. 9.4 illustrates a typical shelf life stability of the *Schizochytrium* sp. oil under frozen conditions where only a minor change was detected over 2 years.

Further information regarding the safety aspects of these and other SCOs is covered in Chapter 15.

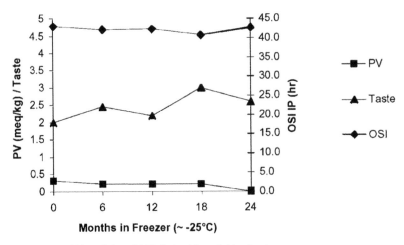

Fig. 9.4. Typical shelf life stability of SCO derived from *Schizochytrium* sp.

2.3 Extraction of Oils from Microalgae

General Considerations

Phototrophic algae have been studied, with respect to their potential to produce several very long-chain polyunsaturated fatty acids (VLC-PUFAs), predominantly EPA and ARA. The major producers of EPA are the diatom *Phaeodactylum tricornutum* (Yongmanitchai & Ward, 1992; Molina Grima et al., 1999), the red alga *Porphyridium cruentum* (Cohen et al., 1988), and the eustigmatophytes *Nannochloropsis* sp. (Seto et al., 1984; Sukenik, 1999) and *Monodus subterraneus* (Cohen, 1994). The red alga mentioned here was, until recently, the only algal source of ARA (Cohen, 1990).

The major impedance preventing the use of microalgae as a source of VLC-PUFAs is the relatively high cost of producing biomass from these microorganisms. Subsequently, little effort has been dedicated to the development of downstream processes for algal SCOs in comparison to alternative sources. Novel panel photobioreactors, currently being tested, are expected to significantly reduce the cost of production.

Under stress conditions (e.g., nitrogen starvation), many microalgae can be induced to accumulate large amounts of oil. However, the accumulated TAGs are mostly constructed of saturated and monounsaturated, with few, if any, PUFAs. When present, PUFAs are predominantly located in the polar membranal lipids (Cohen, 1999). Unfortunately, the content of membrane lipids and, consequently, their fatty acid components, are inherently limited. One of the limitations of fish oils as a source of single PUFAs is the co-occurrence of several PUFAs in the oil, requiring expensive HPLC separations. The same restriction applies to algal oils. Preparative HPLC, however, is the single most expensive component in producing the final product, affecting the cost of the purified product much more than the cost of producing the oil extract itself (Molina Grima et al., 1996).

The differentiation of different lipid classes from each other is, however, much simpler. Thus, the search for promising, PUFA-rich algae should take into consideration not only the absolute content of the PUFA of interest, but also whether or not it is concentrated in specific lipid classes and to what extent other PUFAs are present in these lipids. One possible approach is to search for algae with PUFA-rich TAGs. Indeed, Cohen et al. (2000) hypothesized that some algae, whose habitat is characterized by rapidly changing environmental conditions, can swiftly adapt by mobilizing LC-PUFAs from their TAG to chloroplastic lipids. Based on this hypothesis, they isolated a microalga, identified as the chlorophyte *Parietochloris incisa* (Bigogno et al., 2002). This alga was found to be the richest plant source of ARA. While the alga can withstand very low temperatures, its optimal growth temperature is 25 °C. Under nitrogen starvation, the proportion of AA was close to 60% of total fatty acids and the ARA content was over 20% of dry weight, over 90% of which was deposited in TAG (Khozin-Goldberg et al., 2002). However, the downstream processing of AA from this alga has not yet been studied.

Extraction

Molina Grima et al (1999) recently studied the recovery and separation of algal PUFAs in depth. This section will, thus, only briefly relate aspects covered elsewhere in this review. Direct extraction of wet biomass *P. tricornutum* with 96% ethanol produced almost as many lipids (90%) as those from freeze-dried biomass (96%) (Molina Grima et al., 1996). However, the cell walls in many species that contain potentially valuable oils are quite impermeable, requiring a cell disruption step (e.g., *P. incisa* and the astaxanthin-rich *Haematococcus pluvialis*). Bead mills have been successfully used to disrupt cells of *Scenedesmus obliquus* (Hedenskok & Ebbinghaus, 1972). The classical extraction method for plant lipids utilizes mixtures of chloroform, methanol, and water. However, these solvents are too toxic to be used for nutritional or pharmaceutical purposes. In most algal species, EPA is found in polar lipids, mainly, galactolipids. Unfortunately, the solubility of these lipid classes in most relevant solvents is not very high, resulting in high extraction volumes and incomplete recovery. However, since the final purification of the fatty acids requires their release from the host lipid, a simultaneous saponification/extraction process was shown to be advantageous, with respect to recovery. Biocompatible systems, such as ethanol (96%) and hexane/ethanol (2:5 v/v), have been successfully used for the extraction of fatty acids from a lyophilized biomass of *P. tricornutum* following saponification (Cartens et al., 1996). Similar purities were obtained using this process for the separation of EPA from *Isochrysis galbana* (Robles Modina, 1995) and *P. cruentum* (Giménez Giménez et al., 1998). The yields, however, were significantly lower, reaching 43% and 25%, respectively.

Purification

Reverse-phase chromatography separates fatty acids according to their chain length and the degree of unsaturation (Gunstone et al., 1984). However, this technique could be prohibitively expensive on a large scale. Preliminary separation of the PUFA-rich galactolipids fraction, followed by urea treatment, can reduce cost by increasing output and, subsequently, reducing the cost per run. Cohen and Cohen (1991) demonstrated the potential of this method for the purification of EPA from the galactolipids of *Porphyridium* by successive elution of the transmethylated oil with acetonitrile-water mixtures on C_{18} Sep-Pak filters. More than 80% of the EPA was eluted in three fractions whose EPA content ranged from 85.4% to 93.2%. Yet, when reverse phase chromatography was preceded by urea treatment, 85% of the EPA was recovered at 97% purity. Application of these methods resulted in an EPA concentrate of 97% purity. Similar methods resulted in an ARA concentrate of 80% purity.

Monodus subterraneus is one of the leading algal sources of EPA. Over 70% of cellular EPA is concentrated in the galactolipid monogalactosyldiacylglycerol (MGDG), where it makes up 46% of total fatty acids. As mentioned above, the difference in polarity between lipid classes allows for a much simpler separation that can serve to partially remove some of the other PUFAs. Such separation could be accomplished by washing the lipids over a pad of silica gel. To simulate this method, lipid extracts of *M. subterraneus* were washed over silica gel cartridges with solvents of increasing polarities, and the subsequent fractions were then analyzed (Cohen & Cohen, unpublished data). Transmethylation of the MGDG-rich fraction and urea fractionation of the resulting methyl esters resulted in EPA of 88% purity (Table 9.E). Additional improvements can be expected by incorporating several genetic modifications; for example, the galactolipids of *M. subterraneus* are made of two types of molecular species: eukaryotic-like lipids, with mostly 20:5 at both the *sn*-1 and *sn*-2 positions, and prokaryotic-like lipids, containing shorter (C_{14}–C_{18}) fatty acids in the *sn*-1 position and 20:4 or 20:5 at the *sn*-2 position. These molecular species originate from two different biosynthetic pathways (Khozin-Goldberg et al., 2002). Any modification that will increase the share of eukaryotic-like molecular species at the expense of the prokaryotic-like species would also increase both the EPA content and its ease of purification.

Table 9.E. Fractionation of the Lipids of *Monodus subterraneus* by Stepwise Elution on Silica Gel Cartridges, Transmethylation, and Urea Fractionation. (The process was studied at a small scale only and was not optimized.)

Lipid Fraction	Fatty Acid Composition (% of Total Fatty Acids)										
	14:0	16:0	16:1	16:3	18:0	18:1	18:2	18:3	20:3 n-6	20:4 n-6	20:5 n-3
Total Lipid Extract[a]	6	14	23	2	1	5	1	1	0.5	6	39
5% MeOH Elution[b]	3	9	24	0.6	4	3	1	1	0.4	5	50
Urea Fractionation[c]			0.5				1	1	0.6	7	90

[a]Lipids were extracted by the Bligh and Dyer method (1959). Total fatty acid and EPA content were and 12 and 4.7% of dry weight, respectively.

[b]Neutral lipids were washed away with chloroform. The galactolipids containing fraction were obtained by washing with 5% (v/v) methanol in water. Washes with a higher concentration of methanol yielded other polar lipids with a lower EPA content.

PUFAs, such as ARA, EPA, and DHA, are currently in demand for the treatment of various diseases and their symptoms. However, the inclusion of a PUFA as a drug component would require its purification to over 95%. Different sources would, thus, compete, not only in productivity, concentration, and cost of production, but also by the ease or difficulty of separating the PUFA of interest from other, similar fatty acids. When considering a source of PUFA, one should also consider in which lipids the PUFA is concentrated and what its relative distribution is, with respect to other PUFAs.

References

Bigogno, C.; I. Khozin-Goldberg; S. Boussiba; A. Vonshak; Z. Cohen. Lipid and Fatty Acid Composition of the Green Alga *Parietochloris incisa*. *Phytochemistry* **2002,** *60,* 497–503.

Bligh, E.G.; W.J. Dyer. A Rapid Method for Total Lipid Extraction and Purification. *Can. J. Biochem. Physiol.* **1959,** *37,* 911–917.

Cartens, M.; E. Molina Grima; A. Robles Medina; A. Giménez Giménez; M.J. Ibáñez González. Eicosapentaenoic Acid (20:5n-3) from the Marine Microalga *Phaeodactylum tricornutum*. *J. Am. Oil Chem. Soc.* **1996,** *73,* 1025–1031.

Cohen, Z. The Production Potential of Eicosapentaenoic Acid and Arachidonic Acid of the Red Alga *Porphyridium cruentum*. *J. Am. Oil Chem. Soc.* **1990,** *67,* 916–920.

Cohen, Z. Production Potential of Eicosapentaenoic Acid by *Monodus subterraneus*. *J. Am. Oil Chem. Soc.* **1994,** *71,* 941–945.

Cohen, Z. Production of Polyunsaturated Fatty Acids by the Microalga *Porphyridium cruentum*. *Chemicals from Microalgae*; Z. Cohen, Ed.; Taylor and Francis: London, 1999; pp 1–24.

Cohen, Z.; S. Cohen. Preparation of Eicosapentaenoic Acid (EPA) Concentrate from *Porphyridium cruentum*. *J. Am. Oil Chem. Soc.* **1991,** *68,* 16–19.

Cohen, Z.; A. Vonshak; A. Richmond. Effect of Environmental Conditions of Fatty Acid Composition of the Red Alga *Porphyridium cruentum*: Correlation to Growth Rate. *J. Phycol.* **1988,** *24,* 328–332.

Cohen, Z.; I. Khozin-Goldberg; D. Adlrestein; C. Bigogno. The Role of Triacylglycerols as a Reservoir of Polyunsaturated Fatty Acids for the Rapid Production of Chloroplastic Lipids in Certain Microalgae. *Biochem. Soc. Trans.* **2000,** *28,* 740–743.

De Swaaf M.E.; G. Grobben; G. Eggink; T.C. de Rijk; P. van de Meer; L. Sijtsma. Characterization of Extracellular Polysaccharides Produced by *Crypthecodinium cohnii*. *Appl. Microbiol. Biotechnol.* **2001,** *57,* 395–400.

Engler, C.R.; C.W. Robinson. Effects of Organism Type and Growth Conditions on Cell Disruption by Impingement. *Biotechnol. Lett.* **1981,** *3,* 83–88.

Giménez Giménez, A.; M.J. Ibáñez González; A. Robles Medina; E. Molina Grima; S. García Salas; L. Esteban Cerdán. Downstream Processing and Purification of Eicosapentaenoic (20:5n-3) and Arachidonic Acid (20:4n-6) from the Microalga *Porphyridium cruentum*. *Bioseparation* **1998,** *7,* 89–99.

Gunstone, F.D.; E. Bascetta; C.M. Scrimgeour. The Purification of Fatty Acid Methyl Esters by High Pressure Liquid Chromatography. *Lipids* **1984,** *19,* 801–803.

Hedenskog, G.; L. Ebbinghaus. Reduction of the Nucleic Acid Content of Single-Cell Protein Concentrates. *Biotechnol. Bioeng.* **1972,** *14,* 447–457.

Higashiyama, K.; S. Fujikawa; E.Y. Park; S. Shimizu. Production of Arachidonic Acid by *Mortierella* fungi. *Biotechnol. Bioproc. Eng.* **2002,** *7,* 252–262.

Khozin-Goldberg, I.; C. Bigogno; Z. Cohen. Nitrogen Starvation Induced Accumulation of Arachidonic Acid in the Freshwater Green Alga *Parietochloris incisa. J. Phycol.* **2002,** *38,* 991–994.

Khozin-Goldberg, I.; S. Didi-Cohen; Z. Cohen. Biosynthesis of Eicosapentaenoic Acid (EPA) in the Fresh Water Eustigmatophyte *Monodus subterraneus. J. Phycol.* **2002,** *38,* 745–756.

Kula, M.R.; H. Schütt. Purification of Proteins and the Disruption of Microbial Cells. *Biotechnol. Prog.* **1987,** *3,* 31–42.

Kyle, D.J. Production and Use of a Single Cell Oil Which is Highly Enriched in Docosahexaenoic Acid. *Lipid Technol.* **1996,** *8,* 107–110.

Kyle, D.J. Production and Use of a Single Cell Oil Highly Enriched in Arachidonic Acid. *Lipid Technol.* **1997,** *9,* 116–121.

Molina Grima, E.; A. Robles Medina; A. Giménez Giménez; M.J. Ibáñez González. Gram-Scale Purification of Eicosapentaenoic Acid (EPA, 20:5n-3) from Wet *Phaeodactylum tricornutum* UTEX 640 Biomass. *J. Appl. Phycol.* **1996,** *8,* 359–367.

Molina Grima, E.; F. Garcia Camacho; F.G. Acien Fernandez. Production of EPA from *Phaeodactylum tricornutum. Chemicals from Microalgae*; Z. Cohen, Ed., Taylor and Francis: London, 1999; pp 57–92.

Park, E.Y.; T. Hamanaka; K. Higashayama; S. Fujikawa. Monitoring of Morphological Development of the Arachidonic-Acid-Producing Filamentous Microorganism *Mortierella alpina. Appl. Microbiol. Biotechnol.* **2002,** *59,* 706–712.

Ratledge, C. Microorganisms as Sources of Polyunsaturated Fatty Acids. *Structured and Modified Lipids*; F.D. Gunstone, Ed.; Marcel Dekker: New York, 2001; pp 351–399.

Ratledge, C. Single Cell Oils—A Coming of Age. *Lipid Technol.* **2004,** *16,* 37–41.

Ratledge, C. Microbial Production of γ-Linolenic Acid. *Handbook of Functional Lipids*; C.C. Akoh, Ed.; CRC Press: Boca Raton, FL, 2006; pp 19–45.

Robles Medina, A.; A. Giménez Giménez; F. García Camacho; J.A. Sánchez Pérez; E. Molina Grima; A. Contreras Gómez. Concentration and Purification of Stearidonic, Eicosapentaenoic, and Docosahexaenoic Acids from Cod Liver Oil and the Marine Microalga *Isochrysis galbana. J. Am. Oil Chem. Soc.* **1995,** *72,* 575–583.

Save, S.S.; A.B. Pandit; J.B. Joshi. Use of Hydrodynamic Cavitaion for Large Scale Microbial Cell Disruption. *Trans. I. Chem. Eng.* **1997,** *75,* 41–49.

Seto A.; H.L. Wang; C.W. Hesseltine. Culture Conditions Affect Eicosapentaenoic Acid Content of *Chlorella minutissima. J. Am. Oil Chem. Soc.* **1984,** *61,* 892–894.

Shim, S.M.; C.R. Santerre; J.R. Burgess; D.C. Deardorff. Omega-3 Fatty Acids and Total Polychlorinated Biphenyls in 26 Dietary Supplements. *J. Food Sci.* **2003,** *69,* 2436–2440.

Sinden, K.W. The Production of Lipids by Fermentation within the EEC. *Enzyme Microbial Technol.* **1987,** *9,* 124–125.

Sobus, M.T.; C.E. Holmlund. Extraction of Lipids from Yeast. *Lipids* **1976,** *11,* 341–348.

Streekstra, H. On the Safety of *Mortierella alpina* for the Production of Food Ingredients, Such as Arachidonic Acid. *J. Biotechnol.* **1997,** *56,* 153–165.

Sukenik, A. Production of Eicosapentaenoic Acid by the Marine Eustigmatophyte *Nannochloropsis. Chemicals from Microalgae*; Z. Cohen, Ed.; Taylor and Francis: London, 1999; pp 41–56.

Yongmanitchai, W.; O.P. Ward. Growth and Eicosapentaenoic Acid Production by *Phaeodactylum tricornutum* in Batch and Continuous Culture System. *J. Am. Oil Chem. Soc.* **1992,** *69,* 584–590.

Yuan, C.; J. Wang; Y. Shang; G. Gong; J. Yao; Z. Yu. Production of Arachidonic Acid by *Mortierella alpina* I_{49}–N_{18}. *Food Technol. Biotechnol.* **2002,** *40,* 311–315.

Zhu, M.; P.P. Zhou; L.J. Yu. Extraction of Lipids from *Mortierella alpina* and Enrichment of Arachidonic Acid from the Fungal Lipids. *Biores. Technol.* **2002,** *84,* 93–95.

PART 3

Production of Single Cell Oils Using Photosynthetically Grown Microorganisms

10

Searching for Polyunsaturated Fatty Acid-Rich Photosynthetic Microalgae

Zvi Cohen and Inna Khozin-Goldberg

The Microalgal Biotechnology Laboratory, The Jacob Blaustein Institutes for Desert Research, Ben Gurion University of the Negev, Midreshet Ben Gurion Campus, 84990, Israel

Introduction

Several very long-chain (C_{20}–C_{22}) polyunsaturated fatty acids (VLC-PUFA) are of value for various nutritional and pharmaceutical purposes. Some of these polyunsaturated fatty acids (PUFA) are precursors of different families of prostaglandins and leukotrienes, e.g., arachidonic acid (AA, 20:4n-6) and docosahexaenoic acid (DHA, 22:6n-3)—which are the major PUFA of brain membrane phospholipids, necessary for visual acuity and improved cognitive development of infants (Koletzko et al., 1991; Agostoni et al., 1994). Newborn infants obtain most of these PUFA from breast milk (Hansen et al., 1997), and it was, thus, suggested that the diet of preterm infants who are not breast-fed should be supplemented with AA and DHA (Carlson et al., 1993; Boswell et al., 1991). Indeed, various health authorities now recommend the incorporation of both AA and DHA into baby formulae (Makrides et al., 1995; see also Chapter 16), and the FDA has already approved their combined use. Another PUFA, dihomo-γ-linolenic acid (DGLA, 20:3n-6), displays anti-inflammatory activity and a potential for treating conditions such as atopic eczema, psoriasis, asthma, and arthritis (Fan & Chapkin, 1998).

VLC-PUFA of the n-3 family are abundant in microalgae. For example, *Porphyridium cruentum* (Cohen et al., 1988), *Nannochloropsis* sp. (Seto et al., 1984; Sukenik & Carmeli, 1989), *Phaeodactylum tricornutum* (Yongmanitchai & Ward, 1992; Molina-Grima et al., 1999), and *Monodus subterraneus* (Cohen, 1994) were studied for their potential to produce eicosapentaenoic acid (EPA, 20:5n-3). Likewise, *Crypthecodinium cohnii* (Jiang et al., 1999) and *Chroomonas salina* (Henderson et al., 1992) contain DHA. However, very long chain-n-6 PUFA are relatively rare, and high contents of DGLA are not found in any organism unless it has undergone genetic manipulation (see Chapter 2). AA is almost nonexistent in the lipids of fresh water algae, and, in most marine species, it does not account for more than a few percent of the total fatty acids (TFA) [Table 10.A (Thompson, 1996)].

Table 10.A. Major Fatty Acid Composition of Algae Relatively Rich in Arachidonic Acid.

Species	Major Fatty Acids (% of Total)									
	16:0	16:1	18:1	18:2	18:3 n-6	18:4 n-3	20:4 n-6	20:5 n-3	22:6 n-3	Ref.
Bacillariophyceae										
Thalassiosira pseudonana	10	29		1			14	15		1
Chlorophyceae										
Parietochloris incisa	10	2	16	17	1		43	1		2
Dinophyceae										
Amphidinium carteri	12	1	2	1	3	19	20		24	1
Phaeophyceae										
Desmarestia acculeata	12	2	7	6	10	16	19	19		3
Dictyopteris membranacea	20	1	14	14	11	11	11	9		4
Ectocarpus fasciculatus	17		13	4	15	23	11	13		5
Prasinophyceae										
Ochromonas danica	4		7	26	12	7	8			6
Rhodophyceae										
Gracilaria confervoides	18	3	16	2		1	46			3
Phycodrys sinuosa	22	5	5	1	1		44	2		3
Porphyridium cruentum	34	1	2	12			40	7		7

References: 1. Cobelas & Lechado, 1988; 2. Bigogno et al., 2002b; 3. Pohl & Zurheide, 1979; 4. Hoffman & Eichenberger, 1997; 5. Makewicz et al., 1997; 6. Vogel & Eichenberger, 1992; 7. Cohen, 1990.

Source: Bigogno et al., 2002b.

Occurrence of PUFA-Rich TAG in Microalgae

Many microalgal species were found to contain VLC-PUFA as constituents of their polar lipids; yet, the content of polar lipids is intrinsically limited (Eichenberger & Gribi, 1997; Cohen & Khozin-Goldberg, 2005). Oleaginous algae are able to accumulate neutral lipids, mainly triacylglycerols (TAG), up to 86% of their total cell dry weight in response to environmental stresses, such as nitrogen limitation, salinity, or high temperature (Dubinski et al., 1978; Scragg & Leather, 1988; Roessler, 1990; Cohen, 1999); however, their TAG is composed mainly of saturated and monounsaturated fatty acids (Henderson et al., 1990). In order for microalgae to be used as an economical source of VLC-PUFA, algal strains that can accumulate these PUFA in their TAG must be found.

Cohen (1990) showed that, under nitrogen starvation, *P. cruentum* could be induced to accumulate up to 2.5% AA (dry wt) and 41% of its TFA. The latter had, at the time, the highest content of AA found in any plant source, and it was the only phototrophic microalga to accumulate PUFA in TAG. In contrast, some fungi, especially of the genus *Mortierella*, were shown to accumulate AA up to 60% of their FA (Higashiyama et al., 1998; see also, Chapter 2).

Accumulation of Oil in Microalgae

Oil accumulation in microalgae is a biphasic process. Rapid cell division will continue as long as growth conditions are not limiting. In the lipogenic phase, excess carbon and a growth factor limitation—most often, nitrogen—can result in the accumulation of lipids (Leman, 1997). Imposing nitrogen limitation when light is in excess results in the cessation of growth. Since photosynthetic fixation of carbon continues, the cellular ratio of C/N increases (Mayzaud et al., 1989), and carbon can be channeled into production of non-nitrogenous reserve materials, such as TAG, that serve as a sink for photosynthetically fixed carbon.

According to prevalent interpretations, TAG are accumulated in algae to store carbon and energy to support growth when conditions become favorable (Pohl & Zurheide, 1979; Harwood & Jones, 1989). For this purpose, saturated acyl moieties should be preferred, because they require less energy to produce than PUFA and provide more energy upon oxidation. Energetically, therefore, there is very little sense in producing and accumulating PUFA-rich TAG. Radiolabelling studies of *P. cruentum* have shown that labeled AA, accumulated in TAG, was transferred to the major lipid classes constituting chloroplast membranes, predominantly monogalactosyldiacylglycerol (MGDG) (Khozin et al., 1997; Khozin-Goldberg et al., 2000). Moreover, a mutant of *P. cruentum*, HZ3, which had impaired ability to transfer labels from TAG, showed a decrease in the proportion of 20:5/20:5, the major molecular species of MGDG, and, consequently, a reduced growth at low temperatures (Khozin-Goldberg et al., 2000). (Pairs of numbers that represent the FA separated by a solidus designate the components in the *sn*-1 and *sn*-2 positions, respectively.)

While most algae grow in large bodies of water in which the temperature only changes slowly, *P. cruentum* is found in shallow marshes and wet sands in which the temperature can fluctuate rapidly. The increase in the proportion of EPA in MGDG—especially in the eukaryotic-like component of MGDG (20:5/20:5), observed at low temperatures—may reflect the organism's mode of coping with stress inflicted by sudden temperature changes. Arguably, TAG could be used in a buffering capacity for AA- and EPA-containing DAG and be rapidly mobilized to produce eukaryotic-like (20/20) molecular species of MGDG. The HZ3 mutant is deficient in the eukaryotic pathway, and compensation by enhanced production of prokaryotic-like species is inherently limited.

Algae that reside in ecological niches subject to rapid short-term fluctuations—e.g., temperature, salinity, or nitrogen availability—may require significant and rapid alterations in the fatty acid and molecular species composition of chloroplast membrane lipids. However, under such conditions, the de novo synthesis of PUFA would be too slow. We, thus, hypothesized that, in such algae, TAG may have another role, a buffering capacity, mobilizing PUFA moieties for the construction and rapid adaptation of chloroplastic membranes after sudden changes in environmental conditions (Cohen & Khozin-Goldberg, 2005).

PUFA-Rich TAG Can Be Used as a Reservoir of PUFA for the Modification of Chloroplastic Lipids

In most plants and lower organisms, the degree of unsaturation of membranal FA increases upon temperature reduction (Patterson, 1970). Therefore, adaptation to sudden decreases in temperature must include a mechanism that would enable rapid enhancement of the PUFA content of the membranes. However, the rate of biochemical processes slows down at lower temperatures. Even at room temperature, radiolabelling of *P. cruentum* with acetate has shown that labeled EPA appeared in the chloroplast more than 10 h after the pulse (Khozin et al., 1997), suggesting that algae exposed to rapid changes of temperature would have difficulty increasing their chloroplastic PUFA content at low temperatures by de novo synthesis. The protective role of PUFA against the damaging effect of high intensity light and UV radiation, especially at low temperatures, may also contribute to the ability of the organism to survive and adapt in extreme environments (Whitelam & Codd, 1986). The capacity to store PUFA in TAG would allow the organism to adapt swiftly to the rapidly changing environment. Since high levels of free fatty acids would be toxic to the cell, they must be stored in the form of lipids. Polar lipids are membranal components, and, consequently, accumulation of these lipids is intrinsically limited, which leaves neutral lipids—predominantly, TAG—as the only viable option. We have, therefore, hypothesized that in algae whose natural habitat is characterized by rapid changes in environmental conditions, such as alpine environments, VLC-PUFA-rich TAG might be involved in a buffering capacity for PUFA (Bigogno et al., 2002a).

Isolation and Characterization of *Parietochloris incisa*

In order to test this hypothesis, several algal species were collected from the soil of Mt. Tateyama, Japan (Watanabe et al., 1996). This ecological niche must tolerate a wide ranges of temperatures, from freezing to over 20°C, and light intensity varying from normal to very high, due to snow reflection. Screening of the strains resulted in the identification of the chlorophyte (Trebouxiophyceae) *Parietochloris incisa* as the richest known plant source of AA (Bigogno et al., 2002b). This strain is the first and only reported alga capable of accumulating large quantities of TAG particularly rich in any PUFA. AA is the major FA of *P. incisa*, comprising 34% of the TFA in the logarithmic phase and 43% in the stationary phase (Table 10.B). Other major FA are 16:0, 18:1, and 18:2. Among the minor FA, there are n-6 PUFA, such as 18:3n-6, and 20:3n-6, as well as PUFA of the n-3 family: 16:3n-3, 18:3n-3, and 20:5n-3. Even in logarithmic cultures, TAG was the major lipid class and accounted for 43% of the TFA (Table 10.B). In the stationary phase, its proportion increased even further to 77%. In contrast to most algae, whose TAG are made of mainly saturated and monounsaturated FA, the TAG of *P. incisa* are the major lipid class in which AA is deposited, reaching up to 47% in the stationary phase. Other than AA, TAG also

Table 10.B. Fatty Acids Composition of the Major Lipids of *P. incisa* in the Logarithmic (L) and Stationary (S) Phases.

Lipid Class		Lipid Dist.[a]	Fatty Acid Composition (% of Total Fatty Acids)												
			16:0	16:1 n-11	16:2 n-6	16:3 n-3	18:0	18:1 n-9	18:1 n-7	18:2 n-6	18:3 n-6	18:3 n-3	20:3 n-6	20:4 n-6	20:5 n-3
Biomass	L		13.9	4.7	1.7	4.0	1.7	6.7	5.1	13.2	1.5	10.3	1.2	33.6	1.7
	S		10.1	1.8	1.3	0.9	2.5	12.2	4.2	17.2	0.8	2.0	1.0	42.5	0.7
MGDG[b]	L	22.1	1.9	1.0	8.5	20.5	0.4	3.6	0.8	14.5	0.8	31.9	0.3	13.9	1.1
	S	4.9	3.4	1.2	20.8	11.0	0.5	4.3	0.0	31.4	0.7	18.5	0.2	6.1	0.6
DGDG	L	14.2	16.0	1.1	1.4	1.7	1.9	5.8	4.0	26.0	1.5	18.6	0.7	18.1	1.4
	S	6.8	34.0	0.5	0.2	0.6	4.5	6.9	3.0	31.0	2.4	2.9	0.4	7.6	0.6
PC	L	5.7	29.5	2.7	0.2	0.5	4.2	11.8	2.0	16.0	4.5	2.3	1.9	21.0	1.8
	S	0.7	31.2	1.4	-	-	5.5	8.9	15.4	21.1	2.6	1.4	0.7	9.6	-
DGTS	L	4.0	47.1	1.8	0.3	0.5	7.4	2.0	7.4	9.4	3.6	1.9	0.9	15.2	0.8
	S	4.6	30.0	0.4	2.6	0.6	3.9	5.4	2.8	34.2	3.3	3.5	0.4	10.0	0.6
PE	L	2.9	11.1	1.4	0.2	1.3	3.9	4.2	18.2	4.8	1.2	3.0	4.1	43.2	1.7
	S	0.6	19.8	4.5	0.3	4.8	7.1	4.3	24.7	11.2	1.5	1.8	2.0	14.3	-
TAG	L	42.9	13.3	0.5	tr	0.4	3.7	15.3	6.8	10.4	1.1	1.0	1.5	43.0	1.0
	S	77.1	8.4	0.4	tr	tr	3.1	18.0	4.0	14.1	0.7	0.4	1.1	47.1	0.7

[a]Lipid distribution (% TFA)

[b]Abbreviations: Digalactosyldiacylglycerol, DGDG; digalactosyltrimethylhomoserine, DGTS; monogalactosyldiacylglycerol, MGDG; phosphatidylethanolamine, PE; phosphatidylcholine, PC; triacylglycerol, TAG; undetected, -; trace (<1%), tr.

Source: Bigogno et al., 2002b.

contain 16:0, 18:1, and 18:2. Due to the sharp increase in TAG accumulation in the stationary phase (Table 10.B), the share of cellular AA that was deposited in TAG increased from 60% in the logarithmic phase to 90% in the stationary phase. With the exception of AA, the FA composition of the chloroplastic lipids did not differ greatly from that of typical green algae, such as *Chlorella* (Safford & Nichols, 1970), consisting mostly of C16 and C18 PUFA.

The lipid composition of *P. incisa* is also unusual. The simultaneous presence of diacylglyceryltrimethylhomoserine (DGTS), phosphatidylcholine (PC), and phosphatidylethanolamine (PE) is not very common. DGTS is abundant in many species of green algae, such as *Dunaliella salina*, *Chlamydodmonas reinhardtii*, and *Volvox carteri*; it appears to be located in nonplastidial membranes (Thompson, 1996; Harwood & Jones, 1989). DGTS resembles PC in some aspects and generally appears when PC is either low or absent (Vogel & Eichenberger, 1992). The co-existence of these three lipids also occurs in the EPA-producing eustigmatophytes *Nannochloropsis* sp. (Schneider & Roessler, 1994) and *M. subterraneus* (Nichols & Appleby, 1969). Generally, DGTS appears together with either PC or PE, but not both. For example, *Chlamydodmonas reinhardtii* (Giroud et al., 1988) contains PE and DGTS, while PC and DGTS are found in *D. parva* (Evans et al., 1982) and *D. salina* (Norman & Thompson, 1985). We speculate that the co-occurrence of these lipids in *P. incisa* is related to the production of AA.

Induced Accumulation of AA in P. incisa

To induce nitrogen starvation, cultures of *P. incisa* at the early stationary phase were resuspended and maintained in nitrogen-free medium (Khozin-Goldberg et al., 2002). Electron microscopy revealed that TAG was accumulated in large oil globules (Fig. 10.1). After 14 d, the FA content of the N-free cultures increased from 17 to 36% (dry weight), compared to 25% in the control (Table 10.C). The FA composition demonstrated a sharp increase in the proportion of AA from 40 to 59% of TFA, compared to only 46% in the control. The proportion of AA increased in all lipid classes, but especially in neutral lipids, where it increased to 64%, compared to 51% in the control. Correspondingly, the dry weight content of AA increased from 7 to 21% in the N-free culture and to 11% in the control culture. Neutral lipids, mostly TAG, comprised 87% of TFA, compared to 62% in the control (Table 10.D). Under N-starvation, over 90% of cellular AA was deposited in the TAG. The enhancement in the content of AA under N-starvation was the result of both the increase in the proportion of TAG and the increase of AA in TAG.

The major molecular species of TAG was triarachidonylglycerol, whose share of total TAG was as high as 40%, compared to only 21% in the control (Table 10.E). The total share of the three other molecular species containing two arachidonyl moieties (as well as 18:2, 18:1, or 16:0) amounted to 54%, whereas molecular species containing only one arachidonyl moiety almost disappeared.

The accumulation of FA following nitrogen starvation is driven by the increase in the C/N ratio in the medium. Increasing the availability of carbon could enhance FA

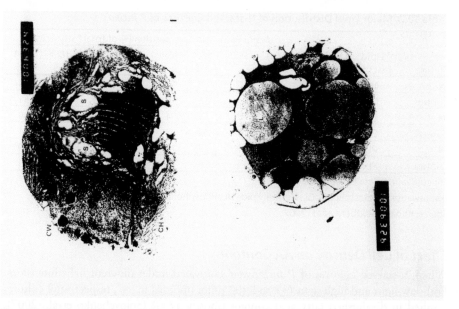

Fig. 10.1. Electron microscopy of *P. incisa*. (Left) Cells cultivated under optimal conditions. (Right) Cells harvested after 14 d of N-starvation. The approximate diameters of the cells are 20μ (left) and 10μ (right).

accumulation even further. We have, thus, studied the effect of the addition of acetate to the medium of N-starved cultures of *P. incisa* (unpublished data). While there was some effect on the FA content, the FA composition was adversely affected. The major increase was noted in the proportion of 18:1, which represents up to a 40% increase, at the expense of AA. Labeling experiments revealed that excess of acetate was used not only for the de novo FA synthesis, but also by the C16 elongase. Apparently, the levels of the obtained oleate were too high, as it competed with AA for incorporation into TAG, rather than being incorporated into phospholipids for further desaturation.

Table 10.C. Effect of Nitrogen Starvation on the Composition and Content of Fatty Acids in *P. incisa*[a].

Culture	Time (d)	Content (% Dry Wt)		Major Fatty Acid Composition (% of Total)						
						18:1 n-9	18:1 n-7	18:2 n-6	18:3 n-6	18:3 n-3	20:4 n-6
		TFA	AA	16:0	18:0	18:1 n-9	18:1 n-7	18:2 n-6	18:3 n-6	18:3 n-3	20:4 n-6
Control	0	16.5	6.6	11	2	11	5	17	1	4	40
Control	14	24.7	11.4	9	2	14	4	17	1	2	46
-N	14	35.8	21.1	9	2	9	6	9	1	1	59

[a]Cultures were resuspended in N-free (-N) or control medium and cultivated in columns for another 14 d.

[b]Abbreviations: Arachidonic acid, AA; total fatty acids, TFA.

Source: Khozin-Goldberg et al., 2002.

Table 10.D. Major Lipid Distribution of N-starved Cultures of *P. incisa*.

Culture	Lipid Fraction	Lipid Share (% TFA)	16:0	18:0	18:1 n-9	18:1 n-7	18:2 n-6	18:3 n-6	18:3 n-3	20:4 n-6
-N	NL	87	8	2	9	4	9	tr	t	64
Control	NL	62	9	1	15	6	10	tr	2	51
-N	GL	10	12	2	5	4	11	1	12	30
Control	GL	19	9	1	5	3	21	tr	15	21
-N	PL	3	26	3	3	14	13	3	3	29
Control	PL	19	17	1	6	11	19	1	7	25

For cultivation details, see Table 10.C.

Abbreviations: Galactolipids, GL; Neutral lipids NL; Phospholipids, PL; trace (<1%), tr.

Source: Khozin-Goldberg et al., 2002.

Effect of Cell Density on AA Content

When N-starved cultures of *P. incisa* were cultivated under different light intensities, both low light and high light (35 and 400 μmol photons m^{-2} s^{-1}, respectively) cultures resulted in diminished fatty acid content (mostly TAG) (Solovchenko et al., 2007). However, under medium light (200 μmol photons m^{-2} s^{-1}), both the fatty content (34% of dry weight) and the proportion of AA (57% of total fatty acids) were higher. Consequently, after 14 d, the AA content of these cultures reached 19% (of dw) (Fig. 10.2). However, the highest productivity of AA (mg L^{-1}) was obtained by cultures maintained on full medium and high light, due to much higher biomass accumulation (Fig. 10.3).

Table 10.E. Major Molecular Species Composition of TAG in Stationary (Control) and Nitrogen Starved (-N) Cultures of *P. incisa*[a].

Molecular Species[a] (% of Total)	Growth Medium	
	Control	-N
20:4/20:4/20:4	21	40
20:4/20:4/18:2	17	16
20:4/20:4/18:1	18	20
20:4/20:4/16:0	11	18
20:4/18:2/16:0	15	2
20:4/18:2/18:1	11	-
20:4/18:1/18:1	2	1
20:4/18:1/18:0	3	-

[a]TAG were separated using reverse phase HPLC. Peaks identified by UV detection were collected, transmethylated, and analyzed by GC. Relative composition was estimated by integrating peak areas obtained using an evaporative light-scattering detector without calibration.

[b]Positional distribution was not determined; the values are given without assignment of any position of the TAG.

Source: Khozin-Goldberg et al., 2002.

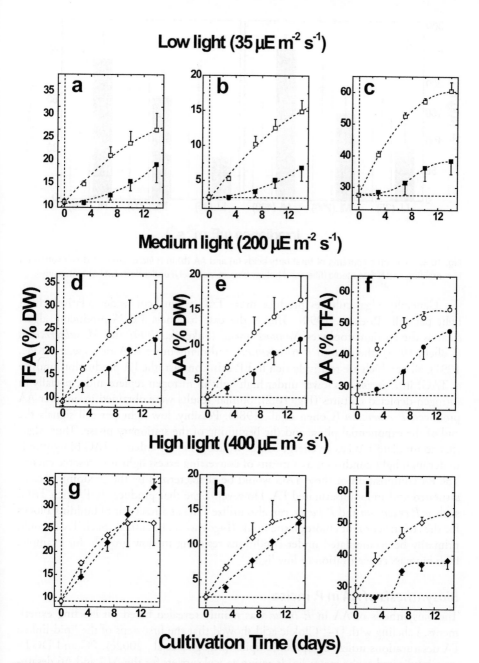

Fig. 10.2. Dynamics of total fatty acids (**a, d, g**), AA (**b, e, h**) content, and AA percentage (**c, f, i**) in *P. incisa* cells grown with (filled symbols) and without (hollow symbols) nitrogen under low (**a-c**), medium (**d-f**), and high (**g-i**) light. Initial biomass density was 1 g L⁻¹ in all cultures. *Source*: Solovchenko et al., 2008.

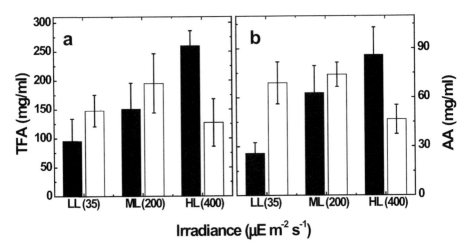

Fig. 10.3. Volumetric contents of total fatty acids (**a**) and AA (**b**) in *P. incisa* after 14 d cultivation on N-containing and N-free media (filled and hollow bars, respectively) under different light intensities.

Generally, algal cultures produce more TAG when kept under relatively high light per cell (Roessler, 1990). This is the case for most PUFA-producing algae, such as the eustigmatophytes *Nannochloropsis* (Sukenik, 1999) and *M. subterraneus* (Cohen, 1994), the diatom *P. tricornutum*, and many other EPA-rich species (Kyle, 1991), all of which are relatively rich in EPA. In contrast, the FA production (mostly as TAG) in *P. incisa* was lower under high light in nitrogen-replete and especially in nitrogen-depleted cultures (Fig. 10.2). Similar results were obtained in another AA producer, *P. cruentum* (Cohen et al., 1988). Possibly, low light per cell signals the end of the exponential phase and the beginning of the stationary phase. Thus, algae that accumulate TAG as an energy source would generally enhance TAG biosynthesis under high light conditions as a means of converting excess light into reserve energy. The FA composition of these TAG would be characterized by the predominance of saturated and monounsaturated FA. However, algae that produce PUFA-rich TAG, such as *P. cruentum* and *P. incisa*, can also utilize TAG as a reservoir of building blocks for the construction of chloroplastic lipids (Bigogno et al., 2002a). Such TAG would primarily be accumulated under conditions resulting in slow growth: for example, high biomass concentration or low light.

Biosynthesis of AA in P. incisa

The biosynthesis of AA in *P. incisa* was mainly revealed based on labeling experiments. Labeling with [1–^{14}C]oleic acid showed that the first steps of the lipid-linked FA desaturations utilize cytoplasmic lipids (Bigogno et al., 2002c). PC and DGTS were implicated as the major lipids acting as acyl carriers for the Δ12 and Δ6 desaturations of oleic acid, which lead sequentially to linoleic (18:2) and γ-linolenic acid (GLA, 18:3n-6). As elongation of 18:3 occurs at its carboxylic end, the acyl groups

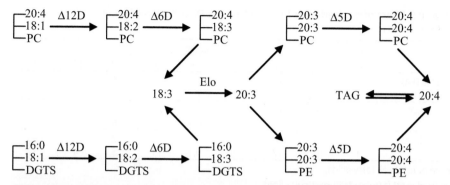

Scheme 10.1. Outline of the biosynthesis of AA in *P. incisa*. The *sn*-1 position in PE also contained significant levels of 18:1n-7. All PUFA are of the n-6 family. Abbreviations: Elo, Elongase.

must be detached from their phospholipid carrier in order to be elongated to 20:3n-6. Since labeled 20:3n-6 was detected first in PE and then in PC, we deduced that these lipids (especially PE) are the most likely substrates for the Δ5 desaturation of 20:3n-6 to AA. The presence of molecular species common to both PC and PE—for example, AA/AA, 18:2/AA, 18:1/AA, and 16:0/AA—further supports this suggestion (Bigogno et al., 2002b).

Galactolipids, mostly MGDG, serve as substrates for the chloroplastic Δ7, n-6, and n-3 desaturases, common in higher plants and many green algae. In the predominant sequence, 18:1/16:0 is sequentially desaturated to the 18:3n-3/16:3n-3 molecular species in a fashion similar to the prokaryotic pathway used by higher plants (Browse & Somerville, 1991) and some green algae (Thompson, 1996). The occurrence of the eukaryotic-like (20/20) molecular species of MGDG and digalactosyldiacylglycerol (DGDG) that contain AA (Bigogno et al., 2002b) suggests that, in *P. incisa*, AA is imported from extra chloroplastic lipids in ways similar to *P. cruentum* (Khozin et al., 1997). An outline of the likely biosynthetic pathways of the PUFA of *P. incisa* is displayed in Scheme 10.1. Interestingly, desaturation of C18 and C20 FA in many algae (Bigogno et al., 2002c) involves one or two lipids, usually PC, PE, or DGTS. In *P. incisa*, it appears that all three lipids are involved at different stages in the desaturation processes.

The inhibitor, salicylhydroxamic acid (SHAM), inhibits both Δ12 and Δ6 desaturations (Bigogno et al., 2002a). In *P. incisa*, SHAM produces a very large increase in 18:1, especially in PC and DGTS, further implicating these lipids as the substrates for the lipid-linked C18 desaturations. Another inhibitor, the substituted pyridazinone SAN 9785, impeded the TAG assembly in *Pavlova lutheri* (Siljegovich & Eichenberger, 1998). In *P. incisa*, Bigogno et al. (2002a) found that the herbicide mentioned previously also inhibited the accumulation of TAG, resulting in a decrease in the share of cell AA that was deposited in TAG from 80 to 52%. Most of the AA that could not be deposited in TAG was exported to the galactolipids MGDG and DGDG. Their respective share of cell AA drastically increased

Table 10.F. Effect of the Herbicide SAN 9785 on the Distribution of AA in Lipids of *P. incisa.*

	SAN 9785	Lipid							
		MGDG	DGDG	SQDG	PG	PC	DGTS	PE	TAG
Lipid (% TFA)[a]	–	7	10	8	2	8	5	2	54
	+	18	21	12	7	7	5	3	25
AA (% TFA)[b]	–	9	12	4	4	21	17	37	40
	+	15	23	1	0.3	43	26	34	58
% of Cell AA[c]	–	2	4	1	0.3	6	3	3	80
	+	10	17	0.5	0.1	11	5	4	52

[a]Share of total cell fatty acids.
[b]AA, % of total fatty acids in particular lipid.
[c]% of total cell AA.

Abbreviations: Digalactosyldiacylglycerol, DGDG; Digalactosyldimethylhomoserine, DGTS; Monogalactosyldiacylglycerol, MGDG; Sulfoquinovosylglycerol, SQDG; Phosphatidylcholine, PC; Phosphatidylethanolamine, PE; Phosphatidylglycerol, PG; Triacylglycerol, TAG.

Source: Bigongo et al., 2002a.

from 2 and 4% to 10 and 17% (Table 10.F). These finding are in keeping with our hypothesis in which these lipids function as the sink and TAG as the source of AA. Presumably, excess AA can utilize the same vehicle to mobilize AA from TAG to galactolipids. Due to this inhibition, the proportion of TAG in *P. incisa* decreased from 54 to 25% of TFA (Table 10.F). However, the proportion of AA in TAG increased sharply from 40 to 58%. The proportion of AA in PC doubled from 21 to 42%, and the percentage of total AA that was deposited in PC increased from 6 to 11%, supporting the suggested role of this lipid as the major donor of AA to TAG. The results of the treatments with SHAM—which inhibited the accumulation of AA, but not of TAG—and of SAN 9785—which inhibited the biosynthesis of TAG, but not that of AA—indicate that these two processes are not necessarily coupled and are independent of each other.

Role of AA in P. incisa

The search for PUFA-rich algal strains was based on our hypothesis that the accumulation of PUFA-rich TAG can serve as a survival strategy in ecological niches characterized by rapidly changing conditions. To assess this hypothesis, we studied the changes in lipid and fatty acid composition during recovery from nitrogen starvation at 24°C. We have followed the redistribution of fatty acid labels in both neutral and polar lipids of *P. incisa* after radiolabelling with oleic acid and recovery from nitrogen starvation at both temperatures (Fig. 10.4). During recovery, the label of TAG continuously decreased and was partially transferred to the chloroplastic lipids MGDG, DGDG, and sulfoquinovosyldiacylglycerol (SQDG). Similarly, in unlabeled cultures, TAG was mainly consumed to support growth; however, there was a significant increase in the content of AA in the chloroplastic lipids: predominantly, MGDG

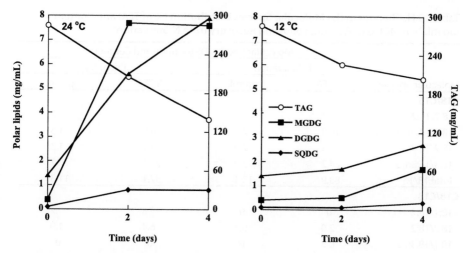

Fig. 10.4. Changes in the volumetric content of arachidonic acid in TAG [O], MGDG (■), DGDG (▲), and SQDG (◆) during recovery from N-starvation at 24°C (left panel) and 12°C (right panel).

(Table 10.G). The content of molecular species of MGDG of the 20/18 and 20/20 types increased from 1.4 to 11.8 µg/mL and from 0.2 to 2.8 µg/mL, respectively (Table 10.H). Molecular species of the 18/16 series were more desaturated but did not increase in content. Likewise, when the growth temperature of an exponentially growing culture of *P. incisa* labeled with [1–¹⁴C]AA was abruptly dropped to 4°C, the

Table 10.G. Major Fatty Acids Composition of Chloroplastic Lipid Classes of *P. incisa* Following Recovery (Rec.) from N-Starvation (Starv.).

Lipid	Conditions	Temp (°C)	% of TFA	TFA (µg mL⁻¹)	16:0	16:2 n-6	16:3 n-3	18:1 n-9	18:0	18:2 n-6	18:3 n-3	20:4 n-6
MGDG	Starv.		1.8	16.0	3.6	23.6	14.6	3.2	0.3	28.9	20.4	2.7
	Rec.	24	4.3	30.5	2.1	4.4	13.3	6.2	0.7	22.5	21.9	22.9
	Rec.	12	2.7	18.3	2.1	1.7	23.2	2.2	0.6	6.7	42.0	16.1
DGDG	Starv.		1.6	14.1	16.7	6.3	1.2	5.8	0.5	42.7	10.5	11.3
	Rec.	24	3.8	26.6	7.1	1.8	0.9	8.8	2.3	37.0	13.9	20.9
	Rec.	12	3.2	21.5	7.3	2.4	3.0	2.4	1.2	12.6	44.5	20.2
SQDG	Starv.		3.2	28.8	62.3	-	-	2.7	0.7	21.1	3.3	0.4
	Rec.	24	4.9	34.8	54.2	-	-	5.7	2.5	17.4	7.9	1.8
	Rec.	12	5.5	37.5	56.8	-	-	1.2	1.6	4.7	16.2	1.4

The header spans: Fatty Acid Composition (% of total) over columns 16:0 through 20:4.

Cultures were resuspended in full medium and grown at 24°C (2 d) or 12°C (4 d).

Abbreviations: Digalactosyldiacylglycerol, DGDG; Monogalactosyldiacylglycerol, MGDG; Sulfo-quinovosyldiacylglycerol, SQDG; trace (<1%), tr.

Source: Khozin-Goldberg et al., 2005.

Table 10.H. Changes in the Molecular Species Distribution (% of TFA) and Content (μg mL⁻¹) in MGDG of *P. incisa*, Following Recovery from N-Starvation[a].

| Molecular Species | Molecular Species Distribution and Content | | | |
| | N-Starvation | | Recovery | |
	% TFA	μg/mL	% TFA	μg/mL
C18/C16				
18:2/16:2	31.3	5.0	2.6	0.8
18:3n-3/16:2	16.0	2.6	4.9	1.5
18:2/16:3n-3	12.5	2.0	4.3	1.3
18:3/16:3n-3	24.2	3.9	25.4	7.7
Total 18/16	83.9	13.4	37.1	11.3
C18/C18				
18:1/18:2	tr	tr	3.8	1.2
18:2/18:2	2.0	0.3	6.4	1.9
18:1/18:3n-3	0.8	tr	tr	tr
18:2/18:3n-3	2.1	0.3	2.8	0.8
18:3n-3/18:3n-3	1.3	0.2	2.4	0.7
Total 18/18	**6.1**	**1.0**	**15.3**	**4.7**
C20/C18				
20:4/18:1	2.4	0.4	10.1	3.1
20:4/18:2	4.4	0.7	21.2	6.5
20:4/18:3n-3	2.1	0.3	7.2	2.2
Total 20/20	**8.9**	**1.4**	**38.5**	**11.8**
C20/C20				
20:4/20:4	**1.0**	**0.2**	**9.0**	**2.8**

[a]Cultures of *P. incisa* were kept on N-free medium for 14 d.
Cultures were resuspended in full medium and cultivated for another 2 d at 24°C. PUFA are of the n-6 family, unless otherwise noted. Positional analysis was not performed.
Abbreviations: total fatty acids, TFA; trace (<1%), tr.
Source: Khozin-Goldberg et al., 2005.

label in TAG, which was mostly associated with AA, was turned over to polar lipids (Bigogno et al., 2002a).

These findings support the hypothesis that one of the roles of AA-rich TAG in *P. incisa* is to serve as a reservoir that can be used to rapidly construct PUFA-rich chloroplastic membranes, especially under the low temperatures at which de novo PUFA synthesis would be very slow. Indeed, it was recently suggested that the sub-cellular role of lipid bodies is much more complex the assumption that they act as relatively inert carbon stores and that they may serve as building blocks for remodeling of polar lipids (Murphy, 2001).

In many green algae and higher plants, chloroplastic lipids are divided into two types: prokaryotic lipids characterized by the presence of C_{16} fatty acids at the *sn*-2 position of their glycerol skeletons (18/16) and eukaryotic lipids containing C_{18} fatty

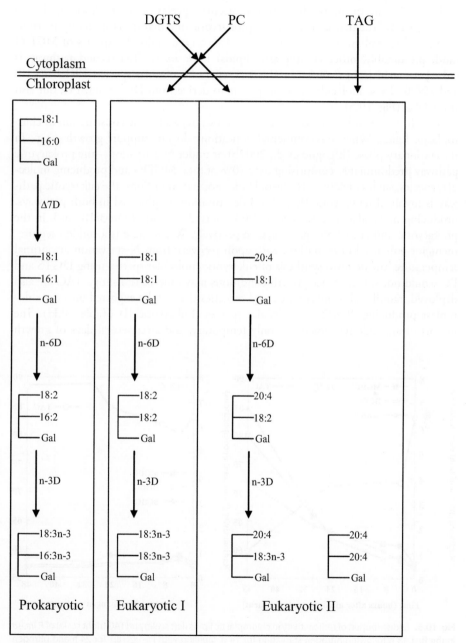

Scheme 10.2. Outline of the biosynthesis of the various types of molecular species of MGDG in *P. incisa*. Unless otherwise noted, PUFA are of the n-6 family.

acids at both positions (18/18). Our findings indicate that in *P. incisa* there are two pathways leading to the production of molecular species of chloroplastic lipids (see Scheme 10.2). The first pathway is the prokaryotic one that in its similarity to 16:3-plants and several green algae gives rise to 18/16 type molecular species of MGDG (and, presumably, other chloroplastic lipids) in *P. incisa*. The second pathway is eukaryotic, which, in *P. incisa*, produces 3 types of molecular species: 18/18, 20/18, and 20/20. These molecular species appears to derive from DGTS and PC, but also from TAG (Fig. 10.5).

We further invoke the existence of several strategies for the construction of chloroplastic lipids. When environmental conditions do not support growth—e.g., in the stationary phase (Bigogno et al., 2002b) or under N-starvation—the prokaryotic pathway predominates, comprising over 70% of total MGDG and producing molecular species, such as 18:2n-6/16:2n-6. Under normal conditions, the eukaryotic pathway is involved (eukaryotic I), and n-3 desaturation is enhanced in both pathways, producing molecular species, such as 18:3n-3/16:3n-3 and 18:3n-3/18:3n-3, in the prokaryotic and eukaryotic pathways, respectively. When there is a sudden requirement for enhanced desaturation, e.g., upon recovery from N-starvation at optimal temperature, the de novo synthesis of eukaryotic molecular species using DGTS and PC would not suffice. Since growth conditions may only be transitory, TAG are also deployed, rapidly releasing AA-rich acyl moieties that can be exported into the chloroplast producing 20:4/20:4 molecular species (Eukaryotic II) (Table 10.H). The import of arachidonyl moieties is only temporary, and after several days of growth

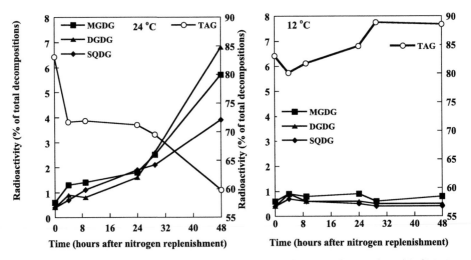

Fig. 10.5. Redistribution of radioactivity in chloroplastic lipids (left scale) and TAG (right scale) of *P. incisa* in the first 48 h following nitrogen starvation for 14 d, labeling with [1-^{14}C]18:1 for 24 h and nitrogen replenishment at 24°C (left panel) or 12°C (right panel); TAG ( ), MGDG (■), DGDG (▲), and SQDG (◆), respectively.

under optimal conditions, chloroplastic AA decreases to its normal level, and prokaryotic molecular species regain their dominance.

Similarly, in another alga, *P. cruentum*, whose ecological niche is characterized by rapid fluctuations, the share of eukaryotic molecular species of MGDG increases from 42% (of total acyl MGDG) at 30°C to 58% at 20°C, and, following N-replenishment, 20/20 molecular species of MGDG are produced by import of AA from TAG (Khozin-Goldberg et al., 2000). Recently, lipid bodies of higher plants were shown to be metabolically active in seeds and other plant organs (Murphy, 2001). Furthermore, Stobart et al. (1997) produced evidence that supports a transacylation mechanism that accounts for TAG turnover in microsomal membranes of developing safflower seeds.

Exposure to high light damages the photosynthetic machinery, particularly at low temperatures; the main site of the damage is the D1 protein of photosystem II. FA desaturation is important for the tolerance of intense light, especially at low temperatures, as it accelerates synthesis of the D1 protein (Gombos et al., 1998). The optimal growth temperature of *P. incisa* is 25°C; however, this alga, which was isolated from the slopes of a mountain in Japan (Watanabe et al., 1996), can withstand temperatures as low as 4°C. By measuring O_2 evolution and variable fluorescence, we found that *P. incisa* is more resistant to low temperatures than *Chlorella*; its resistance is almost as high as that of the cryophyte *Chlamydomonas nivalis* (unpublished).

Production of DGLA

DGLA is a precursor of various biologically active classes, e.g., prostaglandins and leukotrienes, and is, therefore, of potential pharmacological significance, but a lack of sources has prevented its study and, consequently, its use. Whereas higher plants or fungi and algae accumulate other PUFA, DGLA normally occurs only as an intermediate in the biosynthesis of AA; it is not appreciably accumulated in any organism. Furthermore, the conversion of GLA to DGLA in the body is, under certain conditions—e.g., low calcium—significantly diminished, and in such cases, GLA cannot be considered DGLA-equivalent. The only reported source of DGLA is a mutant of the fungus *Mortierella* (Higashiyama et al., 1998; see also Chapter 2).

The high content of AA in *P. incisa* suggests that this organism could be used as a vehicle to produce other PUFA. Thus, we mutated cells of *P. incisa* by treating it with N-methyl-N'-nitro-N-nitrosoguanidine. Since one of the roles of AA in this alga is to provide a high level of unsaturation for membranes at low temperatures, we selected mutants that were defective in their ability to grow under these conditions. Indeed, one of these mutants (P127) is Δ5 desaturase-defective and almost incapable of desaturating DGLA to AA. Comparison of the genomic and mRNA sequences of Δ5 desaturases isolated from the WT and the mutant revealed a single point mutation in a codon encoding for tryptophan, upstream of the HPGG quartet that is highly conserved within a fused cytochrome b5 domain in all cloned Δ5 and Δ6 desaturases

WT   IRRRVYDVTAWVPQHPGGNLIFVKAGR

MUT IRRRVYDVTA —VPQHPGGNLIFVKAGR

Scheme 10.3. Partial amino acid sequences of the Δ5 desaturase in wild type (WT) and mutant (MUT) of *P. incisa*, indicating the location of the mutation.

of different origins. The cytochrome b5 domain functions as an intermediate electron donor in a NADH-dependent acyl-group desaturation (Napier et al., 1999). The mutation resulted in the appearance of a stop codon, thus abolishing the function of the enzyme (see Scheme 10.3, below).

The mutant produces trace amounts of AA and high contents of DGLA, up to 39% of total fatty acids (Table 10.I) and 14% of dry weight, making it a potential source for the production of this pharmaceutically important PUFA. The distribution of lipid classes in the mutant was similar to that of the wild type (data not shown). However, the fatty acid composition showed some interesting differences. In the extrachloroplastic lipids PC, PE and DGTS DGLA substituted AA (Table 10.I). However, in the chloroplastic lipids and TAG, DGLA levels in the mutant were significantly lower than the corresponding levels of AA in the wild type. In TAG, DGLA comprised 27% of AA in comparison to 54% in the wild type. On the other hand, 18:1 increased from 20 to 43%. In the galactolipids MGDG and DGDG, the decrease in DGLA was accompanied by an enhanced n-3 desaturation, which resulted in an increase in 16:3 and 18:3. Indeed, MALDI-TOF mass spectrometry demonstrated that mutant MGDG and DGDG were composed mainly of prokaryotic 18/16 and eukaryotic 18/18 molecular species with a predominance for n-3 desaturated fatty acids (not shown).

We suggest that the mutant reduces the import of DGLA to chloroplastic lipids and compensates for the decreased level of unsaturation in chloroplastic lipids by increasing the production of prokaryotic and eukaryotic type I molecular species (see Scheme 10.3, above) and by augmenting n-3 desaturation. Consequently, the mutant's ability to grow at low temperatures is lower than that of the wild type.

DGLA Productivity

The mutant displayed a decreased proportion of DGLA in comparison to that of AA in the wild type and an increased proportion of C18 fatty acids—predominantly, 18:1 and 18:2. The latter is of special importance since these two fatty acids are mostly accumulated in TAG at the expense of DGLA. These alterations required us to evaluate the effect of environmental conditions on DGLA content and the C18/DGLA ratio. While N starvation enhanced the production of both TFA and the proportion of DGLA, phosphate starvation brought about some increase in the fatty acid content and a reduction of the DGLA proportion. On the other hand, a ten-fold increase in P concentration facilitated biomass production and resulted in higher content of DGLA, both in biomass (% dw) and in culture (mg/L) (Table 10.J). Additionally, we saw a decrease in the (18:1+18:2)/DGLA ratio. The addition of a buffer diminished pH

Table 10.I. Major Fatty Acids Composition of Major Lipid Classes of Wild Type (WT) and Mutant (P127) of *P. incisa* Following 3 d of Nitrogen Starvation.

Lipid Class	Strain	16:0	16:2	16:3 n-3	18:1[a]	18:2	18:3 n-6	18:3 n-3	20:3 n-6	20:4 n-6
MGDG	WT	3.3	18.5	13.5	1.9	26.6	0.8	21.9	tr	11.4
	P127	3.2	8.7	27.8	2.4	14.6	1.9	36.4	3.5	-
DGDG	WT	14.8	3.0	1.0	5.2	39.3	1.1	10.4	0.4	21.4
	P127	15.2	5.0	5.0	5.4	30.2	2.4	25.7	8.6	-
PC	WT	23.0	0.3	1.3	15.1	15.8	6.6	0.8	2.5	28.8
	P127	26.9	0.7	0.2	13.0	12.6	5.5	2.4	30.0	tr
DGTS	WT	28.2	0.8	2.6	9.1	21.0	9.1	0.8	1.0	18.4
	P127	28.1	0.9	1.9	8.3	22.8	12.4	3.3	15.6	tr
PE	WT	8.6	0.3	0.6	25.0	5.2	4.4	0.5	8.3	39.4
	P127	11.1	0.3	0.7	24.1	7.8	3.7	2.0	41.4	tr
TAG	WT	7.5	tr	0.2	20.5	10.3	0.7	0.2	1.1	54.0
	P127	8.5	tr	tr	42.8	12.3	1.4	0.8	27.5	tr

(Header span: **Fatty Acids (% of Total)**)

[a]Total of 18:1n-9 and 18:1n-7.

Source: Cohen & Khozin-Goldberg, unpublished.

fluctuations and further increased the fatty acid content from 34% (at standard phosphate) to 42%, the proportion of DGLA from 28 to 35%, and, consequently, its content from 10 to 15%. Furthermore, the proportion of 18:1 and 18:2 slightly decreased, and, consequently, the (18:1+18:2)/DGLA was diminished from 1.35 to 0.96.

By varying the initial nitrogen content of the N starvation period, we have shown that when the culture was transferred to N-free medium, 18:1 increased sharply, at the expense of DGLA. However, when starvation was started using a medium containing 50 or 25% nitrate (in comparison to the BG11 medium), the (18:1+18:2)/DGLA improved much, and the volumetric content of DGLA was higher due to the higher accumulation of biomass (data not shown).

Table 10.J. Effect of Phosphate Content on Total Fatty Acids and DGLA Content.

P Conc.	Fatty Acid Cont. (% DW)	DGLA Cont. (% DW)	DGLA (% TFA)	18:1+18:2 (% TFA)	Ratio[a]
X1	33.9	9.5	27.9	52.4	1.88
X5	35.7	10.3	29.6	51.4	1.74
X10	36.5	12.4	33.9	47.7	1.41
X10, pH Buffered	41.5	14.7	35.4	44.9	1.32

Cultures were cultivated on nitrogen-free BG-11 medium for 14 d at 0.175 mM (X1), 0.0875 mM (X5), and 1.75 mM (X10) with or without pH buffering using 20 mM Tris-HCl.

[a](18:1+18:2)/DGLA.

Abbreviations: Dry weight, DW; total fatty acids, TFA.

Source: Cohen & Khozin-Goldberg, unpublished.

The combination of light intensity and biomass concentration affects the light per cell. Higher biomass concentration and low (35 µE m^{-2} s^{-1}) light intensities favored higher contents of DGLA (ca. 40%); however, lipid productivity and, consequently, DGLA productivity were negatively affected. Cultures grown in 1 L columns reached the highest biomass content of DGLA (10% of dw) after 14 d of nitrogen starvation at a light intensity of 180 µE m^{-2} s^{-1} and a biomass density of 1 g L^{-1}. The highest culture content of DGLA of 755 mg L^{-1} was obtained, however, under similar nitrogen starvation and high light intensity (450 µE m^{-2} s^{-1}) at a biomass density of 2 g L^{-1}. We expect the combination of growth and nutritional conditions to result in even higher contents and productivities of DGLA.

Conclusion

According to prevailing theories (Pohl & Zurheide, 1979; Harwood & Jones, 1989), plants accumulate TAG to store energy. Indeed, the TAG of most algal species contain mostly saturated and monounsaturated FA; these FA are less complicated to produce and provide more energy. Our studies of the red microalga, *P. cruentum*, have shown that VLC-PUFA accumulated in TAG can be utilized to biosynthesize the eukaryotic-like molecular species of chloroplastic lipids, especially at low temperatures. We hypothesized that some algae whose habitat is characterized by rapidly changing environmental conditions can swiftly adapt to these conditions by mobilizing VLC-PUFA from their TAG to chloroplastic lipids. Based on this hypothesis, a chlorophyte microalga, *P. incisa*, was isolated. This alga is the richest plant source of the nutraceutically valuable PUFA AA. While this alga can withstand very low temperatures, its optimal growth temperature is 25°C where the maximal accumulation of AA occurs at relatively high biomass concentration and, especially, under nitrogen starvation. Under these conditions, the proportion of AA can reach 60% TFA, and the AA exceeds 20% dry weight. This is, to the best of our knowledge, the highest reported content of any PUFA—let alone, AA—in algae. Radiolabelling studies have shown that labeled AA was transferred from TAG to polar lipids upon a sudden reduction in temperature; this transfer indicates that TAG of *P. incisa* may indeed have a role as a depot of AA that can be incorporated into the membranes, enabling the organism to quickly respond to stress induced by low temperatures. Further elucidation of the mechanisms that result in the accumulation of AA in TAG could lead to the identification of other algal species rich in LC-PUFA.

The finding that most AA of *P. incisa* is deposited in TAG is of practical value, since TAG are the preferred chemical form to introduce AA into baby formulae. The capacity for AA accumulation makes *P. incisa* one of the best candidates for large-scale production of AA. *P. incisa* could be also utilized to produce high purity AA for pharmaceutical purposes. The high PUFA content of this organism drove us to select a Δ5 desaturase-deficient mutant that accumulates DGLA, rather than AA.

Acknowledgments

This research was supported in part under Grant No. TA-MOU-00-C20–013, U.S.-Israel Cooperative Development Research Program, Economic Growth, U.S. Agency for International Development. The authors would like to acknowledge the dedicated technical assistance of Ms. S. Didi-Cohen.

References

Agostoni, C.; E. Riva; R. Bellu; S. Trojan; D. Luotti; M. Giovannini. Effects of diet on the lipid and fatty acid status of full term infants at 4 months. *J. Am. Clin. Nutr.* **1994**, *13*, 658–664.

Bigogno, C.; I. Khozin-Goldberg; Z. Cohen. Accumulation of arachidonic acid and triacylglycerols in the microalga *Parietochloris incisa* (Chlorophyceae). *Phytochemistry* **2002a**, *60*, 135–143.

Bigogno, C.; I. Khozin-Goldberg; S. Boussiba; A. Vonshak; Z. Cohen. Lipid and fatty acid composition of the green alga *Parietochloris incisa*. *Phytochemistry* **2002b**, *60*, 497–503.

Bigogno, C.; I. Khozin-Goldberg; D. Adlerstein; Z. Cohen. Biosynthesis of arachidonic acid in the oleaginous microalga *Parietochloris incisa* (Chlorophyceae): Radiolabeling studies. *Lipids* **2002c**, *37*, 209–216.

Boswell, K.; E.K. Koskelo; L. Carl; S. Galza; D.J. Hensen; K.D. Williams; D.J. Kyle. Preclinical evaluation of single cell oils that are highly enriched with arachidonic acid and docosahexaenoic acid. *Food Chem. Toxicol.* **1996**, *34*, 585–593.

Browse, J.; C. Somerville. Glycerolipid synthesis: Biochemistry and regulation. *Ann. Rev. Plant Physiol. Plant Mol. Biol.* **1991**, *42*, 467–506.

Carlson, S.E.; S.H. Werkman; J.M. Peeples; R.J. Cooke; E.A. Tolley. Arachidonic acid status correlates with first year growth in preterm infants. *Proc. Natl. Acad. Sci.* **1993**, *90*, 1073–1077.

Cobelas, M.A.; J.Z. Lechado. Lipids in microalgae. A review. I. Biochemistry, *Grasas y Aceites* **1988**, *40*, 118–145.

Cohen, Z. The production potential of eicosapentaenoic and arachidonic acids by the red alga *Porphyridium cruentum*. *J. Am. Oil Chem. Soc.* **1990**, *67*, 916–920.

Cohen, Z. Production of eicosapentaenoic acid by the alga *Monodus subterraneus*. *J. Am. Oil Chem. Soc.* **1994**, *71*, 941–946.

Cohen, Z. Production of polyunsaturated fatty acids by the microalga *Porphyridium cruentum*. *Production of Chemicals by Microalgae*; Z. Cohen, Ed.; Taylor and Francis: London, 1999; pp 1–24.

Cohen, Z.; A. Vonshak; A. Richmond. Effect of environmental conditions of fatty acid composition of the red alga *Porphyridium cruentum*: Correlation to growth rate. *J. Phycol.* **1988**, *24*, 328–332.

Cohen, Z.; I. Khozin-Goldberg. Searching for PUFA-rich microalgae. *Single Cell Oils*, 1st edn.; Z. Cohen C. Ratledge, Eds.; American Oil Chemists' Society: Champaign, IL, 2005; 53–72.

Dubinsky, Z.; T. Berner; S. Aaronson. Potential of large-scale algal culture for biomass and lipid production in arid lands. *Biotech. Bioeng. Symp.* **1978**, *8*, 51–68.

Eichenberger, W.; C. Gribi. Lipids of *Pavlova lutheri*: Cellular site and metabolic role of DGCC. *Phytochemistry* **1997**, *45*, 1561–1567.

Evans, R.W.; M. Kates; M. Ginzburg; B.Z. Ginzburg. Lipid composition of the halotolerant algae, *Dunaliella parva* Lerche and *Dunaliella tertiolecta*. *Biochim. Biophys. Acta* **1982**, *712*, 186–195.

Fan, Y-Y.; R.S. Chapkin. Importance of dietary γ-linolenic acid in human health and nutrition. *J. Nut.* **1998**, *128*, 1411–1414.

Giroud, C.; A. Gerber; W. Eichenberger. Lipids of *Chlamydomonas reinhardtii*. Analysis of molecular species and intracellular sites of biosynthesis. *Plant Cell Physiol.* **1988**, *29*, 587–595.

Gombos, Z.; E. Kanervo; N. Tsvetkova; T. Sakamoto; E.M. Aro; N. Murata. Genetic enhancement of the ability to tolerate photoinhibition by introduction of unsaturated bonds into membrane glycerolipids. *Plant Phys.* **1998**, *115*, 551–559.

Hansen, J.; D. Schade; C. Harris; K. Merkel; D. Adamkin; R. Hall; M. Lim; F. Moya; D. Stevens; P. Twist. Docosahexaenoic acid plus arachidonic acid enhance preterm infant growth. *Prostaglandins, Leukotriens, Essential Fatty Acids* **1997**, *57*, 196.

Harwood, J.; L. Jones. Lipid metabolism in algae. *Adv. Bot. Res.* **1989**, *16*, 2–52.

Henderson, R.J.; E.E. Mackinlay. Radiolabeling studies of lipids in the marine cryptomonad *Chroomonas salina* in relation to fatty acid desaturation. *Plant Cell Physiol.* **1992**, *33*, 395–406.

Henderson, R.J.; E.E. Mackinaly; P. Hodgson; J.L. Harwood. Differential effects of the substituted pyridazinone herbicide Sandoz 9785 on lipid composition and biosynthesis in photosynthetic and non-photosynthetic marine microalgae. II. Fatty acid composition. *J. Exp. Bot.* **1990**, *41*, 729–736.

Higashiyama, K.; T. Yaguchi; K. Akimoto; S. Fujikawa; S. Shimizu. Enhancement of arachidonic acid production by *Mortierella alpina* 1S-4. *J. Am. Oil Chem. Soc.* **1998**, *75*, 1501–1505.

Hoffman, M.; W. Eichenberger. Lipid and fatty acid composition of the marine brown alga *Dictyopteris membranacea*. *Plant Cell Physiol.* **1997**, *389*, 1046–1052.

Jiang, Y.; F. Chen; S.Z. Liang. Production potential of docosahexaenoic acid by the heterotrophic marine dinoflagellate *Crypthecodinium cohnii*. *Proc. Biochem.* **1999**, *34*, 633–637.

Khozin, I.; D. Adlerstein; C. Bigogno; Y.M. Heimer; Z. Cohen. Elucidation of the biosynthesis of EPA in the microalga *Porphyridium cruentum* II: Radiolabeling studies. *Plant Physiol.* **1997**, *114*, 223–230.

Khozin-Goldberg, I.; Z.Y. Hu; D. Adlerstein; S. Didi Cohen; Y.M. Heimer; Z. Cohen. Triacylglycerols of the red microalga *Porphyridium cruentum* participate in the biosynthesis of eukaryotic galactolipids. *Lipids* **2000**, *5*, 881–889.

Khozin-Goldberg, I.; C. Bigogno; Z. Cohen. Nitrogen starvation induced accumulation of arachidonic acid in the freshwater green alga *Parietochloris incisa*. *J. Phycol.* **2002**, *38*, 991–994.

Khozin-Goldberg, I.; P. Shrestha; Z. Cohen. Mobilization of arachidonyl moieties from triacylglycerols into chloroplastic lipids following recovery from nitrogen starvation of the microalga *Parietochloris incisa*. *Biochim. Biophys. Acta* **2005**, *1738*, 63–71.

Koletzko, B.; M. Braun. Arachidonic acid and early human growth: Is there a relation? *Ann. Nutr. Metabol.* **1991**, *35*, 128–131.

Kyle, D.J. Specialty oils from microorganisms. *Biotechnology of Plant Fats and Oils*; J. Rattray, Ed.; American Oil Chemists' Society: Champaign, IL, 1991; pp 130–143.

Leman, J. Oleaginous microorganisms: An assessment of the potential. *Adv. Appl. Microbiol.* **1997**, *43*, 195–243.

Makewicz, A.; C. Gribi; W. Eichenberger. Lipids of *Ectocarpus fasciculatus* (Phaeophyceae). Incorporation of [1–^{14}c]oleate and the role of TAG and MGDG in lipid metabolism. *Plant Cell Physiol.* **1997**, *38*, 952–960.

Makrides, M.; M. Neumann; K. Simmer; J. Pater; R. Gibson. Are long-chain polyunsaturated fatty acids essential in infancy? *Lancet* **1995**, *345*, 1463–1468.

Mayzaud, P.; J.P. Chanut; R.G. Ackman. Seasonal changes of the biochemical composition of marine particulate matter with special reference to fatty acids and sterols. *Mar. Ecol. Prog. Ser.* **1989**, *56*, 189–204.

Molina-Grima, E.; F. Garcia Camacho; F.G. Acien Fernandez. Production of EPA from *Phaeodactylum tricornutum*. *Chemicals from Microalgae*; Z. Cohen, Ed.; Taylor and Francis: London, 1999; pp 57–92.

Murphy, D.J. The biogenesis and functions of lipid bodies in animals, plants and microorganisms. *Prog. Lipid Res.* **2001**, *40*, 325–438.

Napier, J.A.; O. Sayanova; P. Sperling; E. Heinz. A growing family of cytochrome b5-domain fusion proteins. *Trends Plant Sci.* **1999**, *4*, 2–4.

Nichols, B.W.; R.S. Appleby. The distribution of arachidonic acid in algae. *Phytochemistry* **1969**, *8*, 1907–1915.

Norman, H.; G.A. Thompson Jr. Quantitative analysis of *Dunaliella salina* diacylglyceroltrimethylhomoserine and its individual molecular species by high performance liquid chromatography. *Plant Sci.* **1985**, *42*, 83–87.

Patterson, G.W. Effect of culture temperature on fatty acid composition of *Chlorella sorokiniana*. *Lipids* **1970**, *5*, 597–600.

Pohl, P.; F. Zurheide. Fatty acids and lipids of marine algae and the control of their biosynthesis by environmental factors. *Marine Algae in Pharmaceutical Science*; H.A. Hoppe, T. Levring, Y. Tanaka, Eds.; Walter de Gruyter: Berlin, 1979; pp 473–523.

Roessler, P.G. Environmental control of glycerolipid metabolism in microalgae: Commercial implications and future research directions. *J. Phycol.* **1990**, *26*, 393–399.

Safford, R.; B.W. Nichols. Positional distribution of fatty acids in monogalactosyldiglyceride fractions from leaves and algae. *Biochim. Biophys. Acta* **1970**, *210*, 57–64.

Schneider, J.C.; Roessler, P. Radiolabeling studies of lipids and fatty acids in *Nannochloropsis* (Eustigmatophyceae), an oleaginous marine alga. *J. Phycol.* **1994**, *30*, 594–598.

Scragg, A.H.; R.R. Leather. Production of fats and oils by plant and algal cell cultures. *Single Cell Oil*; R.S. Moreton, Ed.; Longman: UK, 1988; pp 71–98.

Seto, A.; H.L. Wang; C.W. Hesseltine. Culture conditions affect eicosapentaenoic acid content of *Chlorella minutissima*. *J. Am. Oil Chem. Soc.* **1984**, *61*, 892–894.

Siljegovich-Hänggi, N.; W. Eichenberger. Effect of the substituted pyridazinone san 9785 on the lipid and fatty acid biosynthesis in *Pavlova lutheri* (Haptophyceae). *Advances in Plant Lipid*

Research; J. Sanchez, E. Cerda-Olmedo, E. Martinez-Force, Eds.; Universidad de Sevilla: Seville, 1998; pp 259–261.

Solovchenko, A.E.; I. Khozin-Goldberg; S. Didi-Cohen; Z. Cohen; M.N. Merzlyak. Effects of light intensity and nitrogen starvation on growth, total fatty acids and arachidonic acid in the green microalga *Parietochloris incisa*. *J. Appl. Phycol.* **2007**, *20*, 245–251.

Stobart, K.; M., Mancha; M. Lenman; A. Dahlqvist; S. Stymne. Triacylglycerols are synthesized and utilized by transacylation reactions in microsomal preparations of developing sunflower (*Carthamus tinctorius* l.) seeds. *Planta* **1997**, *203*, 58–56.

Sukenik, A. Production of eicosapentaenoic acid by the marine eustigmatophyte *Nannochloropsis*. *Chemicals from Microalgae*; Z. Cohen, Ed.; Taylor and Francis: London, 1999; pp 41–56.

Sukenik, A.; Y. Carmeli. Regulation of fatty acid composition by irradiance level in the eustigmatophyte *Nannochloropsis sp. J. Phycol.* **1989**, *25*, 686–692.

Thompson, G.A. Lipids and membrane function in green algae. *Biochim. Biophys. Acta* **1996**, *1302*, 17–45.

Vogel, G.; W. Eichenberger. Betaine lipids in lower plants. biosynthesis of DGTS and dgta in *Ochromonas danica* (Chrysophyceae) and the possible role of DGTS in lipid metabolism. *Plant Cell Physiol.* **1992**, *33*, 427–436.

Watanabe, S.; S. Hirabashi; S. Boussiba; Z. Cohen; A. Vonshak; A. Richmond. *Parietochloris incisa* comb. nov. (Trebuxiophyceae, Chlorophyta). *Physiol. Res.* **1996**, *44*, 107–108.

Whitelam G.C.; G.A. Codd. Damaging effects of light on microorganisms. *Microbes in Extreme Environments*; R.A. Herbert, G.A. Codd, Eds.; Academic Press: London, 1986; pp 129–169.

Yongmanitchai, W.; O.P. Ward. Growth and eicosapentaenoic acid production by *Phaeodactylum tricornutum* in batch and continuous culture system. *J. Am. Oil Chem. Soc.* **1992**, *69*, 584–590.

11

Carotenoid Production Using Microorganisms

Michael A. Borowitzka

Algae R & D Center, School of Biological Sciences and Biotechnology, Murdoch University, Murdoch, WA 6150 Australia

Introduction

A wide variety of carotenoids are biosynthesized by plants, algae, fungi, and bacteria (Goodwin, 1980; Liaaen-Jensen & Egeland, 1999). They appear to play a range of roles in these organisms, especially light-harvesting in plants and protecting cells from oxidative damage. A number of carotenoids, such as β-carotene, astaxanthin, lutein, and lycopene, are good antioxidants in lipid phases, functioning as free radical scavengers or singlet oxygen quenchers. Singlet oxygen ($^1O_2$), a free radical, is known to damage DNA. Carotenoids act by absorbing the energy of the singlet oxygen into the carotenoid chain, which leads to the degradation of the carotenoid molecule, but protects other molecules or tissues from damage (Woodall et al., 1997). Carotenoids can also prevent the chain reaction production of free radicals initiated by the degradation of polyunsaturated fatty acids, thus preventing the accelerated degradation of lipid membranes. These properties of carotenoids are used to explain the epidemiological and experimental studies that indicate that dietary carotenoids inhibit the onset of a range of diseases, such as cataracts, age-related macular degeneration, multiple sclerosis, arteriosclerosis, and some cancers (Osganian, 2003; Tapiero et al., 2004; Hosokawa et al., 2009). Dietary carotenoids also play an important role in animal pigmentation and nutrition (Liñán-Cabello, 2002). The characteristic colour of salmonid flesh is due mainly to astaxanthin and canthaxanthin (Schiedt, 1986). Similarly, the pigmentation of certain crustaceans, such as prawns and crabs, is due to astaxanthin. β-Carotene has been shown to enhance fertility in cattle (Aréchiga, 1998) and pigs (Kyrakis, 2000), as well as crustaceans (Liñán-Cabello, 2002).

"Natural" carotenoids for applications in human and animal nutrition are extracted from a number of natural plant sources, and, in the last 25 years, several algal, fungal, and yeast sources have also been developed as commercial sources of β-carotene and astaxanthin. Microbial sources of other carotenoids, such as lycopene, lutein, zeaxanthin, and canthaxanthin, are also being developed.

β-Carotene

Dunaliella salina

The halophilic, unicellular, biflagellate green alga *Dunaliella salina* (Dunal) Teodoresco (sometimes also called *D. bardawil*) is the richest natural source of the carotenoid β-carotene, containing up to 14% of the dry weight as β-carotene. Other species of *Dunaliella* such as *D. parva* produce smaller amounts of carotenoids or extremely little e.g. *D. viridis* and D. *tertiolecta* (Borowitzka & Siva, 2007). The β-carotene is accumulated as droplets in the chloroplast stroma of the alga, especially under the following conditions: high temperature, high salinity, high irradiance, and low nitrogen. In *D. salina*, the β-carotene occurs as a mixture of the *cis* (mainly *9-cis*) and *trans* isomers, whereas synthetic β-carotene is only obtained in the form of the *trans* isomer. The formation of *cis* isomers in *Dunaliella* is favored by low irradiances (Ben-Amotz et al., 1988; Orset & Young, 2000); however, there also appear to be some differences between strains (Jimenez & Pick, 1994).

Commercial production commenced in Australia, Israel, and the United States in the 1980s, and, since that time, *Dunaliella* production plants have also been established in India and Inner Mongolia (Borowitzka & Borowitzka, 1988b; Schlipalius, 1991; Borowitzka, 1994; Ben-Amotz, 2004). The Australian plants at Hutt Lagoon, Western Australia, and Whyalla, South Australia, operated by Cognis Nutrition and Health produce the bulk of the world's natural β-carotene; the American production plant is no longer operational.

Dunaliella salina is a photoautotroph, although it is able to utilize organic C sources, such as glycerol and acetate, to some extent in light (Suárez et al., 1998; Mojaat et al., 2008). The rate of carotenoid accumulation is a function of the total irradiance received by the cells (Ben-Amotz & Shaish, 1992), and the maximum content of carotenoids produced is a function of the salinity (Borowitzka et al., 1990).

All species of *Dunaliella* have a specific requirement for sodium, and *D. salina* has the broadest reported NaCl tolerance range of any organism: from about 10 to 35% (w/v) NaCl (= saturation) (Borowitzka & Siva, 2007). The optimum salinity for growth is about 22% NaCl, whereas the optimum salinity for carotenogenesis is greater than 30% NaCl (Borowitzka et al., 1990). The actual salinity used for commercial culture is usually greater than 30% NaCl—i.e., 10 times seawater salinity—because of the need to prevent the growth of predators, such as brine shrimp and the amoeba *Fabrea salina* (Post et al., 1983), and avoid overgrowth by the non-carotenogenic *D. viridis* (Borowitzka & Borowitzka, 1989).

The optimum growth temperature for *D. salina* is between 21 and 40°C and is dependent on the strain and irradiance (Gibor, 1956; Borowitzka, 1981); the growth temperature range, however, is from less than 0°C to over 40°C (Wegmann et al., 1980; Siegel et al., 1984). Higher temperatures stimulate carotenogenesis, and the high temperatures (>35°C) and high irradiances experienced during the summer months at the *D. salina* production plant at Hutt Lagoon, Western Australia, result in maximum β-carotene productivity at that time of the year.

The solubility of CO_2 in the high salinity brines used to culture these algae is very low, and it is further reduced at the high temperatures and high pH normally encountered. The solubility of CO_2 at 15°C in 2M NaCl is 1.02 g L^{-1}, and at 40°C it is only 0.53 g L^{-1} (Lazar et al., 1983); therefore, the addition of inorganic carbon, either as CO_2 or HCO_3^-, stimulates growth, provided that there is no precipitation of carbonates.

The best nitrogen source for *D. salina* is nitrate as ammonium salts, urea and nitrite are less effective (Borowitzka & Borowitzka, 1988a). Care must be taken to control culture pH as nitrite uptake can lead to alkalinisation and ammonium uptake to acidification of the medium, and also, potentially, cell death. Ammonium also inhibits carotenoid formation. The optimum phosphate concentration for growth is approximately 0.02 g L^{-1} K_2HPO_4, and higher concentrations can inhibit growth (Gibor, 1956). Like other algae, *D. salina* requires Mg^{2+}, Ca^{2+}, K, chelated iron, and various trace elements—especially, Mn, Zn, Co, and Cu—for growth. High sulphate concentrations above 2 mM are required for optimum cell division, and higher concentrations enhance carotenogenesis (Ben-Amotz & Avron, 1989). No vitamin requirement has been demonstrated, and the optimum pH for growth is between pH 7–9 (Borowitzka & Borowitzka, 1988b).

Commercial production of *D. salina* for β-carotene is carried out either in extremely large (several hundreds of Ha in area), unmixed, open ponds, as in Australia, or, more intensively, in smaller, paddle wheel-mixed, raceway ponds, as in Israel. Cultivation in closed systems, such as tubular photobioreactors, has also been attempted (García-González et al., 2005), however none of these systems has, as yet, proven commercially viable (Borowitzka & Borowitzka, 1989; Ben-Amotz, 2004).

Open pond culture can be likened to agriculture, as the cultures are exposed to the daily and seasonal vicissitudes of weather (e.g., changes in temperature, light, rainfall, and wind), as well as potential contaminants from the environment. Cultures may be operated in either batch or continuous mode, with the latter being the significantly more economical option. A two-stage process has been considered. In this process the algae are first grown at lower salinity to achieve maximum biomass, followed by a high-salinity stage for maximum carotenoid accumulation; however this process has been found to be uneconomical (Borowitzka et al., 1984). An alternative, two-stage culture process has been described by Ben Amotz (1995) in which the algae are first grown in a nitrate-rich medium, to maximize biomass production, and then diluted with nitrate-deficient medium for the β-carotene accumulation stage. This culture regime increased β-carotene productivity by about 50%.

There is little published information on the actual productivity of commercial *D. salina* operations. Ben Amotz (1999) states that, in raceway ponds that are 20 cm deep, annual average productivities of about 200 mg β-carotene m^{-2} day^{-1} are achieved, so a 50 000 m^2 plant produces 10 kg β-carotene day^{-1}. The average cell concentration of β-carotene is about 5% of the dry weight, and the average concentration of algae in the ponds ranges from about 0.1 g dry wt L^{-1} in unmixed ponds to up to 1.0 g L^{-1} in raceway ponds.

The harvesting of algae and the extraction and purification of the β-carotene represent major cost areas for the production of algal β-carotene; therefore, the exact details of the processes used by the various producers are closely guarded. The harvesting of *D. salina* is more difficult than for most other microalgae. *Dunaliella salina* is a naked single cell, approximately 20 x 10 μm in size, neutrally or positively buoyant in a high specific gravity, high viscosity brine. Cell densities in the ponds are relatively low, between 0.1 to 0.5 g of dry weight·L^{-1}; thus, very large volumes of medium have to be processed. These properties have led to the development of a wide range of harvesting or pre-concentration methods to be used before more conventional harvesting methods, which include high pressure filtration using diatomaceous earth as a filter medium (Ruane, 1974b); exploitation of salinity-dependent buoyancy properties in stationary and moving gradients (Bloch et al., 1982); exploitation of the phototactic and gyrotactic responses of the algae (Kessler, 1982, 1985); salinity-dependent hydrophobic adhesion properties of *Dunaliella* cells; flocculation (Sammy, 1987); deep bed filtration (Guelcher & Kanel, 1998); or bubble columns (Guelcher & Kanel, 1999). Methods currently used in commercial production include centrifugation or flocculation and flotation.

Extraction of β-carotene can be achieved using several methods. Conventional solvent extraction is efficient (Ruane, 1974a; Ohishi et al., 1999; Bailey et al., 2002): however, the presence of solvent residues in the final product is not acceptable for most markets. More acceptable extraction methods involve the use of hot vegetable oil (Nonomura, 1987) or supercritical carbon dioxide extraction (Cygnarowicz & Seider, 1990; Mendes, 2003). Alternatively, the algal biomass is drum-dried, stabilized, and sold as a β-carotene-rich nutritional supplement for human nutrition or as an additive for animal feed, such as prawn feed. Unlike the synthetic product, the algal product also contains small amounts of other carotenoids such as zeaxanthin. The relative proportion of the *cis* and *trans* isomers in the final product can be manipulated during the extraction and formulation process (Schlipalius, 1997).

A new alternative culture process has been described in which the cells are grown in a two-phase photobioreactor consisting of an aqueous and a biocompatible organic (dodecane) phase (Hejazi et al., 2002). In this system, β-carotene is continuously removed ("milked") from the algal cells without impairing their viability, achieving β-carotene productivities of up to 2.45 mg.m^{-2}.day^{-1}. Although this productivity is two to three times higher that the productivity in current commercial open pond cultures, it remains to be seen whether the process is commercially viable.

Blakeslea trispora

Several fungi of the order Mucorales synthesize β-carotene;they include *Phycomyces blakesleeanus*, *Choaneophora curcurbitarum*, and *Blakeslea trispora*. Industrial production of β-carotene from fungi has focused on *B. trispora* because of its inherently higher β-carotene content and the ability to grow well in standard fermentations of liquid media. Current commercial producers are DSM and Vitaene. Industrial production using *B. trispora* involves the separate culture of (+) and (-) mating strains, followed by the joint fermentation of the two strains in media rich in hydrocarbons (kerosene), carbohydrates, vegetable oils, and chemical additive (Ciegler, 1965; Ninet & Renaut, 1979). This process gives yields of 0.3% of dry weight of β-carotene or about 3 g·L⁻¹. The process has been markedly improved over time with carotenoid contents of up to 20% of dry weight reported (Finkelstein, 1995). β-Carotene is accumulated in the mycelium, and synthesis is stimulated by trisporic acids, such as β-ionone, by a positive feedback control (Lampila, 1985). In a mutant strain of *B. trispora* grown with β-ionone addition, biomass containing 30 mg β-carotene.g⁻¹ dry weight can be produced with controlled oxygen addition and control of the age of the vegetative growth phase in 5-6 days (Costa Perez, 2003). Light has also been shown to stimulate carotenoid synthesis, although the degree of photo-induction of carotenogenesis varies with light wavelength and the length of exposure time (Lampila, 1985). Jeong et al. (1999) have postulated that this light effect might be caused by oxidative stress through the production of free oxygen radicals, and they have shown that hydrogen peroxide also increases β-carotene production.

Blakeslea is unusual amongst the Mucorales in that carotene biosynthesis is not inhibited by the end product β-carotene, as is the case in *P. blakesleeanus* (Murillo, 1976; Mehta & Cerdá-Olmedo, 1995). *Blakeslea* and *Phycomyces* also differ in their response to blue light and many chemicals (Lampila, 1985; Cerdá-Olmedo, 2001), and, in *Phycomyces,* the highest carotenoid contents are achieved on solid substrates or quiet (unmixed) liquid media.

Astaxanthin

Haematococcus pluvialis

The freshwater green flagellate alga *Haematococcus pluvialis* can accumulate up to about 6% (approximately 2-4% in commercial cultures) of dry weight as astaxanthin, making it the best natural source of this carotenoid. In nature and in culture, the alga exists in three cell forms: (1) a motile, naked, biflagellate stage; (2) a non-motile, naked palmella stage; and (3) a non-motile, thick-walled aplanospore stage. The astaxanthin is accumulated in droplets in the perinuclear cytoplasm in the aplanospore stage of the life cycle and occurs as the $3S, 3'S$ isomer in the form of free astaxanthin, as well as mono- and di-esters with $C_{16:0}$, $C_{18:0}$, $C_{20:0}$, and $C_{18:1}$ fatty acids (Grung et al., 1992). Commercial production commenced in the mid-1990s in the United States (Hawaii) and Sweden, and later in Israel and India.

The optimum growth temperature is between 15-20°C, and, in batch culture, the motile stage predominates during the logarithmic phase of growth, whereas the carotenogenic aplanospores are formed during stationary phase. This change in cell morphology—from rapidly dividing, shear sensitive flagellated cells to the extremely robust, non-motile aplanospores—means that, generally, a two stage, batch growth process is preferred (Cysewski & Lorenz, 2004; García-Malea et al., 2008). In this process, conditions are first optimized for the growth of the non-carotenogenic flagellate cells to achieve the maximum biomass, followed by a stage, in which the cells are "stressed" to induce aplanospore formation and maximize astaxanthin accumulation.

Haematococcus pluvialis can grow photoautotrophically, as well as mixotrophically or heterotrophically, using organic C sources, such as acetate (Droop, 1955; Borowitzka et al., 1991; Gong & Chen, 1997), although heterotrophically grown cells grow more slowly and produce only small amounts of astaxanthin in the dark (Kobayashi et al., 1997). In mixotrophic growth, acetate and pyruvate addition enhances growth rate, cell yield, and astaxanthin formation (Borowitzka et al., 1991; Harker et al., 1996). There is significant variability between strains, however, in their response to acetate and their ability to use other carbon sources, such as glucose, glycerol, and glycine (Borowitzka, 1992). The maximum astaxanthin content of the cells is also strain dependent (Lee & Soh, 1991). Strain selection is, therefore, an important factor in the development of a commercial process for this alga. Increased astaxanthin production in several strains has also been achieved through mutagenesis (Chen et al., 2003) and genetic engineering (Steinbrenner & Sandmann, 2006), but better results have been obtained through the isolation and screening of new strains (Borowitzka, unpublished results).

Carotenogenesis is stimulated by a wide range of "stress" factors, such as nutrient limitation, especially N-limitation; high light; high temperatures; or NaCl addition (Borowitzka et al., 1991; Boussiba & Vonshak, 1991; Fan et al., 1994; Cordero et al., 1996): in fact almost any factor that inhibits growth. Boussiba and Vonshak (1991) were able to demonstrate that the inhibition of cell division by the addition of vinblastin promoted astaxanthin accumulation. Other studies also have shown that reactive oxygen species (ROS) enhance astaxanthin synthesis by an, as yet, unknown mechanism (Kobayashi et al., 1993; Lu, 1998; Boussiba, 2000). The light regulation of carotenogenesis in *Haematococcus* is apparently under photosynthetic redox control. The plastoquinone pool appears to function as a redox sensor, and reduction of the plastoquinone pool leads to the transcriptional activation of most, if not all, genes involved in astaxanthin biosynthesis (Steinbrenner, 2003).

Current commercial processes first produce the algal biomass in closed photobioreactors, such as bubble columns, plate reactors, or tubular photobioreactors, under optimal conditions in a nutrient-replete medium. This stage is then followed by a carotenoid-accumulation step under reduced N and P conditions in a high light environment, either in raceway ponds or in tubular photobioreactors. This whole growth cycle takes several weeks (Boussiba et al., 1997; Olaizola, 2000; Cysewski & Lorenz, 2004). The final product is a spray-dried powder of mechanically or enzymatically broken and stabilised, astaxanthin-rich aplanospores that can be used as a pigmenter for salmonid fish (Sommer et al., 1991, 1992) or as a human nutritional supplement (Guerin et al., 2003).

Phaffia rhodozyma/Xenophyllomyces dendrorhous

The basidomycetous yeast, *Phaffia rhodozyma/Xenophyllomyces dendrorhous*, is unusual amongst carotenoid containing yeasts in that it has the ability to vigorously ferment sugars. On the basis of life history studies and sequence analysis of rDNA IGS and ITS regions, there appear to be distinct anamorphic (asexual) species, *P. rhodozyma*, and a teleomorphic species (producing both sexual and asexual spores), *X. dendrorhous* (Kuscera, 1998; Fell, 1999). This yeast produces unesterified astaxanthin, predominantly in the 3*R*, 3′*R* form and has been extensively studied as a commercial source of this carotenoid because of very high cell densities of >50 g dry weight L^{-1} when grown in a fermentor (Echavarri-Erasun, 2002). Cell yield and carotenoid production are decreased at high sugar concentrations, and, therefore, fed-batch fermentation is used to maximize productivity (Ho, 1999). Elicitors prepared from various fungi, as well as plant extracts, have also been shown to increase both biomass and carotenoid yields (Wang et al., 2006; Kim et al., 2007).

The main limitations of commercial utilization are the low levels of astaxanthin found in wild-type isolates and the thick wall and capsule of the yeast, which must be ruptured to ensure bioavailability of the astaxanthin. However, several companies have developed astaxanthin-hyperproducing strains that produce >1% of dry weight of astaxanthin by chemical mutagenesis and screening, and genetic engineering methods are also being used to enhance astaxanthin production (Chun, 1992; Visser, 2003; Kim et al., 2005). An (2001) has achieved improved growth and carotenogenesis in the carotenoid hyperproducing mutant stain 2A2N in media containing tricarboxilic acid cycle intermediates. These findings suggest that the use of media containing molasses (Haard, 1988), corn steep liquor (Kesava, 1998), or corn wet-milling co-products (Hayman, 1995) can be used in conjunction with grape juice (Myers, 1994) as a TCA cycle intermediates source for the increased production of biomass with a high carotenoid content. A recent review by Frengova and Beshkova (2009) provides a detailed description of factors affecting carotenoid formation and biomass production in *Phaffia rhodozyma/Xenophyllomyces dendrorhous*.

A range of mechanical, enzymatic, and chemical methods have also been developed to break the yeast's cell wall and improve the bioavailability of astaxanthin (An & Choi, 2003; Storebakken et al., 2004), including a two-stage process where, in the second stage, the yeast is grown in mixed culture with *Bacillus circulans*, a bacterium with high cell wall lytic activity (Fang & Wang, 2002).

Lutein

In recent years, there has been extensive research on the production of lutein as this carotenoid has been shown to have a range of health effects, such as the delay of age-related cataracts and macular degeneration (Chiew & Taylor, 2007). The major commercial source of lutein is marigold flowers (*Tagetes* spp), which contain only about 0.03% lutein (Piccaglia et al., 1998).

Potential algal sources of lutein with a higher lutein content than marigolds include the green algae *Muriellopsis* sp., *Chlorella zofingiensis* (Del Campo, 2000), *C. prototbecoides* (Shi et al., 2000), *C. acidophila* (Garbayo et al., 2008), and *Scenedesmus almeriensis* (Sánchez et al., 2008) (see Table 11.A). *Muriellopsis* sp. has been grown successfully outdoors in both tubular photobioreactors and in open ponds with a

Table 11.A. Carotenoid Content and Yield for a Selected Range of Microorganisms[a].

Organism	Principal Carotenoid(s) Accumulated	Maximum Carotenoid Content Reported (mg g⁻¹ dry wt)	Carotenoid Yield (mg.L⁻¹)
ALGAE			
Dunaliella salina	β-carotene	~500 (-1400)	~5 - 50
Haematococcus pluvialis	(3S,3'S) astaxanthin	~300 (-600)	
Chlorococcum sp.	astaxanthin (canthaxanthin) adonirubin, 4'-hydroxy echinenone)	8	45
Neochloris wimmeri	astaxanthin	19.3	
Protosiphon botryoides	astaxanthin	14.3	
Chlorella zofingiensis	astaxanthin	6.8	24.8
Coeloastrella striolata	astaxanthin	47.5	
Trentepohlia aurea	β-carotene	21	
Chlorella sorokiniana	lutein (β-carotene)	4.3	730
Chlorella prototcoides	lutein	6.5	136
Chlorella acidophila	Lutein (β-carotene)	20.2	
Murelliopsis sp.	lutein	5.5	35
Chlorococcum citriforme	lutein	7.4	38
Scenedesmus almeriensis	lutein	5.3	
FUNGI & YEASTS			
Xanthophyllomyces dendrorhous/ Phaffia rhodozyma	(3R,3'R) astaxanthin	>10	~5
Blakeslea trispora	β-carotene	17	
	lycopene	41	
Phycomyces blakesleeanus	β-carotene		
Mycobacterium aurum (mutant)	lycopene	7.4	16.2
BACTERIA			
Brevibacterium KY-4313	canthaxanthin	0.6	1-2

[a] Some of the data have been recalculated from the information supplied in the articles.

lutein content of about 5-6 mg g^{-1} dry weight (Del Campo et al., 2001; Blanco et al., 2007).

The food yeast *Candida albicans* has also been genetically engineered to produce up to 7.8 mg lutein g^{-1} (Shimada, 1998).

Other Carotenoids and Other Organisms

Carotenogenesis is widespread amongst microorganisms, and a number of species are being studied as potentially new industrial sources of carotenoids. In addition, the cloning of carotenoid pathways provides a potential tool for engineering overproducing strains or creating carotenoid producers from non-producing species (Schmidt-Dannert, 2000).

Many species of green algae, other than *Haematococcus pluvialis*, are known to produce high concentrations of astaxanthin as their major carotenoid; they include *Chlorococcum* sp., *Neochloris wimmeri*, *Chlorella zofingiensis*, and *Protosiphon botryoides* (Table 11.A). Recently, a new bacterial genus has been isolated that contains about 0.13 mg g^{-1} (*3S, 3'S*)-astaxanthin (Tsubokura, 1999). A mutant of *Dunaliella salina* has also been isolated that produces up to 6 mg g^{-1} zeaxanthin (Jin et al., 2003). The yeast *Rhodotorula glutinis* produces the carotenoids, torulene, torularhodin, γ-carotene, and β-carotene, in various proportions (Frengova & Beshkova, 2009), and Bhosale and Gadre (2001) have isolated a β-carotene-overproducing mutant of this yeast that contained up to 2 mg β-carotene g^{-1} (82% of total carotenoids). Ninet and Renaud (1979) reported yields of 10-40 mg L^{-1} of zeaxanthin in a *Flavobacterium* sp. and yields of up to 0.9 g L^{-1} of mostly unidentified xanthophylls, with isorenieratene (an aromatic carotene) predominating in the actinomycete *Streptomyces mediolani*. The potential of these non-photosynthetic bacteria as commercial carotenoid sources remains to be fully explored. Various metabolic engineering approaches have also been used to create new carotenoid-producing organisms or to enhance carotenoid production. For example, the bacterium *Escherichia coli* has been engineered to produce lycopene (Alper et al., 2006), and the yeast *Saccharomyces cereviceae* (Verwaal et al., 2007) has been engineered to produce β-carotene. As yet, however, the carotenoid yields are too low for commercial application.

Conclusion

Several species of microorganisms have been successfully commercialized in the last twenty-five years as natural sources of the carotenoids β-carotene and astaxanthin. New sources of these carotenoids, as well as lutein and lycopene, are in advanced stages of development. Genetic engineering also provides a powerful tool for understanding the carotenoid biosynthetic pathways and their control, and this information can be used to optimize production systems. Genetic engineering has also been used to create overproducing strains and for the transformation of non-carotenogenic species into carotenoid producers (Das et al., 2007). At this stage, this information is of limited commercial interest as the natural carotenoid market reacts adversely to any GMO product. However, this response may change in the future, and it is clear that commercial carotenoid biosynthesis using microorganisms has a great and colorful future.

References

Abe, K.; H. Hattori; M. Hirano. Accumulation and Antioxidant Activity of Secondary Carotenoids in the Aerial Microalga *Coelastrella striolata* Var. *multistrata*. *Food Chem.* **2007**, *100*, 656–661.

Abe, K.; N. Nishimura; M. Hirano. Simultaneous Production of ß-Carotene, Vitamin E and Vitamin C by the Aerial Microalga *Trentepohlia aurea*. *J. Appl. Phycol.* **1999**, *11*, 331–336.

Alper, H.; K. Miyaoku; G. Stephanopoulus. Characterization of Lycopene-Overproducing *E. coli* Strains in High Cell Density Fermentations. *Appl. Microbiol. Biotechnol.* **2006**, *72*, 968–974.

An, G.H. Improved Growth of the Red Yeast, *Phaffia rhodozyma\Xanthophyllomyces dendrorhous*, in the Presence of Tricarboxylic Acid Cycle Intermediates. *Biotech. Lett.* **2001**, *23*, 1005–1009.

An, G.H.; E.S. Choi. Preparation of the Red Yeast, *Xanthophyllomyces dendrorhodus*, as Feed Additive with Increased Availability of Astaxanthin. *Biotech. Lett.* **2003**, *25*, 767–771.

Aréchiga, C.F. Effect of Injection of ß-Carotene or Vitamin E and Selenium on Fertility of Lactating Dairy Cows. *Theriogenology* **1998**, *50*, 65–76.

Bailey, D.T.; R.J. Daughenbauh; R. Arslanian; L.A. Kaufmann; S.L. Richheimer; Z.Z. Liu; J.M. Piffarerio; C.J. Kurtz. High Purity Beta-Carotene and Process for Obtaining the Same. U.S. Patent 0082459, 2002.

Ben-Amotz, A. New Mode of *Dunaliella* Diotechnology: Two-Phase Growth for ß-Carotene Production. *J. Appl. Phycol.* **1995**, *7*, 65–68.

Ben-Amotz, A. Production of Beta-Carotene from *Dunaliella*. *Chemicals from Microalgae*; Z. Cohen, Ed.; Taylor & Francis: London, 1999; pp 196–204.

Ben-Amotz, A. Industrial Production of Microalgal Cell-Mass and Secondary Products—Major Industrial Species: *Dunaliella*. *Microalgal Culture: Biotechnology and Applied Phycology*; A. Richmond, Ed.; Blackwell Science: Oxford, 2004; pp 273–280.

Ben-Amotz, A.; A. Lers; M. Avron. Stereoisomers of ß-Carotene and Phytoene in the Alga *Dunaliella bardawil*. *Pl. Physiol.* **1988**, *86*, 1286–1291.

Ben-Amotz, A.; M. Avron. The Biotechnology of Mass Culturing *Dunaliella* for Products of Commercial Interest. *Algal and Cyanobacterial Biotechnology*; R.C. Cresswell; T.A.V. Rees; N. Shah, Eds.; Longman Scientific & Technical: Harlow, 1989; pp 91–114.

Ben-Amotz, A.; A. Shaish. ß-Carotene Diosynthesis. *Dunaliella: Physiology, Biochemistry, and Biotechnology*; M. Avron, A. Ben-Amotz, Eds.; CRC Press: Boca Raton, 1992; pp 205–216

Beutner, S.B. Quantitative Assessment of Antioxidant Properties of Natural Colourants and Phytochemicals: Carotenoids, Flavenoids, Phenols and Indigoids. The Role of ß-Carotene in Antioxidant Function. *J. Sci. Food. Agric.* **2001**, *81*, 559–568.

Bhosale, P.; R.V. Gadre. Optimization of Carotenoid Production from Hyper-Producing *Rhodotorula glutinis* Mutant 32 by a Factorial Approach. *Lett. Appl. Microbiol.* **2001**, *33*, 12–16.

Blanco, A.M.; J. Moreno; J.A. Del Campo; J. Rivas; M.G. Guerrero. Outdoor Cultivation of Lutein-Rich Cells of *Muriellopsis* sp. in Open Ponds. *Appl. Microbiol. Biotechnol.* **2007,** *73,* 1259–1266.

Bloch, M.R.; J. Sasson; M.E. Ginzburg; Z. Goldman; B.Z. Ginzburg; N. Garti; A. Perath. Oil Products from Algae. U.S. Patent 4,341,038, 1982.

Borowitzka, L.J. The Microflora. Adaptations to Life in Extremely Saline Lakes. *Hydrobiologia* **1981,** *81,* 33–46.

Borowitzka, L.J. Commercial Pigment Production from Algae. *Algal Biotechnology in the Asia-Pacific Region*; S.M. Phang, K. Lee, M.A. Borowitzka, B. Whitton, Eds.; Institute of Advanced Studies, University of Malaya: Kuala Lumpur, 1994; pp 82–84.

Borowitzka, L.J.; M.A. Borowitzka; T. Moulton. The Mass Culture of *Dunaliella*: From Laboratory to Pilot Plant. *Hydrobiologia,* **1984,** *116/117,* 115–121.

Borowitzka, L.J.; M.A. Borowitzka. Industrial Production: Methods and Economics. *Algal and Cyanobacterial Biotechnology*; R.C. Cresswell, T.A.V. Rees, N. Shah, Eds.; Longman Scientific: London, 1989; pp 294–316.

Borowitzka, M.A. Comparing Carotenogenesis in *Dunaliella* and *Haematococcus*: Implications for Commercial Production Strategies. *Profiles on Biotechnology*; T.G. Villa, J. Abalde, Eds.; Universidade de Santiago de Compostela: Santiago de Compostela, 1992; pp 301–310.

Borowitzka, M.A.; L.J. Borowitzka. Limits to Growth and Carotenogenesis in Laboratory and Large-Scale Outdoor Cultures of *Dunaliella salina*. *Algal Biotechnology*; T. Stadler, J. Mollion, M.C. Verdus, Y. Karamanos, H. Morvan, D. Christiaen, Eds.; Elsevier Applied Science: Barking, 1988a; pp 371–381.

Borowitzka, M.A.; L.J. Borowitzka. *Dunaliella. Micro-algal Biotechnology*; M.A. Borowitzka, L.J. Borowitzka, Eds.; Cambridge University Press: Cambridge, 1988b; pp 27–58.

Borowitzka, M.A.; L.J. Borowitzka; D. Kessly. Effects of Salinity Increase on Carotenoid Accumulation in the Green Alga *Dunaliella salina*. *J. Appl. Phycol.* **1990,** *2,* 111–119.

Borowitzka, M.A.; J.M. Huisman; A. Osborn. Culture of the Astaxanthin-Producing Green Alga *Haematococcus pluvialis* 1. Effects of Nutrients on Growth and Cell Type. *J. Appl. Phycol.* **1991,** *3,* 295–304.

Borowitzka, M.A.; C.J. Siva. The Taxonomy of the Genus *Dunaliella* (Chlorophyta, Dunaliellales) with Emphasis on the Marine and Halophilic Species. *J. Appl. Phycol.* **2007,** *19,* 567–590.

Boussiba, S. Carotenogenesis in the Green Alga *Haematococcus pluvialis*: Cellular Physiology and Stress Response. *Physiol. Plant.* **2000,** *108,* 111–117.

Boussiba, S.; A. Vonshak. Astaxanthin Accumulation in the Green Alga *Haematococcus pluvialis*. *Pl. Cell. Physiol.* **1991,** *32,* 1077–1082.

Boussiba, S.; A.Vonshak; Z. Cohen; A. Richmond. A Procedure for Large-Scale Production of Astaxanthin from *Haematococcus*. PCT Patent 9,728,274, 1997.

Bubrick, P. Production of Astaxanthin from *Haematococcus*. *Biores. Technol.* **1991,** *38,* 237–239.

Cerdá-Olmedo, E. Production of Carotenoids with Fungi. *Biotechnology of Vitamins, Pigments and Growth Factors*; E.J. Vandamme, Ed.; Elsevier Applied Science: London, 1989; pp 27–42.

Cerdá-Olmedo, E. *Phycomyces* and the Biology of Fight and Colour. *FEMS Microbiol. Rev.* **2001,** *25,* 503–512.

Chen, Y.; D. Li; W. Lu; J. Xing; B. Hui; Y. Han. Screening and Characterization of Astaxanthin-Hyperproducing Mutants of *Haematococcus pluvialis*. *Biotech. Lett.* **2003,** *25,* 527–529.

Chiew, C.J.; A. Taylor. Nutritional Antioxidants and Age-Related Cataract and Macular Degeneration. *Exp. Eye Res.* **2007,** *84,* 229–245.

Chun, S.B. Strain Improvement of *Phaffia rhodozyma* by Protoplast Fusion. *FEMS Microbiology Letters* **1992,** *93,* 221–226.

Ciegler, A. Microbial Carotenogenesis. *Adv Appl Microbiol,* **1965,** *7,* 1–34.

Cordero, B.; A. Otero; M. Patino; B.O. Arredondo; J. Fabregas. Astaxanthin Production from the Green Alga *Haematococcus pluvialis* with Different Stress Conditions. *Biotech. Lett.* **1996,** *18,* 213–218.

Costa Perez, J. Method of Producing Beta-Carotene by Means of Mixed Culture Fermentation Using (+) and (-) Strains of *Blakeslea trispora*. European Patent Application 1,367,131, 2003.

Cygnarowicz, M.L.; W.D. Seider. Design and Control of a Process to Extract Beta-Carotene with Supercritical Carbon Dioxide. *Biotech. Prog.* **1990,** *6,* 82–91.

Cysewski, G.R.; R.T. Lorenz. Industrial Production of Microalgal Cell-Mass and Secondary Products—Species of High Potential: *Haematococcus*. *Microalgal Culture: Biotechnology and Applied Phycology;* A. Richmond, Ed.; Blackwell Science: Oxford, 2004; pp 281–288.

Das, A.; S.H. Yoon; S.H. Lee; J.Y. Kim; D.K. Oh; S.W. Kim. An Update on Microbial Carotenoid Production: Application of Recent Metabolic Engineering Tools. *Appl. Microbiol. Biotechnol.* **2007,** *77,* 505–512.

Del Campo, J.A. Carotenoid Content of Chlorophycean Microalgae: Factors Determining Lutein Accumulation in *Muriellopsis* sp. (Chlorophyta). *J. Biotech.* **2000,** *76,* 51–59.

Del Campo, J.A.; H. Rodríguez; J. Moreno; M.A. Vargas; J. Rivas; M.G. Guerrero. Lutein Production by *Muriellopsis* sp. in an Outdoor Tubular Photobioreactor. *J. Biotech.* **2001,** *85,* 289–295.

Droop, M.R. Carotogenesis in *Haematococcus pluvialis*. *Nature* **1955,** *175,* 42.

Echavarri-Erasun, C. Fungal Carotenoids. *Applied Mycology and Biotechnology;* G.G. Khachtourians, Ed.; Agriculture and Food Production; Elsevier: Amsterdam, 2002; Vol. 2, 45–85

Fan, L.; A. Vonshak; S. Boussiba. Effect of Temperature and Irradiance on Growth of *Haematococcus pluvialis* (Chlorophyceae). *J. Phycol.* **1994,** *30,* 829–833.

Fang, T.J.; J.M. Wang. Extractibility of Astaxanthin in a Mixed Culture of an Overproducing Mutant of *Xanthophyllomyces dedrorhous* and *Bacillus circulans* in Two-Stage Batch Fermentation. *Proc. Biochem.* **2002,** *37,* 1235–1245.

Fell, J.W. Separation of Strains of the Yeasts *Xanthophyllomyces dedrorhous* and *Phaffia rhodozyma* Based on rDNA IGS and ITS Sequence Analysis. *J. Indust. Microbiol. Biotechnol.* **1999,** *23,* 677–681.

Finkelstein, M. *Blakeslea trispora* Mated Culture Capable of Increased Beta-Carotene Production. U.S. Patent 5,422,247, 1995.

Frengova, G.I.; D.M. Beshkova. Carotenoids from *Rhodotorula* and *Phaffia*: Yeast of Biotechnological Importance. *J. Indust. Microbiol. Biotechnol.* **2009,** *36,* 163–180.

Garbayo, I.; M. Cuaresma; C. Vílchez; J.M. Vega. Effect of Abiotic Stress on the Production of Lutein and ß-Carotene by *Chlamydomonas acidophila*. *Proc. Biochem.* **2008,** *43,* 1158–1161.

García-González, M.; J. Moreno; J.C. Manzano; F.J. Florencio; M.G. Guerrero. Production of *Dunaliella salina* Biomass Rich in 9-*cis*-ß-Carotene and Lutein in a Closed Tubular Photobiore-actor. *J. Biotech.* **2005**, *115*, 81–90.

García-Malea, M.; F.G. Acíen; E. Del Río; J.M. Fernández; M.C. Cerón; M.G. Guerrero; E. Molina-Grima. Production of Astaxanthin by *Haematococcus pluvialis*: Taking the One-Step System Outdoors. *Biotech. Bioeng.* **2008**, *102*, 651–657.

Gibor, A. The Culture of Brine Algae. *Biol. Bull.* **1956**, *3*, 223–229.

Gong, X.; F. Chen. Optimisation of Culture Medium for Growth of *Haematococcus pluvialis*. *J. Appl. Phycol.* **1997**, *9*, 437–444.

Goodwin, T.W. *Biochemistry of Carotenoids*. Plants; Chapman and Hall: London, 1980; Vol.1.

Grung, M.; F.M.L. D'Souza; M.A. Borowitzka; S. Liaaen-Jensen. Algal Carotenoids 51. Secondary Carotenoids 2. *Haematococcus pluvialis* Aplanospores as a Source of (3*S*,3'*S*)-Astaxanthin Esters. *J. Appl. Phycol.* **1992**, *4*, 165–171.

Guelcher, S.A.; J.S. Kanel. Method for Deep Bed Filtration of Microalgae. W.O. Patent 9,828,404, 1998.

Guelcher, S.A.; J.S. Kanel. Method for Dewatering Microalgae with a Bubble Column. U.S. Patent 5,910,254, 1999.

Guerin, M.; M.E. Huntley; M. Olaizola. *Haematococcus* Astaxanthin: Applications for Human Health and Nutrition. *Trends in Biotechnology,* **2003**, *21*, 210–216.

Haard, N.F. Astaxanthin Formation by the Yeast {\i Phaffia rhodozyma} on Molasses. *Biotech. Lett.* **1988**, *10*, 609–614.

Harker, M.; A.J. Tsavalos; A.J. Young. Factors Responsible for Astaxanthin Formation in the Chlorophyte *Haematococcus pluvialis*. *Biores. Technol.* **1996**, *55*, 207–214.

Hayman, G.T. Production of Carotenoids by *Phaffia rhodozyma* Grown on a Medium Composed of Corn Wet-Milling Co-products. *J. Indust. Microbiol.* **1995**, *14*, 389–395.

Hejazi, M.A.; C. de Lamaliere; J.M.S. Rocha; M. Vermue; J. Tramper; R.H. Wijffels. Selective Extraction of Carotenoids from the Microalga *Dunaliella salina* with Retention of Viability. *Biotech. Bioeng.* **2002**, *79*, 29–36.

Ho, K.P. Growth and Carotenoid Production of *Phaffia rhodozyma* in Fed-Batch Cultures with Different Feeding Methods. *Biotech. Lett.* **1999**, *21*, 175–178.

Hosokawa, M.; T. Okada; N. Mikami; I. Konishi; K. Miyashita. Bio-Functions of Marine Carotenoids. *Food Sci. Biotech.* **2009**, *18*, 1–11.

Jeong, J.C.; I.Y. Lee; S.W. Kim; Y.H. Park. Stimulation of ß-Carotene Synthesis by Hydrogen Peroxide in *Blakeslea trispora*. *Biotech. Lett.* **1999**, *21*, 683–686.

Jimenez, C.; U. Pick. Differential Stereoisomer Compositions of ß,ß-Carotene in Thylakoids and in Pigment Globules in *Dunaliella*. *J. Plant Physiol.* **1994**, *143*, 257–263.

Jin, E.; B. Feth; A. Melis. A Mutant of the Green Alga *Dunaliella salina* Constitutively Accumulates Zeaxanthin under All Growth Conditions. *Biotech. Bioeng.* **2003**, *81*, 115–124.

Johnson, E.A. *Phaffia rhodozyma*: Colorful Odyssey. *Internat. Microbiol.* **2003**, *6*, 169–174.

Kerr, S.C. Factors Enhancing Lycopene Production by a New *Mycobacterium aurum* Mutant. *Biotech. Lett.* **2004**, *26*, 103–108.

Kesava, S.S. An Industrial Medium for Improved Production of Carotenoids from a Mutant Strain of *Phaffia rhodozyma. Bioproc. Eng.* **1998,** *19,* 165–170.

Kessler, J.O. Algal Cell Harvesting. U.S. Patent 4,324,067, 1982.

Kessler, J.O. Hydrodynamic Focusing of Motile Algal Cells. *Nature* **1985,** *313,* 208–210.

Kim, J.H.; S.W. Kang; S.W. Kim; H.I. Chang. High-Level Production of Astaxanthin by *Xanthophyllomyces dendrorhous* Mutant JH1 Using Statistical Experimental Designs. *Biosci. Biotech. Biochem.* **2005,** *69,* 1743–1748.

Kim, S.K.; J.H. Lee; C.H. Lee; Y.C. Yoon. Increased Carotenoid Production in *Xanthophyllomyces dendrorhous* G267 Using Plant Extracts. *J. Microbiol.* **2007,** *45,* 128–132.

Kobayashi, M.; T. Kakizono; S. Nagai. Enhanced Carotenoid Diosynthesis by Oxidative Stress in Acetate-Induced Cyst Cells of a Green Unicellular Alga, *Haematococcus pluvialis. Appl. Env. Microbiol.* **1993,** *59,* 867–873.

Kobayashi, M.; Y. Kurimura; Y. Tsuji. Light-Independent, Astaxanthin Production by the Green Microalga *Haematococcus pluvialis* under Salt Stress. *Biotech. Lett.* **1997,** *19,* 507–509.

Kuscera, J. Homothallic Life Cycle in the Diploid Red Yeast *Xanthophyllomyces dendrorhous. Antonie Van Leeuwenhoek* **1998,** *73,* 163–168.

Kyrakis, S.C. Effect of ß-Carotene on Health Status and Performance of Sows and Their Litter. *J. Anim. Physiol. Anim. Nutr.* **2000,** *83,* 150–157.

Lampila, L.E. A Review of Factors Affecting Biosynthesis of Carotenoids by the Order Mucorales. *Mycopathologica* **1985,** *90,* 65–80.

Lazar, B.; A. Starinsky; A. Katz; E. Sass; S. Ben-Yaakov. The Carbonate System in Hypersaline Solutions: Alkalinity and $CaCO_3$ Solubility of Evaporated Seawater. *Limnol. Oceanogr.* **1983,** *28,* 978–986.

Lee, Y.K.; C.W. Soh. Accumulation of Astaxanthin in *Haematococcus lacustris* (Chlorophyta). *J. Phycol.* **1991,** *27,* 575–577.

Liaaen-Jensen, S.; E.S. Egeland. Microalgal Carotenoids. *Chemicals from Microalgae*; Z. Cohen, Ed.; Taylor & Francis: London, 1999; pp 145–172

Liñán-Cabello, M.A. Bioactive Roles of Carotenoids and Retinoids in Crustaceans. *Aquacult. Nutr.* **2002,** *8,* 299–309.

Liu, B.H. Secondary Carotenoids Formation by the Green Alga *Chlorococcum* sp. *J. Appl. Phycol.* **2000,** *12,* 301–307.

Lu, F.V. Does Astaxanthin Protect *Haematococcus* against Light Damage? *Z. Naturforsch.* **1998,** *53c,* 93–100.

Marcos, A.T. Lycopene Production Method. European Patent Application 1,184,464, 1920.

Matsukawa, R. Antioxidants from Carbon Dioxide Fixing *Chlorella sorokiniana. J. Appl. Phycol.* **2000,** *12,* 263–267.

Mehta, B.J.; E. Cerdá-Olmedo. Mutants of Carotene Production in *Blakeslea trispora. Appl. Microbiol. Biotechnol.* **1995,** *42,* 836–838.

Mendes, R.L. Supercritical Carbon Dioxide Extraction of Compounds with Pharmaceutical Importance from Microalgae. *Inorganica Chimica Acta* **2003,** *356,* 328–334.

Mojaat, M.; J. Pruvost; A. Foucault; J. Legrand. Effect of Organic Carbon Sources and Fe^{2+} Ions on Growth and ß-Carotene Accumulation by *Dunaliella salina*. *Biochem. Eng. J.* **2008,** *39,* 177–184.

Murillo, F.J. Regulation of Carotene Synthesis in *Phycomyces*. *Mol. Gen. Genet.* **1976,** *148,* 19–24.

Myers, P.S. Astaxanthin Production by a *Phaffia rhodozyma* Mutant on Grape Juice. *World J. Microbiol. Biotech.* **1994,** *16,* 178–183.

Ninet, L.; J. Renaut. Carotenoids. *Microbial Technology*; H.J. Peppler, Ed.; Academic Press: NY, 1979; pp 529–544.

Nonomura, A.M. Process for Producing a Naturally Derived Carotene-Oil Composition by Direct Extraction from Algae. U.S. Patent 4,680,314, 1987.

Ohishi, N.; T. Suzuki; K. Yagi. Composition Containing 9-cis Beta-Carotene in High Purity and Method of Obtaining the Same. Canada Patent 2,234,332, 1999.

Olaizola, M. Commercial Production of Astaxanthin from *Haematococcus pluvialis* Using 25,000-liter Outdoor Photobioreactors. *J. Appl. Phycol.* **2000,** *12,* 499–506.

Orosa, M.; J.F. Valero; C. Herrero; J. Abalde. Comparison of the Accumulation of Astaxanthin in *Haematococcus pluvialis* and Other Green Microalgae under N-Starvation and High Light Conditions. *Biotech. Lett.* **2001,** *23,* 1079–1085.

Orset, S.; A.J. Young. Exposure to Low Irradiances Favors the Synthesis of 9-*cis*-ß,ß-Carotene in *Dunaliella salina* (Teod.). *Pl. Physiol.* **2000,** *122,* 609–617.

Osganian, S.K. Dietary Carotenoids and Risk of Coronary Artery Disease in Women. *Am. Jf. Clin. Nutr.* **2003,** *77,* 1390–1399.

Piccaglia, R.; M. Marotti; S. Grandi. Lutein and Lutein Ester Content in Different Types of *Tagetes patula* and *T. erecta*. *Industrial Crops Production* **1998,** *8,* 45–51.

Post, F.J.; L.J. Borowitzka; M.A. Borowitzka; B. Mackay; T. Moulton. The Protozoa of a Western Australian Hypersaline Lagoon. *Hydrobiologia* **1983,** *105,* 95–113.

Ruane, M. Extraction of Caroteiniferous Material from Algae. Australian Patent 487,018, 1974a.

Ruane, M. Recovery of Algae from Brine Suspensions. Australian Patent 486,999, 1974b.

Sammy, N. Method for Harvesting Algae. Australian Patent 70,924, 1987.

Sánchez, J.F.; J.M. Fernández; F.G. Acíen; A. Rueda; J. Pèrez-Parra; E. Molina. Influence of Culture Conditions on the Productivity and Lutein Content of the New Strain *Scenedesmus almeriensis*. *Proc. Biochem.* **2008,** *43(3981987),* 405.

Schiedt, K.V. Astaxanthin and Its Metabolites in Wild Rainbow Trout (*Salmo gairdneri* R.). *Comp. Biochem. Physiol. B - Comp. Biochem* . **1986,** *83,* 9–12.

Schlipalius, L. The Extensive Commercial Cultivation of *Dunaliella salina*. *Biores. Technol.* **1991,** *38,* 241–243.

Schlipalius, L.E. High *cis* Beta-Carotene Composition. U.S. Patent 5,612,485, 1997.

Schmidt-Dannert, C. Engineering Novel Carotenoids in Microorganisms. *Curr. Op. Biotech.* **2000,** *11,* 255–261.

Shi, X.M.; Z.H. Zhang; F. Chen. Heterotrophic Production of Biomass and Lutein by *Chlorella protothecoides* on Various Nitrogen Sources. *Enz. Microb. Technol.* **2000,** *27,* 312–318.

Shi, X.M.; Y. Jiang; F. Chen. High-Yield Production of Lutein by the Green Microalga *Chlorella prototothecoides* in Heterotrophic Fed-Batch Culture. *Biotech. Prog.* **2002**, *18(7231997)*, 727.

Shimada, H. Increased Carotenoid Production by the Food Yeast *Candida utilis* through Metabolic Engineering of the Isoprenoid Pathway. *Appl. Env. Microbiol.* **1998**, *64*, 2676–2680.

Siegel, B.Z.; S.M. Siegel; T. Speitel; J. Waber; R. Stoeker. Brine Organisms and the Question of Habitat-Specific Adaptation. *Origins of Life* **1984**, *14*, 757–770.

Sommer, T.R.; F.M.L. D'Souza; N.M. Morrissy. Pigmentation of Adult Rainbow Trout, *Oncorhynchus mykiss*, Using the Green Alga *Haematococcus pluvialis*. *Aquaculture* **1992**, *106*, 63–74.

Sommer, T.R.; W.T. Potts; N.M. Morrissy. Utilization of Microalgal Astaxanthin by Rainbow Trout (*Oncorhynchus mykiss*). *Aquaculture* **1991**, *94*, 79–88.

Steinbrenner, J. Light Induction of Carotenoid Biosynthesis Genes in the Green Alga *Haematococcus pluvialis*: Regulation by Photosynthetic Redox Potential. *Pl. Mol. Biol.* **2003**, *52*, 343–356.

Steinbrenner, J.; G. Sandmann. Transformation of the Green Alga *Haematococcus pluvialis* with a Phytoene Desaturase for Accelerated Astaxanthin Biosynthesis. *Appl. Env. Microbiol.* **2006**, *72*, 7477–7484.

Storebakken, T.; M. Sørensen; B. Bjerkeng; J. Harris; P. Monahan; S. Hiu. Stability of Astaxanthin from Red Yeast, *Xanthophyllomyces dedrorhous*, during Feed Processing: Effects of Enzymatic Cell Wall Disruption and Extrusion Temperature. *Aquaculture* **2004**, *231*, 489–500.

Suárez, G.; T. Romero; M.A. Borowitzka. Cultivo de la Microalga {\i Dunaliella salina} en Medio Orgánico. *Anais do IV Congresso Latino-Americano, II Reuniao Iberio-Americana, VII Reuniao Brasileira de Ficologia*; E.J. de Paula, M. Corediro-Marino, D.P. Santos, E.M. Plastino, M.T. Fujii, N.S. Yokoya, Eds.; Sociedade Ficologica de America Latina e Caribe & Sociatade Brasiliera de Ficologia: Caxambu, 1998; pp 371–382.

Tapiero, H.; D.M. Townsend; K.D. Tew. The Role of Carotenoids in the Prevention of Human Pathologies. *Biomed. Pharmacol.* **2004**, *58*, 100–110.

Tsubokura, A. Bacteria for Production of Carotenoids. U.S. Patent 5,858,761, 1999.

Verwaal, R.; J. Wang; J. Meijnen; H. Visser; G. Sandmann; J.A. van den Berg; A.J.J. van Ooyen. High-Level Production of Beta-Carotene in *Saccharomyces cerevisiae* by Successive Transformation with Carotenogenic Genes from *Xanthophyllomyces dendrorhous*. *Appl. Env. Microbiol.* **2007**, *73*, 4342–4350.

Visser, H. Metabolic Engineering of the Astaxanthin-Biosynthetic Pathway of *Xanthophyllomyces dedrorhous*. *FEMS Yeast Research* **2003**, *4*, 221–231.

Wang, W.; L. Yu; P. Zhou. Effects of Different Fungal Elicitors on Growth, Total Carotenoids and Astaxanthin Formation by *Xanthophyllomyces dendrorhous*. *Biores. Technol.* **2006**, *97*, 26–31.

Wegmann, K.; A. Ben-Amotz; M. Avron. Effect of Temperature on Glycerol Retention in the Halotolerant Algae *Dunaliella* and *Asteromonas*. *Pl. Physiol.* **1980**, *66*, 1196–1197.

Woodall, A.A.; S.W. Lee; R.J. Weesie; M.J. Jackson; G.Britton. Oxidation of Carotenoids by Free Radicals: Relationship between Structure and Reactivity. *Biochim. Biophys. Acta* **1997**, *1336*, 33–42.

Zhang, D.H. Two-Step Process for Ketocarotenoid Production by a Green Alga, *Chlorococcum* sp. Strain MA-1. *Appl. Microbiol. Biotechnol.* **2001**, *55*, 537–540.

Zhang, D.H.; Y.-K. Lee. Ketocarotenoid Production by a Mutant of *Chlorococcum* sp. in an Outdoor Tubular Photobioreactor. *Biotech. Lett.* **1999**, *21*, 7–10.

PART 4

Single Cell Oils for Biofuels

12

Survey of Commercial Developments of Microalgae as Biodiesel Feedstock

Marguerite Torrey

American Oil Chemists' Society, Urbana, Illinois, USA

Introduction

Over the past half-century, many nations have come to realize that petroleum, as a transportation fuel, is a finite resource. Efforts to develop alternative fuels during that period have seemed to ebb and flow with changes in politics and the global economy. OPEC (Organization of the Petroleum Exporting Countries) restrictions on oil shipments in 1973 and again in 1979 forced nations to consider how to produce their own fuels, thus increasing their national security, and along with this response came an expanded interest in the development of biofuels. For example, as a response to the OPEC embargo, Brazil moved to develop an alternative fuel from its plentiful sugarcane, producing ethanol as a transportation fuel. Other nations considered how to use their own agricultural products to develop liquid fuels that would more nearly resemble petroleum products.

Out of that came the development of what is now termed biodiesel (Knothe, 2004), a fuel made from animal fats, waste cooking fats and oils, and vegetable oils–such as soybean, rapeseed, corn, and palm oils. (In this work, biodiesel is taken to mean triglycerides that have been transesterified with methanol to produce fatty acid methyl esters (FAME) or with ethanol to produce fatty acid ethyl esters. There are some who also lump into this term "renewable diesel," or hydrocarbons generated from plant material.)

Microalgae—the microscopic single-cell plants that cloud poorly maintained swimming pools and aquaria and form scums on eutrophic bodies of water—have been considered a potential alternative source of fuel oil for at least 50 years (Sheehan et al., 1998), but little concerted effort was expended as long as petroleum was relatively inexpensive. Rising oil costs over the course of the past decade have shifted the interest of fuel producers to generating transportation fuels from agricultural commodities, such as corn (ethanol) and soybean, rapeseed, and palm oil (biodiesel). Globally rising food prices in much of 2008 were initially blamed on the use of these food crops as feedstocks for biofuels. Start-up companies and multinationals that were already exploring the possibilities of microalgae leaped at the possibility that they could sidestep the food vs. fuel conundrum and be the first to produce commercial quantities of oil from algae at a reasonable cost.

Algae—Pros and Cons

Proponents have put forward a number of reasons to consider algae as a biofuel feedstock (Table 12.A), and Hu et al. (2008) elaborated on some of the more important ones. (i) With few exceptions, microalgae are not generally a human food source. Thus, growing

Table 12.A. Comparison of Open Ponds/Raceways with Photobioreactors[a].

Parameter or Issue	Open Ponds/Raceways	Photobioreactors (PBR)
Space required	High	Low for PBR itself
Water loss	Very high; can lead to salt precipitation	Low
CO_2 loss	Can be high	Low
O_2 concentration	Usually low, owing to continuous outgassing	Builds up in a closed system. O_2 must be removed to prevent inhibition of photosynthesis and photo-oxidation
Temperature	Quite variable; function of pond depth	Cooling often required, achieved by spraying water on PBR or immersing tubes in cooling baths
Shear	Low (gentle mixing)	High (fast and turbulent flows needed for good mixing, pumping through gas exchangers)
Cleaning	No issue	Required (algal wall growth reduces light intensity, but cleaning causes abrasion, limiting lifetime of PBR)
Risk of contamination	High	Low
Quality of biomass	Variable	Reproducible
Concentration of biomass	Low: 0.1–0.5 g/L	High: 2–8 g/L
Production flexibility	Only a few species possible, difficult to switch	High: switching possible
Process control and reproducibility	Limited: flow speed, mixing, temperature (by pond depth)	Possible with limits
Weather dependence	High (light intensity, temperature, precipitation)	Medium (light intensity, cooling requirements)
Start-up	6–8 weeks	2–4 weeks

[a]Modified from Carlsson et al. (2007).

algae for fuel does not compete with production of human food. (ii) Some strains of microalgae can produce ten times more oil per hectare per year than oilseed plants, such as soybeans. (iii) Some strains, under carefully defined conditions, produce as much as 50% of their dry cell weight as oil. (iv) Algal growth requirements are simple: carbon dioxide, nutrients, light, and temperatures high enough above 0 °C that they promote cell division. (v) Besides growing in fresh water, some oil-producing algal strains can grow in saline, brackish, or highly contaminated water, thus avoiding the diversion of water fit for drinking and irrigation into the production of algae. Some projects have investigated the possibility of growing algae in sewage or industrial effluents. (vi) Algae can be grown on suboptimal land, such as in deserts or near industrial sites, where food crops would not thrive. (vii) Under optimal conditions, algae can grow rapidly, doubling their population in as little as three and a half hours (Chisti, 2007).

There are drawbacks to algae as feedstock, however. (i) During periods of darkness, the cells continue to respire, breaking down as much as 25% of the biomass produced during the lighted period (Ratledge & Cohen, 2008). Thus, net yield can be considerably less than gross yield. (ii) Lower temperatures reduce production; if algae are to be grown at ambient temperatures, then the geographic areas where they will thrive are confined to the tropics and certain areas of the temperate zone. Growing algae in water originating from geothermal sources could lessen this restriction. (iii) Growing algae on suboptimal land, such as in the desert, leads to issues regarding the availability of an enriched source of CO_2 to feed to the algae and the logistics of transporting algal oils and the fuels produced from them to the consumer. (iv) Algae grown in open settings, such as ponds or lagoons, are subject to predation and to the introduction of and competition with algae that produce little oil. (v) In a system containing high concentrations of algal biomass, self-shading reduces the attainment of maximal growth. (vi) CO_2 transfer across the air-water interface must be optimized to obtain maximal growth. (vii) Higher levels of oil production in cells are offset by lower rates of cell growth (Sheehan et al., 1998; Ratledge & Cohen, 2008). For oil accumulation to occur, there must be nutrient limitation, either of silicon (for diatoms) or nitrogen, so that excess carbon will be held as storage lipids instead of being incorporated into new cells.

History

From 1978 to 1996, the U.S. Department of Energy's Office of Fuels Development funded a program to identify economic sources of natural oil that could be used in place of petroleum. One component of this effort, the Aquatic Species Program (ASP), originated from an idea, proposed in the 1950s, of using algae as an oil feedstock (Sheehan et al., 1998). In the latter part of its existence, the ASP concentrated on the possibility of growing high lipid-content algae in ponds using waste CO_2 from coal-fired power plants as a carbon source.

The ASP isolated about 3,000 strains of algae that could grow under severe conditions (e.g., pH, temperature, nutrients) and produce sufficient quantities of oil.

Further experiments reduced the list to about 300 strains, most of which were members of the Chlorophyceae (green algae) and Bacillariophyceae (diatoms).

The ASP carried out its first field experiments in open ponds in California and Hawaii and then moved to ponds in Roswell, New Mexico. At the latter site, careful control of pH and physical conditions demonstrated how to achieve very efficient (>90%) utilization of CO_2. Despite these promising preliminary results, the Department of Energy terminated the ASP in 1996. In the Program's final report, Sheehan et al. (1998) said, "Even with aggressive assumptions about biological productivity, we project costs for biodiesel which are two times higher than current petroleum diesel fuel costs." At the time of that report, crude oil traded for $20 per barrel, and technology costs were much higher than they are in 2009, after adjusting for inflation.

The question of algae as an alternative fuel source resurfaced with some urgency around 2005 for at least three reasons: (1) concern about global warming, brought on by the burning of fossil fuels; (2) the need for national energy security, that is, the need for nations to be independent of the vagaries of foreign petroleum sources; and (3) the need for a fuel that cannot also serve a major source of food for humans. A number of companies that have entered the race to produce commercial quantities of algae-based fuels are listed in Table 12.B, and a number of cooperative projects also are taking place between companies on this list and large corporations, such as Chevron, Royal Dutch Shell, Boeing, Raytheon, UOP (a Honeywell Company), and General Electric.

In 2009, three principal methods for growing algae to produce triglycerides for FAME are being explored with major investment capital. The idea of open ponds or lagoons is the oldest and the most cost effective, but closed photobioreactors (PBR: algal growth systems based on tubular, flat plate, or other designs) circumvent some of the disadvantages associated with ponds. Both ponds and PBR depend on algal incorporation of CO_2 in the process of photosynthesis:

$$x \; H_2O \text{ and } y \; CO_2 + \text{other nutrients} \rightarrow \text{carbohydrates} + \text{proteins} + \text{lipids},$$

to produce lipids and oils that can be processed into fuel. In the third method, algae ferment organic compounds, in the absence of light, to produce oil.

Open Ponds

The ASP (Sheehan et al., 1998) was among the first efforts to grow large quantities of algae in open-air raceway ponds (Fig. 12.1). These closed-loop recirculation channels are formed from concrete or compacted earth (Chisti, 2007). The channels may be further coated or lined with white plastic to increase light reflectivity. Channels are typically about 0.3 m or less in depth to allow the algae adequate exposure to light. Some systems incorporate a paddlewheel to provide mixing and circulation for nutrients and to keep cells suspended. Baffles may be placed within the flow channel to guide flow around bends and aid in the mixing process. In ASP experiments, nutrients were added continuously in make-up water, and algae were harvested for processing into oil and by-products. One of the largest raceway systems constructed to date,

Table 12.B. Companies Currently Developing Algae as a Feedstock for Biodiesel[a].

Company Other[b]	Location[c]	Internet Address	Growth Mode[d]
Abpe Pty Ltd	Perth, Australia	www.apbe.com.au	PBR
Algae @ Work	Boulder, Colorado	www.algaeatwork.com	PBR
A2BE Carbon Capture LLC			
Algaecake Technologies Corp.	Tempe, Arizona	www.algaecake.com	PBR
Algae Floating Systems	South San Francisco, California	www.algaefloatingsystems .com	PBR
AlgaeLink N.V.	Roosendaal, The Netherlands	www.algaelink.com	PBR
AlgaeVenture Systems	Marysville, Ohio	www.algaevs.com	PBR
Algaewheel	Indianapolis, Indiana	www.algaewheel.com	O/PBR
Aquaflow Bionomic Corp.	Nelson, New Zealand	www.aquaflowgroup.com	O
Aquatic Energy LLC	Lake Charles, Louisiana	www.aquaticenergy.com	O
Aurora Biofuels	Alameda, California	www.aurorabiofuels.com	O
BARD (Biofuel Advance Research and Development) Algae	Philadelphia, Pennsylvania/ Painesville Township,Ohio	www.bardalgae.com	PBR
Bayer Technology Services	Leverkusen, Germany	www.bayertechnology.com	F
BFS Biopetróleo	San Vicente del Raaspeig (Alicante), Spain	www.biopetroleo.com	PBR
BioAlgene	Seattle, Washington	www.bioalgene.com	O
BioCentric Energy Inc.	Huntington Beach, California	www.biocentricenergy.com	PBR
BioMax Fuels	Laverton, North Victoria, Australia	www.energetix.com.au/ fuelalgae.html	PBR
Bionavitas	Redmond, Washington	www.bionavitas.com	O, PBR
Blue Marble Energy	Seattle, Washington	www.bluemarbleenergy.net	O
Bodega Algae, LLC	Jamaica Plain, Massachusetts	www.bodegaalgae.com	PBR
Canadian Pacific Algae	Nanaimo, British Columbia, Canada	http://canadianpacificalgae.com	O
Carbon Capture Corporation	La Jolla, California	www.carbcc.com	O
Carbon Trust	London, United Kingdom	www.carbontrust.co.uk	O

(Cont. on next page)

Table 12.B., cont. Companies Currently Developing Algae as a Feedstock for Biodiesel[a].

Company Other[b]	Location[c]	Internet Address	Growth Mode[d]
Cellana Royal Dutch Shell	Hawaii	www.cellana.com	O
Circle Biodiesel and Ethanol Corp.	San Marcos, California	www.circlebio.com	PBR
Culturing Solutions Inc.	St. Petersburg, Florida	www.culturingsolutions.com	PBR
Diversified Energy XL Renewables	Gilbert, Arizona	www.diversified-energy.com www.xlrenewables.com	Hybrid PBR
Dynamic Biogenics	Sacramento, California	www.dynamicbiogenics.com	PBR
ExxonMobil Research & Engineering	Clinton, New Jersey Fairfax, Virginia	www.exxonmobil.com/algae	GE/O/PBR/F
Galp Energia	Lisbon, Portugal	www.galpenergia.com	
General Atomics	San Diego, California	www.ga.com	
Global Green Solutions JV with Valcent Vertigro System	Vancouver, Canada	www.globalgreensolutionsinc.com	PBR
GreenFuel Technologies	Cambridge, Massachusetts	www.greenfuelonline.com	O, PBR
Greenshift Corporation Veridium Corp.	New York, New York	www.greenshift.com	PBR
Green Star Products	Chula Vista, California	www.greenstarusa.com	Hybrid PBR
HR Biopetroleum	Hawaii	www.hrbp.com	PBR, O
Ingrepro B.V.	Borculo, Netherlands	www.ingrepro.nl	O, PBR
International Energy	Newark, New Jersey	www.internationalenergyinc.com	PBR
Inventure Chemical	Tacoma, Washington	www.inventurechem.com	PBR
Kai BioEnergy	Del Mar, California	www.kaibioenergy.com	O
Kent BioEnergy Corporation	San Diego, California	http://kentbioenergy.com	O
Live Fuels Inc.	Menlo Park, California	www.livefuels.com	O
OriginOil	Los Angeles, California	www.originoil.com	PBR
PetroAlgae LLC	Melbourne, Florida	www.petroalgae.com	O

(Cont. on next page)

Table 12.B., cont. Companies Currently Developing Algae as a Feedstock for Biodiesel.[a]

Company Other[b]	Location[c]	Internet Address	Growth Mode[d]
Petrosun Inc.	Scottsdale, Arizona	www.petrosuninc.com	O
Petrosun Biofuels			
Phycal LLC	Highland Heights, Ohio	www.phycal.com	O
Phyco₂	Santa Maria, California	http://phyco2.us/home.html	PBR
Plankton Power	Wellfleet, Massachusetts	www.planktonpower.com	Modified O, PBR
Primafuel Inc.	Signal Hill, California	www.primafuel.com	PBR
Royal Dutch Shell	The Hague, The Netherlands	http://royaldutchshellplc.com/=2algae	O,PBR
RWE (Renewed World Energies) Corp.	Georgetown, South Carolina	www.rwenergies	PBR
Sapphire	San Diego, California	www.sapphireenergy.com	O
		www.greencrudeproduction.com	
SarTec	Anoka, Minnesota	www.sartec.com	PBR
SBAE Industries NV/SA	Waarschoot, Belgium	www.sbae-industries.com	PBR
Seambiotic	Tel Aviv, Israel	www.seambiotic.com	O
Solazyme Inc.	San Francisco, California	www.solazyme.com	F
Solena	Washington, DC	www.solenagroup.com	PBR
Solix Biofuels Inc.	Fort Collins, Colorado	www.solixbiofuels.com	PBR
Solray Energy	Christchurch, New Zealand	www.solrayenergy.co.nz	
SunEco Energy	Chino, California	www.sunecoenergy.com	O
Sunrise Ridge Algae, Inc.	Katy, Texas	www.sunrise-ridge.com	O
Synthetic Genomics	LaJolla, California	www.syntheticgenomics/com	GE/O/PBR/F
Valcent Products Inc.	El Paso, Texas	www.valcent.net	PBR
Global Green Solutions			
W² Energy Corp.	Carson City, Nevada	www.w2energy.com	PBR
XL Renewables	Phoenix, Arizona	www.xlbiorefinery.com	Hybrid PBR

[a]Modified from Torrey (2008).

[b]"Other" refers to company names representing joint ventures, cooperative arrangements, subsidiary relationships, etc.

[c]All companies are located within the United States, except where noted.

[d]Abbreviations: F, fermentation; GE, genetically engineered; O, open pond; PBR, photobioreactor.

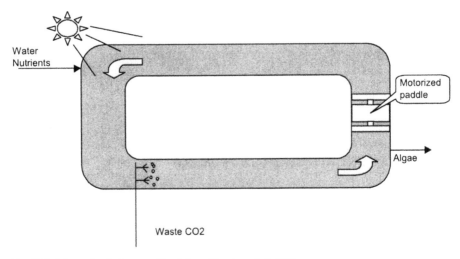

Fig. 12.1. Schematic of a Raceway Pond. From Sheehan et al., 1998.

owned by Earthrise Nutritionals (Irvine, California, USA; www.earthrise.com) occupies 400,000 m² (Spolaore et al., 2006) and is used to produce *Spirulina*, a blue-green alga, as a dietary supplement. Seven companies currently focused on growing algae for biofuel in some sort of open system are considered in the rest of this section. Throughout this chapter, access to information was frequently limited by proprietary considerations.

Aquaflow Bionomic Corporation Ltd.

Located in Nelson, New Zealand, this company, established in 2005, has set itself the task of being "the first company in the world to economically produce biofuel from wild algae harvested from open-air environments, to market it, and meet the challenge of increasing demand" (Aquaflow Bionomic Corp., accessed Jun 2008a). Aquaflow has been harvesting the algal scum that grows on nutrient-rich water sources such as sewage treatment ponds, dewatering it, and processing it into biodiesel. The company has worked only with wild algae, rather than culturing specific strains, citing the expense of the latter.

In December 2006, the possibilities of Aquaflow's processes were demonstrated when David Parker, the New Zealand Minister of Energy, test-drove an unmodified, standard Land Rover, fueled by a B5 blend (5% biodiesel, 95% mineral diesel mix) that incorporated Aquaflow's algal biodiesel, around the Parliament Building in Wellington. Subsequent analysis of the engine showed little variation in performance between B5 and pure diesel, except for lowered emissions, a positive feature (Aquaflow Bionomic Corp., accessed May 2009).

Aquaflow has been conducting algae-harvesting experiments on the Blenheim sewage treatment ponds, which cover over 60 ha and serve about 27,000 people. In

March 2008, a company press release said it had achieved commercial-scale, continuous harvesting of tons of wild algae per day (Aquaflow Bionomic Corp., accessed Jun 2008b). Aquaflow claims (Aquaflow Bionomic Corp., accessed Jan 2009a) it was able to harvest algae successfully even in New Zealand's South Island (latitude 41–42°S) winters.

Aquaflow and UOP-Honeywell signed a Memorandum of Understanding in October 2008, agreeing to work together to commercialize algal biofuel technology and use the growth of algae to sequester CO_2 emitted by fossil fuel-burning power plants (Aquaflow Bionomic Corp., accessed Nov 2008). This agreement was followed by an announcement in December 2008 that UOP's proprietary hydroprocessing technology had been used to convert wild algae to synthetic paraffinic kerosene, which, when blended with petroleum-based kerosene, can be used to power aircraft (Aquaflow Bionomic Corp., accessed Jan 2008b).

PetroSun Inc.

This company, located in Scottsdale, Arizona, along with its subsidiary PetroSun Biofuels, Inc., is developing algae farms in warmer areas of the United States, including Louisiana, Texas, Arizona, and Mississippi, as well as Mexico and Central America. The company has acquired a 1100-acre (450-ha) network of saltwater ponds near Harlingen, Texas, which is a few kilometers inland from the Gulf of Mexico, as part of its plan to grow algae in water not needed for human consumption or irrigation. PetroSun indicated it would grow native microalgal strains, so as not to disrupt local ecosystems if any inadvertent transfers of microalgae from company ponds to other bodies of water should occur. A press release in March 2008 (PetroSun, accessed Jun 2008) claimed that these ponds annually could produce a minimum of 4.4 million gallons of algal oil (16.6 million liters) and 110 million pounds of biomass (50,000 metric tons). The latter could then be processed into ethanol, animal feed, fertilizer, and other products.

In late 2008, PetroSun announced its interest in converting as much as 80,000 acres (32,000 ha) of former commercial catfish ponds in Mississippi, Alabama, Louisiana, and Arkansas into algal farm systems (PetroSun, accessed Nov 2008). As well as basic rent, pond owners were to be offered royalty income, based on the production of algal oil and algal biomass, and a potential future benefit from a carbon credit program. The company required commitments of 320 acres (130 ha) within the same area; people interested in joining the program, but with less land, could partner with a neighbor (Biodiesel Now, accessed Mar 2009). PetroSun claimed that ponds under its management in the Gulf Coast region had an annual potential production rate of 2,000 gallons of algae oil per acre (19,000 L/ha) (LeBlanc, 2009).

The company's announced intention is to ship algal oil to fuel refineries for processing. As of February 2009, the company had released no information regarding its success in producing algal oil, even on a bench-scale basis.

Seambiotic Ltd.

Like Aquaflow Bionomics, Seambiotic Ltd., headquartered in Tel Aviv, Israel, is striving to develop a portfolio of products and services derived from algae grown in ponds open to the air. The company's goals are to lower air pollution, to lessen global warming, and to produce algae with high concentrations of lipids and carbohydrates that can be processed into biodiesel and bioethanol, respectively (Seambiotic Ltd., accessed Jan 2009).

Since 2003, Seambiotic has been developing technology to grow marine microalgae in raceway ponds near the Israeli Electric Company power plant (IEC), located on the shore of the Mediterranean Sea, near Ashkelon (Seambiotic Ltd., accessed Jan 2009). Flue gases from the smokestack of the coal-burning power station are passed through a desulfurizing scrubbing system (thus, concentrating the CO_2) and then transferred through pipes to Seambiotic's open (raceway) ponds. Diffusers are used to dissolve the gases in the water, and gentle mixing keeps the cells suspended.

The company, to date, has cultivated selected species of microalgae, including *Nannochloropsis* and *Skeletonema*. The National Renewable Energy Laboratory (Boulder, Colorado, USA) analyzed *Nannochloropsis* grown in this system and found a lipid content of 37% on an ash-free, dry weight basis, according to a December 2008 posting on the company website (Seambiotic Ltd., accessed Jan 2009). At its 1000 m^2 pilot-scale site, which includes eight raceway ponds, the company can achieve 23 $g/m^2/day$ of algae. Seambiotic's scientific advisor and algal growth expert, Ami Ben-Amotz, said the company shipped 3 tons of *Nannochloropsis* to biofuel manufacturers in 2008 (Grant, 2009). At a typical oil content of 30% oil per gram of biomass, 3 tons would approximate 816 kg of algal oil, which could be converted to 100–200 gallons (400–800 L) of biofuel.

In June 2008, Seambiotic and Inventure Chemical Technology (Gig Harbor, Washington, USA) entered a joint venture to build a commercial plant to produce algae for conversion to biodiesel, a fermentable sugar solution, and a concentrated protein solution (Inventure Chemicals, accessed Jan 2009). Inventure was considered a good match with Seambiotic, as it already had a record of being able to produce biodiesel meeting ASTM standards from algal feedstock.

Seambiotic (accessed Jan 2009) announced in January 2009 that it was scaling up from a pilot-plant stage by establishing a 5-ha open-pond algal farm in collaboration with the IEC. Production was expected to increase later in 2009. Plans were to send lipids from the algae grown at this location to biodiesel manufacturers, carbohydrates to bioethanol producers, and proteins to makers of nutraceuticals (Grant, 2009).

Because the company claims its expertise is in growing algae, not in making biodiesel, it plans to establish fully integrated microalgal refinery operations by partnering with strategic fuel companies and electric plants to develop sites. Seambiotic expects to market these operations outside of Israel because of the scarcity, within the country, of large swathes of land near the sea, which, at its present location, provide make-up water for the ponds (Grant, 2009).

PetroAlgae

Another company planning to site its algae-growing ponds near fossil fuel-burning power plants is PetroAlgae of Melbourne, Florida, USA. The "ponds" they use are long, plastic tubes, about 5 ft wide by 200 ft long (1.5 m × 61 m), open to the air (Anderson, 2008). Stock solutions of algae that are selected for each specific location and that are indigenous to the region (Lane, accessed May 2009) are first grown in the laboratory, then transferred to the ponds, which contain a nutrient solution. CO_2 is bubbled in, not only providing nutrition to what the company is calling "micro-crops" (PetroAlgae, accessed May 2009) but also aiding mixing. Additionally, paddle-wheels promote uniform mixing (Stephens, accessed Feb 2009).

The company has been experimenting with twelve main strains of algae, some of which are licensed from Arizona State University. None of them are genetically modified (Lane, accessed Feb 2009). These strains reportedly have a lipid content of 30–40%.

In much of 2008, the company was claiming (Anderson, 2008; Garner, accessed Jun 2008; Kanellos, accessed Jun 2008) its system would be able to yield 10,000–14,000 gallons of algal oil per acre (100,000–130,000 L/ha). By early 2009, how-ever, PetroAlgae had nearly completed and was marketing a scaleable algal production system that it said was capable of producing 5,000–6,000 gallons/acre (50,000–60,000 L/ha), annually. In May 2009, the company said its systems were able to generate 40 g of biomass per square meter per day, or 400 kg/ha/day (Lane, accessed May 2009).

A company spokesperson has said that algae can be harvested in just two days after inoculation into the PetroAlgae system (Garner, accessed Jun 2008) and, thereaf-ter, can be harvested daily, or even more frequently (PetroAlgae, accessed May 2009). Lane (accessed Feb 2009) first reported that PetroAlgae was concentrating the algae to 50% water, then extracting the resultant paste with hexane to isolate the algal oil. In a later report (Lane, accessed May 2009), Chief Executive Officer John Scott indi-cated the company is separating the protein for use as an animal feed concentrate and is refining the oils and carbohydrates from the remainder to produce "a spectrum of distillates that is weighted toward the diesel side of the refinery." The company is still testing extraction parameters, however, to optimize costs, yields, and flow rates.

In February 2009, Lane (accessed Feb 2009) reported each unit of PetroAlgae's system was scaled to 0.5 acre (0.2 ha), with production systems based on multiples of 0.5 acre. The minimum area needed to produce a viable quantity of algal oil is esti-mated to be several hundred acres, and the ideal is 12,500 acres (5,000 ha).

Situating a PetroAlgae-licensed system within less than 10 miles (16 km) of a fossil fuel-burning power plant should allow for the economical transport of CO_2 to the algal growth ponds (Lane, accessed Feb 2009). The company has developed a proprietary process to concentrate power plant emissions, which average 12% CO_2, to more than 50% CO_2 before they are added to the ponds; at the same time, sulfur dioxides and mercury are removed.

PetroAlgae entered its first licensing agreement, in March 2009, with GTB Power Enterprise, Ltd. (Los Angeles, California, USA). GTB will use PetroAlgae's technology to construct and operate ten facilities at which algae will be grown and harvested for the production of oil and biomass. These will be located in the People's Republic of China, Taiwan, and Yonaguni (a Japanese island). Each facility will consist of 5,000 ha of land and will be constructed in increments of 500 ha (Yahoo! Finance, accessed Mar 2009). PetroAlgae does not expect that it will reach a production rate of 100 million gallons (400 million liters) of fuel per year by the end of 2010 with its licensees (Lane, accessed May 2009), but it does anticipate that licensees will be able to generate algae within a few months of initiating a project.

HR Biopetroleum

Hawaii-based HR Biopetroleum has developed a two-stage system, involving both PBR and open ponds, for growing proprietary strains of algae on a commercial scale. The company contends its technology removes some of the drawbacks of open ponds that were enumerated earlier in this chapter, particularly the need for rapid growth so as to be economically viable and the need to avoid contamination with undesirable algal species (HR Biopetroleum, accessed Jan 2009).

The company uses unique, naturally-occurring algal strains, licensed from the University of Hawaii, that grow rapidly and produce high quantities of oil under its culture conditions. Algae are initially grown in a PBR, consisting of large temperature- and pH-controlled tubes, connected in parallel, where they are exposed to sunlight and kept in suspension. The HR Biopetroleum system has much in common with that described by Huntley and Redalge (2006).

In the second stage, algae are transferred from the PBR to a sunlit, open-pond system (a recirculating raceway, lined with plastic) where the cells are exposed to a nutrient-depleted culture medium. Only enough nutrients are transferred with the fresh culture to allow cell division to continue for one day (HR Biopetroleum, accessed Jan 2009). [Huntley and Redalge (2006) suggest that oil content can reach 25% of dry weight by the second day after inoculation.] Such limitation stimulates oil production. Depending on the product desired, the company says (HR Biopetroleum, accessed Jan 2009) that it then harvests, cleans, and prepares the pond for a new production cycle on the second or third day after inoculation. HR Biopetroleum harvests its algal crop by gravitational settling into a slurry. Excess water is removed, and then centrifugation is used to further concentrate the algae. Other techniques, discussed later in this chapter, may be more economical than centrifugation. The resultant biomass is dried, and oil and other by-products can be extracted by proprietary procedures.

The company summarizes its approach as follows (HR Biopetroleum, accessed Jan 2009): "The coupled system minimizes cost. In a coupled system, photobioreactors provide a continuous source of single-species culture in ample quantity to inoculate the open ponds, allowing the batch cultures in open ponds to exhaust the nutrient supply in a short time, thus avoiding the perils of contamination by other species."

HR Biopetroleum and Royal Dutch Shell formed a joint venture company, Cellana, in 2008 to construct a pilot facility to demonstrate the former's algal production technology and its commercial viability (HR Biopetroleum, accessed Jan 2009).

Green Star Products, Inc.

This company, headquartered in San Diego, California, carried out field experiments on algal growth in 2007 at a demonstration facility in Hamilton, Montana, USA, and claimed these had been some of the larger algal demonstration projects to date (Green Star Products, accessed Jun 2008). The company describes its system as a hybrid: Algae are grown in an enclosed, controlled environment, similar to a PBR, but, within that environment, they are contained in ponds that are kept in motion by paddlewheels. Enclosing the system provides the following benefits: (i) reduction of the amount of light reaching the algae, thus preventing oxidation and sun bleaching on high-sunlight days; (ii) reduction of the rate of evaporation and exclusion of precipitation in the form of rain and snow, thus exerting control over salinity; and (iii) minimization of contamination by unwanted organisms and windblown particles.

In Phase I of the Montana demonstration, variables necessary to algal production were controlled—temperature, salinity, evaporation, and pH—and pond construction costs were established. Make-up water came from well water at the site, and additional CO_2, derived from the generator that powered the circulation pumps, was fed to the algae (Stella, 2009). A "micronutrient booster growth formula" enhanced growth several-fold (Green Star Products, accessed Oct 2008). Flow rates and mixing rates were evaluated.

Phase II concentrated on growing a laboratory-tested algal strain that produced a 21% oil content (this percentage is lower than what many algal entrepreneurs consider to be a value at which a profit can be made). Unexpectedly, the selected strain was able to tolerate sustained ambient temperatures as high as 46 °C for several hours at a time. Cells larger than 2 μm were harvested, and smaller cells were returned to the growth medium to continue growing and reproducing. Regarding the harvest, the company reported it had been unprepared for "the exponential growth of the algae" and had to modify the pumping system in place to avoid damaging the algal cells.

In Phase III, Green Star Products grew algae at its Montana site under winter conditions, including temperatures below −18 °C and snowfall that reached 14 inches (360 mm). In addition to adding CO_2 from the generator to the system as an algal nutrient, heat from the generator was discharged into the algal water, providing some warmth for their growth. During this several-month experiment, temperatures within the reactor were 60–65 °F (15.6–18.3 °C), providing sufficient heat to melt snow falling on the enclosure (Stella, 2009). Algae continued to grow despite lower light levels, shorter days, and lower temperatures. The company proposes that CO_2 from coal burning could be used to as a carbon source for algae in its system.

According to Sims (accessed Feb 2009), Green Star has licensed a technology that eliminates the need to dry the harvested algae before extracting the oil. Instead, oil is continuously stripped from the algae, and the rest of the biomass is broken down into carbohydrates, proteins, and other components.

Production figures have not been released to the public, but the company predicts it will produce commercial quantities by August 2009 at an undisclosed facility in the Pacific Northwest (Stella, 2009).

Sapphire Energy

Venture capitalists have directed over $100 million in funding to Sapphire Energy in La Jolla, California, USA (Sapphire Energy, accessed May 2009b). The company is using algae to produce hydrocarbon compounds, or what it terms "green crude" (Sapphire Energy, accessed May 2009a). This product is not, strictly speaking, derived from single-celled organisms. Rather, green crude represents an algal product that can be refined to produce gasoline, diesel, and jet fuel, all of which can be used in existing pipelines, refineries, cars, trucks, and planes without the preliminary chemical modifications that are necessary for converting FAME into biodiesel.

Sapphire Energy received nearly $1 million in March 2009 (American Fuels, accessed May 2009) in the U.S. Federal Omnibus Appropriations Bill to fund an algae-to-fuel demonstration project, to be sited on 20 acres (8 ha) in Portales, New Mexico. The algae, which have been genetically engineered to produce hydrocarbons, will be grown in brackish water in raceway ponds open to the air (Kanellos, accessed May 2009a). Genome researchers help pinpoint the algae best suited to a site. According to Kanellos (accessed May 2009a), Tim Zenk, the vice president of corporate affairs for Sapphire, indicated that the company has minimized the danger of rogue algal blooms escaping to the surrounding area as well as out-competition by natural local strains because "we will optimize [the algae] to live only in certain conditions."

Sapphire has already used algae to make a gasoline that met fuel quality standards, achieving a 91 octane rating (Blumenthal, accessed May 2009). In April 2009, the company announced it will produce 1 million gallons (4 million liters) of diesel and jet fuel per year by 2011; by 2018, they anticipate generating 100 million gallons annually, and up to 1 billion gallons of fuel per year by 2025 (Sapphire Energy, accessed May 2009c).

Photobioreactors

As a consequence of the problems associated with growing algae in open-air ponds (Table 12.A), entrepreneurs are searching for ways to use PBR to produce algae, with the hope that, whereas ponds may work in some locations, perhaps PBR should work in many more. Benemann (accessed Nov 2008) noted that little information has been released yet on the detailed designs, operations, yields, and other parameters that are important when evaluating the potential for commercial algal production with either open systems or PBR. On the other hand, academic and government laboratories have produced hundreds of publications on algal cultivation with small-scale PBR, leading to "a perception that the latter are the better and more promising systems" (Benemann, accessed Nov 2008). Investment in PBR systems is certainly riskier than in pond systems, as the following paragraphs will demonstrate.

In its most general form, a PBR is a closed, often tubular, system in which algae are grown. Enclosing the system permits essentially single-species culture for prolonged periods. The solar collector for a tubular PRB is often an array of straight, transparent growth tubes, oriented either vertically or horizontally. Another design that has been investigated is the construction of a PBR as a tubular helix, which appears somewhat like a coil spring or a Slinky® toy. Tube diameters are designed to allow adequate light penetration into the dense cultures they contain, maximizing algal productivity. Turbulent flow must be maintained within the tubes to prevent settling out.

GreenFuel Technologies Corporation

Headquartered in Cambridge, Massachusetts, USA, GreenFuel Technologies Corporation was founded in 2001 with the goal of growing algae on CO_2 derived from flue gases, smokestacks, and fermentations to produce biomass for processing into biofuel and feed.

The company has been testing its ideas in movable, greenhouse-like structures that contain a series of racks from which are suspended five-foot-long plastic algae-growing tubes (GreenFuel Technologies Corporation, accessed Jan 2009). The algae inoculated into these tubes are chosen for their innate growth rate, their ability to thrive under the CO_2 content of the emissions being fed to them, their favorable overall composition (lipids, carbohydrates, protein), and their ability to grow in specific climatic conditions.

According to the company website (accessed Jan 2009), gases are introduced into the PBR, in which algae are suspended in a nutrient medium. Once growth has been initiated, part of the medium is withdrawn continuously and then dewatered in two stages. In the first, the algal biomass is concentrated 10–30 times, and, in the second, enough additional water is removed so that a "cake" is formed. The GreenFuel protocol then calls for the algal cake to be processed into oil, delipidated meal, and dried, whole algae. The company does not plan to produce its own biodiesel; instead, it plans to market its algal oil as feedstock to companies already in the business of making FAME for fuel.

The company website states that GreenFuel "has successfully installed its systems" at six facilities that burn gas, coal, and oil and discharge gases containing 5–30% CO_2. These have been small-scale operations, and no commercial quantities of algal oil have been produced. Indeed, in a 2007 attempt to scale up by 100 times from an earlier test model at the 1,060-megawatt Arizona Public Service Redhawk site (Phoenix, Arizona, USA), algae grew more prolifically in the Arizona sun than GreenFuel had anticipated. The organisms clogged the system and died, forcing operators to shut it down (Stipp, accessed Jan 2009). The company concluded that trying to scale up a small test system by a factor of 100, all at once, led to the malfunction (Waltz, 2009).

The company learned from the experience and, in October 2008, announced the second phase of a joint project with Aurantia, SA (Madrid, Spain) to develop and scale up algal farming technologies in the Iberian peninsula (GreenFuel Technologies, accessed Oct 2008). The first phase began in December 2007 at the Holcim cement plant, near Jerez, Spain, where GreenFuel successfully grew a variety of naturally occurring algal strains, using Holcim flue gases. In the second phase, the algae are to be inoculated, grown, and harvested from a 100-m^2 prototype, vertical, thin-film PBR. After land is acquired, Aurantia-GreenFuel intends to construct a 1,000-m^2 algal greenhouse and harvesting facilities, adjacent to the cement plant, while continuing to evaluate the influence of the plant's flue gases on algal growth rates. Ultimately, the companies predict they can scale up production to 100 ha of algal greenhouses, producing 25,000 tons of algal biomass yearly.

Perhaps because GreenFuel was among the earliest companies to attempt to develop technology to grow algae commercially for fuel using a PBR system, its operations have been subjected to more scrutiny than later arrivals. Dimitrov (accessed Feb 2009a) concluded that GreenFuel's methods are "not economically feasible . . . . Fundamental thermodynamic constraints make it impossible for such approach[es] to be commercially viable for fuel prices below $800 per barrel." The company quickly protested that its methods are sound (Dimitrov, accessed Feb 2009b), but Dimitrov maintained that his conclusion was correct. John Benemann, coauthor of the final report on the National Renewable Energy Laboratory and the ASP (Sheehan et al., 1998), also expressed misgivings about GreenFuel's technologies (Rapier, accessed Feb 2009).

Note added in proofing: On May 13, 2009, Duncan McIntyre, a spokesperson for Polaris Venture Partners, that had invested in GreenFuel, announced that the company was shutting down after running out of money (LaMonica, accessed May 2009). He also said investors were exploring ways to sell GreenFuel's intellectual properties and assets.

AlgaeLink N.V.

Dutch manufacturer AlgaeLink N.V. is cultivating algae for processing into biodiesel in high-impact, 93%-clear PMMA (polymethylmethacrylate) tubes containing an ultraviolet light inhibitor at its home location in Roosendaal, The Netherlands. The tubes are either 480 or 2000 m long, with a diameter of 250 mm (AlgaeLink, accessed Feb 2009) and wall thicknesses of 200–400 microns (AlgaeLink, accessed Jun 2008). The tubes are mounted horizontally inside the shelter of greenhouses, either in straight lines or as a sequence of linked U-tubes (Fig. 12.2). Because the systems are modular, more tubes can be added to the system, as needed.

Nutrients and acidity levels in the system are regulated, and pumps intermix algae with CO_2 and nutrients. The company says its patented technology has overcome problems of shading in these tubes. It also points out that it has patented an internal cleaning system for the growth tubes that prevents clogging and permits non-stop harvesting (Henley, accessed Feb 2009). In March 2008, AlgaeLink indicated (accessed

Fig. 12.2. Photobioreactor as Produced by AlgaeLink N.V. Courtesy of AlgaeLink N.V., Roosendaal, The Netherlands.

Jun 2008) that it harvests algae by filtration or centrifugation, but it is less precise in describing how it separates the cells into their chemical components. The company also says it has a patent pending on an oil extraction system that does not use chemicals, does not require that the algae be dried, and does not use an oil press.

The company suggests the material remaining after oil removal can be sold as feed for fish and oyster farms. Alternatively, the remainder can be used as a source of vitamins; nutritional supplements; antioxidants, such as α-carotene and astaxanthin; or sterols, for use in pharmaceutical products.

AlgaeLink manufactures equipment for growing algae to sell to interested third parties. The company is presently installing algae-growing tubes in the open air, without shelter, in the south of Spain and in China (Henley, accessed Feb 2009). The company claims it can supply turnkey algal cultivation plants producing 1–250 tons of dry algae/day. However, the company has no installations yet producing on a commercial scale (Deckers, 2009).

Valcent Products

A joint venture with Global Green Solutions (Vancouver, Canada), Valcent Products is growing algae in a vertical system of 10-foot (3.3-m) long, segmented plastic bags that are closed to the air at its facility in El Paso, Texas, USA. Vertigro, the name for its

Fig. 12.3. Vertigro Algae Biofuel System from Valcent Products. Courtesy of Valcent Products, El Paso, Texas, USA.

patented vertical bioreactor system, is designed to increase the available surface area to expose cells to the sunlight.

In the Vertigro system, algae-containing water is initially held in an underground tank, so as to maintain a fairly constant temperature (Jones, accessed Jun 2008). The process is initiated when a pump lifts the algae-containing water to a holding tank 3 m above the top of the plastic bags (Fig. 12.3). Algal water then moves by gravity through horizontal flow paths (connected at alternate ends by vertical tubular spaces) within clear plastic sheets. Light shines on the algae as they pass through the tubes, and photosynthesis occurs. At the bottom of the sheets, a collection chamber feeds back into the underground tank, where the O_2 produced by photosynthesis is stripped out, more CO_2 is added, and the cycle begins again.

According to Jones (accessed Jun 2008), once the concentration of algae reaches a predetermined level, say, 1.5 grams per liter, harvesting can begin. Over a 24-hour period, half of the fluid is skimmed to remove the algae, and the remaining water is returned to the tank. By matching the skimming rate to the rate at which algae grow back to their original density, the system becomes a continuous process, so long as light and CO_2 are available to generate algae (and oil).

With respect to algae, Valcent is involved only in growing naturally occurring, indigenous species; that is, the company is an algae farmer (Mitra, accessed Feb 2009). The company has no plans at present to invest in manipulating algae genetically to increase oil production. Once oil has been extracted from the harvested cells, the company business plan is to sell the oil to refineries equipped to make biodiesel.

Valcent has yet to achieve the production of commercial quantities of algae. In a November 2008 interview (Mitra, accessed Feb 2009), Valcent's Chief Executive Officer and President Glen Kertz said the company was conducting experiments with a 30-panel reactor and planned to scale up further by building a 100-panel reactor, covering one-eighth of an acre (0.05 ha). Still further in the future, there are plans are for a 1–5 acre (0.4–2 ha) pilot facility.

On January 13, 2009, Valcent laid off nineteen employees at its El Paso facility, or about half of its staff (Kolenc, accessed February 2009). Maintenance staff were left in place to manage the algal research and development facilities (Valcent, accessed Feb 2009).

Note added in proofing: Company President Glen Kertz stepped down in April (Reuters News Agency, accessed Apr 2009), and Valcent/Global Green Solutions are still determining how to proceed with their algal business (Johnston, 2009).

Solix Biofuels

Like Valcent, Solix Biofuels (Fort Collins, Colorado, USA) is developing a PBR system based on growing algae in suspended bags. Solix considered a number of configurations to attempt to optimize the light impinging on the PBR—a traditional glass box bioreactor (very long, very high, small width), ranks of horizontally oriented glass tubes, and raceways (Wilson, accessed May 2009). Company experiments showed that diffuse light, reaching into a tubular array, produced better yields than direct light, for example, shining on a raceway system. The system they chose to optimize is a series of bags into which CO_2 and nutrients are circulated. The bags are immersed in and supported by water, which helps control temperature.

Solix is a spin-off and technology partner of Colorado State University (CSU; Fort Collins, USA). Solix seed funds were used to sponsor research by CSU faculty and graduate students to identify algal species with the best potential to grow at a large scale and produce high yields of fuel and chemical feedstocks, and to develop the technology to bring the process to commercial scale (Solix Biofuels, accessed May 2009).

In November 2008, Solix announced it had raised $10.5 million in outside funding that it would use to build an algal biofuel facility near Durango, Colorado (RedOrbit.com, accessed May 2009). Located on a 10-acre (4 ha) site, it will be built in two phases. The first phase, to be completed by May 2010, will consist of 4 acres of PBR for growing algae and 1 acre for a laboratory. The second phase is to be a 5-acre expansion that will allow the pilot facility to produce on a commercial scale. Investors have committed $5 million for construction of the pilot plant.

In September 2008, the company's chief technology officer, Bryan Willson, anticipated being able to achieve an algal productivity of 30–40 g/m²/day by the second quarter of 2009 (Willson, accessed May 2009). In late June 2009, Willson amended this prediction to say the company was producing about 2,500 gallons per acre (23,000 L/ha) and that it expected to reach cost parity with $80/barrel of oil in 3–4 years (Lane, accessed July 2009).

OriginOil

In its exploration of how to profit from algal production with a PBR, OriginOil (Los Angeles, California, USA) is working to optimize the light to which it exposes the algae, rather than depending solely on sunlight. In a pond with actively growing algae, sunlight only reaches to a depth of a few centimeters because of self-shading by algal cells. That is, growth occurs only in a superficial layer. To circumvent this problem, OriginOil has trademarked its Helix BioReactor, which features a rotating, vertical shaft with very low energy lights, arranged in a helix or spiral pattern and producing a strobe effect as the shaft spins through the algae. Thus, instead of a single growth layer, the result is a theoretically unlimited number of growth layers (OriginOil, accessed May 2009a). The lighting elements on the shaft are engineered to produce light waves of red and blue frequencies, rather than the full spectrum of sunshine, to optimize the photosynthetic process (Kram, accessed Feb 2009). The company claims biomass within this system can double in as little as a few hours.

OriginOil is promoting the use of a method that it calls Quantum Fracturing at two points during the process of growing algae for their oil content. In the initial stage, quantum fracturing breaks "water, carbon dioxide, and other nutrients . . . at very high pressure to create a slurry of micron-sized nutrition bubbles" that are then discharged to the algal culture, held at ambient pressure in the Helix BioReactor (OriginOil, accessed May 2009b). This step is taken to enhance the speed of nutrient distribution and to avoid aeration.

The other point at which quantum fracturing is used is in harvesting. The creation of micron-sized bubbles, combined with low-power microwaves, disrupts the algal cell walls, thus releasing the oil. The company home page presents a video of the course of gravitational separation of quantum-fractured cells of *Nannochloropsis oculata* and the oil they contained (oil floats to the top, biomass sinks to the bottom) over a period of one hour (OriginOil, accessed May 2009). Oil can then be skimmed off and refined, and the remaining biomass can be collected from the bottom for processing into other products, all without the use of added chemicals, heavy machinery, or initial dewatering.

According to Schill (accessed May 2009), OriginOil intends to distribute the technology it has developed for harvesting algae and extracting oil separately from its technology for growing algae, which it is still improving. In May 2009, OriginOil and Desmet Ballestra (Zaventem, Belgium) announced formation of a partnership to begin to commercialize OriginOil's oil extraction systems (Environmental Protection, accessed May 2009). This agreement may allow energy efficiency gains of 90% in certain configurations (The Bioenergy Site, accessed May 2009). Once the company has erected a pilot plant and started to sell equipment and/or know-how for oil extraction, OriginOil plans to perfect its PBR system (Kanellos, accessed May 2009b).

In an October 2008 press release, OriginOil said, based on its prototype work, that it expected to be able to achieve the production of 20–25 g dry algal matter/L over a 24–36 hour period (Origin Oil, accessed Oct 2008). OriginOil subsequently signed a Cooperative Research and Development Agreement with the U.S. Department of Energy's Idaho National Laboratory to validate and commercially scale the company's technology (Sims, accessed May 2009).

Fermentation

Moving from growing algae photosynthetically in outdoor ponds to enclosed PBR significantly increases costs and complexity of the process. A major issue is providing adequate light for maximal photosynthetic growth. Several companies are attempting to circumvent these issues by dispensing with the need for sunlight.

Solazyme

California-based Solazyme is developing processes based on the observation that some microalgal species are able to switch from photosynthetic to heterotrophic growth. The company contends that growing algae heterotrophically, on glucose or other carbon sources, in the dark, forces them to produce more oil than they would in the light (Bullis, accessed Jun 2008). That is, nullifying the photosynthetic process allows other metabolic processes that convert sugar into oil to become more active. Also, the algae can grow to higher concentrations in the dark than they can under sunlight, because they are no longer light-limited by shading and day length, as they are when growing photosynthetically.

Solazyme has been genetically modifying strains of marine algae and growing them in stainless steel fermentors. The company has investigated alterations in light-harvesting genes, chlorophyll biosynthesis genes, and signaling genes, as well as synthetic genes containing unnatural codons, to maximize triglyceride production (Waltz, 2009).

The oil produced by algae in Solazyme's process has been extracted and further processed into a range of fuels, including biodiesel, renewable diesel, and jet fuel. Chevron Technology Ventures, a division of Chevron USA Inc., is cooperating with Solazyme to develop and test algal biodiesel (Solazyme, accessed Jun 2008). So far, Solazyme has announced a biodiesel, Soladiesel BD™, which, it says, exceeds the requirements of both the ASTM biodiesel standard D6751 and the European standard EN 14214. In June 2008, the company also announced a renewable diesel, Soladiesel

RD™, which meets both ASTM D975 specifications and the ASTM ultra-low sulfur diesel standards (Solazyme, accessed Aug 2008). According to a company press release (Solazyme, accessed Nov 2008), both fuels have been successfully road-tested in their unblended forms (100% algal diesel) for thousands of miles in standard, unmodified diesel engines. The company also announced that it has produced an algae-based jet fuel that meets all 11 of the tested criteria (ASTM D1655) for Jet A-1 fuel.

Another angle that Solazyme is pursuing to increase the economic potential for growing algae to produce biodiesel is to feed glycerol (a by-product of the manufacture of biodiesel from triglycerides), by itself or in combination with other feedstocks, to a second microorganism—such as an oleaginous yeast, a fungus, or another alga. This would allow for the fermentative production of oil, which would be processed into biodiesel (Trimbur et al., 2009).

The company expects annual production of algal biofuel to reach tens to thousands of gallons by the end of 2009 (Grant, 2009) and 100 million gallons by 2012 or 2013 (Lane, accessed Apr 2009).

Bayer Technology

In Germany, Bayer Technology Services is considering the economic possibilities of growing algae in a "deep-dark-tank" (DDT) system, in which algae are grown on sugars from biomass (sugar, starch, cellulose) and from which animal feed and oil are harvested. Such a facility would require considerably less land than open ponds and far less capital than current PBR concepts (Steiner, accessed Feb 2009).

A DDT approach would allow more flexibility than a PBR, particularly with respect to having a choice of feedstocks and avoiding seasonal variations. Calculations tend to indicate that a DDT system would be more energy efficient than a facility producing bioethanol. Furthermore, the technology may enable decentralized production. On the other hand, the technology clearly derives no benefit at all from the high photosynthetic capabilities of algae.

Synthetic Genes

Founded in 2005 by J. Craig Venter, of human genome fame, Synthetic Genomics Inc. (SGI; La Jolla, California, USA) is using genetic and genomic techniques to devise ways to acquire fuel from algae. The company is modifying genes to create new secretion pathways so that algae can continuously expel the oil they synthesize, thus making collection of the oil easier (Waltz, 2009). In 2008, Venter said the algae with which the company is working have been genetically engineered to use sunlight and CO_2 to synthesize C_8, C_{10}, and larger lipids (Ladd, accessed Jan 2009). Because the algae secrete these molecules continuously, they can be regarded as chemical factories. Venter pointed out that, by supplying the algae with concentrated CO_2, the process could benefit the environment by converting CO_2 into usable products.

When asked when SGI would have commercial quantities of transportation fuels available, Venter replied that SGI has a goal of having "multiple things on the market" within five years (Ladd, accessed Jan 2009). As of January 2009, capital to build a pilot plant was still being sought (Zimmer, accessed Jan 2009).

The Future

Interest in algae as a source of fatty acids for the production of biodiesel/renewable diesel is high, and news coverage of advances appears in a number of venues (Table 12.C), but the fact remains that no one has yet shown that they can cheaply and reliably transform "pond scum" into a fuel in quantities that would significantly affect the consumption of petroleum-based fuels. As Benemann (accessed Nov 2008) pointed out, "There is no guarantee that a sufficiently low-cost process can actually be engineered. On the other hand, there are no clear 'show-stoppers' that would suggest that either the biological or engineering R&D cannot be eventually successful."

Table 12.C. Sources of Business News About Algae as a Feedstock for Biodiesel.

Organization	Website	Comment
Algal Biomass Organization	www.algalbiomass.org	Many backers in aviation industry
European Algal Biomass Association	www.eaba-association.eu	
National Algae Organization	www.nationalalgaeassociation.org	Predominantly, small entrepreneurs
National Biodiesel Board	www.biodiesel.org	
Biobased News	www.biobasednews.com	Bi-weekly e-newsletter
Biodiesel Magazine	www.biodieselmagazine.com/	Published monthly
Biodiesel News	www.biobasednews.com/news/biodiesel	
World Grain's Biofuels Business e-newsletter	www.biofuelsbusiness.com	Available print and digital
Biofuels Digest	www.biofuelsdigest.com	Updated Monday–Friday
Biofuels International	www.biofuels-news.com	Hard copy published 10 times per year
Biofuels Journal	www.biofuelsjournal.com	
Checkbiotech	http://checkbiotech.org	Updated Monday–Friday
Greentech Media	www.greentechmedia.com	Weekly e-newsletter
LexisNexis	www.lexis.com	Aggregates news from around the world Monday–Friday
National Renewable Energy Laboratory	www.nrel.gov	
Oilgae	www.oilgae.com	

References

AlgaeLink. http://www.algaelink.com/commercial-cultivation-plants.htm (accessed Feb 2009).

AlgaeLink Home Page. http://www.algaelink.com (accessed Jun 2008).

American Fuels. Sapphire Energy Algae to Fuel Demonstration Project Receives Funding. [Online] Mar 15, 2009. http://www.americanfuels.info/2009/03/sapphire-energy-algae-to-fuel. html (accessed May 2009).

Anderson, S. Biofuels Profile: PetroAlgae. *Biofuels Bus.* Sept **2008,** 46–50.

Aquaflow Bionomic Corporation Home Page. http://www.aquaflowgroup.com (accessed Jun 2008a).

Aquaflow Bionomic Corporation. Aquaflow Makes Crucial Algae Biofuel Breakthroughs. Press Release [Online] Mar 31, 2008. http://www.aquaflowgroup.com (accessed Jun 2008b).

Aquaflow Bionomic Corporation. Aquaflow and US Refinery Leader (UOP) Join Forces on Renewable Fuels. Press Release [Online] Oct 30, 2008. http://www.aquaflowgroup.com (accessed Nov 2008).

Aquaflow Bionomic Corporation. Prospectus and Investment Statement. [Online] Oct 30, 2008. http://www.aquaflowgroup.com (accessed Jan 2009a).

Aquaflow Bionomic Corporation. Aquaflow Wild Algae Converted to Key Jet Fuel Component. Press Release [Online] Dec 15, 2008. http://www.aquaflowgroup.com (accessed Jan 2009b).

Aquaflow Bionomic Corp. World First Wild Algae Bio-diesel Test Drive. Press Release [Online] Dec 15, 2006. http://www.aquaflowgroup.com/ScoopWorld1stwildalgaebio-dieseltestdrive151206.htm (accessed May 2009).

Benemann, J.R. Opportunities and Challenges in Algae Biofuels Production. Prepared for Algae World 2008, Singapore, November 17–18, 2008. [Online] http://www.futureenergyevents.com/algae/whitepaper (accessed Nov 2008).

Biodiesel Now. Weary Catfish Farmers Becoming Landlords of Algae. [Online] http://www.biodieselnow.com/forums/t/24247.aspx (accessed Mar 2009).

The Bioenergy Site. OriginOil in Partnership with Desmet Ballestra. [Online] May 5, 2009. http://www.thebioenergysite.com/news/3645/originoil-in-partnership-with-desmet-ballestra (accessed May 2009).

Blumenthal, L. Boeing Brews Up Algae, a Promising Fuel of the Future. *The News Tribune* [Online] http://www.thenewstribune.com/news/local/story/560281.html (accessed May 2009).

Bullis, K. Fuel from Algae. *Technol. Rev.* [Online] Feb 22, **2008.** http://www.technologyreview.com/business/20319/page2 (accessed Jun 2008).

Carlsson, A.S.; J.B. van Beilen; R. Möller; D. Clayton. *Micro- and Macro-algae: Utility for Industrial Application*; CPL Press: Newburg, Berkshire, UK, 2007, pp. 1–87.

Chisti, Y. Biodiesel from Algae. *Biotechnol. Adv.* **2007,** *25,* 294–306.

Deckers, S. AlgaeLink, Roosendaal, The Netherlands. Personal communication, 2009.

Dimitrov, K. GreenFuel Technologies: A Case Study for Industrial Photosynthetic Energy Capture. [Online] March 2007. http://www.nanostring.net/Algae/CaseStudy.pdf (accessed Feb 2009a).

Dimitrov, K. GreenFuel Technologies: A Case Study for Industrial Photosynthetic Capture— Follow-up Discussion. [Online] 2007. http://www.nanostring.net/Algae/CaseStudyFollowup.pdf (accessed Feb 2009b).

Environmental Protection. OriginOil Teams Up with Desmet Ballestra for Algae Market. [Online] May 12, 2009. http://www.eponline.com/articles/72010 (accessed May 2009).

Garner, B. Turning Algae into Oil. [Online] http://www.wptv.com/mostpopular/story. aspx?content_id=ed49a7ea-a5a5-40de-be86-d28361fcce63 (accessed Jun 2008).

Grant, B. Future Oil. *The Scientist* **2009,** *23 (2),* 36–41.

GreenFuel Technologies Corporation. GreenFuel Algae CO_2 Recycling Project with Aurantia Enters Second Phase at Spanish Cement Plant. Press Release [Online] Oct 21, 2008. http://www.greenfuelonline.com (accessed Oct 2008).

GreenFuel Technologies Corporation Home Page. http://www.greenfuelonline.com (accessed Jan 2009).

Green Star Products. Complete Algae Demonstration Report. [Online] May 2008. http://www.greenstarusa.com/news/08-05-09.html (accessed Jun 2008).

Green Star Products. Green Star One Step Closer to Marketing Algae Booster: GSPI Receives Independent Test Confirmation of Algae Booster Effectiveness. Press Release [Online] Oct 13, 2008. http://www.greenstarusa.com/news/08-10-13.html (accessed Oct 2008).

Henley, P. In Bloom: Growing Algae for Fuel. *BBC News* [Online] Oct 9, 2008. http://news.bbc.co.uk/2/hi/science/nature/7661975.stm (accessed Feb 2009).

HR Biopetroleum Home Page. http://www.hrbp.com (accessed Jan 2009).

Hu, Q.; M. Sommerfeld; E. Jarvis; M. Ghiradi; M. Posewitz; M. Seibert; A. Darzins. Microalgal Triacylglycerols as Feedstocks for Biofuel Production: Perspectives and Advances. *Plant J.* **2008,** *54,* 621–639.

Huntley, M.E.; D.G. Redalge. CO_2 Mitigation and Renewable Oil from Photosynthetic Microbes: A New Appraisal. *Mitigation and Adaptation Strategies for Global Change* **2006,** *12,* 573–608.

Inventure Chemical Technology. Inventure Chemical and Seambiotic Enter Joint Venture to Build Commercial Algae to Biofuel Plant in Israel. Press Release [Online] Jun 18, 2008. http://www.inventurechem.com/news5.html (accessed Jan 2009).

Johnston, H. Global Green Solutions, Vancouver, Canada. Personal communication, 2009.

Jones, W.D. The Power of Pond Scum: Biodiesel and Hydrogen from Algae. *IEEE Spectrum Online* [Online] April 2008. http://www.spectrum.ieee.org/apr08/6175 (accessed Jun 2008).

Kanellos, M. Life on Mars: The Secret Ingredient for Biofuel? Greentech Media. [Online] May 26, 2008. http://www.greentechmedia.com/articles/life-on-mars-the-secret-ingredient-for-biofuel-935.html (accessed Jun 2008).

Kanellos, M. Inside Sapphire's Algae-Fuel Plans. Greentech Media. [Online] Oct 13, 2008. http://www.greentechmedia.com/green-light/post/inside-sapphires-algae-fuel-plans-646 (accessed May 2009a).

Kanellos, M. A Tool Kit for Algae Growers to Demo Soon. Greentech Media. [Online] May 27, 2009. http://www.greentechmedia.com/green-light/post/a-tool-kit-for-algae-growers (accessed May 2009b).

Knothe, G. The History of Vegetable Oil-Based Diesel Fuels. *The Biodiesel Handbook*; G. Knothe, J. Van Gerpen, J. Krahl, Eds.; AOCS Press: Urbana, IL, 2004; pp 4–16.

Kolenc, V. Valcent Restructures, Lays Off 19: More Layoffs Could Follow. *El Paso Times* [Online] Jan 13, 2009. http://www.elpasotimes.com/business/ci_11438333 (accessed Feb 2009).

Kram, J.W. OriginOil Presents at Algae Conference. *Biodiesel Magazine* [Online] Aug 23, 2008. http://biodieselmagazine.com/article.jsp?article_id=2585 (accessed Feb 2009).

Ladd, C. 10 Big Questions for Maverick Geneticist J. Craig Venter on America's Energy Future. *Popular Mechanics* [Online] Jul 30, 2008. http://www.popularmechanics.com/blogs/science_news/4275738.html (accessed Jan 2009).

LaMonica, M. Algae Front-Runner GreenFuel Shuts Down. CNET News [Online] http://news.cnet.com/8301-11128_3-10239916-54.html (accessed May 2009).

Lane, J. "A Vertically Integrated, Scaleable, 5,000 Gallon/Acre Algae Fuel System": A Biofuels Digest Special Report on PetroAlgae. *Biofuels Digest* [Online] Feb 25, 2009. http://biofuelsdigest.com/blog2/2009/02/25/a-vertically-integrated-scaleable-licensable-5-6000-gallon-per-acre-algae-to-energy-production-system-a-biofuels-digest-special-report-on-petroalgae (accessed Feb 2009).

Lane, J. Coskata, Qteros and Cobalt Biofuels Project "100 Mgy Production by 2012"; Solazyme Says It Will Reach the 100 Mgy Mark by "2012 or 2013." *Biofuels Digest* [Online] Apr 21, 2009. http://biofuelsdigest.com/blog2/2009/04/21/coskata-qteros-and-cobalt-biofuels-project-100-mgy-production-by-2012-solazyme-says-it-will-reach-the-100-mgy-mark-by-2012-or-2013 (accessed Apr 2009).

Lane, J. PetroAlgae Debuts New Drop-in Fuel Process for Green Diesel from Algae; Revenues This Year: Digest Exclusive Update. *Biofuels Digest* [Online] May 18, 2009. http://biofuelsdigest.com/blog2/2009/05/18/petroalgae-debuts-new-drop-in-fuel-process-for-green-diesel-from-algae-revenues-this-year-doubles-algal-production-rate-digest-exclusive-update (accessed May 2009).

Lane, J. Algae Pioneer Solix Closes $16.8 Million Series A Financing with New Chinese Investor. *Biofuels Digest* [Online] Jul 1, 2009. http://biofuelsdigest.com/blog2/2009/07/01/algae-pioneer-solix-closes-168-million-series-a-financing-with-new-chinese-investor (accessed Jul 2009).

LeBlanc Jr, G.M. PetroSun Inc., Scottsdale, Arizona, USA. Personal communication, 2009.

Mitra, S. Engineering Algae-Based Bio Fuel: Valcent CEO Glen Kertz (Part 3). [Online] Nov 21, 2008. http://www.sramanamitra.com/2008/11/21/engineering-renewable-energy-valcent-ceo-glen-kertz-part-3 (accessed Feb 2009).

OriginOil. Veteran Algae Scientist Praises OriginOil's Technology Results. Press Release [Online] Oct 1, 2008. http://www.originoil.com/company-news/veteran-algae-scientist-praises-originoils-technology-results.html (accessed Oct 2008).

OriginOil Home Page. http://www.originoil.com (accessed May 2009).

OriginOil. Helix BioReactor™. http://www.originoil.com/technology/helix-bioreactor.html (accessed May 2009a).

OriginOil. Quantum Fracturing™. http://www.originoil.com/technology/quantum-fracturing.html (accessed May 2009b).

PetroAlgae. PetroAlgae Chairman Issues Special Letter to Shareholders. Press Release [Online] Mar 14, 2009. http://www.petroalgae.com/press.php (accessed May 2009).

PetroSun, Inc. PetroSun Issues Algae-to-Biofuels Corporate Updates. Press Release [Online] Mar 25, 2008. http://biz.yahoo.com/iw/080324/0378475.html (accessed Jun 2008).

PetroSun, Inc. PetroSun Biofuels Proposes New Crop to Catfish Farmers at the Mississippi Biomass Council Meeting. Press Release [Online] Nov 14, 2008. http://www.marketwire.com/press-release/Petrosun-Inc-920681.html (accessed Nov 2008).

Rapier, Robert. The Man Who Wrote the Book on Algal Biodiesel. *The Oil Drum* [Online] May 17, 2007. http://www.theoildrum.com/node/2541 (accessed Feb 2009).

Ratledge, C.; Z. Cohen. Microbial and Algal Oils: Do They Have a Future for Biodiesel or As Commodity Oils? *Lipid Technol.* **2008,** *20,* 155–160.

RedOrbit.com. Solix Biofuels Completes $10.5 Million Series A Funding Round, Receives Commitment for Additional $5 Million to Build Biofuels Pilot Plant in Southwest Colorado. [Online] Nov 11, 2008. http://www.redorbit.com/news/business/1599409/solix_biofuels _completes_105_million_series_a_funding_round_receives/ (accessed May 2009).

Reuters News Agency. Valcent Products Inc.: Progress Report. [Online] http://www.reuters.com/ article/pressRelease/idUS197487+09-Apr-2009+MW20090409 (accessed Apr 2009)

Sapphire Energy. Company Backgrounder. 2008. [Online] http://www.sapphireenergy.com/pdfs/ sapphire%20Company%20Backgrounder.pdf (accessed May 2009a).

Sapphire Energy. Sapphire Energy Builds Investment Syndicate to Fund Commercialization of Green Crude Production. Press Release [Online] Sept 17, 2008. http://www.sapphireenergy. com/press_release/4 (accessed May 2009b).

Sapphire Energy. Algae-Based Fuel Projected to be Commercial-Ready in Three Years. Press Release [Online] Apr 16, 2009. http://www.sapphireenergy.com/press_release/11 (accessed May 2009c).

Schill, S.R. OriginOil Achieves Rapid Algae Oil Extraction. *Biodiesel Magazine* [Online] May 2009. http://www.biodieselmagazine.com/article.jsp?article_id=3417 (accessed May 2009).

Seambiotic Ltd. Home Page. http://www.seambiotic.com (accessed Jan 2009).

Sheehan, J.; T. Dunahay; J. Benemann; P. Roessler. A Look Back at the U.S. Department of Energy's Aquatic Species Program: Biodiesel from Algae. National Renewable Energy Laboratory: Golden, Colorado, USA, 1998; Report #NREL/TP-580-24190.

Sims, B. Green Star Products Acquires Algae License. *Biodiesel Magazine* [Online] Apr 2008. http://www.biodieselmagazine.com/article.jsp?article_id=2189 (accessed Feb 2009).

Sims, B. OriginOil Signs Agreement with U.S. DOE. *Biodiesel Magazine* [Online] Mar 2009. http://www.biodieselmagazine.com/article.jsp?article_id=3253 (accessed May 2009).

Solazyme. Solazyme and Chevron Technology Ventures Enter into Biodiesel Feedstock Development and Testing Agreement. Press Release [Online] Jan 22, 2008. http://www.solazyme.com/ news080122_2.shtml (accessed Jun 2008).

Solazyme. Solazyme Produces First Algal-Based Renewable Diesel to Pass American Society for Testing and Materials D-975 Specifications. Press Release [Online] Jun 11, 2008. http://www. solazyme.com/news080611.html (accessed Aug 2008).

Solazyme. Solazyme Showcases World's First Algal-Based Renewable Diesel at Governor's Global Climate Summit. Press Release [Online] Nov 19, 2008. http://www.solazyme.com/news081119. shtml (accessed Nov 2008).

Solix Biofuels Home Page. http://www.solixbiofuels.com (accessed May 2009).

Spolaore, P.; C. Joannis-Cassan; E. Duran; A. Isambert. Commercial Applications of Microalgae. *J. Biosci. Bioeng.* **2006,** *101,* 87–96.

Steiner, U. Biofuels' Cost Explosion Necessitates Adaptation of Process Concepts. Presented at the European White Biotechnology Summit, Frankfurt, Germany, May 21–22, 2008. [Online] http://

www.mstonline.de/mikrosystemtechnik/mst-fuer-energie/algen/SteinerWhiteBiotechSummit08
.pdf (accessed Feb 2009).

Stella, J. Green Star Products, Chula Vista, California, USA. Personal communication, 2009.

Stephens, H.A. Fellsmere Farm's Algae May Help Bring Diesel Fuel Prices Down. *TC Palm* [Online] Feb 6, 2009. http://www.tcpalm.com/news/2009/feb/06/fellsmere-farms-algae-may -help-bring-diesel-fuel-p/ (accessed Feb 2009).

Stipp, D. The Next Big Thing in Energy: Pond Scum? *Fortune* [Online] **2008,** *157(8),* 142. http:// money.cnn.com/2008/04/14/technology/perfect_fuel.fortune/index.htm (accessed Jan 2009).

Torrey, M. Algae in the Tank. *inform* **2008,** *19,* 432–437.

Trimbur, D.E.; C.-S. Im; H.F. Dillon; A.G. Day; S. Franklin; A. Coragliotti. Glycerol Feedstock Utilization for Oil-Based Fuel Manufacturing. U.S. Patent Application 20090004715, January 1, 2009.

Valcent Products Inc. Valcent Announces Moving Verticrop™ High Density Growing System to Its U.K. Subsidiary. Press Release [Online] Jan 27, 2009. http://www.valcent.net/s/NewsReleases. asp?ReportID=336047 (accessed Feb 2009).

Waltz, E. Biotech's Green Gold? *Nat. Biotechnol.* **2009,** *27,* 15–18.

Willson, B. Low Cost Photobioreactors for Algal Biofuel Production & Carbon Capture. [Online] Presented Sept 18, 2008. http://www.netl.doe.gov/publications/proceedings/08/H2/pdfs/ Solix%20Carbon%20Recycling,%209-18-08.pdf (accessed May 2009).

Yahoo! Finance. Form 8-K for PetroAlgae Inc. 11-Mar-2009, Entry into a Material Definitive Agreement. [Online] http://biz.yahoo.com/e/090311/palg.ob8-k.html (accessed Mar 2009).

Zimmer, C. The High-Tech Search for a Cleaner Biofuel Alternative. *Environment 360* [Online] Jan 5, 2009. http://www.e360.yale.edu/content/feature.msp?id=2106 (accessed Jan 2009).

·• 13 •·

Algae Oils for Biofuels: Chemistry, Physiology, and Production

Michael A. Borowitzka

Algae R & D Center, School of Biological Sciences and Biotechnology, Murdoch University, Murdoch, WA 6150 Australia

Introduction

We have long known that microalgae can contain high levels of lipids and sugars, and they have, therefore, often been proposed as potential sources of renewable biodiesel and bioethanol. In response to the oil crisis in the 1970s, extensive research on the use of algae to produce biodiesel was undertaken at the Solar Energy Research Institute in Colorado, USA (see Sheehan et al., 1998 for summary), who concluded that, although technically feasible, biodiesel production from microalgae was economically unsustainable at that time. Interest in algae as a source of renewable biofuels has been reinvigorated recently by increasing oil prices and the need to reduce CO_2 emissions because of the threat of global warming.

Microalgae are attractive as sources of renewable biofuels—e.g., biodiesel, bioethanol, and H_2—not only because of their high lipid and/or sugar content and high areal productivity, but also because they can be grown on land using saline water and on land which that is not suitable for agriculture. Algae-based biofuels are also expected to reduce CO_2 emissions. Furthermore, as will be shown in this chapter, the great diversity of microalgae provides opportunities for the selection of species and strains with particular fatty acid profiles with advantages for the production of biofuel with specific properties. In the longer term, microalgae present the option of genetic modification to enhance lipid productivity and/or modify the fatty acid profile (Courchesne et al., 2009).

Lipids and Fatty Acids

The total oil and fat content of microalgae ranges from about 1 to 70% of ash-free dry weight[1], although contents higher than 40% are only very rarely observed, and then only

---

[1] Lipid contents should always be reported on the basis of ash-free dry weight, rather than purely dry weight, as the ash content of microalgae is variable and can be quite high, especially in marine and hypersaline species or species that have siliceous or calcareous cell wall components, such as diatoms, some chrysophytes, and the coccolithophorid algae.

in algae that have been in the stationary phase of growth for some time (Borowitzka, 1988). Microalgae lipids are generally esters of glycerol and fatty acids with a chain-length of C_{14} to C_{22}; they may be saturated or unsaturated. In the eukaryotic algae triacylglycerols are the most common storage lipids and may constitute up to 80% of the total lipid fraction (Tornabene et al., 1983). The other major algal membrane lipids are sulphonoquinovosyl diglyceride (SL); monogalactosyl diglyceride (MGDG): digalactosyl diglyceride (DGDG), found mainly in the chloroplast; and phosphatidyl glycerol (PG) and phosphatidyl ethanolamine (PE), found mainly in the plasma membrane and the endoplasmic membrane systems (Guschina & Harwood, 2006).

Hydrocarbons

Microalgae generally contain less than 5% of dry weight as hydrocarbons, with the exception of the green alga, *Botryococcus braunii*, in which contents of up to 61% of dry weight as hydrocarbons have been reported (Metzger et al., 1985). The highest contents, however, are known only from field samples, and the levels are lower in cultured *Botryococcus* (Wake & Hillen, 1981; Templier et al., 1984). The bulk of the *Botryococcus* hydrocarbon (c. 95%) is located outside of the cells in the colony matrix and occluded globules. Three chemical "races" of *B. braunii* have been identified: (1) the A-race, which produces essentially *n*-alkadiene and triene hydrocarbons, odd-carbon-numbered from C_{23} to C_{33}; (2) the B-race that produces C_{30}-C_{37} triterpenoid hydrocarbons, the botryococcenes, and C_{34} methylated squalenes; and (3) the L-race that producs a single tetraterpenoid hydrocarbon: lycopadiene. Some sixty strains of *B. braunii* have been analysed for their hydrocarbon content and composition so far (see Metzger & Largeau, 1999 for partial list), and hydrocarbon contents range from 0.4% to 61% of dry weight for A-race strains, about 9% to 40% for B-race strains, and 0.1% to 8% for L-race strains (Metzger & Largeau, 2005). In the "Berkeley strain" of *B. braunii*, the botryococcene fraction consists of ten compounds (C_nH_{2n-10}; n =30-34) that makes up about 25–40% of its dry weight. About 7% of the botryoccocenes, mainly C_{30} to $C_{32,}$ are located in the cells. The external (colonial matrix) pool contains >99% of the C_{33} and C_{34} compounds, as well as lower chain-length botryococcenes (Wolf et al., 1985).

Algae other than *Botryococcus* do not accumulate their lipids and hydrocarbons extracellularly, but rather as part of the cell membrane systems and as oil droplets within the cytoplasm or chloroplasts.

Triacylglycerols, Fatty Acids and Biodiesel Production

Biodiesel is prepared from algal lipids by esterifying free fatty acids or transesterifying triacylglycerol fatty acids by reacting them with an alcohol, usually methanol or ethanol; other alcohols, such as propanol, butanol, isopropanol, *tert*-butanol, branched alcohols and octanol, can be used. Because of the cost, methanol is the most widely used alcohol as opposed to the more expensive ethanol. Methanol is also more reactive, and the fatty acid methyl esters (FAMEs) are more volatile than the fatty acid ethyl esters (FAEEs) produced with ethanol. FAEEs also have slightly higher viscosities and slightly lower cloud and pour points than the corresponding FAMEs (Bozbas, 2008). However, it must be remembered that methanol is largely produced from non-renewable fossil fuel feedstocks, whereas ethanol can be produced from renewable feedstocks by fermentation.

There are four main synthetic approaches for biodiesel production:

1. Base-catalysed transesterification (by far the most widely used method at this time, mainly because of low costs), which proceeds at a relatively high rate at low temperatures (Demirbas, 2003). Most commonly alkalis, such as NaOH or KOH are used although alkoxides, such as sodium methoxide (NaOMe), are more efficient catalysts, and their use is becoming more common.

2. Acid-catalysed transesterification with simultaneous esterification of free fatty acids (Fukuda et al., 2001; Meher et al., 2006). Sulphuric acid, phosphoric acid, hydrochloric acid, and sulphonic acid are the acids typically used. This process is rarely employed because of the slower reaction rate and the greater corrosiveness of the acids.

3. Non-catalytic conversion via transesterification and esterification under supercritical alcohol conditions (Kudsiana & Saka, 2001; Demirbas, 2003).

4. Lipase enzyme-catalysed transesterification (Ranganathan et al., 2008; Robles-Medina et al., 2009). Although lipases have a number of advantages, their high cost means that their application is highly restricted.

More recently, processes using metal oxide-base catalysts (microparticulate porous zirconia, titania, and alumina) at high pressure and high temperature have been developed (McNeff et al., 2008).

The fatty acid composition of the algal oil has important effects on the efficiency of the transesterification process. For example, the presence of free fatty acids results in the formation of soaps during base-catalysed transesterification process, a concomitant reduction in yield, and an increase in the downstream processing and water use to remove these soaps. Thus, oils with high free fatty acid contents are better transesterified using the acid-catalysed process. It has been suggested that, for algal oils with a high free fatty acid content, a two-stage process be used (Robles-Medina et al., 2009); in the first step, acid catalysts are used to convert the free fatty acids to methyl esters, followed by a base catalyst process that converts the remaining triacylglycerols to methyl esters. Nagle and Lemke (1990) compared acid- and base-catalysed transesterification using lipids extracted from the diatom *Chaetoceros muelleri*. They achieved a maximum yield of 10 mg of fatty acid methyl esters (FAME) from 250 mg lipids using hydrochloric acid-methanol and only 3.3 mg of FAME using NaOH as a catalyst under the same conditions. The relatively high content of free fatty acids in lipid extracts of some algae, however, may be an artifact of extraction. Berge et al. (1995), using the diatom *Skeletonema costatum*, found that when the harvested cells were immediately treated with boiling water to deactivate the lipases before lipid extraction, they could detect no free fatty acids in the lipid extract, unlike when these fatty acids were extracted using the Bligh and Dyer (1959) method without the boiling water treatment.

Combined extraction and esterification methods have also been developed. For example, Belarbi et al. (2000) developed a simultaneous extraction and esterification method using a slurry (≈82% water by weight) of either the diatom *Phaeodactylum tricornutum* or the green alga *Monodus subterraneus* that obtained a 77.5% yield of FAME by transesterification with methanol and acetyl chloride by heating in a boiling water bath for 120 min at 2.5 atm.

The nature of the component fatty esters produced largely determines the properties of the biodiesel (Knothe, 2005). Thus, oils with a high content of unsaturated fatty acids result in a biodiesel that is less viscous and has a greater *cloud point* (the temperature at which the fuel becomes cloudy, due to solidification) and *pour point* (the temperature at which fuel stops flowing). This property makes the biodiesel more suitable for cold conditions; however, it is also more prone to oxidation and has a lower *cetane index* (related to the ignition delay time and combustion quality of the fuel). Conversely, oils with a high proportion of long chain fatty acids (> C_{18}) have a higher cetane index, lower cloud and pour points, and are more viscous (Knothe, 2005). The position of the double bonds in unsaturated fatty acids also affects the oxidative stability of biodiesel. Thus, esters of linoleic acid (double bonds at $\Delta 9$ and $\Delta 12$) oxidise more slowly than esters of linolenic acid (double bonds at $\Delta 9$, $\Delta 12$ and $\Delta 15$) (Frankel, 1998). Unsaturation also may decrease the lubricity of the biodiesel and may contribute to gum formation in the engine.

The different algal phyla, and even species, vary markedly in the fatty acid composition of their lipids (Table 13.A), thus providing an opportunity to select strains with a preferred fatty acid profile in order to manufacture fuels with different properties. It must also be remembered, though, that many of the oleaginous species—especially diatoms, cryptomonads and eustigmatophytes—also have significant contents of long-chain polyunsaturated fatty acids (Brown et al., 1997) that will affect the utility of their oils for transesterification to produce biofuels. The European standards for biodiesel for vehicle use (EN14214) and for heating oil (EN 14213) limit the content of FAMEs with four or more double bonds to a maximum of 1% mol (Knothe, 2006). As the lipids of many oleaginous microalgae have a high content of highly polyunsaturated fatty acids, such as eicosapentaenoic acid (C20:5*n*-3) and docosahexaenoic acid (C22:6*n*-3), they are likely not to meet the European biodiesel standards. However, the content of fatty acids with more than four double bonds can still be reduced by partial catalytic hydrogenation of the oil (Dijkstra, 2006).

Table 13.A. Relative Fatty Acid Composition of the Major Microalgae Classes of Interest for Biofuels.[a,b]

Fatty Acid and Double Bond Position		Algal Class[c]							
		CY	RHO	CRY	BAC	CHR	HAP	DIN	CHL
14:0		4	1	2	2	2	1	2	1
16:0		6	3	3	3	2	2	4	3
16:1	9	5	1	2	5	3	3	3	2
16:2	6,9	2	·	Tr	1	1	1	1	1
16:2	9,12	(5)[d]	·	1	1	1	1	1	1
16:3	6,9,12	·	·	1	2	2	2	2	2
16:4	6,9,12,15	·	·	1	1	·	·	·	2
17:1		1	·	·	·	·	·	·	·
18:0		3	·	1	1	1	1	1	1
18:1	9	4	2	1	1	2	3	2	5
18:2	9,12	4	3	2	1	3	2	1	4
18:3	6,9,12	3	2	1	1	2	Tr	1	3
18:3	9,12,15	4	1	3	·	1	2	1	4
18:4	6,9,12,15	3	1	5	1	1	2	2	2
20:0		·	·	1	·	·	·	1	1
20:1	11	1	·	2	·	1	·	·	2
20:2	8,11	·	·	·	1	Tr	Tr	1	1
20:3	8,11,14	·	1	·	1	1	1	·	1
20:4	5,8,11,14	·	4	1	1	1	1	1	1
20:4	8,11,14,17	·	·	1	1	1	1	2	·
20:5	5,8,11,14,17	·	2	2	3	3	3	·	1
22:0		·	·	·	·	·	·	·	1
22:1	1	·	·	·	1	·	1	1	1
22:5	4,7,10,13,16	·	·	1	Tr	·	·	·	1
22:5	7,10,13,16,19	·	·	·	·	·	2	1	1
22:6	4,7,10,13,16,19	·	·	·	·	·	1	3	·

[a]Modified and updated from Borowitzka (1988).
[b]Values in the table are: · = not reported; Tr = Trace; 1 = up to 10% of total fatty acids; 2 = up to 20%; etc.
[c]The classes are (with number of strains examined): CY = Cyanobacteria (53); RHO = Rhodophyceae [23]; CRY = Cryptophyceae (10); BAC = Bacillariophyceae (13); CHR = Chrysophyceae (6); HAP = Haptophyceae (5); DIN = Dinophyceae (8); CHL = Chlorophyceae (21).
[d]Present in only a few of the strains examined.

Algal Physiology, Lipid Content, and Fatty Acid Composition

The total lipid content and the lipid and fatty acid composition change with growth conditions, often in a taxon-specific, and even species-specific, manner. In many species, conditions that reduce the growth rate (e.g., nutrient limitation, suboptimal irradiance, temperature, or salinity—often referred to by many workers as "stress" conditions) will cause the lipid content to increase, especially in the form of triacylglycerols, which act as storage lipids (Borowitzka, 1988; Roessler, 1990). Similarly, in batch cultures, the total lipid content of many species has been found to increase as the culture "ages" and reaches stationary phase (Table 13.B). However, because of the reduction in growth rate, due to nutrient limitation, the lipid productivity declines. On the other hand, some species, such as *Dunaliella tertiolecta*, *Tetraselmis* sp., *Biddulphia aurita*, *Synedra ulna*, and *Nannochloris atomus*, (Shifrin & Chisholm, 1981; Reitan et al., 1994) reduce the total lipid content with nutrient limitation.

Changes in the growth environment (e.g., irradiance, temperature, salinity) may also affect the fatty acid composition. Furthermore, the fatty acid fraction of the total lipids in marine microalgae is about 40% (range 32–53%) (Cohen, 1986; Reitan et al., 1994). This relatively low fraction of fatty acids occurs because the algae contain high levels of unusual fatty acids and also have a high content of non-fatty acids lipids (Kayama et al., 1989) (Table 13.C).

Table 13.B. Change in Total Lipid Content with Culture Age or between Log and Stationary Phase in Cultures of Microalgae.

Species	Change in Lipid (% of dry weight)	Reference
Chlorella vulgaris	22-28	Collyer & Fogg, 1955
Scenedesmus obliquus	19-32	Piorreck et al., 1984
Botryococcus braunii	23-34	Belcher, 1968
Euglena gracilis	24-65	Piorreck & Pohl, 1984
Amphora sp.	11.7-15.7	Barclay et al., 1985
Navicula pelliculosa	14.5-19	Piorreck & Pohl, 1984
Phaeodactylum tricornutum	24-29	Piorreck & Pohl, 1984
Phaeodactylum tricornutum	23-35	Chrismadha & Borowitzka, 1994
Thalassiosira pseudonana	7.8-21.6	Fisher & Schwarzenbach, 1978
Isochrysis galbana	10-20	Fidalgo et al., 1998

Table 13.C. Lipid Class Composition of N-nutrient-replete *Phaeodactylum tricornutum.*[a]

Lipid Class	Content (pg cell^{-1})	% Total Lipids
Aliphatic hydrocarbons	0.09	3.5
Sterol esters & wax esters	0.04	1.6
Methyl esters & other short-chain esters	0.02	0.8
Triacylglycerides	0.11	4.3
Free fatty acids	0.21	8.2
Free aliphatic alcohols	0.06	2.4
Free sterols	0.06	2.4
Diacylglycerides	0.03	1.2
Acetone-mobile polar lipids[b]	0.95	37.3
Phospholipids & other acetone-immobile lipids	0.98	38.3
Total	2.55	100

[a]Modified from Parrish & Wangersky, (1987)
[b]Includes Chlorophyll *a*

Nutrients (N, P Si)

As pointed out above, the decreased growth rate in most microalgae, due to nitrogen limitation, results in an increased content of total lipids (Borowitzka, 1988; Takagi et al., 2000; Tonon et al., 2002; Rodolfi et al., 2009; Griffiths & Harrison, 2009). Similarly, the inhibition of growth in diatoms, due to silicon limitation, also leads to an increased lipid content (Coombs et al., 1967; Taguchi et al., 1987). The source of nitrogen also affects lipid content and fatty acid distribution in some algae. For example, Li et al. (2008) found that the freshwater green alga *Neochloris oleoabundans,* cultivated on nitrate, contained about twice as much lipid on a dry weight basis than when grown on urea or ammonium. They also observed the highest lipid productivity of 0.133 g L^{-1} d^{-1} at 0.5 mM nitrate concentration in cultures that were also bubbled with 5% v/v CO_2-containing air. Fidalgo et al. (1998) also found that urea (compared with nitrate and nitrite) gave a higher lipid content in early stationary phase cultures of *I. galbana* and that these cultures also had the highest proportion of PUFAs.

Phosphorous limitation has also been demonstrated to increase lipid content in several species, such as *Monodus subterraneus* (Khozin-Goldberg & Cohen, 2006), *I. galbana, Pavlova lutheri, Phaeodactylum tricornutum,* and *Chaetoceros* sp. The increase in lipid content mainly results from increased triacyclycerol (TAG) synthesis (Lombardi & Wangersky, 1991; Reitan et al., 1994). In *Nannochloris atomus, Tetraselmis* sp., and *Dunaliella tertiolecta,* however, P-limitation decreased lipid content on a dry weight basis (Siron et al., 1989; Reitan et al., 1994). In the marine microalgae studied by Reitan et al. (1994), they found that P-limitation increased the relative content of 16:0 and 18:1 fatty acids and reduced the relative content of 18:4n-3, 20:5n-3 and 22:6n-3 fatty acids. Interestingly, in a marine *Chlorella vulgaris*, supplementation with Fe also increased the neutral lipid content as determined by Nile Red fluorescence (Liu et al., 2008).

Carbon Supply

Most microalgae in dense cultures are carbon limited, and the addition of CO_2 enhances the growth rate and biomass productivity; if the CO_2 or the associated shift in the culture pH does not affect lipid content, lipid productivity will also be enhanced. In a study on the effect of CO_2 on lipid content, Becker and Venkataraman (1982) found that increasing the CO_2 supply increased the total lipid content slightly in *Scenedesmus acutus*. Hu and Gao (2003) obtained similar results with *Nannochloropsis* sp., and Widjaja et al. (2009) found that the addition of CO_2 resulted in higher lipid productivity in *Chlorella vulgaris*. Adding bicarbonate to the medium also increased the total fatty acid content in exponentially growing *Pavlova lutheri* (Guihéneuf et al., 2009). In *Dunaliella viridis*, the addition of CO_2 increased the total lipid content only under N-limiting conditions (Gordillo et al., 1998). On the other hand, in *D. salina*, a one-day increase in CO_2 concentration from 2% to 10% increased the total fatty acids on a dry weight basis by 30% (Muradyan et al., 2004). This increase was due mainly to de novo fatty acid synthesis, while their elongation and desaturation was inhibited, leading to an increase in the relative content of saturated fatty acids at high levels of CO_2. However, CO_2 addition does not increase the lipid content of all algae (Raghavan et al., 2008).

Several studies have also examined the effect of the addition of an organic carbon source, usually acetate, in light (i.e., mixotrophic growth). The addition of acetate in the light either had no effect on lipid content in some species (e.g., *Pavlova lutheri* (Guihéneuf et al., 2009) or had a slight inhibitory effect (e.g., *Nitzschia communis* (Dempster & Sommerfeld, 1998). Unfortunately, few of these studies first determined whether the algal cell could actually take up acetate, and this oversight may be one reason for the variable results.

Irradiance

Irradiance has profound effects on both lipid content and fatty acid composition of microalgae. In *Pavlova lutheri*, grown at high C concentrations, increasing the irradiance from 20 to 340 µ mol photons $m^{-2} s^{-1}$ increased the lipid content of the cells by over 50% (Guihéneuf et al., 2009). Similarly, increased irradiance increased the total fatty acid content of *Monodus subterraneus* (Lu et al., 2001) and *Nannochloropsis* (Sukenik et al., 1989).

In general, low irradiances lead to the formation of polar lipids, especially those associated with the thylakoid membranes, whereas high irradiances result in a decrease in polar lipids and an increase of neutral storage lipids, mainly TAGs (Sukenik et al., 1989; Fabregas et al., 2004). Day length also affects the fatty acid profile and overall lipid productivity (Tzovenis et al., 2003). Increased UV radiation has also been demonstrated to affect the fatty acid composition of diatoms (Liang et al., 2006).

Salinity

Suboptimal salinities ("salt stress") were found to increase the chloroform fraction containing TAGs and free fatty acids in *Botryococcus braunii*, but decreased the lipid content in *Isochrysis* sp. (Ben-Amotz et al., 1985). Chelf (1990) also found that increasing salinity decreased neutral lipid accumulation in the diatoms *Chaetoceros muelleri* and *Navicula saprophila*. In *M. subterraneus*, increasing salinity increased lipid content from 34% to 46% (Iwamoto & Sato, 1986), and, in an "A"-race strain of *B. braunii*, increasing salinity from 34 to 85 mM NaCl resulted in approximately a two-fold increase in the relative proportion of palmitic and oleic acid (Ranga Rao et al., 2007); salinity apparently does not affect the total lipid content (Vasquez-Duhalt & Arrendondo-Vega, 2001).

Temperature

Decreasing the temperature generally results in increased fatty acid unsaturation (Morgan-Kiss et al., 2006), and increasing the temperature in increased saturation (Renaud et al., 1995). The effect of temperature on total lipid content is difficult to interpret. For example, in *Ochromonas danica* and *Nannochloropsis salina* (Boussiba et al., 1987), increasing temperature increased the lipid content, whereas temperature had little effect on the lipid content of *Chlorella sorokiniana* (Patterson, 1970).

Biochemistry

Despite much research on the lipid content and fatty acid composition of microalgae, the biochemical basis of lipid accumulation under nutrient-deficient conditions and other environmental conditions has received little investigation. Where it has been investigated, the focus has often been on the synthesis of long-chain polyunsaturated fatty acids and not on the synthesis of TAGs, which contain mainly saturated and monounsaturated fatty acids. Recently Riekhof et al. (2005) have used genetic analysis of the *Chlamydomonas reinhardtii* genome to construct a hypothetical pathway for glycerolipid biosynthesis, and further metabolomics- and proteomics-assisted genome annotation (May et al., 2008) of the ever-increasing number of algae whose genomes have been partially or wholly sequenced (Grossman, 2007) will assist in establishing the metabolic pathways that will help researchers develop methods to enhance lipid formation.

Lipid accumulation under N-deficient conditions has been attributed, in part, to the fact that storage lipids and most membrane lipids do not contain nitrogen and, therefore, continue to be synthesized in N-deficient cells, while the synthesis of N-containing compounds, such as proteins and nucleic acids, is markedly reduced.

Roessler (1988a, 1988b), studying the diatom *Cyclotella cryptica* with radiotracer experiments, found that, within 4 h of silicon deficiency, the percentage of carbon newly assimilated into lipids doubled, whereas the percentage of C going to storage carbohydrate (chrysolaminarin) decreased by 50%. He also showed a concomitant slow conversion of pre-existing non-lipid compounds into lipids. He calculated that about 55-68% of the lipids produced by the diatom during the first 12 h of silicon starvation were newly synthesized, and the rest of these lipids resulted from the conversion process. The bulk of the lipids produced were TAGs. Acetyl-CoA carboxylase, which may catalyse the rate-limiting step of fatty acid biosynthesis, doubled in activity within 4 h after the onset of silicon deficiency.

TAG synthesis requires large amounts of ATP and NADPH (for example, the synthesis of a C18 fatty acid requires approximately 24 NADPH), and it is, therefore, not surprising that TAG synthesis is enhanced under high light conditions and/or conditions when the energy produced by photosynthesis exceeds that needed for cell growth and division, which can occur under nutrient limiting conditions. In synchronized cultures of *Chlorella ellipsoidea* (Otsuka & Morimura, 1966), neutral lipids are produced primarily during the growth phase in light and consumed during cell division in the dark. If cell division is inhibited due to nutrient deficiency or high pH, lipid accumulation may conceivably occur, as observed by Guckert and Cooksey (1990). Interestingly, no cell cycle changes in lipid content have been observed in synchronised cultures of *Navicula pelliculosa* or *Oocystis polymorpha* (Darley, 1976; Shifrin & Chisholm, 1981).

Strain Selection

There are many aspects that need to be considered when selecting algae strains for biofuels production. Obviously, high lipid productivity (i.e., lipid content x growth rate) and a suitable fatty acid profile are very important. However, biological and physical attributes, relating to suitability for culture in an extremely large-scale culture system— including temperature tolerance, oxygen sensitivity, salinity range, and shear sensitivity—are equally important to consider when trying to achieve reliable high productivities in an open pond or a closed photobioreactor system (Borowitzka, 1998). Finally, for economic reasons, the ease of harvesting and dewatering the cells and extracting the lipids is important and needs to be considered when selecting strains.

It is not only important to select the right species, but also some gains in productivity and lipid quality may be achieved through strain selection. For example, Liang et al (2005) screened sixty clones isolated from a single culture of the diatom *Cylindrotheca fusiformis* and found that, under identical growth conditions, the total lipid content ranged from 7.3 to 23.4%, with a similar variation in the fatty acid distribution of the lipids. Alonso et al (1992a, 1992b) also found marked variation in the lipids and fatty acids of *Isochrysis galbana*. Clearly, this observed variation must be due to genetic differences between the clones, but it is not known whether these differences arose during the long period in culture of the strain or because the original strain was not derived from a single cell, but from a number of cells (Alonso et al., 1994). Significant differences between different isolates of the same species have often been observed (e.g., Shaw et al., 1989; Borowitzka, 1992; Rodolfi et al., 2009). These findings highlight the importance of both species selection and strain selection.

Production

The production of microalgae for biofuels presents a rather new paradigm to the commercial production of microalgae—the scale of production desired is at least one order of magnitude greater that that for any fine chemical, nutraceutical, or food for aquaculture application, and the final product has a significantly lower unit value. This difference presents the commercial algae culture industry with an enormous challenge to reduce production costs. There is an ongoing debate regarding whether algal culture for biofuels should be in "open" systems, such as raceway ponds, or "closed" systems, such as tubular or plate photobioreactors. Hybrid systems that use a combination of "closed" and "open" systems also are possible (Pushparaj et al., 1997). Almost all currently operating commercial algae culture plants use raceway ponds, with the exception of the extensive open pond systems used for the production of *D. salina* in Australia and the tubular photobioreactors that are used for the production of *H. pluvialis* in Israel and *Chlorella* in Germany (Borowitzka, 2005). The reason for the prevalence of open ponds is that their open pond culture is significantly cheaper than culture in closed photobioreactors (Borowitzka, 1999b). Experience has shown that many algae species can be grown reliably in open ponds for extended periods with minimum contamination. Similarly, many species can be grown in closed photobioreactors, but many others cannot, due to problems with turbulence or fouling of the inside surfaces of the photobioreactor. However, open pond algae plants require a larger land area than closed photobioreactor systems and, generally, have a larger requirement for water because of greater evaporative losses. Closed photobioreactors, on the other hand, have a higher capital cost for construction and have a higher energy requirement for culture circulation. They may also require cooling in high irradiance conditions (Borowitzka, 1999a).

For commercial-scale culture, the process must be reliable, easily managed, and have high lipid productivity over the whole year. However, almost all proposed culture systems are necessarily located outdoors to use natural daylight as the energy source. One important consequence of this choice is that the algal culture is exposed to a continuously changing and dynamic environment with varying irradiance, temperature, and O_2 concentration, all of which affect algal productivity, as well as lipid content and fatty acid composition. Direct extrapolation from laboratory-scale results or short-term outdoor experiments, therefore, is extremely difficult or impossible. Furthermore, the algae cultures must be operated in continuous culture mode, as batch culture is significantly more expensive.

The data on algal productivity, especially lipid productivity, in scientific literature are mainly from short-term, small-scale experiments, usually carried out in the laboratory, and are often difficult to interpret because of the range of units used. Griffiths and Harrison (2009) have recently attempted to summarize this data. There are very few published data on the productivity of algal cultures outdoors over long periods (>3 months). However, from those that are available, some general conclusions can be reached. The best annual average biomass productivities achieved so far are in outdoor raceway ponds with a depth of 20 to 30 cm are about 20-23 g dry wt m^{-2} d^{-1} (i.e., about 0.1 g dry wt L^{-1} d^{-1}); productivity under high irradiance and warm temperature conditions in summer often exceeds 30-35 g dry wt m^{-2} d^{-1}, but falls sharply in winter due to shorter day length, reduced irradiances, and lower temperatures, even in mild climates, such as that found in Perth, Western Australia (Fig. 13.7). At a biomass productivity of 20 g dry wt m^{-2} d^{-1} and with a cell lipid content of 30% of dry weight, an annual lipid productivity of 21.9 t ha^{-1} can be calculated. Thus, climate represents one key criterion for the location of an algae production plant when seeking to achieve the best average annual productivities. Sites for production plants should be located in regions with the highest possible average annual solar irradiation (i.e., close to the equator, with low cloud cover) and a suitable temperature range. Furthermore, the production plant must be located near a suitable

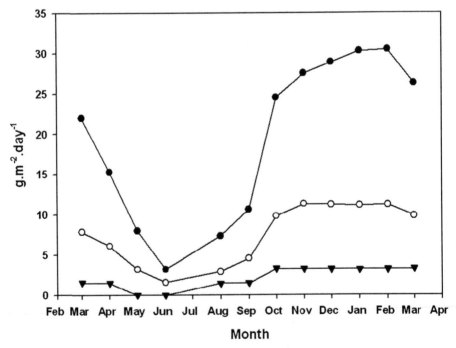

Fig. 13.1. Productivity of the calcareous haptophyte *Pleurochrysis carterae*, grown outdoors in a 20 cm deep raceway pond in semi-continuous culture in Perth, Western Australia, over the whole year, [●] = dry weight productivity; [O] = lipid productivity; [▼] = CaCO$_3$ productivity (data from Moheimani & Borowitzka, 2006).

water source and, in order to achieve the high productivities required, a source of cheap CO_2 such as from a power station or a cement plant.

As with all microalgal processes, harvesting, dewatering, and further downstream processing are very significant components of the overall production cost (Borowitzka, 1999b; Molina Grima et al., 2003). The method of harvesting is species-dependent and is also influenced by the culture system used. Unfortunately, most species of interest are too small for simple filtration; therefore, the most likely and lowest-cost first step in harvesting is flocculation, followed by flotation or settling. Depending on the lipid extraction process, further dewatering step(s) may also be necessary.

In order to recoup some of the costs of producing algal lipids for biodiesel production, in the interest of producing the algal lipids economically, several options have been proposed (see Chisti, 2007 for a discussion of some of these options). First, sugars, as well as lipids for biodiesel production, can be extracted and fermented to produce bioethanol. For the biomass remaining after extraction (mainly proteins), several options have been suggested, such as its use as animal feed, digestion to produce methane, and/or pyrolysis to produce an oil-like fuel source and char. Other proposed options are to apply the so-called "biorefinery" concept and extract valuable products, such as carotenoids, long chain polyunsaturated fatty acids, sterols, or bioactive peptides (Borowitzka, 1995, 1999c) and use the remaining lipids for biodiesel production. Whether this process is possible and economical remains to be determined.

Conclusion

The production of biodiesel from microalgae has long been proposed, and, in the last 20 years, there has been extensive research carried out. However, the high cost of algae production means that, to date, it has only been possible to commercialize relatively high-value microalgae products and not a low-value product, such as lipids for biofuels. The many attractive features of microalgae have led to a significant effort in the past few years to try and develop markedly lower cost production processes in order to make algal oil for biodiesel commercially viable. This task remains extremely challenging, and it will still take some time before algae biofuels become a commercial reality.

References

Alonso, D.L.; E.M. Grima; J.A.S. Perez; J.L.G. Sanchez; F.G. Camacho. Isolation of Clones of *Isochrysis galbana* Rich in Eicosapentaenoic Acid. *Aquaculture* **1992a,** *102,* 363–371.

Alonso, L.; E.M. Grima; J.A.S. Perez; J.L.G. Sanchez; F.G. Camacho. Fatty Acid Variation among Different Isolates of a Single Strain of *Isochrysis galbana*. *Phytochemistry* **1992b,** *31,* 3901–3904.

Alonso, D.L.; C.I.S. Delcastillo; J.L.G. Sanchez; J.A.S. Perez; F.G. Camacho. Quantitative Genetics of Fatty Acid Variation in *Isochrysis galbana* (Prymnesiophyceae) and *Phaeodactylum tricornutum* (Bacillariophyceae). *J. Phycol.* **1994,** *30,* 553–558.

Barclay, B.; N. Nagle; K. Terry; P. Roessler. Collecting and Screening Microalgae from Shallow, Inland Saline Habitats. SERI/CP-23-2700, **1985**, 52–68. SERI Aquatic Species Review.

Becker, E.W.; L.V. Venkataraman. *Biotechnology and Exploitation of Algae—The Indian Approach*; German Agency for Tech. Co-op.: Eschborn, 1982.

Belarbi, E.H.; E. Molina; Y. Chisti. A Process for High Yield and Scalable Recovery of High Purity Eicosapentaenoic Acid Esters from Microalgae and Fish Oil. *Enz. Microb. Technol.* **2000**, *26*, 516–529.

Belcher, J.H. Notes on the Physiology of *Botryococcus braunii. Kützing. Arch. Mikrobiol.* **1968**, *61*, 335–346.

Ben-Amotz, A.; T.G. Tornabene; W.H. Thomas. Chemical Profiles of Selected Species of Microalgae with Emphasis on Lipids. *J. Phycol.* **1985**, *21*, 72–81.

Berge, J.P.; J.P. Gouygou; J.P. Dubacq; P. Durand. Reassessment of Lipid Composition of the Diatom, *Skeletonema costatum. Phytochemistry* **1995**, *39*, 1017–1021.

Bligh, E.G. ; W.J. Dyer. A Rapid Method of Total Lipid Extraction and Purification. *Can. J. Biochem. Physiol.* **1959**, *37*, 911–917.

Borowitzka, M.A. Fats, Oils and Hydrocarbons. *Micro-algal Biotechnology*; M.A. Borowitzka; L.J. Borowitzka, Eds.; Cambridge University Press: Cambridge, 1988; pp 257–287.

Borowitzka, M.A. Comparing Carotenogenesis in *Dunaliella* and *Haematococcus*: Implications for Commercial Production Strategies. *Profiles on Biotechnology*; T.G. Villa; J. Abalde, Eds.; Universidade de Santiago de Compostela: Santiago de Compostela, 1992; pp 301–310.

Borowitzka, M.A. Microalgae as Sources of Pharmaceuticals and Other Biologically Active Compounds. *J. Appl. Phycol.* **1995**, *7*, 3–15.

Borowitzka, M.A. Limits to Growth. *Wastewater Treatment with Algae*; Y.S. Wong; N.F.Y. Tam, Eds.; Springer-Verlag: Berlin, 1998; pp 203-226.

Borowitzka, M.A. Commercial Production of Microalgae: Ponds, Tanks, Tubes and Fermenters. *J. Biotech.* **1999a**, *70*, 313–321.

Borowitzka, M.A. Economic Evaluation of Microalgal Processes and Products. *Chemicals from Microalgae*; Z. Cohen, Ed.; Taylor & Francis: London, 1999b; pp 387–409.

Borowitzka, M.A. Pharmaceuticals and Agrochemicals from Microalgae. *Chemicals from Microalgae*; Z. Cohen, Ed.; Taylor & Francis: London, 1999c; pp 313–352.

Borowitzka, M.A. Culturing Microalgae in Outdoor Ponds. *Algal Culturing Techniques*; R.A. Anderson, Ed.; Elsevier Academic Press: London, 2005; pp 205–218. CP;T.G. J.

Boussiba, S.; A. Vonshak; Z. Cohen; Y. Avissar; A. Richmond. Lipid and Biomass Production by the Halotolerant Microalga *Nannochloropsis salina. Biomass* **1987**, *12*, 37–47.

Bozbas, K. Biodiesel as an Alternative Motor Fuel: Production and Policies in the European Union. *Ren. Sust. Energy Rev.* **2008**, *12*, 542–552.

Brown, M.R.; S. W. Jeffrey; J.K. Volkman; G.A. Dunstan. Nutritional Properties of Microalgae for Mariculture. *Aquaculture* **1997**, *151*, 315–331.

Chelf, P. Environmental Control of Lipid and Biomass Production in Two Diatom Species. *J. Appl. Phycol.* **1990**, *2*, 121–129.

Chisti, Y. Biodiesel from Microalgae. *Biotech. Adv.* **2007**, *25*, 294–306.

Chrismadha, T.; M.A. Borowitzka. Growth and Lipid Production of *Phaeodactylum tricornutum* in a Tubular Photobioreactor. *Algal Biotechnology in the Asia-Pacific Region*; S.M. Phang ; Y.K. Lee; M.A. Borowitzka ; B.A. Whitton, Eds.; Institute of Advanced Studies, University of Malaya: Kuala Lumpur, 1994; pp 122–129.

Cohen, Z. Products from Microalgae. *CRC Handbook of Microalgal Mass Culture*; A. Richmond, Ed.; CRC Press: Boca Raton, 1986; pp 421–454.

Collyer, D.M.; G.E. Fogg. Studies on Fat Accumulation by Algae. *J. Exp. Bot.* **1955,** *6,* 256–275.

Coombs, J.; W.M. Darley; O. Holm-Hansen; B.E. Volcani. Studies on the Biochemistry and Fine Structure of Silica Shell Formation in Diatoms. Chemical Composition of *Navicula pelliculosa* during Silicon-Starvation Synchrony. *Pl. Physiol.* **1967,** *42,* 1601–1606.

Courchesne, N.M.D.; A. Parisien; B. Wang; C.Q. Lan. Enhancement of Lipid Production using Biochemical, Genetic and Transcription Factor Engineering Approaches. *J. Biotech.* **2009,**

Darley, W.M. Studies on the Biochemistry and Fine Structure of Silica Shell Formation in Diatoms. Division Cycle and Chemical Composition of *Navicula pelliculosa* during Light-Dark Synchronized Growth. *Planta* **1976,** *130,* 159–167.

Demirbas, A. Biodiesel Fuels from Vegetable Oils via Catalytic and Non-catalytic Supercritical Alcohol Transesterifications and Other Methods: A Survey. *Energy Conversion & Management* **2003,** *44,* 2093–2109.

Dempster, T.A.; M. Sommerfeld. Effects of Environmental Conditions on Growth and Lipid Accumulation in *Nitzschia communis* (Bacillariophyceae). *J. Phycol.* **1998,** *34,* 712–721.

Dijkstra, A.J. Revisiting the Formation of *Trans* Isomers during Partial Hydrogenation of Triacylglycerol Oils. *Eur. J. Lipid Sci. Technol.* **2006,** *108,* 249–264.

Fabregas, J.; A. Maseda; A. Dominguez; A. Otero. The Cell Composition of *Nannochloropsis* sp. Changes under Different Irradiances in Semicontinuous Culture. *World Journal of Microbiology & Biotechnology* **2004,** *20,* 31–35.

Fidalgo, J.P.; A. Cid; E. Torres; A. Sukenik; C. Herrero. Effect of Nitrogen Source and Growth Phase on Proximate Biochemical Composition, Lipid Classes and Fatty Acid Profile of the Marine Microalga *Isochrysis galbana*. *Aquaculture,* **1998,** 105–116.

Fisher, N.S.; R.P. Schwarzenbach. Fatty Acid Dynamics in *Thalassiosira pseudonana* (Bacillariophyceae). Implications for Physiological Ecology. *J. Phycol.* **1978,** *34,* 143–150.

Frankel, E.N. *Lipid Oxidation*; The Oily Press: Dundee, 1998.

Fukuda, H.; A. Kondo; H. Noda. Biodiesel Fuel Production by Transesterification of Oils. *J. Biosci. Bioeng.* **2001,** *92,* 405–416.

Gordillo, F.J.L.; M. Goutx; F.L. Figueroa; F.X. Niell. Effect of Light Intensity, CO_2 and Nitrogen Supply on Lipid Class Composition of *Dunaliella viridis*. *J. Appl. Phycol.* **1998,** *10,* 135–144.

Griffiths, M.J.; S.T.L. Harrison. Lipid Productivity as a Key Characteristic for Choosing Algal Species for Biodiesel Production. *J. Appl. Phycol.* **2009,**

Grossman, A.R. In the Grip of Algal Genomics. *Adv. Exp. Med. Biol.* **2007,** *616,* 54–76.

Guckert, J.B.; K.E. Cooksey. Triglyceride Accumulation and Fatty Acid Profile Changes in *Chlorella* (Chlorophyta) during High pH-induced Cell Cycle Inhibition. *J. Phycol.* **1990,** *26,* 72–79.

Guihéneuf, F.; V. Mimouni; L. Ulmann; G. Tremblin. Combined Effects of Irradiance Level and

Carbon Source on Fatty Acid and Lipid Class Composition in the Microalgae *Pavlova lutheri* Commonly Used in Aquaculture. *J. Exp. Mar. Biol. Ecol.* **2009**, *369*, 136–143.

Guschina, I.A.; J.L. Harwood. Lipids and Lipid Metabolism in Eukaryotic Algae. *Prog. Lipid Res.* **2006**, *45*, 160–186.

Hu, H.; K. Gao. Optimisation of Growth and Fatty Acid Composition of a Unicellular Marine Picoplankton, *Nannochloropsis* sp., with Enriched Carbon Sources. *Biotech. Lett.* **2003**, 25, 421–425.

Iwamoto, H.; S. Sato. Production of EPA by Freshwater Unicellular Algae. *JAOCS* **1986**, *63*, 434–438.

Kayama, M.; S. Araki; S. Sato. Lipids of Marine Plants. *Marine Biogenic Lipids, Fats, and Oils*; R.G. Ackman, Ed.; CRC Press: Boca Raton, 1989; pp 3–48.

Khozin-Goldberg, I.; Z. Cohen. The Effect of Phosphate Starvation on the Lipid and Fatty Acid Composition of the Fresh Water Eustigmatophyte *Monodus subterraneus*. *Phytochemistry* **2006**, *67*, 696–701.

Knothe, G. Dependence of Biodiesel Fuel Properties on the Structure of Fatty Acid Alkyl Esters. *Fuel Proc. Technol.* **2005**, *86*, 1059–1070.

Knothe, G. Analyzing Biodiesel: Standards and Other Methods. *JAOCS* **2006**, *83*, 823–833.

Kudsiana, D.; S. Saka. Methyl Esterification of Free Fatty Acids of Rapeseed Oil as Treated in Supercritical Methanol. *Fuel* **2001**, *80*, 225–231.

Li, Y.; M. Horsman; B. Wang; N. Wu; C.Q. Lan. Effect of Nitrogen Sources on Cell Growth and Lipid Accumulation of Green Alga *Neochloris oleoabundans*. *Appl. Microbiol. Biotechnol.* **2008**, *81*, 629–636.

Liang, Y.; J. Beardall; P. Heraud. Effect of UV Radiation on Growth, Chlorophyll Fluorescence and Fatty Acid Composition of *Phaeodactylum tricornutum* and *Chaetoceros muelleri* (Bacillariophyceae). *Phycologia* **2006**, *45*, 605–615.

Liang, Y.; K. Mai; S. Sun. Differences in Growth, Total Lipid Content and Fatty Acid Composition among 60 Clones of *Cylindritheca fusiformis*. *J. Appl. Phycol.* **2005**, *17*, 61–65.

Liu, Z.Y.; G.C. Wang; B.C. Zhou. Effect of Iron on Growth and Lipid Accumulation in *Chlorella vulgaris*. *Biores. Technol.* **2008**, *99*, 4717–4722.

Lombardi, A.T.; P.J. Wangersky. Influence of Phosphorus and Silicon on Lipid Class Production by the Marine Diatom *Chaetoceros gracilis* Grown in Turbidostat Cage Cultures. *Mar. Ecol. Prog. Ser.* **1991**, *77*, 39–47.

Lu, C.; K. Rao; D. Hall; A. Vonshak. Production of Eicosapentaenoic Acid (EPA) in *Monodus subterraneus* Grown in a Helical Tubular Photobioreactor as Affected by Cells Density and Light Intensity. *J. Appl. Phycol.* **2001**, *13*, 517–522.

May, P.; S. Wienkoop; S. Kempa; B .Usadel; N. Christian; J. Ruprecht; J. Weiss; L. Recuenco-Munoz; O. Ebenhöh; W. Weckwerth; et al. Metabolomics- and Proteomics-Assisted Genome Annotation and Analysis of the Draft Metabolic Network of *Chlamydomonas reinhardtii*. *Genetics* **2008**, *179*, 157–166.

McNeff, C.V.; L.C. McNeff; B .Yan; D.T. Nowlan; M. Rasmussen; A.E. Gyberg; B.J. Krohn; R.L. Fedie; T.R. Hoye. A Continuous Catalytic System for Biodiesel Production. *Appl. Catalysis A* **2008**, *343*, 39–48.

Meher, L.C.; D. Vidya Sagar; S.N. Naik. Technical Aspects of Biodiesel Production by Transesterification—A Review. *Ren. Sust. Energy Rev.* **2006,** *10,* 248–268.

Metzger, P.; C. Berkaloff; E. Casadevall; A. Coute. Alkadiene- and Botryococcene-Producing Races of Wild Strains of *Botryococcus braunii. Phytochemistry* **1985,** *24,* 2305–2312.

Metzger, P.; C. Largeau. Chemicals from *Botryococcus braunii. Chemicals from Microalgae*; Z. Cohen, Ed.; Taylor & Francis: London, 1999; pp 205–260.

Metzger, P.; C. Largeau. *Botryococcus braunii*: A Rich Source for Hydrocarbons and Related Ether Lipids. *Appl. Microbiol. Biotechnol.* **2005,** *66,* 486–496.

Moheimani, N.R.; M.A. Borowitzka. The Long-tTerm Culture of the Coccolithophore *Pleurochrysis carterae* (Haptophyta) in Outdoor Raceway Ponds. *J. Appl. Phycol.* **2006,** *18,* 703–712.

Molina Grima, E.; E.H. Belarbi; F.G. Áacién Fernandez; A. Robles Medina; Y. Chisti. Recovery of Microalgal Biomass and Metabolites: Process Options and Economics. *Biotech. Adv.* **2003,** *20,* 491–515.

Morgan-Kiss, R.M.; J.C. Priscu; T. Pocock; L. Gudynaite-Savitch; N.P.A. Huner. Adaptation and Acclimation of Photosynthetic Microorganisms to Permanently Cold Environments. *Microbiol. Mol. Biol. Rev.* **2006,** *70,* 222–252.

Muradyan, E.A.; G.L. Klyachko-Gurvich; L.N. Tsoglin; T.V. Sergeyenko; N.A. Pronina Changes in Lipid Metabolism during Adaptation of the *Dunaliella salina* Photosynthetic Apparatus to High CO_2 Concentration. *Russ. J. Pl. Physiol.* **2004,** *51,* 53–62.

Nagle, N.; P. Lemke. Production of Methyl Ester Fuel from Microalgae. *Appl. Biochem. Biotechnol.* **1990,** *24/25,* 355–361.

Otsuka, H.; Y. Morimura Changes in Fatty Acid Composition of *Chlorella ellipsoidea* during Its Cell Cycle. *Plant Cell. Physiol.* **1966,** *7,* 663–670.

Parrish, C.C.; P.J. Wangersky. Particulate and Dissolved Lipid Classes in Cultures of *Phaeodactylum tricornutum* Grown in Cage Culture Turbidostats with a Range of Nitrogen Supply Rates. *Mar. Ecol. Prog. Ser.* **1987,** *35,* 119–128.

Patterson, G.W. Effect of Culture Temperature on Fatty Acid Composition of *Chlorella sorokiniana. Lipids* **1970,** *5,* 579–600.

Piorreck, M.; K.-H. Baasch; P. Pohl. Biomass Production, Total Protein, Chlorophylls, Lipids and Fatty Acids of Freshwater Green and Blue-green Algae under Different Nitrogen Regimes. *Phytochemistry* **1984,** *23,* 207–216.

Piorreck, M.; P. Pohl. Formation of Biomass, Total Protein, Chlorophylls, Lipids and Fatty Acids in Green and Blue-green Algae during One Growth Phase. *Phytochemistry* **1984,** *23,* 217–223.

Pushparaj, B.; E. Pelosi; M.R. Tredici; E. Pinzani; R. Materassi. An Integrated Culture System for Outdoor Production of Microalgae and Cyanobacteria. *J. Appl. Phycol.* **1997,** *9,* 113–119.

Raghavan, G.; C.K. Haridevi; C.P. Gopinathan. Growth and Proximate Composition of the *Chaetoceros calcitrans* f. *pumilus* under Different Temperature, Salinity and Carbon Dioxide Levels. *Aquacult. Res.* **2008,** *39,* 1053–1058.

Ranga Rao, A.; C. Dayanandra; R. Sarada; T.R. Shamala; G.A. Ravishankar. Effect of Salinity on Growth of Green Alga *Botryococcus braunii* and Its Constituents. *Biores. Technol.* **2007,** *98,* 560–564.

Ranganathan, S.V.; S.L. Narasimhan; K. Muthukumar. An Overview of Enzymatic Production of Biodiesel. *Biores. Technol.* **2008,** *99,* 3975–3981.

Reitan, K.I.; J.R. Rainuzzo; Y. Olsen. Effect of Nutrient Limitation on Fatty Acid and Lipid Content of Marine Microalgae. *J. Phycol.* **1994,** *30,* 972–979.

Renaud, S.M.; H.C. Zhou; D.L. Parry; L.V. Thinh; K.C. Woo. Effect of Temperature on the Growth, Total Lipid Content and Fatty Acid Composition of Recently Isolated Tropical Microalgae *Isochrysis* sp, *Nitzschia closterium, Nitzschia paleacea,* and Commercial Species *Isochrysis* sp (clone T iso). *J. Appl. Phycol.* **1995,** *7,* 595–602.

Riekhof, W.R.; B.B. Sears; C. Benning. Annotation of Genes Involved in Glycerolipid Biosynthesis in *Chlamydomonas reinhardtii:* Discovery of the Betaine Lipid Synthase BTA1$_{Cr}$. *Eukaryotic Cell* **2005,** *4,* 242–252.

Robles-Medina, A.; P.A. González-Moreno; L. Esteban-Cerdan; E. Molina-Grima. Biocatalysis: Towards Even Greener Biodiesel Production. *Biotech. Adv.* **2009,** 398–408.

Rodolfi, L.; G.C. Zitelli; N. Bassi; G. Padovani; N. Biondi; G. Bonini; M.R. Tredeci. Microalgae for Oil: Strain Selection, Induction of Lipid Synthesis and Outdoor Mass Cultivation in a Low-Cost Photobioreactor. *Biotech. Bioeng.* **2009,** *102,* 100–112.

Roessler, P.G. Changes in the Activities of Various Lipid and Carbohydrate Biosynthetic Enzymes in the Diatom *Cyclotella cryptica* in Response to Silicon Deficiency. *Arch. Biochem. Biophys.* **1988a,** *267,* 521–528.

Roessler, P.G. Effects of Silicon Deficiency on Lipid Composition and Metabolism in the Diatom *Cyclotella cryptica. J. Phycol.,* **1988b,** *24,* 394–400.

Roessler, P.G. Environmental Control of Glycerolipid Metabolism in Microalgae—Commercial Implications and Future Research Directions. *J. Phycol.* **1990,** *26,* 393–399.

Shaw, P.M.; G.J. Jones; J.D. Smith; R.B. Johns. Intraspecific Variations in the Fatty Acids of the Diatom *Skeletonema costatum. Phytochemistry* **1989,** *28,* 8111 –8115.

Sheehan, J.; T. Dunahay; J. Benemann; P. Roessler. A Look Back at the U.S. Department of Energy's Aquatic Species Program—Biodiesel from Algae; National Renewable Energy Laboratory:, Golden, Colorado, 1998.

Shifrin, N.S.; S.W. Chisholm. Phytoplankton Lipids: Interspecific Differences and Effects of Nitrate, Silicate and Light-Dark Cycles. *J. Phycol.* **1981,** *17,* 374–384.

Siron, R.; G. Giusti; B. Berland. Changes in the Fatty Acid Composition of *Phaeodactylum tricornutum* and *Dunaliella tertiolecta* during Growth and under Phosphorous Deficiency. *Mar. Ecol. Prog. Ser.* **1989,** *55,* 95–100.

Sukenik, A.; Y. Carmeli; T. Berner. Regulation of Fatty Acid Composition by Irradiance Level in the Eustigmatophyte *Nannochloropsis* sp. *J. Phycol.* **1989,** *25,* 686–692.

Taguchi, S.; J.A. Hirata; E.A. Laws. Silicate Deficiency and Lipid Synthesis in Marine Diatoms. *J. Phycol.* **1987,** *23,* 260–267.

Takagi, M.; K. Watanabe; K. Yamaberi; T. Yoshida. Limited Feeding of Potassium Nitrate for Intracellular Lipid and Triglyceride Accumulation of *Nannochloris* sp. UTEX LB1999. *Appl. Microbiol. Biotechnol.* **2000,** *54,* 112–117.

Templier, J.; C. Largeau; E. Casadevall. Mechanism on Non-isoprenoid Hydrocarbon Biosynthesis in *Botryococcus braunii. Phytochemistry* **1984,** *23,* 1017–1028.

Tonon, T.; D. Harvey; T.R. Larson; I.A. Graham. Long Chain Polyunsaturated Fatty Acid Production and Partitioning to Triacylglycerols in Four Microalgae. *Phytochemistry* **2002,** *61,* 15–24.

Tornabene, T.G.; G. Holzer; S. Lien; N. Burris. Lipid Composition of the Nitrogen Starved Green Alga *Neochloris oleoabundans. Enz. Microb. Technol.* **1983,** *5,* 435–440.

Tzovenis, I.; N. De Pauw; P. Sorgeloos. Optimisation of T-TSO Biomass Production Rich in Essential Fatty Acids II. Effect of Different Light Regimes on the Production of Fatty Acids. *Aquaculture* **2003,** *216,* 223–242.

Vasquez-Duhalt, R.; B.O. Arrendondo-Vega. Haloadaptation in the Green Alga *Botryococcus braunii* (Race A). *Phytochemistry* **2001,** *30,* 2919–2925.

Wake, L.V.; L.W. Hillen. Nature and Hydrocarbon Content of Blooms of the Alga *Botryococcus braunii* Occurring in Australian Freshwater Lakes. *Aust. J. Mar. Freshwat. Res.* **1981,** *32,* 353–367.

Widjaja, A.; C.C. Chien; Y.H. Ju. Study of Increasing Lipid Production from Freshwater Microalgae *Chlorella vulgaris. Journal of the Taiwan Institute of Chemical Engineers* **2009,** *40,* 13–20.

Wolf, F.R.; A.M. Nonomura; J.A. Bassham. Growth and Branched Hydrocarbon Production in a Strain of *Botryococcus braunii* (Chlorophyta). *J. Phycol.* **1985,** *21,* 388–396.

14

Production of Lipids for Biofuels Using Bacteria

Daniel Bröker,[a] Yasser Elbahloul,[b] and Alexander Steinbüchel[a]

[a]Institut für Molekulare Mikrobiologie und Biotechnologie, Westfälische Wilhelms-Universität Münster, D-48149 Münster, Germany; [b]Faculty of Science-Botany Dept., Alexandria University, 21511-Moharam Bey Alexandria, Egypt

Abstract

There is an increasing demand for biofuels from renewable resources especially, for crude oil, because of rising energy prices induced by a shortage of fossil resources and increasing environmental awareness. People want to retain their mobility, but not at the expense of massive environmental, economical, and social problems. Thus, the production of biofuels seems to be a promising solution. This chapter will deal with the bacterial production of lipids for biofuels. Lipid-based substitutes, such as methyl and ethyl esters of fatty acids, are commonly designated as biodiesel; these esters exhibit several positive characteristics, such as biodegradability, non-toxicity, low sulfur content, and the absence of aromatic compounds. At present, a wide variety of vegetable oils are predominantly used as renewable resources for chemically alkali-catalyzed or enzymatically-catalyzed transesterification reactions yielding biodiesel. Instead of using vegetable oils, oils with microbial origins that occur in the cells as storage compounds could be used. In this chapter, we will focus on in vivo lipid biosynthesis and production in bacteria, also introducing bacterial acyltransferases—the key enzymes for the biosynthesis of triacylglycerols and wax esters. We will review attempts to establish the formation of fatty acid ethyl esters in a recombinant *Escherichia coli* strain and discuss the perspectives of this so-called "microdiesel."

Introduction

Biotechnological production of bacterial storage compounds, such as polyhydroxyalkanoic acid (PHA), has increased, especially due to the fact that these PHAs are biodegradable (Steinbüchel, 1996, 2001; Steinbüchel & Hein, 2001). One prominent example is the copolymer poly(3-hydroxybutyrate-co-3-hydroxyvalerate) [poly(3HB-co-3HV)], which was commercially developed by Imperial Chemical Industries (ICI) and sold under the tradename Biopol˚. Beside these well-investigated bacterial storage compounds, other intracellular storage compounds, such as cyanophycin, triacylglycerols (TAG) and wax esters (oxoesters of primary long-chain fatty acids and primary long-chain fatty alcohols) have become of scientific and biotechnological interest (Sallam et al., 2009; Stöveken & Steinbüchel, 2008). The production of alternative fuels has attracted wide attention due to the public's perception of petroleum reserves as diminishing; the environmental consequences of exhaust gases from petroleum-based fuels; energy prices, especially for crude oil; and heightened environmental awareness. Thus, intensive research on biofuels, notably on biodiesel, has been performed (Ma & Hanna, 1999). Biodiesel possesses several positive characteristics—e.g., its biodegradability, non-toxicity, low sulfur content, absence of aromatic compounds, and environmental-friendliness—as it is obtained from renewable resources. In general, biodiesel is produced by the transesterification (Fig. 14.1A) or esterification (Fig. 14.1B) of TAG or of free fatty acids, respectively , with short chain alcohols, such as methanol, yielding fatty acid methyl esters (FAME) and glycerol, a side product. The lengths of the acid moieties depend on the biological sources of the lipids often representing TAG. In several studies, use of TAG from different sources as an alternative fuel for diesel engines was described (Fukuda et al., 2001; Ma & Hanna, 1999; Srivastava & Prasad, 2000).

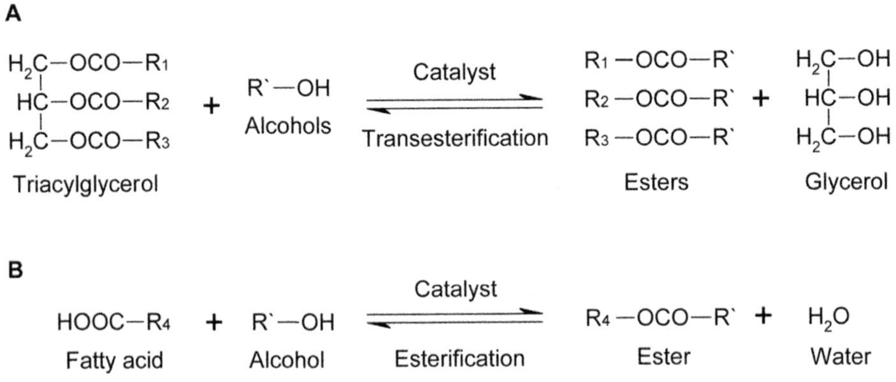

A

$$H_2C-OCO-R_1$$
$$HC-OCO-R_2 \quad + \quad R'-OH \quad \xrightleftharpoons{\text{Catalyst}} \quad R_1-OCO-R' \quad H_2C-OH$$
$$H_2C-OCO-R_3 \quad \quad \text{Alcohols} \quad \text{Transesterification} \quad R_2-OCO-R' \quad + \quad HC-OH$$
$$R_3-OCO-R' \quad H_2C-OH$$

Triacylglycerol   Esters   Glycerol

B

$$HOOC-R_4 \quad + \quad R'-OH \quad \xrightleftharpoons{\text{Catalyst}} \quad R_4-OCO-R' \quad + \quad H_2O$$

Fatty acid   Alcohol   Esterification   Ester   Water

Fig. 14.1. Biodiesel production by transesterification (A) or esterification (B) of TAG or free fatty acids, respectively, with short chain alcohols. R_1 to R_4 represent the fatty acid side chains, whereas R' indicates the alcohol side chains.

Besides wax esters and steryl esters, TAG occur as neutral storage lipids in plants, animals, yeast, fungi, and bacteria (Alvarez & Steinbüchel, 2002). In plants, TAG are the dominating intracellular storage lipids present in seeds (Murphy & Vance, 1999). Interestingly, jojoba seeds store lipids as long-chain wax esters (Yermanos, 1975). The main components of intracellular lipid bodies in animals are TAG and steryl esters, predominantly occuring in hepatocytes and adipocytes (Murphy & Vance, 1999). Lipid bodies accumulated in yeasts and filamentous fungi are similar to those from plants and animals (Holdsworth & Ratledge, 1991; Zweytick et al., 2000). Bacteria also store significant amounts of lipids as TAG and wax esters (Alvarez et al., 2000; Kalscheuer et al., 2007).

Thus, biodiesel can be subdivided according to the source of the oil and to the catalysis of the transesterification process into biodiesel: (i) processes based on vegetable, animal, microbial or waste oil, and fats and a chemical transesterification process performed by petrochemical methanol (Fig. 14.2A), (ii) processes based on vegetable, animal, microbial or waste oil, and fats; a biologically catalyzed transesterification process using lipases (Du et al., 2007; Iso et al., 2001; Orcaire et al., 2006) or whole cells (Li et al., 2007a); and biological or petrochemical alcohol (Fig. 14.2B); and (iii) microbial biodiesel production, including synthesis of alcohol for the transesterification process (Fig. 14.2C). The latter type is now known as microdiesel and has become an issue of great interest since Kalscheuer et al. (2006) reported on its formation.

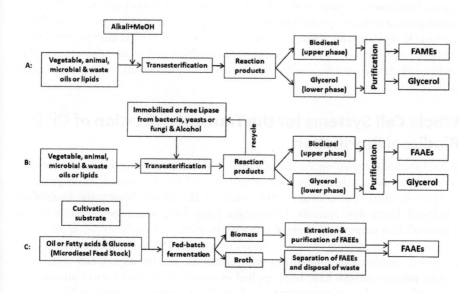

Fig. 14.2. Different types of biodiesel classified according to the sources of lipids and alcohol used in the transesterification process. We distinguish between biodiesels based on (A) vegetable, animal, microbial, or waste oil and fats, and a chemical transesterification process performed by petrochemical methanol; (B) vegetable, animal, microbial, or waste oil and fats, and a biologically catalyzed transesterification process, using lipases or whole cells and biological or petrochemical alcohol; and (C) microbial biodiesel production, including synthesis of alcohol for the transesterification process.

Enzymatic Transesterification of Oil for Biodiesel Production

The use of extracellular lipases has been extensively studied for the production of biodiesel with the aim of alleviating the problems associated with alkali catalysis, such as the generation of waste glycerol and the purification of biodiesel. Different sources for lipases were investigated for the transesterification of triglycerides with short-chain alcohols to produce alkyl esters (Adamczak & Bednarski, 2004; Hama et al., 2006; Nakashima et al., 1988, 1990; Shimada et al., 1999; Tamalampudi et al., 2007). Shimada et al. (1999) reported on the stepwise addition of methanol to oil using a lipase obtained from *Candida antarctica* (Novozym 435) immobilized on an acrylic resin. Because lipase inhibition by methanol is avoided by this strategy, this immobilized system, used for over 50 cycles, yielded a conversion of 95% methyl ester. In another study, Watanabe et al. (2000) used Novozym 435 in a batch system for over 100 days without significant reduction in yield and, thus, increased the methyl ester content to about 93%. The application of nonspecific lipases isolated from *Candida rugosa*, *Pseudomonas cepacia*, and *P. fluorescens* showed relatively high conversion rates in comparison to regio-specific lipases, such as the lipase from *Rhizopus oryzae*. When the *sn*-1(3)-regio-specific lipase from *R. oryzae* was used, methyl ester moieties were obtained in positions *sn*-1 and *sn*-3, but not in position *sn*-2 of the triglycerides. This finding reflects the need for nonspecific lipases in the production of biodiesel (Kaieda et al., 1999)

Whole Cell Systems for the Transesterification of Oil for Biodiesel Production

Several studies reported on the use of varieties of bacteria, yeast, and fungi as whole-cell biocatalysts to enhance the cost effectiveness of the transesterification processes (Ban et al., 2001; Fujita et al., 2002; Narita et al., 2006). Among the established whole-cell biocatalyst systems, filamentous fungi have arisen as the most robust whole-cell biocatalyst for industrial transesterification and methanolysis applications of plant oils (Atkinson et al., 1979; Nakashima et al., 1988, 1989). Heterologously-expressed membrane-bound lipases with enhanced activity on the cell surfaces of certain microorganisms have been applied in various transesterification processes. A yeast cell surface display system for lipase from *R. oryzae* was developed by Matsumoto et al. (2002). A main advantage of using surface-expressed lipases is their easy access to the substrates during alcoholysis, which renders pretreatment of the catalyst cells unnecessary and, thus, decreases production costs. Other studies dealt with the permeabilization of intracellular, lipase-producing yeast strains with isopropyl alcohol, which may further increase the efficiency of whole-cell biocatalysts (Kondo et al., 2000).

However, the transesterification reaction by whole cell catalysis is considered to be more time consuming in comparison with in vitro processes catalyzed by immobilized lipases. An outstanding reaction rate for a transesterification reaction using immobilized lipases has been obtained by Watanabe et al. (2000) and Samukawa et al. (2000) with Novozym 435 in continuous operation (7 h with 92–94% methyl ester content) and fed-batch operations (3.5 h, methyl oleate pretreatment with 87% methyl ester content), respectively. In contrast, the reaction rate of whole-cell biocatalysts using a batch reaction was very low, and the process required periods of more than 72 h (Matsumoto et al., 2002) or 165 h (Matsumoto et al., 2001), respectively. Investigations into more efficient whole-cell biocatalysts are in progress. Tamalampudi et al. (2007) reported on the potential of recombinant filamentous *Aspergillus oryzae* containing a lipase-encoding gene from *C. antarctica* as an efficient whole-cell biocatalyst, its immobilization in biomass support particles (BSPs), and its considerably facilitating applications in industrial biocatalysis, both in aqueous and non-aqueous media. Moreover, advances in the heterologous expression of nonspecific lipases, such as lipases from *C. antarctica* and *P. cepacia*, as well as of methanol-tolerant lipases might lead to the development of recombinant whole-cell biocatalysts and allow even more efficient transesterification of plant oils (Fukuda et al., 2001).

Nevertheless, a wide variety of renewable vegetable oils have been predominantly used for chemically alkali-catalyzed or enzymatically catalyzed transesterification reactions to biodiesel (Vasudevan & Briggs, 2008). The commercial alkali-catalyzed transesterification needs a high temperature (160–180°C), whereas higher costs for lipases, the most frequently used enzymatic catalysts, are countered by the fact that they save energy (Metzger & Bornscheuer, 2006). Major costs of biodiesel production are associated with the cost of the oil substrate (Shah et al., 2004; Tamalampudi et al., 2008). Moreover, the use of renewable vegetable oils provokes social problems, because the cultivation of these feedstocks for biodiesel production often competes with the cultivation of food and feed. The first challenge for microdiesel production is to find suitable microbial strains that are capable of overproducing triacylglycerols or wax esters de novo (Stöveken & Steinbüchel, 2008). In recent years, many lipid-accumulating microorganisms have been studied for single cell oil (SCO) production, especially for use in the production of biodiesel. Numerous oleaginous bacteria, fungi, yeasts, and microalgae have been reported to grow and accumulate significant amounts of lipids, similar to vegetable oil (Aggelis et al., 1995; Aggelis & Sourdis, 1997; Li et al., 2007b; Papanikolaou et al., 2002, 2004; Ratledge, 2004), methyl-esters, and soaps (Metzger & Largeau, 2005), used as sole carbon and energy sources. In this context, bacteria using non-food products as carbon sources for conversion into TAG have the potential to be used as alternative production organisms for biodiesel.

Occurrence and Function of Neutral Storage Lipids in Bacteria

Bacteria accumulating large amounts of TAG mainly belong to the actinomyces group, notably *Mycobacterium* sp. (Akao & Kusaka, 1976; Barksdale & Kim, 1977), *Nocardia* sp. (Alvarez et al., 1997a), *Rhodococcus* sp. (Alvarez et al., 1996), *Dietzia* sp. (Alvarez & Steinbüchel, 2002), *Gordonia* sp. (Alvarez & Steinbüchel, 2002), *Micromonospora* sp. (Hoskisson et al., 2001), and *Streptomyces* sp. (Olukoshi & Packter, 1994). TAG accumulation has also been described for members of the Gram-negative genus *Acinetobacter*, though the amounts are small in comparison to the accumulated wax esters (Makula et al., 1975; Scott & Finnerty, 1976; Singer et al., 1985). TAG are stored in spherical lipid bodies in these bacteria; their numbers and sizes depend on the respective species, growth phase, and cultivation conditions (Table 14.A). Among these species, the *Rhodococcus opacus* strain PD630 is able to store TAG of up to 89% of the total cellular dry weight, and its cells are almost completely filled with several lipid bodies exhibiting diameters ranging from 50 to 400 nm (Fig. 14.3) (Alvarez et al., 1996). These lipid bodies were mainly composed of TAG (87%), diacylglycerols (≈5%), free fatty acids (≈5%), phospholipids (PLs) (1.2%), and proteins (0.8%). The main components of the TAG were hexadecanoic acid (36.4% of total fatty acid content), octadecenoic acid (19.1%), and considerable amounts of odd-numbered fatty acid residues like heptadecanoic acid (11.4%) and heptadecenoic acid (10.6%) (Alvarez et al., 1996).

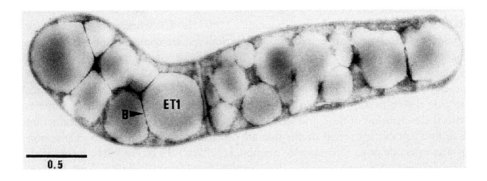

Fig. 14.3. Electron microscopic picture of lipid bodies in cells of *R. opacus* strain PD630. The cell contained several electron transparent lipid bodies (ET1) exhibiting a thin boundary layer (B). The bar represents a length of 0.5 µm. The photograph is reproduced from Alvarez et al. (1996).

Table 14.A. Occurrence of Triacylglycerols (TAG) in bacteria. For each bacterium, possible carbon sources for conversion into TAG and the obtained TAG contents are indicated. In case of different possible carbon sources, the TAG contents for cultivation with the carbon source are indicated in bold (modified according to Alvarez & Steinbüchel, 2002).

Bacterium	Carbon Source	TAG Content	Reference
Gram-Positive Bacteria			
Dietza maris	Acetate, hexadecane	19.2% (cdw)[a]	Alvarez, 2003
Gordonia amarae	Gluconate, hexadecane	6.1% (cdw)[a]	Alvarez, 2003
Micromonospora echinospora	Glucose	8.0% (cdw)[a]	Hoskisson et al., 2001
Mycobacterium avium	Palmitic acid	5.0% (cdw)[a]	Barksdale & Kim, 1977
M. ratisbonense	Squalene	n.r.	M. Berekaa & A. Steinbüchel, unpublished data
M. smegmatis	Complex medium	n.r.	Akao & Kusaka, 1976
M. tuberculosis	Glycerol-containing medium	n.r.	Barksdale & Kim, 1977
Nocardia asteroides	Gluconate, pentadecane, hexadecane, gas-oil	12.2% (cdw)[a]	Alvarez, 2003
N. corallina	Glucose, hexadecane, valerate	23.9% (cdw)[a]	Alvarez et al., 1997a
N. globerula	Gluconate, hexadecane, pristane	18.6% (cdw)[a]	Alvarez et al., 2001; Alvarez, 2003
N. restricta	Gluconate, hexadecane	19.3% (cdw)[a]	Alvarez, 2003
Rhodococcus erythropolis	Gluconate, pentadecane, hexadecane, valerate	21.0% (cdw)[a]	Alvarez et al., 1997a; Alvarez, 2003
R. fascians	Glucose, pentadecane, hexadecane, valerate	18.1% (cdw)[a]	Alvarez et al., 1997a; Alvarez, 2003
R. opacus	Gluconate, fructose, acetate, citrate, sucinate, propionate, valerate, phenylacetate, olive oil, phenyldecane, n-alkanes	87.0% (cdw)[a]	Alvarez et al., 1996, 1997a, 2003

(Cont. on next page)

Table 14.A., continued

Bacterium	Carbon Source	TAG Content	Reference
R. ruber	Glucose, acetate, citrate, valerate, pentadecane, hexadecane	26.0% (cdw)[a]	Alvarez et al., 1997a
Rhodococcus sp. strain 20	Gluconate, hexadecane	8.1% (cdw)[a]	Alvarez, 2003
Streptomyces coelicolor	Complex medium	84 mg/ml medium[b]	Olukoshi & Packter, 1994; Karandikar et al., 1997
S. lividans	Complex medium	125 mg/ml medium[b]	Olukoshi & Packter, 1994
S. albulus	Complex medium	56 mg/ml medium[b]	Olukoshi & Packter, 1994
S. griseus	Complex medium	93 mg/ml medium[b]	Olukoshi & Packter, 1994
Gram-Negative Bacteria			
Acinetobacter calcoaceticus strain BD413	Hexadecane	4.0% (cdw)[a]	Reiser & Somerville, 1997
A. lwoffi	Hexadecane + ethanol	16 µg/mg protein[b]	Vachon et al., 1982
Acinetobacter sp. strain H01-N	Hexadecane, hexadecanol	1.9% (cdw)[a]	Makula et al., 1975; Scott & Finnerty, 1976; Singer et al., 1985
Acinetobacter sp. strain 211	Acetate, ethanol, olive oil, hexadecanol, heptadecane	25.0% (cdw)[a]	Alvarez et al., 1997b
Pseudomonas aeruginosa strain 44T1	Glucose, n-alkanes, olive oil	38.0% (cdw)[a]	De Andrès et al., 1991
Other Bacteria			
Nostoc commune	Complex medium	n.r.	Taranto et al., 1993

Abbreviations and symbols: *cdw*, cellular dry weight; *n.r.*, not reported; [a] total amounts of cellular fatty acids; [b] total amounts of TAG.

Wax esters are predominantly accumulated by *Acinetobacter* sp. cultivated on alkanes and aromatic hydrocarbons (Fixter & Fewson, 1974; Fixter & McCormack, 1976; Gallagher, 1971; Scott & Finnerty, 1976) and also, in lesser amounts, by *Moraxella* sp. (Bryn et al., 1977), *Micrococcus* sp. (Russell & Volkman, 1980), and *Alcanivorax* sp. (Bredemeier et al., 2003). Actinomycetes belonging to the genera *Corynebacterium* (Bacchin et al., 1974), *Mycobacterium* (Wang et al., 1972), and *Nocardia* (Raymond & Davis, 1960) have also been described as accumulating wax esters. Isoprenoid wax esters have been identified in the *Marinobacter* species, e.g., in *M. hydrocarbonoclasticus* DSM 8798, grown in marine sediment materials where there is an abundance of recalcitrant acyclic isoprenoid alcohols, such as farnesol and phytol, that are derived from (bacterio) chlorophyll molecules (Holtzapple & Schmidt-Dannert, 2007; Rontani et al., 1997, 1999; 2003).

Among these species, *Acinetobacter baylyi* strain ADP1 and *A. calcoaceticus* strain HO1-N are able to store wax esters of about 25% of their total cellular dry weight, but here only one or several wax ester bodies can be observed per cell that exhibit an average diameter of 200 nm (Scott & Finnerty, 1976; Stöveken et al., 2005; Wältermann et al., 2005). Chemical analysis reveals that these wax ester bodies consist of hexadecylpalmitate (85.6%), hexadecanol (4.8%), and PLs (9.6%) (Singer et al., 1985). Interestingly, some species of *Acinetobacter* (Dewitt et al., 1982; Makula et al., 1975; Singer et al., 1985) and *Alcanivorax* (Bredemeier et al., 2003; Kalscheuer et al., 2007) synthesize extracellular wax esters from alkanes, and investigations seeking to understand the function of these extracellular wax esters and the mechanisms of their export are in progress (Manilla-Perez et al., unpublished).

In bacteria, neutral storage lipids serve as carbon and energy storage compounds, which can be mobilized in the absence of a carbon source and the presence of other essential supplements (Alvarez et al., 2000). For example, TAG can be hydrolyzed to glycerol and fatty acids by lipases. Glycerol is then phosphorylated, oxidized to dihydroxyacetone phosphate, and further catabolized in the glycolytic pathway. In general, fatty acids are converted to coenzyme A (CoA) esters and oxidized in the β-oxidation pathway, yielding acetyl-CoA. The latter is further catabolized in the tricarboxylic acid cycle (TCA) or used in anabolism. Thus, lipid accumulation is advantageous for natural survival in nutrient-poor habitats. It was considered that free fatty acids have a membrane damaging potential, and their incorporation into non-toxic storage compounds protects the cells from high-cellular concentrations of these molecules (Alvarez et al., 2001; Wältermann & Steinbüchel, 2005). Furthermore, an adaption to aridity was proposed, as oxidation of the hydrocarbon chains of the lipids under conditions of dehydration would generate considerable amounts of water, helping the cells to survive (Alvarez et al., 2004). As pathogenic bacteria, e.g., *M. tuberculosis*, are also able to accumulate TAG, a probable influence of TAG metabolism in pathogenesis is discussed (Daniel et al., 2004; Garton et al., 2002).

Biosynthesis of Neutral Storage Lipids in Bacteria and Involved Enzymes

Bacteria and eukaryotes store carbon and energy as neutral storage lipids, such as TAG and wax esters, which are composed of glycerol esterified to three fatty acids (Fig. 14.4A) or fatty alcohols esterified to fatty acids, respectively (Fig. 14.4B). Fatty acid de novo synthesis is catalyzed by the fatty acid synthetase complex, using acetyl-CoA and malonyl-CoA as substrates and NADPH as reductant. After acetate and malonate have been transferred from CoA to the sulfhydryl group of the acyl carrier protein (ACP), fatty acid de novo synthesis occurs. During this two-stage process, two carbons are added to the carboxyl end of the growing fatty acid chain by the synthetase complex. Synthesis is repeated until the proper chain length of the growing fatty acid, carried by ACP, has been reached, yielding fatty acids with a certain chain length (Magnuson et al., 1993). These fatty acids have to be attached to glycerol arising from reduction of the glycolytic intermediate dihydroxyacetone phosphate or to fatty alcohols resulting from reductive and oxidative pathways using acyl-CoA and hydrocarbons as precursors, yielding TAG or wax esters, respectively.

In contrast to eukaryotes, where different specialized enzyme classes are responsible for catalyzing the transfer of long chain acyl residues, which yields TAG and wax esters (Athenstaedt & Daum, 2006; Yen et al., 2008), in bacteria predominantly one class of acyltransferases catalyzing the last step of either TAG and wax ester biosynthesis occurs (Alvarez et al., 2008; Arabolaza et al., 2008; Daniel et al., 2004; Holtzapple & Schmidt-Dannert, 2007; Kalscheuer & Steinbüchel, 2003; Kalscheuer et al., 2003, 2007; Wältermann et al., 2006). This enzyme class possesses an extraordinarily low substrate specificity and accepts a broad range of various substances as alternative acceptor molecules, allowing biosynthesis of such structurally different compounds as TAG and wax esters. Due to their catalyzed reactions, members of this enzyme class are designated as wax ester synthase/acyl coenzyme A:diacylglycerol acyltransferases (WS/DGAT) (Fig. 14.4) (Kalscheuer & Steinbüchel, 2003; Stöveken et al., 2005; Uthoff et al., 2005). In vitro experiments demonstrated that purified WS/DGAT from *A. baylyi* strain ADP1 also exhibited acyl-CoA:monoacylglycerol acyltransferase (MGAT) activity and that the exposed hydroxy groups in *sn*-1 and *sn*-3 positions of acylglycerols were preferred in comparison to those in the *sn*-2 position. WS/DGAT acylated linear alcohols, ranging from ethanol to triacontanol, exhibited the highest activity towards medium-chain-length alcohols from tetradecanol to octadecanol. Enzyme activity decreased with increasing chain length. Primary alcohols were preferred in comparison to secondary alcohols. To determine the specificity of the WS, acyl-CoA thioesters with a carbon chain length of the acyl moieties from C2 to C20 were tested. The highest activities were obtained for carbon chain lengths from C14 to C18. Acylation of cyclic and aromatic alcohols, such as cyclohexanol or phenylethanol, was also catalyzed by WS/DGAT, confirming the unspecificity of the latter (Stöveken et al., 2005).

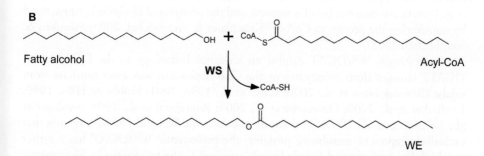

Fig. 14.4. In vivo reactions catalyzed by bacterial wax ester synthase/acyl coenzyme A:diacylglycerol acyltransferases (WS/DGAT). (A) diacylglycerol transferase reaction; (B) wax ester synthase reaction. Abbreviations: DAG, diacylglycerol; DGAT, diacylglycerol transferase; Acyl-CoA, acyl-coenzyme A; TAG, triacylglycerol; WS, wax ester synthase.

Further bacterial acyltransferases of the WS/DGAT family have been identified, -e.g., from *M. tuberculosis* strain H37Rv (Daniel et al., 2004), *A. borkumensis* strain SK2 (Kalscheuer et al., 2007), *Marinobacter hydrocarbonoclusticus* DSM8798 (Holtzapple & Schmidt-Dannert, 2007; Rontani et al., 1999), *R. opacus* strain PD630 (Alvarez et al., 2008), and *Streptomyces coelicolor* (Arabolaza et al., 2008). All of these strains exhibited a broad substrate range; however, they showed preferences for different substrates. Another family of long chain acyltransferases, containing the so-called polyketide-associated proteins (PAP), was also characterized. PAP exclusively occur in mycobacteria and *S. coelicolor*, in addition to WS/DGAT (Onwueme et al., 2004). They catalyze reactions similar to those of the WS/DGATs, such as acylation of alcohols of different chain length, as well as secondary or tertiary alcohols by PAPA5 (Onwueme et al., 2004) or the subsequent palmitoylation of trehalose 2-sulfate by PAPA1 and PAPA2 from *M. tuberculosis* (Kumar et al., 2007).

Almost all members of the WS/DGAT family, and also the PAP from mycobacteria and *S. coelicolor*, contain a highly conserved motif (HHXXXDG, amino acids 133-138 of AtfA from *A. baylyi* strain ADP1), which is suggested to be involved in fatty acyl-CoA binding or catalytically participates in the acyltransferase reaction (Daniel et al., 2004; Kalscheuer & Steinbüchel, 2003; Stöveken & Steinbüchel, 2008). It was suggested that the conserved histidine residue deprotonates the hydroxyl group of the fatty-alcohol or the DAG, respectively, acting as a base catalyst, which would then enable a nucleophilic attack on the thioester bond of the fatty acyl-CoA molecule (Fig. 14.4A). An oxoester bond is formed, and the protonated histidine is regenerated by transferring the proton to CoA-S⁻ (Kalscheuer & Steinbüchel, 2003; Stöveken & Steinbüchel, 2008).

Interestingly, WS/DGAT exhibit no sequence homology to the DGAT1 and DGAT2 families from eukaryotes or the CoA-dependent wax ester synthase from jojoba (Bouvier-Navé et al., 2000; Cases et al., 1998, 2001; Hobbs & Hills, 1999; Lardizabal et al., 2000; Onwueme et al., 2004; Routaboul et al., 1999; Sandager et al., 2002; Zou et al., 1999). In contrast to eukaryotic WS and DGAT enzymes that exclusively represent membrane proteins, the prokaryotic WS/DGAT has a rather amphiphilic character and is only loosely attached to the membrane by electrostatic interactions (Cao et al., 2003; Cases et al., 1998; Lardizabal et al., 2000; Kamisaka et al. 1997; Stöveken et al., 2005; Wältermann & Steinbüchel, 2005).

Investigations into Bacterial Neutral Lipid Production

In bacteria, neutral storage lipids serve as carbon and energy storage compounds and are only accumulated under certain cultivation conditions. *Rhodococcus opacus* and *S. lividans* contain few TAG when cultivated in complex media with a high content of carbon and nitrogen. The lipid content and the number of TAG bodies significantly increased when the cells were cultivated in a mineral salt medium with a low nitrogen-to-carbon ratio, which produced its maximum yield in the late stationary growth phase (Packter & Olukoshi, 1995; Wältermann et al., 2005). Voss and Steinbüchel

(2001) investigated the fermentative production of TAG from the low-cost carbon sources—sugar beet molasses and sucrose—using *R. opacus* strain PD630. Fed-batch fermentations at 30 L and 500 L in stirred tank bioreactors at 30°C yielded approximately 37.5 g of cell dry matter (CDM) per liter, with 52% TAG, and 18.4 g of CDM per liter, with 38.4% TAG, respectively (Voss & Steinbüchel, 2001). Mona et al. (2008) studied lipid accumulation for *Gordonia* sp. DG in comparison with the *R. opacus* strain PD630, using different agro-industrial residuals or wastes like sugar cane molasses, carob, and orange waste as carbon sources. The use of orange waste as a carbon source resulted in the maximum lipid content for both strains and an increased lipid unsaturation, with 18:3 as major unsaturated fatty acids for *R. opacus* and 22:0 and 6:0 as dominant fatty acids for *Gordonia* sp. DG (Mona et al., 2008). However, no account was taken in this work of the possible content of lipids or fatty acids in the original feedstock material, and, in addition, the authors reported lipid levels of over 95% of the cell dry wt, which, if substantiated, would be the highest value ever recorded for any oleaginous microorganism. Thus, the production of lipids using wild-type strains depends on the used strain, the cultivation conditions, the carbon source being used, and the nitrogen-to-carbon ratio. To establish biotechnological production in industry, comparative optimization studies that examine different suitable strains and certain cultivation conditions have to be performed. Another approach explores the metabolic engineering of suitable bacteria using respective genes that code for enzymes involved in the well-understood lipid biosynthesis (Kalscheuer et al., 2006).

Engineered *E. coli* Strain for Microdiesel Production

E. coli does not produce fatty acid ethyl esters (FAEE) by its natural metabolism but forms only ethanol, among other fermentation products, anaerobically during mixed acid fermentation. Ethanol is synthesized from acetyl-CoA via two sequential NADH-dependent reductions that are catalyzed by a multifunctional alcohol dehydrogenase (*adhE* gene product) (Goodlove et al., 1989; Kessler et al., 1992). However, ethanol levels that naturally occur in *E. coli* under anaerobic conditions are probably not sufficient to support the formation of significant amounts of FAEE in the transesterification process. Genes encoding the NADH-oxidizing system from *Zymomonas mobilis* were heterologously expressed in *E. coli* (Ingram et al., 1987; Alterthum & Ingram, 1989; Kalscheuer et al., 2006). The NADH-oxidizing pathway in *Z. mobilis* comprises two reactions catalyzed by two different enzymes. The pyruvate decarboxylase (PDC, EC 4.1.1.1) catalyzes the non-oxidative decarboxylation of pyruvate yielding acetaldehyde and CO_2. Two alcohol dehydrogenase isozymes (ADH, EC 1.1.1.1) catalyze the reduction of acetaldehyde to ethanol during fermentation, accompanied by the oxidation of NADH to NAD (Wills et al., 1981; Neale et al., 1986). Substantial amounts of ethanol are produced under aerobic conditions, which are sufficient for the transesterification process and the formation of FAEE. As we have already mentioned, microbial FAEE biosynthesis for microdiesel production is based on the exploitation of the extraordinarily low substrate specificity of WS/DGAT (the *atfA* gene product) from the *A. baylyi* strain ADP1.

Thus, Kalscheuer et al. (2006) constructed a plasmid, designated as pMicrodiesel, which contained three genes—*atfA*, *pdc*, and *adhB*—under the control of two *lacZ* promoters to ensure the effective transcription and expression of WS/DGAT, from the *A. baylyi* strain ADP1, pyruvate decarboxylase, and alcohol dehydrogenase from *Z. mobilis*, respectively, in *E. coli*. These metabolically engineered strains produced large amounts of ethanol and simultaneously expressing WS/DGAT, providing an unusual, alternative substrate for this acyltranferase. The latter used this ethanol as an acyl group acceptor, despite the fact that it operated at a lower efficiency than the natural substrate, glycerol, and yielded considerable amounts of FAEEs. Fig. 14.5 illustrates a strategy for the in vitro biosynthesis of FAEEs from sugars.

Pilot-Scale Cultivation of the Microdiesel Strain

FAEE biosynthesis, using the engineered *E. coli* strain, is strictly dependent on the presence of sodium oleate in the medium (Kalscheuer et al., 2006). The medium was, therefore, supplemented with 0.2% (w/v) sodium oleate as source of fatty acids and 2% (w/v) glucose for ethanol production. During the aerobic fed-batch cultivation of the engineered *E. coli* strain, FAEE increased throughout the cultivation and reached 1.3 g/L after 72 h with a final cellular dry biomass of 4.9 g/L. This finding corresponds to a FAEE content of 26% (w/w). The formed FAEE were accumulated intracellularly, and no significant extracellular FAEE were detected in the cell-free supernatant (Kalscheuer et al., 2006).

Recently, Elbahloul and Steinbüchel (unpublished data) cultivated *E. coli* harboring pMicrodiesel in a 20 L fed-batch bioreactor. Cell densities of about 60 g/L, containing up to 25% (w/w) FAEE, were obtained. An interesting aspect of this study was the use of glycerol as carbon source. Since glycerol is considered a byproduct or residual of biodiesel production, its utilization in microdiesel production will be of great interest in the future. After achieving the appropriate amount of growth during the fermentation process, glucose and oleic acid or sodium oleate were added to the mineral salts medium to initiate the accumulation of FAEE. Glucose is converted to ethanol, due to the activity of the alcohol dehydrogenase, and oleic acid is transesterified by the activity of WS/DGAT. The downstream processing of microdiesel starts with the harvest of cells by centrifugation. Afterwards, the cells were dried and extracted by an organic solvent, such as acetone, which is subsequently removed from the extraction mixture, yielding microdiesel. The process for microdiesel production is still under development and will be further simplified (Elbahloul & Steinbüchel, unpublished data). Analyses by gas chromatography (GC) and by GC/MS revealed the presence of 16:0, 16:1, 18:0, and 18:1 ethyl esters in cells cultivated on complex mediums and fed with glucose and oleate (Kalscheuer et al., 2006). However, if the cells were cultivated on glycerol and fed with glucose and oleate, they produced a microdiesel composed of 10:0, 11:0, 12:0, and 14:0 ethyl esters, in addition to 16:0, 16:1, 18:0, and 18:1 ethyl esters (Elbahloul & Steinbüchel, unpublished data).

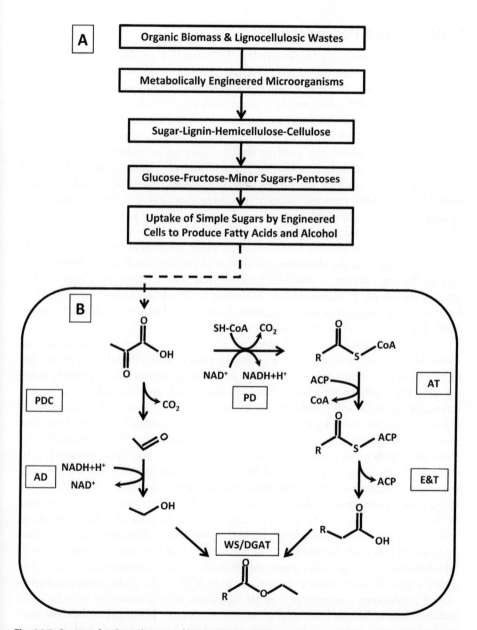

Fig. 14.5. Strategy for the utilization of lignocellulosic wastes and organic biomass for the production of fatty acids ethyl esters (FAEEs). Part A: conversion of wastes into simple sugars. Part B: simple sugars can be metabolized by engineered microbial strains to overproduce fatty acids and alcohol to produce FAEEs by the activity of the heterologously expressed WS/DGAT. Abbreviations: PDC, Pyruvate Decarboxylase; AD, Alcohol Dehydrogenase; PD, Pyruvate Dehydrogenase; ACP, Acyl Carrier Protein; AT, Acetyl Transacylase; E&T, Elongation cycles followed by termination of fatty acid synthesis; WS/DGAT, Wax Ester Synthase/Acyl-CoA:Diacylglycerol Acyltransferase.

Outlook and Perspectives

Current research activities for improving the microbial production of diesel-like fuels are focusing on the use of raw materials, the process design, and the quality of the produced diesel. The major cost in biotechnological processes comes from the feedstocks (above 60%), followed by expenditures for processing.

Thus, one aspect of biodiesel production that we must continue to examine is the source of the oil substrate. An important feature of microdiesel is the possibility of producing it without the need to feed bacteria an oil substrate. Attempts to overproduce fatty acids by engineering the pathway of fatty acid biosynthesis and degradation are critical for the production of biodiesel, microdiesel, and for other lipid. Lu et al. (2008) reported on genotypic alterations aiming at the overproduction of fatty acids in *E. coli*, which were achieved by four different gentical modifications. The endogenous *fadD* gene, encoding an acyl-CoA synthetase, was knocked out to block fatty acid degradation. Furthermore, a plant thioesterase was heterologously expressed to increase the abundance of shorter-chain fatty acids which improve the fuel quality. In addition, a acetyl-CoA carboxylase was heterologously expressed to increase the supply of malonyl-CoA. Moreover, feedback inhibition was released by heterologous expression of an endogenous thioesterase (Davis et al., 2000; Lu et al., 2008; Voelker & Davies, 1994; Yuan et al., 1995).

Another important aspect of biodiesel production is the source of cheap and renewable substrates available for the biomass production of biocatalysts. Engineering known industrial strains of bacteria, yeasts, and fungi to express a set of genes required for utilizing complex cellulosic, hemicellulosic, and polysaccharide wastes and renewable resources will enable sustainable processes for microdiesel production. At the same time, the use of lignocellulosic biomass will not compete with food and feed production from oily crops and other vegetables (Antoni et al., 2007). Lignocellulosic materials represent a promising option as a feedstock for lipid production by engineered microorganisms (Fig. 14.5). Generally, the use of sucrose-containing waste materials, such as cane or beet molasses, allows us to produce microbial biomass at the lowest costs, compared with starchy materials (mainly grains), which are used as food material. Voss & Steinbüchel (2001) produced cells of a *R. opacus* strain with a TAG content of about 39% (wt/wt) at pilot scale cultivation from sugar beet molasses and sucrose. Nevertheless, the availability and low cost of a wide range of lignocellulosic materials offer many possibilities for the development of microbial diesel. Lignocellulosic biomass comprises about 50% of the global biomass, and its annual production was estimated at 10-50 billion tons (Claassen et al., 1999). In addition, lignocellulosic biomass is a resource that can be processed in different ways to produce many other products, such as synthesis gas, methanol, hydrogen, and electricity (Chum & Overend, 2001).

Acknowledgments

We thank H.M. Alvarez for his authorization to reproduce the electron microscopic picture of lipid bodies in cells of *R. opacus* strain PD630.

References

Adamczak, M.; W. Bednarski. Enhanced Activity of Intracellular Lipases from *Rhizomucor meihi* and *Yarrowia lipolytica* by Immobilization on Biomass Support Particles. *Process Biochem.* **2004,** *39,* 1347–1361.

Aggelis, G.; J. Sourdis. Prediction of Lipid Accumulation Degradation in Oleaginous Microorganisms Growing on Vegetable Oils. *Antonie van Leeuwenhock* **1997,** *72,* 159–165.

Aggelis, G.; M. Komaitis; S. Papanikolaou; G. Papadopoulos. A Mathematical Model for the Study of Lipid Accumulation in Oleaginous Microorganisms. Lipid Accumulation during Growth of *Mucor circinelloides* CBS172-27 on a Vegetable Oil. *Gracas y Aceites* **1995,** *46,* 169–173.

Akao, T.; T. Kusaka. Solubilization of Diglyceride Acyltransferase from Membrane of *Mycobacterium smegmatis. J. Biochem. (Tokyo)* **1976,** *80,* 723–728.

Alterthum, F.; L.O. Ingram. Efficient Ethanol Production from Glucose, Lactose, and Xylose by Recombinant *Escherichia coli. Appl. Environ. Microbiol.* **1989,** *55,* 1943–1948.

Alvarez, H.M. Relationship between β-oxidation Pathway and the Hydrocarbon-Degrading Profile in Actinomycetes Bacteria. *Int. Biodeterior. Biodegrad.* **2003,** *52,* 35–42.

Alvarez, H.M.; A. Steinbüchel. Triacylglycerols in Prokaryotic Microorganisms. *Appl. Microbiol. Biotechnol.* **2002,** *60,* 367–376.

Alvarez, H.M.; F. Mayer; D. Fabritius; A. Steinbüchel. Formation of Intracytoplasmic Lipid Inclusions by *Rhodococcus opacus* Strain PD630. *Arch. Microbiol.* **1996,** *165,* 377–386.

Alvarez, H.M.; R. Kalscheuer; A. Steinbüchel. Accumulation of Storage Lipids in Species of *Rhodococcus* and *Nocardia* and Effect of Inhibitors and Polyethylene Glycol. *Fett. Lipid* **1997a,** *99,* 239–246.

Alvarez, H.M.; O.H. Pucci; A. Steinbüchel. Lipid Storage Compounds in Marine Bacteria. *Appl. Microbiol. Biotechnol.* **1997b,** *47,* 132–139.

Alvarez, H.M.; R. Kalscheuer; A. Steinbüchel. Accumulation and Mobilization of Storage Lipids by *Rhodococcus opacus* PD630 and *Rhodococcus ruber* NCIMB 40126. *Appl. Microbiol. Biotechnol.* **2000,** *54,* 218–223.

Alvarez, H.M.; M.F. Souto; A. Viale; O.H. Pucci. Biosynthesis of Fatty Acids and Triacylglycerols by 2,6,10,14-Tetramethyl Pentadecane Grown Cells of *Nocardia globerula* 432. *FEMS Microbiol. Lett.* **2001,** *200,* 195–200.

Alvarez, H.M.; R.A. Silva; A.C. Cesari; A.L. Zamit; S.R. Peressutti; R. Reichelt; U. Keller; U. Malkus; C. Rasch; T. Maskow; et al. Physiological and Morphological Responses of the Soil Bacterium *Rhodococcus opacus* PD630 to Water Stress. *FEMS Microbiol. Ecol.* **2004,** *50,* 75–86.

Alvarez, A.F.; H.M. Alvarez; R. Kalscheuer; M. Wältermann; A. Steinbüchel. Cloning and Characterization of a Gene Involved in Triacylglycerol Biosynthesis and Identification of Additional Homologous Genes in the Oleaginous Bacterium *Rhodococcus opacus* PD630. *Microbiology* **2008,** *154,* 2327–2335.

Antoni, D.; W. Zverlov; W.H. Schwarz. Biofuels from Microbes. *Appl Microbiol Biotechnol.* **2007,** *77,* 23–35.

Arabolaza, A.; E. Rodriguez; S. Altabe; H. Alvarez; H. Gramajo. Multiple Pathways for Triacylglycerol Biosynthesis in *Streptomyces coelicolor. Appl. Environ. Microbiol.* **2008,** *74,* 2573–2582.

Athenstaedt, K.; G. Daum. The Life Cycle of Neutral Lipids: Synthesis, Storage and Degradation. *Cell. Mol. Life Sci.* **2006,** *63,* 1355–1369.

Atkinson, B.; G.M. Black; P.J.S. Lewis; A. Pinches. Biological Particles of Given Size, Shape, and Density for Use in Biological Reactors. *Biotechnol. Bioeng.* **1979,** *21,* 193–200.

Bacchin, P.; A. Robertiello; A. Viglia. Identification of *N*-decane Oxidation Products in *Corynebacterium* Cultures by Combined Gas Chromatography Mass Spectrometry. *Appl. Microbiol.* **1974,** *28,* 737–741.

Ban, K.; M. Kaieda; T. Matsumoto; A. Kondo; H. Fukuda. Whole-cell Biocatalyst for Biodiesel Fuel Production Utilizing *Rhizopus oryzae* Cells Immobilized within Biomass Support Particles. *Biochem. Eng. J.* **2001,** *8,* 39–43.

Barksdale, L.; K.-S. Kim. Mycobacterium. *Bacteriol. Rev.* **1977,** *41,* 217–372.

Bouvier-Navé, P.; P. Benveniste; P. Oelkers; S.L. Sturley; H. Schaller. Expression in Yeast and Tobacco of Plant cDNAs Encoding Acyl CoA: Diacylglycerol Acyltransferase. *Eur. J. Biochem.* **2000,** *267,* 85–96.

Bredemeier, R.; R. Hulsch; J.O. Metzger; L. Berthe-Corti. Submersed Culture Production of Extracellular Wax Esters by the Marine Bacterium *Fundibacter jadensis. Mar. Biotechnol.* **2003,** *5,* 579–583.

Bryn, K.; E. Jantzen; K. Bovre. Occurrence and Patterns of Waxes in Neisseriaceae. *J. Gen. Microbiol.* **1977,** *102,* 33–43.

Cao, J.; P. Burn; Y. Shi. Properties of the Mouse Intestinal Acyl-CoA:monoacylglycerol Acyltransferase, MGAT2. *J. Biol. Chem.* **2003,** *278,* 25657–25663.

Cases, S.; S.J. Smith; Y.W. Zheng; H.M. Myers; S.R. Lear; E. Sande; S. Novak, C. Collins; C.B. Welch; A.J. Lusis; et al. Identification of a Gene Encoding an Acyl CoA:diacylglycerol Acyltransferase, a Key Enzyme in Triacylglycerol Synthesis. *Proc. Natl. Acad. Sci. USA* **1998,** *95,* 13018–3023.

Cases, S.; S.J. Stone; P. Zhou; E. Yen; B. Tow; K.D. Lardizabal; T. Voelker; R.V. Farese. Cloning of DGAT2, a Second Mammalian Diacylglycerol Acyltransferase, and Related Family Members. *J. Biol. Chem.* **2001,** *276,* 38870–38876.

Chum, H.L.; R.P. Overend. Biomass and Renewable Fuels. *Fuel Processing Technol.* **2001,** *71,* 187–195.

Claassen, P.A.M.; J.B. van Lier; A.M. López Contreras; E.W.J. van Niel; L. Sijtsma; A.J.M. Stams; S.S. de Vries; R.A. Weusthuis. Utilisation of Biomass for the Supply of Energy Carriers. *Appl. Microbiol. Biotechnol.* **1999,** *52,* 741–755.

Daniel, J.; C. Deb; V.S. Dubey; T.D. Sirakova; B. Abomoelak; H.R. Morbidoni; P.E. Kolattakudy. Induction of a Novel Class of Diacylglycerol Acyltransferases in *Mycobacterium tuberculosis* as It Goes into a Dormancy-like State in Culture. *J. Bacteriol.* **2004,** *186,* 5017–5030.

Davis, M.S.; J. Solbiati; J.E. Cronan. Overproduction of Acetyl-CoA Carboxylase Activity Increases the Rate of Fatty Acid Biosynthesis in *Escherichia coli. J. Biol. Chem.* **2000,** *275,* 28593–28598.

De Andrès, C.; M.J. Espuny; M. Robert; M.E. Mercade; A. Manresa; J. Guinea. Cellular Lipid Accumulation by *Pseudomonas aeruginosa* 44T1. *Appl. Microbiol. Biotechnol.* **1991,** *35,* 813–816.

Dewitt, S.; J.L. Ervin; D. Howesorchison; D. Dalietos; S.L. Neidleman; J. Geigert. Saturated and Unsaturated Wax Esters Produced by *Acinetobacter* sp. HO1-N Grown on C16-C20 n-alkanes. *J. Am. Oil Chem. Soc.* **1982,** *59,* 69–74.

Du, W.; D.H. Liu; L.L. Li; L.M. Dai. Mechanism Exploration during Lipase-mediated Methanolysis of Renewable Oils for Biodiesel Production in a Tert-Butanol System. *Biotechnol. Prog.* **2007,** *23,* 1087–1090.

Fixter, L.M.; C.A. Fewson. Accumulation of Waxes by *Acinetobacter calcoaceticus* NCIB-8250. *Biochem. Soc. Trans.* **1974,** *2,* 944–945.

Fixter, L.M.; J.G. McCormack. Effect of Growth-Conditions on Wax Content of Various Strains of *Acinetobacter*. *Biochem. Soc. Trans.* **1976,** *4,* 504–505.

Fujita, Y.; S. Takahashi; M. Ueda; A. Tanaka; H. Okada; Y. Morikawa; T. Kawaguchi; M. Arai; H. Fukuda; A. Kondo. Direct and Efficient Production of Ethanol from Cellulosic Material with a Yeast Strain Displaying Cellulolytic Enzymes. *Appl. Environ. Microbiol.* **2002,** *68,* 5136–5141.

Fukuda, H.; A. Kondo; H. Noda. Biodiesel Fuel Production by Transesterification of Oils. *J. Biosci. Bioeng.* **2001,** *92,* 405–416.

Gallagher, I.H.C. Occurrence of Waxes in *Acinetobacter*. *J. Gen. Microbiol.* **1971,** *68,* 245–247.

Garton, N.J.; H. Christensen; D.E. Minnikin; R.A. Adegbola; M.R. Barer. Intracellular Lipophilic Inclusions of Mycobacteria *in vitro* and *in sputum*. *Microbiology* **2002,** *148,* 2951–2958.

Goodlove, P.E.; P.R. Cunningham; J. Parker; D.P. Clark. Cloning and Sequence Analysis of the Fermentative Alcoholdehydrogenase-encoding Gene of *Escherichia coli*. *Gene* **1989,** *85,* 209–214.

Hama, S.; S. Tamalampudi; T. Fukumizu; K. Miura; H. Yamaji; A. Kondo; H. Fukuda. Lipase Localization in *Rhizopus oryzae* Cells Immobilized within Biomass Support Particles for Use as Whole-Cell Biocatalysts in Biodiesel-Fuel Production. *J. Biosci. Bioeng.* **2006,** *101,* 328–333.

Hobbs, D.H.; M.J. Hills. Expression and Characterization of Diacylglycerol Acyltransferase from *Arabidopsis thaliana* in Insect Cell Cultures. *FEBS Lett.* **1999,** *452,* 145–149.

Holdsworth, J.E.; C. Ratledge. Triacylglycerol Synthesis in the Oleaginous Yeast *Candida curvata* D. *Lipids* **1991,** *26,* 111–118.

Holtzapple, E.; C. Schmidt-Dannert. Biosynthesis of Isoprenoid Wax Ester in *Marinobacter hydrocarbonoclasticus* DSM 8798: Identification and Characterization of Isoprenoid Coenzyme A Synthetase and Wax Ester Synthases. *J. Bacteriol.* **2007,** *189,* 3804–3812

Hoskisson, P.A.; G. Hobbs; G. P. Sharples. Antibiotic Production, Accumulation of Intracellular Carbon Reserves, and Sporulation in *Micromonospora echinospora* (ATCC 15837). *Can. J. Microbiol.* **2001,** *47,* 148–152.

Ingram, L.O.; T. Conway; D.P. Clark; G.W. Sewell; J.F. Preston. Genetic Engineering of Ethanol Production in *Escherichia coli*. *Appl. Environ. Microbiol.* **1987,** *53,* 2420–2425.

Iso, M.; B. Chen; M. Eguchi; T. Kudo; S. Shrestha. Production of Biodiesel Fuel from Triglycerides and Alcohol using Immobilized Lipase. *J. Mol. Catal. B: Enzyme* **2001,** *16,* 53–58.

Kaieda, M.; T. Samukawa; T. Matsumoto; K. Ban; A. Kondo; Y. Shimada; H. Noda; F. Nomoto; K. Ohtsuka; E. Izumoto; et al. Biodiesel Fuel Production from Plant Oil Catalyzed by *Rhizopus oryzae* Lipase in a Water-Containing System without an Organic Solvent. *J. Biosci. Bioeng.* **1999**, *88*, 627–631.

Kalscheuer, R.; A. Steinbüchel. A Novel Bifunctional Wax Ester Synthase/Acyl-CoA:diacylglycerol Acyltransferase Mediates Wax Ester and Triacylglycerol Biosynthesis in *Acinetobacter calcoaceticus* ADP1. *J. Biol. Chem.* **2003**, *287*, 8075–8082.

Kalscheuer, R.; T. Stölting; A. Steinbüchel. Microdiesel: *Escherichia coli* Engineered for Fuel Production. *Microbiology* **2006**, *152*, 2529–2536.

Kalscheuer, R.; S. Uthoff; H. Luftmann; A. Steinbüchel. In vitro and in vivo Biosynthesis of Wax Diesters by an Unspecific Bifunctional Wax Ester Synthase/Acyl-CoA: Diacylglycerol Acyltransferase from *Acinetobacter calcoaceticus* ADP1. *Eur. J. Lipid Sci. Technol.* **2003**, *105*, 578–584.

Kalscheuer, R.; T. Stöveken; U. Malkus; R. Reichelt; P.N. Golyshin; J.S. Sabirova; M. Ferrer; K.N. Timmis; A. Steinbüchel. Analysis of Storage Lipid Accumulation in *Alcanivorax borkumensis*: Evidence for Alternative Triacylglycerol Biosynthesis Routes in Bacteria. *J. Bacteriol.* **2007**, *189*, 918–928.

Kamisaka, Y.; S. Mishra; T. Nakahara. Purification and Characterization of Diacylglycerol Acyltransferase from the Lipid Body Fraction of an Oleaginous Fungus. *J. Biochem.* **1997**, *121*, 1107–1114.

Karandikar, A.; G.P. Sharples; G. Hobbs. Differentiation of *Streptomyces coelicolor* A3(2) under Nitrate-limited Conditions. *Microbiology* **1997**, *143*, 3581–3590.

Kessler, D.; W. Herth; J. Knappe. Ultrastructure and Pyruvate Formate-Lyase Radical Quenching Property of the Multienzymic AdhE Protein of *Escherichia coli*. *J. Biol. Chem.* **1992**, *267*, 18073–18079.

Kondo, A.; Y. Liu; M. Furuta; Y. Fujita; T. Matsumoto; H. Fukuda. Preparation of High-Activity Whole-Cell Biocatalyst by Permeabilization of Recombinant Yeasts with Alcohol. *J. Biosci. Bioeng.* **2000**, *89*, 554–558.

Kumar, P.; M.W. Schelle; M. Jain; F.L. Lin; C.J. Petzhold; M.D. Leavell; J.A. Leary; J.S. Cox; C.R. Bertozzi. PapA1 and PapA2 Are Acyltransferases Essential for the Biosynthesis of the *Mycobacterium tuberculosis* Virulence Factor Sulfolipid-1. *Proc. Natl. Acad. Sci. USA* **2007**, *104*, 11221–11226.

Lardizabal, K.D.; J.G. Metz; T. Sakamoto; W.C. Hutton; M.R. Pollard; M.W. Lassner. Purification of a Jojoba Embryo Wax Synthase, Cloning of Its cDNA, and Production of High Levels of Wax in Seeds of Transgenic *Arabidopsis*. *Plant Physiol.* **2000**, *122*, 645–655.

Li, W.; W. Du; D.H. Liu. *Rhizopus oryzae* IFO 4697 Whole Cell Catalyzed Methanolysis of Crude and Acidified Rapeseed Oils for Biodiesel Production in Tert-Butanol System. *Process Biochem.* **2007a**, *42*, 1481–1485.

Li, Y.H.; Z.B. Zhao; F.W. Bai. High-Density Cultivation of Oleaginous Yeast *Rhodosporidium toruloides* Y4 in Fed-Batch Culture. *Enzyme Microb. Technol.* **2007b**, *41*, 312–317.

Lu, X.; V. Harmit; C. Khosla. Overproduction of Free Fatty Acids in *E. coli*: Implications for Biodiesel Production. *Metab. Eng.* **2008**, *10*, 333–339.

Ma, F.R.; M.A. Hanna. Biodiesel Production: A Review. *Bioresour. Technol.* **1999**, *70*, 1–15.

Magnuson, K.; S. Jackowski; C.O. Rock; J.E. Cronan Jr. Regulation of Fatty Acid Biosynthesis in *Escherichia coli. Microbiol. Rev.* **1993,** *57,* 522–542.

Makula, R.A.; P.J. Lockwood; W.R. Finnerty. Comparative Analysis of Lipids of *Acinetobacter* Species Grown on Hexadecane. *J. Bacteriol.* **1975,** *121,* 250–258.

Matsumoto, T.; S. Takahashi; M. Kaieda; M. Ueda; A. Tanaka; H. Fukuda; A. Kondo. Yeast Whole-Cell Biocatalyst Constructed by Intracellular Overproduction of *Rhyzopus oryzae* Lipase is Applicable to Biodiesel Fuel Production. *Appl. Microbiol. Biotechnol.* **2001,** *57,* 515–520.

Matsumoto, T.; H. Fukuda; M. Ueda; A. Tanaka; A. Kondo. Construction of Yeast Strains with High Cell Surface Lipase Activity by Using Novel Display Systems Based on the Flo1p Flocculation Functional Domain. *Appl. Environ. Microbiol.* **2002,** *68,* 4517–4522.

Metzger, J.O.; U. Bornscheuer. Lipids as Renewable Resources: Current State of Chemical and Biotechnological Conversion and Diversification. *Appl. Microbiol. Biotechnol.* **2006,** *71,* 13–22.

Metzger, P.; C. Largeau. *Botryococcus braunii* a Rich Source for Hydrocarbons and Related Ether Lipids. *Appl. Microbiol. Biotechnol.* **2005,** *66,* 486–496.

Mona, K.G.; H.O. Sanaa; M.A. Linda. Single Cell Oil Production by *Gordonia* sp. DG Using Agro-industrial Wastes. *World J. Microbiol. Biotechnol.* **2008,** *24,* 1703–1711.

Murphy, D.J.; J. Vance. Mechanism of Lipid-Body Formation. *Trends Biol. Sci.* **1999,** *24,* 109–115.

Nakashima, T.; H. Fukuda; S. Kyotani; H. Morikawa. Culture Conditions for Intracellular Lipase Production by *Rhizopus chinensis* and Its Immobilization within Biomass Support Particles. *J. Ferment. Technol.* **1988,** *66,* 441–448.

Nakashima, T.; H. Fukuda; Y. Nojima; S. Nagai. Intracellular Lipase Production by *Rhizopus chinensis* Using Biomass Support Particles in a Circulated Bed Fermentor. *J. Ferment. Bioeng.* **1989,** *68,* 19–24.

Nakashima, T.; S. Kyotani; E. Izumoto; H. Fukuda. Cell Aggregation as a Trigger for Enhancement of Intracellular Lipase Production by a *Rhizopus* Species. *J. Ferment. Bioeng.* **1990,** *70,* 83–89.

Narita, J.; K. Okano; T. Tateno; T. Tanino; T. Sewaki; M.H. Sung; H. Fukuda; A. Kondo. Display of Active Enzymes on the Cell Surface of *Escherichia coli* Using PgsA Anchor Protein and Their Application to Bioconversion. *Appl. Microbiol. Biotechnol.* **2006,** *70,* 564–572.

Neale, A.D.; R.K. Scopes; J.M. Kelly; R.E.H. Wettenhall. The Two Alcohol Dehydrogenases of *Zymomonas mobilis*: Purification by Differential Dye Ligand Chromatography, Molecular Characterization and Physiological Role. *Eur. J. Biochem.* **1986,** *154,* 119–124.

Olukoshi, E.R.; N.M. Packter. Importance of Stored Triacylglycerols in *Streptomyces*-Possible Carbon Source for Antibiotics. *Microbiology* **1994,** *140,* 931–943.

Onwueme, K.C.; J.A. Ferreras; J. Buglino; C.D. Lima; L.E.N. Quari. Mycobacterial Polyketide-associated Proteins Are Acyltransferases: Proof of Principle with *Mycobacterium tuberculosis* PapA5. *Proc. Natl. Acad. Sci. USA* **2004,** *101,* 4608–4613.

Orcaire, O.; P. Buisson; A.C. Pierre. Application of Silica Aerogel Encapsulated Lipases in the Synthesis of Biodiesel by Transesterification Reactions. *J. Mol. Catal. B: Enzym.* **2006,** *42,* 106–113.

Packter, N.M.; E.R. Olukoshi. Ultrastructural Studies of Neutral Lipid Localisation in *Streptomyces. Arch. Microbiol.* **1995,** *164,* 420–427.

Papanikolaou, S.; I. Chevalot; M. Komaitis; I. Marc; G. Aggelis. Single Cell Oil Production by *Yarrowia lipolytica* Growing on an Industrial Derivative of Animal Fat in Batch Cultures. *Appl. Microbiol. Biotechnol.* **2002,** *58,* 308–312.

Papanikolaou, S.; M. Komaitis; G. Aggelis. Single Cell Oil (SCO) Production by *Mortierella isabellina* Grown on High-Sugar Content Media. *Bioresour. Technol.* **2004,** *95,* 287–291.

Ratledge, C. Fatty Acid Biosynthesis in Microorganisms Being Used for Single Cell Oil Production. *Biochimie* **2004,** *4,* 1–9.

Raymond, R.L.; J.B. Davis. *n*-Alkane Utilization and Lipid Formation by a *Nocardia. Appl. Microbiol.* **1960,** *8,* 329–334.

Reiser, S.; C. Somerville. Isolation of Mutants of *Acinetobacter calcoaceticus* Deficient in Wax Ester Synthesis and Complementation of One Mutation with a Gene Encoding a Fatty Acyl Coenzyme A Reductase. *J. Bacteriol.* **1997,** *179,* 2969–2975.

Rontani, J.F.; M.J. Gilewicz; V. Michotey, T.L. Zheng; P.C. Bonin; J.C. Bertrand. Aerobic and Anaerobic Metabolism of 6,10,14-trimethylpentadecan-2-one by a Denitrifying Bacterium Isolated from Marine Sediments. *Appl. Environ. Microbiol.* **1997,** *63,* 636–643.

Rontani, J.F.; P.C. Bonin; J.K. Volkman. Biodegradation of Free Phytol by Bacterial Communities Isolated from Marine Sediments under Aerobic and Denitrifying Conditions. *Appl. Environ. Microbiol.* **1999,** *65,* 5484–5492.

Rontani, J.F.; A. Mouzdahir; V. Michotey; P. Caumette; P. Bonin. Production of a Polyunsaturated Isoprenoid Wax Ester during Aerobic Metabolism of Squalene by *Marinobacter squalenivorans* sp. nov. *Appl. Environ. Microbiol.* **2003,** *69,* 4167–4176.

Routaboul, J.M.; C. Benning; N. Bechtold; M. Caboche; L. Lepiniec. The TAG1 Locus of *Arabidopsis* Encodes for a Diacylglycerol Acyltransferase. *Plant Physiol. Biochem.* **1999,** *37,* 831–840.

Russell, N.J.; J.K. Volkman. The Effect of Growth Temperature and Wax Ester Composition in the Psychrophilic Bacterium *Micrococcus cryophilus* ATCC 15174. *J. Gen. Microbiol.* **1980,** *118,* 131–141.

Sallam, A.; A. Steinle; A. Steinbüchel. Cyanophycin: Biosynthesis and Applications. *Microbial Production of Biopolymers and Polymer Precursors—Application and Perspectives*; B.A.H. Rehm, Ed.; Caister Academic Press: Norfolk UK, 2009; pp 79–99.

Samukawa, T.; M. Kaieda; T. Matsumoto; K. Ban; A. Kondo; Y. Shimada; H. Noda; H. Fukuda. Pretreatment of Immobilized *Candida antarctica* Lipase for Biodiesel Fuel Production from Plant Oil. *J. Am. Oil Chem. Soc.* **2000,** *90,* 180–183.

Sandager, L.; M.H. Gustavsson; U. Stahl; A. Dahlqvist; E. Wiberg; A. Banas; M. Lenman; H. Ronne; S. Stymne. Storage Lipid Synthesis is Non-Essential in Yeast. *J. Biol. Chem.* **2002,** *277,* 6478–6482.

Scott, C.C.L.; W.R. Finnerty. Characterization of Intracytoplasmic Hydrocarbon Inclusions from Hydrocarbon-Oxidizing *Acinetobacter* species HO1-N. *J. Bacteriol.* **1976,** *127,* 481–489.

Shah, S.; S. Sharma; M.N. Gupta. Biodiesel Preparation by Lipase-Catalyzed transesterification of Jatropha Oil. *Energy Fuels* **2004,** *18,* 154–159.

Shimada, Y.; Y. Watanabe; T. Samukawa1; A. Sugihara; H. Noda; H. Fukuda; Y. Tominaga. Conversion of Vegetable Oil to Biodiesel Using Immobilized *Candida antarctica* Lipase. *J. Am. Oil Chem. Soc.* **1999,** *77,* 789–793.

Singer, M.E.; S.M. Tyler; W.R. Finnerty. Growth of *Acinetobacter* sp. Strain HO1-N on *n*-Hexadecanol: Physiological and Ultrastructural Characteristics. *J. Bacteriol.* **1985,** *162,* 162–169.

Srivastava, A.; R. Prasad. Triglyceride Based Diesel Fuels. *Renew. Sust. Ener. Rev.* **2000,** *4,* 111–133.

Steinbüchel, A. PHB and Other Polyhydroxyalkanoic Acids. *Biotechnology,* 2nd edn.; H.J. Rehm, G. Reed, A. Pühler, P. Stadler, Eds.; Wiley-VCH: Weinheim, 1996; Vol. 6, pp 403–464.

Steinbüchel, A. Perspectives for Biotechnological Production and Utilization of Biopolymers: Metabolic Engineering of Polyhydroxyalkanoate Biosynthesis Pathways as a Successful Example. *Macromol. Biosci.* **2001,** *1,* 1–24.

Steinbüchel, A.; S. Hein. Biochemical and Molecular Basis of Polyhydroxyalkanoic Acids in Microorganisms. *Biopolyesters;* A. Steinbüchel, W. Babel, Eds.; Adv. Biochem. Eng./Biotechnol. 2001, 71, 81–123.

Stöveken, T.; A. Steinbüchel. Bacterial Acyltransferases as an Alternative for Lipase-Catalyzed Acylation for the Production of Oleochemicals and Fuels. *Angew. Chem. Int. Ed. Engl.* **2008,** *47,* 3688–3694.

Stöveken, T.; R. Kalscheuer; U. Malkus; R. Reichelt; and A. Steinbüchel. The Wax Ester Synthase/ Acyl-coenzyme A:diacylglycerol Acyltransferase from *Acinetobacter* sp. Strain ADP1: Characterization of a Novel Type of Acyltransferase. *J. Bacteriol.* **2005,** *187,* 1869–1376.

Tamalampudi, S.; M.M. Talukder; S. Hama; T. Tanino; Y. Suzuki; A. Kondo; H. Fukuda. Development of Recombinant *Aspergillus oryzae* Whole-Cell Biocatalyst Expressing Lipase-Encoding Gene from *Candida antarctica. Appl. Microbiol. Biotechnol.* **2007,** *75,* 387–395.

Tamalampudi, S.; M.R. Talukder; S. Hama; T. Numata; A. Kondob; H. Fukuda. Enzymatic Production of Biodiesel from Jatropha Oil: A Comparative Study of Immobilized Whole-Cell and Commercial Lipases as a Biocatalyst. *Biochem. Eng. J.* **2008,** *39,* 185–189.

Taranto, P.A.; T.W. Keenan; M. Potts. Rehydration Induces Rapid Onset of Lipid Biosynthesis in Desiccated *Nostoc commune* (Cyanobacteria). *Biochim. Biophys. Acta.* **1993,** *1168,* 228–237.

Uthoff, S.; T. Stöveken; N. Weber; K. Vosmann; E. Klein; R. Kalscheuer; A. Steinbüchel. Thio Wax Ester Biosynthesis Utilizing the Unspecific Bifunctional Wax Ester Synthase/Acyl- CoA: diacylglycerol Acyltransferase of *Acinetobacter* sp. Strain ADP1. *Appl. Environ. Microbiol.* **2005,** *71,* 790–796.

Vachon, V.; J.T. McGarrity; C. Breuil; J.B. Armstrong; D.J. Kushner. Cellular and Extracellular Lipids of *Acinetobacter lwoffi* during Growth on Hexadecane. *Can. J. Microbiol.* **1982,** *28,* 660–666.

Vasudevan, P. T.; M. Briggs. Biodiesel Production—Current State of the Art and Challenges. *J. Ind. Microbiol. Biotechnol.* **2008,** *35,* 421–430.

Voelker, T.A.; H.M. Davies. Alteration of the Specificity and Regulation of Fatty Acid Synthesis of *Escherichia coli* by Expression of Plant Medium-Chain Acyl-Acyl Carrier Protein Thioesterase. *J. Bacteriol.* **1994,** *176,* 7320–7327.

Voss, I.; A. Steinbüchel. High Cell Density Cultivation of *Rhodococcus opacus* for Lipid Production at a Pilot-Plant Scale. *Appl. Microbiol. Biotechnol.* **2001,** *55,* 547–555.

Wältermann, M.; A. Steinbüchel. Neutral Lipid Bodies in Prokaryotes: Recent Insights into Structure, Formation, and Relationship to Eukaryotic Lipid Depots. *J. Bacteriol.* **2005,** *187,* 3607–3619.

Wältermann, M.; A. Hinz; H. Robenek; D. Troyer; R. Reichelt; U. Malkus; H.J. Galla; R. Kalscheuer; T. Stöveken; P. von Landenberg; et al. Mechanism of Lipid Body Formation in Bacteria: How Bacteria Fatten Up. *Mol. Microbiol.* **2005,** *55,* 750–763.

Wältermann, M.; T. Stöveken; A. Steinbüchel. Key Enzymes for Biosynthesis of Neutral Lipid Storage Compounds in Prokaryotes: Properties, Function and Occurrence of Wax Ester Synthase/Acyl-CoA:diacylglycerol Acyltransferases. *Biochimie* **2006,** *89,* 230–242.

Wang, L.; H.K. Schnoes; K. Takayama; D.S. Goldman. Synthesis of Alcohol and Wax Ester by a Cell-Free System in *Mycobacterium tuberculosis. Biochim. Biophys. Acta* **1972,** *260,* 41–48.

Watanabe, Y.; Y. Shimada; A. Sugihara; H. Noda; H. Fukuda; Y. Tominaga. Continuous Production of Biodiesel Fuel from Vegetable Oil Using Immobilized *Candida antarctica* Lipase. *J. Am. Oil Chem. Soc.* **2000,** *77,* 355–360.

Wills, C.; P. Kratofil; D. Londo; T. Martin. Characterization of the Two Alcohol Dehydrogenases of *Zymomonas mobilis. Arch. Biochem. Biophys.* **1981,** *210,* 775–785.

Yen, C.L.; S.J. Stone; S. Koliwad; C. Harris; R.V. Farese Jr. Thematic Review Series: Glycerolipids. DGAT Enzymes and Triacylglycerol Biosynthesis. *J. Lipid Res.* **2008,** *49,* 2283–2301.

Yermanos, D.M. Composition of Jojoba Seed during Development. *J. Am. Oil Chem. Soc.* **1975,** *52,* 115–117.

Yuan, L.; T.A. Voelker; D.J. Hawkins. Modification of the Substrate Specificity of an Acyl-Acyl Carrier Protein Thioesterase by Protein Engineering. *Proc. Natl. Acad. Sci. USA.* **1995,** *92,* 10639–10643.

Zou, J.T.; Y.D. Wie; C. Jako; A. Kumar; G. Selvaraj; D.C. Taylor. The *Arabidopsis thaliana* TAG1 Mutant Has a Mutation in a Diacylglycerol Acyltransferase Gene. *Plant J.* **1999,** *19,* 645–653.

Zweytick, D.; K. Athenstaedt; G. Daum. Intracellular Lipid Particles of Eukaryotic Cells. *Biochim. Biophys. Acta Rev. Biomembranes* **2000,** *1469,* 101–120.

PART 5

Safety and Nutrition of Single Cell Oils

15

Safety Evaluation of Single Cell Oils and the Regulatory Requirements for Use as Food Ingredients

Alan S. Ryan, Sam Zeller, and Edward B. Nelson
Martek Biosciences Corporation, 6480 Dobbin Road, Columbia, MD 21045

Introduction

Safety refers to a "reasonable certainty of no harm" based on an intellectual concept, not an inherent property of a substance. Safety assessment is an ongoing process, not an absolute determination based on a single point in time. When considering the safety of a product, tolerability is also important. Tolerability refers to the capacity a subject has to consume a substance without the occurrence of an adverse event or the appearance of an increased sensitivity to this substance. Three processes are often used when considering the safety or tolerability of single cell oils (SCOs). The first involves risk assessment. Risk assessment is based on existing nonclinical data from past *in vitro* laboratory or animal studies on the product, compiling data from previous clinical testing or marketing of the product in a country whose population is relevant to the U.S. population, or undertaking new preclinical studies designed to provide the evidence necessary to indicate the safety of administering the product to humans. During preclinical testing, the product's potentially toxic effects are evaluated through *in vitro* and *in vivo* laboratory testing on animals. The second and third processes involve risk management (regulation and control) and risk communication.

In the United States, the Food and Drug Administration (FDA) has the primary responsibility of regulating new food ingredients. Manufacturers may propose the addition of new food ingredients in the U.S. by either filing a food additive petition with the FDA to request a formal pre-market review or making a "generally recognized as safe" (GRAS) determination. It is important to note that it is the use of a substance, rather than the substance itself, that is eligible for the GRAS determination. In the European Union (EU), Australia, and New Zealand, the concept of "novel foods" has been developed and integrated into existing regulations . Manufacturers wishing to obtain novel foods approval for new food ingredients typically approach the competent authority to consult on the appropriate approach (e.g., novel food application) to obtain pre-market clearance.

The safety of SCOs has been evaluated and discussed over the past decades in numerous forums, including in published articles, in presentations at scientific conferences, by panels of individuals qualified by their training and experience to evaluate the safety of food ingredients, and by regulatory bodies, around the world. A large number of nonclinical and clinical studies have demonstrated the safe use of DHASCO® [docosahexaenoic acid (DHA) oil derived from *Crypthecodinium cohnii*] and ARASCO® [arachidonic acid (ARA) oil derived from *Mortierella alpina* in infant formulas]. These oils have been successfully commercialized for use in the infant formula sold in over seventy-five countries. Published studies also document the nonclinical safety evaluation of DHA oil derived from *Schizochytrium* sp. (DHASCO-S), *Ulkenia* sp. (DHA45-oil), and ARA-rich oil from a species of *Mortierella alpina* (SUNTGA40S). These preclinical and clinical studies, described previously by Zeller (2005), were used to support safe use of SCOs in a number of food, beverage, and infant formula applications.

It is noteworthy that many studies have evaluated the nutritional and clinical efficacy of SCO. In many of these studies, however, there is no mention of safety and/or tolerability. Foods defined as GRAS are, by definition, safe; therefore, the lack of data for safety and tolerability in clinical efficacy studies is understandable. Likewise, when a study does not include a control product or placebo for comparison, assessment of any adverse events is difficult. For these reasons, there is often a lack of information on safety and tolerability in studies that have considered the nutritional and clinical effects of SCOs. However, this situation will likely change in the U.S., because the FDA now requires the reporting of serious adverse events in subjects using a dietary supplement, a practice that began in 2008. The FDA requires that all dietary supplement distributors include an identifying address or phone number on the label of the product so that customers or health professionals can report a serious adverse event.

The intent of this chapter is to review general aspects of the safety and tolerability of SCOs, specifically, DHA from DHASCO®, DHASCO-S, and DHA45-oil and ARA from ARASCO® and SUNTGA40S. Recent reviews and comments that describe the safety of DHA (Lien, in press) and ARA (Calder, 2007) are also available.)This updated version of our previously published chapter (Zeller, 2005), which described preclinical data in detail, focuses on human studies in infants and adults. A discussion is also presented on the regulatory pathway for obtaining pre-market clearance of SCOs for use as food ingredients in the U.S. and as novel food ingredients in the EU, Australia and New Zealand, and Canada, citing some of the more recent published approvals for these oils.

Safety Evaluation

Safety evaluation of food ingredients, including SCOs, is based on a reasonable assumption that such ingredients do not cause any harm. Sections 201(s), 201(z),

409, and 412 of the Federal Food, Drug, and Cosmetic Act (FDCA) and the FDA's Toxicological Principles for the Safety Assessment of Direct Food Additives and Color Additives Used in Food, also known as the Redbook (Office of Food Additive Safety, 2001, 2004), form part of the basis to assess the safety of food ingredients. The FDA Redbook was prepared to assist in the design of protocols for animal studies conducted to test the safety of food ingredients and includes detailed guidelines for testing the effects of food ingredients.

There are two routes for obtaining regulatory clearance for a food ingredient in the United States. A food additive petition process requires pre-market review and approval by the FDA, whereas, under the GRAS notification proposed rule, a manufacturer can determine that a substance is GRAS if there is a consensus among qualified scientific experts about its safety under the intended conditions of use. The manufacturer may then notify the FDA, and, if the agency has no questions, a letter of no objection is issued. The main difference between a food additive petition and a GRAS notification is that a food additive petition places the responsibility of declaring that a substance is safe and approved under the conditions of use with the regulatory agency, while the GRAS notification process places the responsibility of demonstrating that a substance is GRAS and, therefore, safe under the conditions of use with the manufacturer. The procedures for both the GRAS notification and the food additive petition are intended to ensure the safety, not the efficacy, of the proposed ingredient. The majority of SCOs, utilized as new food ingredients, are likely to follow the GRAS process to enable marketing in the U.S.

Safety Assessment Approach

SCOs are relatively unique due to the nontraditional nature of the source organism used for their production and, perhaps, certain compositional attributes; therefore, flexibility is important when determining overall safe use in a given application. However, there are a few common technical elements to take into consideration. Chemical and physical characterization of the product is important as safety considerations often revolve around what is known regarding the product and its individual components. In addition to the characterization of major constituents, minor components of complex products derived from SCOs must be examined to determine the potential for toxicity, including the possibility for naturally occurring toxins (from the source organism), heavy metals, and hazardous levels of pathogenic microorganisms, as well as potential by-products, formed from the degradation of certain pathways or introduced from production processing. Knowledge of chemical composition/structure, along with the product's intended uses and levels of exposure, is important in determining an appropriate level of toxicological assessment. The FDA has prepared recommending guidelines for the minimum toxicity tests necessary for safety evaluation of food additives, based on levels of concern. Concern levels, as determined by the agency, are "relative measures of the degree to which the use of an additive may present a hazard to human health" (Office of Food Additive Safety,

2004). The concern level is based on the extent of human exposure (dose) and the toxicological effects the additive may have on biological systems. There are three broad levels of concern, with Level Three representing the highest probable risk to human health Level One the lowest probable risk, and Level Two acting as the intermediate level.

Safety of Source Organisms

The use of SCOs for human consumption, including their use in infant formulas (both preterm and term), has a relatively short history, but, with respect to certain compositional attributes, SCOs may be considered similar to various plant oils. The main difference between SCOs and plant oils, which is also the source of their "novelty" in certain regulatory jurisdictions, is related to the source organism used for their production. The following section discusses some of the commercially available SCOs, along with a description of their source organisms, with references to published articles that support the safety of SCOs.

DHASCO is the trade name used by Martek Biosciences Corporation (Martek) to describe the docosahexaenoic acid (DHA) SCO produced by the microalga *Crypthecodinium cohnii*. This single-celled, heterotrophic organism is a marine algal species that has been extensively studied in both laboratory and commercial environments. *C. cohnii* is a member of the Dinophyta (dinoflagellates), a distinct phylum of unicellular eukaryotic microalgae that consists of approximately 2,000 species (Van den Hoek et al., 1995). *C. cohnii* is non-pathogenic in man and animals, and the organism does not produce any toxins. There are a small number of photosynthetic species of Dinophyta that are known to produce a group of closely related toxins (Steidinger & Baden, 1984), which are passed through the food chain via zooplankton and can contaminate fish and bivalves. However, these toxin-producing species are few in number, and there are no known heterotrophic species of the Dinophyta that are toxin producers. In the many reports of *C. cohnii* in culture over the last 30 years, there has never been any indication that *C. cohnii* produces any toxin, nor is it related to any toxin-producing species (Dodge, 1984).

ARASCO is the trade name used by Martek for the arachidonic acid (ARA) SCO produced through a fermentation process that uses the common soil fungus *Mortierella alpina* (see also Chapter 5). *M. alpina* has been studied at length and is not pathogenic to humans, nor does it produce mycotoxins harmful to humans or animals (Streekstra, 1997). *Mortierella* species have been well studied for many years in both laboratory and commercial environments, and their morphology, biochemistry, and physiology is well documented. *Mortierella alpina* has also been described in Japanese publications and patents as a potential source of ARA, and, as a consequence, it has been the subject of recent investigations (see also Chapter 2). In none of this recent work, consistent with earlier studies, has there ever been any report of pathogenicity or toxigenicity to humans or animals by *M. alpina* (Domsch et al., 1980; Scholer et al., 1983). SUNTGA40S is the trade name used by Suntory

Limited for a source of ARA-rich oil. SUNTGA40S is derived from fermentation of the fungus *Mortierella alpina*. This fungus is not pathogenic for humans and has not been reported to produce mycotoxins.

DHASCO-S is Martek's trade name for DHA oil derived from the heterotrophic fermentation of the marine alga *Schizochytrium* sp. *Schizochytrium* sp. is a thraustochytrid and a member of the Chromista kingdom (Stramenopilia), which includes golden algae, diatoms, yellow-green algae, haptophyte and cryptophyte algae, and oomycetes. There are no reports of this organism producing toxic chemicals, nor is it pathogenic. The two toxic compounds known to be produced in Chromista (to which *Schizochytrium* sp. belongs) are largely restricted to two genera (domoic acid in *Pseudonitzschia* and prymnesin in *Prymnesium* spp.) that are in a separate class and phylum, respectively, from the thraustochytrids. No evidence of the two toxic compounds produced in the Chromista—namely, domoic acid and *Prymnesium* toxin—was found in *Schizochytrium* sp. algae using chemical and biological assays. Chemical and biological analysis of the production strain confirmed the absence of common algal toxins (Food and Drug Administration, 2004).

DHA45-oil is described (Kroes et al., 2003; Kiy et al., 2005) as a refined oil produced through a fermentation process that uses a strain of the marine protist, *Ulkenia* sp. This organism is stated to be a member of the non-pathogenic and non-toxigenic family of Thraustochytriaceae (Kroes et al., 2003).

Safety of the SCO Components

Evaluation of the overall safety of a SCO involves a review of the safety of the identified components of this oil. SCOs are typically comprised of fatty acids found esterified to glycerol (e.g., triglycerides) and may contain minor amounts of other lipid classes (e.g., steryl esters, free sterols, or carotenoids). Identified fatty acids present in commercialized SCOs have been described as components of normal human diets or metabolites of fatty acids. Sterols identified in SCOs are commonly found in several traditional food sources, including animal fat, vegetable oil, and human milk, or are part of a human's normal metabolic pathway of cholesterol biosynthesis. A safe history of use of the individual fatty acids and sterols is further supported as a result of their abundant natural presence in food; the small quantities expected to be consumed; extensive knowledge of the absorption, distribution, metabolism and excretion in mammalian species; and published safety information on the specific fatty acids, sterols, and similar compounds.

Infant Formula and Human Milk Clinical Studies with SCOs

SCOs have been added to commercial infant formula since 1995, and their safety profile is well established. These formulas are now available in more than seventy-five countries worldwide, including the U.S., since February 2002, and Canada, since January 2003. It is estimated that 50 million infants have received formulas containing SCOs with no known or reported pattern of adverse events associated with the consumption of any of these formulas. Almost all preterm infants (a vulnerable population) and 90% of term infants in the U.S. are fed infant formulas with SCOs. The American Academy of Pediatrics, Committee on Nutrition (2009) notes that "the body of literature suggests that LCPUFAs (long-chain polyunsaturated fatty acids, e.g., DHA and ARA) are important for growth and development of infants." The committee also notes that LCPUFAs appear to be safe (American Academy of Pediatrics, Committee on Nutrition, 2009). The safety profile related to the use of DHA and ARA in formulas designed for preterm infants is especially important because failure-to-thrive is considered a serious adverse event. Consensus recommendations, on behalf of the World Association of Perinatal Medicine, The Early Nutrition Academy, and the Child Health Foundation, suggest the use of an infant formula that provides DHA at levels between 0.2 and 0.5 weight percent of total fat and the minimum amount of ARA equivalent to the contents of DHA (Koletzko et al., 2008). The American Dietetic Association and Dietitians of Canada published similar recommendations (Kris-Etherson & Innis, 2007). Some of the infant formulas commercially available contain about 35 mg of DHA (DHASCO) per 100 kcal and 70 mg of ARA (ARASCO) per 100 kcal. An average 7 kg infant consumes about 800-900 kcal of formula, or up to 315 mg of DHA and 630 mg ARA per day. This amount would be equivalent to over 3,000 mg of DHA/day and 6,300 mg of ARA/day in a 70 kg adult.

Considering one of the most vulnerable populations—preterm infants—two recent studies demonstrate the efficacy and safety of SCOs added to infant formulas. In a study of 361 preterm infants who were randomized into three feeding groups (control; 17 mg DHASCO/100 kcal + 34 mg ARASCO/100 kcal; 17 mg fish-DHA/100 kcal + 34 mg ARASCO/100 kcal) until 92 weeks, postmenstrual age (PMA), with follow-up to 118 weeks PMA, the body weight of the algal-DHA group was significantly greater than the control group and the fish-DHA group at 118 weeks PMA (Clandinin et al., 2005). Additionally, the algal-DHA group was significantly longer than the fish-DHA group at 92 weeks PMA (Clandinin et al., 2005). Supplementation did not increase morbidity or the incidence of adverse events. Recently, the safety of DHASCO and ARASCO has been demonstrated in a study of 141 preterm infants with birth weights <1500 g who were administered 32 mg of DHA and 31 mg of ARA per 100 L of human milk, which was added directly to human milk, versus a control group (Henriksen et al., 2008). Supplementation started one week after birth and lasted until discharge (9 weeks of life). At 6 months of age, there were no differences in growth between groups. Preterm infants who received DHA and ARA performed better on the problem solving subscore of the Ages and Stages Questionnaire (Henriksen et al., 2008). There were no significant differences in adverse events between the two groups. A list of studies that considered DHASCO and ARASCO supplementation of infant formula is provided in Appendix I.

DHA is always present in human milk, but its concentration is influenced by maternal diet (Yuhas et al., 2006). For this reason, it is now recommended that pregnant and lactating women should aim to achieve an average dietary intake of at least 200 mg DHA/day, either from food or supplements (Koletzko et al., 2007). In pregnant and lactating women, intakes of up to 1 g/day of DHA have been used in clinical trials without and significant adverse events. While levels of ARA have not been clearly defined or recommended for pregnant and lactating women, 400 mg/day of ARA (ARASCO), in combination with 320 mg/day of DHA and 80 mg/day of EPA, was administered for 8 weeks without any reported adverse events (Wesler et al., 2008).

Children and Special Population Studies

To evaluate the effect that DHA (DHASCO-S) had on cognitive functions in healthy preschool children, 400 mg/day of DHA was administered to 175 four-year-old children for four months (Ryan & Nelson, 2008). Regression analysis yielded a statistically significant (p < 0.018) positive association between the blood level of DHA and higher scores on the Peabody Picture Vocabulary Test, a test of listening comprehension and vocabulary acquisition. The investigational product was well tolerated. No subject in either treatment group experienced a serious adverse event.

An ongoing, long-term study of X-linked retinitis pigmentosa in 44 young men (mean age=16 years) reported four-year safety data in subjects who received either 400 mg/day of algal-DHA or a placebo (Wheaton et al., 2003). All adverse events were minor and equally distributed between the algal-DHA and placebo groups. Long-term algal-DHA supplementation did not compromise plasma antioxidant capacity, platelet aggregation, liver function enzyme activity, or plasma lipoprotein content.

Thirty-six adults (18 to 65 years of age) who met the medical criteria for major depressive disorder, without psychotic features, were administered 2 g/day of algal-DHA or placebo for 6 weeks (Marangell et al., 2003). Changes in depression scores did not differ between groups after treatment. Almost half (n=14) of the subjects in the algal-DHA group reported a fishy aftertaste. However, there were no serious adverse events reported in either group. In a pilot study that considered women with bipolar disorder, four of the ten women were administered 3.4 g/day of algal-DHA for 4 months (Marangell et al., 2006). The study reported no significant effects of algal-DHA on bipolar disorder, but treatment with algal-DHA was well tolerated. In another pilot study, thirty-five adults with major depressive disorder were randomized to receive either 1, 2, or 4 g/day of algal-DHA for 12 weeks (Mischoulon et al., 2008). Algal-DHA was effective for depression at lower doses (1 g/day). Treatment with algal-DHA was well tolerated, and no significant dose-related adverse events were reported.

Nineteen patients with cystic fibrosis were administered 50 mg/kg.day of algal-DHA or a placebo for 6 months (Lloyd-Still et al., 2001). Treatment with alga-DHA did not affect levels of plasma antioxidants, nor were there any treatment-related serious adverse events, including changes in liver enzyme levels or lung function. In a smaller study with five cystic fibrosis patients, who received 70 mg/kg.day of algal-DHA for 6 weeks, no treatment-related serious adverse events were reported (Brown et al., 2001).

The effects of algal-DHA (DHASCO, DHASCO-S, *Ulkenia* sp.) on triglyceride (TG) levels and related cardiovascular risk factors have been recently summarized (Ryan et al., 2009). Sixteen studies considered subjects with both normal and elevated TG levels, including those with persistent hypertriglyceridemia (HTG) treated with concomitant 3-hydroxy-3-methylglutaryl coenzyme A (HMG-CoA) reductase inhibitor (statin) therapy. At doses of 1 to 2 g/day, algal-DHA significantly lowered TG levels (up to 26%) when administered either alone or in combination with statins. The reduction in TG levels was markedly greater in hypertriglyceridemic subjects than in normal subjects. Algal-DHA modestly increased plasma levels of both HDL and LDL. The increased plasma level of LDL was associated with a shift of lipoprotein particle size towards larger, less atherogenic sub-fractions. In some subjects, blood pressure and heart rate were significantly reduced. Algal-DHA was well tolerated and no serious adverse events were reported.

Bleeding Potential

Although prolongation of bleeding time was reported in early omega-3 fatty acid studies using fish oil, no evidence of clinically important bleeding has been reported in pooled data or in patients undergoing coronary artery bypass or coronary angioplasty (Harris, 2007). The effects of a fish-enriched diet or dietary supplements of fish oil or algal-DHA on platelet aggregation and homeostatic factors were evaluated in healthy men (Agren et al., 1996, 1997). The results indicated that a fish diet and fish oil, but not algal-DHA, inhibited in vitro platelet aggregation, but hemostatic factors were not affected by the n-3 fatty acid supplementation. Sanders et al. (2006) also reported no significant differences between the algal-DHA group and the placebo group in serum plasma VII antigen, FVII activated, fibrinogen, and von Willebrand factor. Supplements of 1.62 g/day of algal-DHA in healthy vegetarians also failed to reduce collagen- and ADP-induced aggregation or collagen-stimulated thromboxane A_2 release (Conquer & Holub, 1996). Compared with baseline values, higher doses of algal-DHA (6 g/day) administered to healthy men for 90 days reduced urinary thromboxane B_2 excretion (Ferretti et al., 1998). However, the prothrombin time, activated thromboplastin time, and the antithrombin-III levels before and after supplementation with 6 g/day of algal-DHA did not differ significantly (Nelson et al., 1997a). Additionally, the in vivo bleeding times did not show any significant difference during and after algal-DHA supplementation (Nelson et al., 1997a). It appears that, even at high doses, DHA administered alone does not significantly affect clinical bleeding. Based on their consideration of studies that evaluated bleeding time, the FDA has affirmed the use of up to 3 g/day of DHA and EPA as safe for use in food (Food and Drug Administration, 1997a).

Tolerability

The use of fish oil is often associated with "fishy" burps or other mild gastrointestinal complaints. In one study, Schwellenbach et al. (2006) compared the safety and tolerability of 1000 mg/day of DHA (DHASCO-S) and fish oil (1252 mg/day of DHA+EPA) in patients with coronary artery disease and elevated TG levels. A significantly greater proportion of subjects in the fish oil group reported a fishy taste. Arterburn et al. (2007) compared the bioequivalence of DHASCO with DHASCO-S at doses of 200, 600, and 1,000 mg/day for 28 days in healthy individuals. Both SCOs produced equivalent DHA levels in plasma phospholipids and erythrocytes. At each dose, compared with subjects who received DHASCO-S, the incidence of eructation was significantly higher in subjects who received DHASCO.

ARA

Unlike DHA, ARA is used primarily in infant formulas. Calder (2007) suggests that ARA appears to be an important component of infant formulas, and, via this mode of application, may be helpful in growth and development.

A review of literature by Calder (2007) indicates that studies on blood lipids, platelet reactivity, and bleeding time appear to confirm that a significant increase in ARA intake by healthy adults, up to 1.5 g/day, is unlikely to have any adverse events, but the effect of increased intake in diseased individuals is not known. A recent study by Kusumoto et al. (2007) demonstrated that there was no effect of supplemental ARA (ARASCO) (838 mg/day) on blood pressure, serum lipid and glucose concentrations, or serum markers of liver function. These findings are consistent with an earlier study by Nelson et al. (1997b) who demonstrated that adding 1.5 g/day of ARASCO for 50 days produced no observable changes in blood coagulation and thrombotic tendencies in healthy adults. However, larger doses of ARA (6 g/day as an ethyl ester) (Seyberth et al., 1975) increased ex vivo platelet aggregation, which prompted concern about the potential for adverse pro-thrombotic action; the study was stopped early after 3 weeks. However, no changes in blood pressure, sodium balance, renal creatinine clearance, or other laboratory measures were reported.

Allergies Associated with DHA and ARA

DHASCO, DHASCO-S, DHA45-oil, SUNTGA40S, and ARASCO are produced by algal fermentation and do not contain any fish components. These SCOs are considered by the FDA to be "highly refined oils" that are not associated with allergic reactions (The Threshold Working Group of the FDA, 2005). In general, edible oils can be derived from major food allergens, such as soybeans and peanuts, and may contain variable levels of protein. The FDA states that "the consumption of highly refined oils derived from major food allergens by individuals who are allergic to the source food does not appear to be associated with allergic reactions" (The Threshold Working Group of the FDA, 2005).

Marketed Experience and Clinical Exposure Data

In addition to infant formula, DHA, as provided in DHASCO-S, has been added as an ingredient to a variety of food products and nutritional supplements. Since 1997,

DHASCO-S oil has been commercially produced and marketed in countries outside of the U.S. as a natural product. Within the U.S., DHASCO-S is marketed as a dietary supplement for adults, including pregnant and lactating women. Various trade names are used, including *Life's* DHA™, Neuromins®, Neuromins PL Omega Gold®, and Expecta® Lipil® capsules. To date, Martek has sold over 45 million capsules for consumption.

Exposure data for selected and completed controlled clinical trials using DHASCO are listed in Table 15.A. The Martek-sponsored study, monitored according to Good Clinical Practice (GCP) guidelines, is denoted by an asterisk. Other studies, including Investigator-Initiated Studies (IIS), with or without Martek monitoring, used DHASCO or DHASCO-S in oil or capsule form. For these clinical studies, Martek released product only after receiving an Investigational Review Board (IRB) or Institutional Ethics Committee (IEC) approval of the study protocol. For some studies, investigators purchased product and conducted their studies without Martek's knowledge until publication. While Martek endeavored to obtain information, especially safety data, it was not possible to determine the completeness or accuracy of the data from such studies, most of which were not compliant with GCP guidelines.

Table 15.A. DHASCO (DHA) Exposure Table from Selected Studies.

Investigator/ First Author	Subjects	Condition	DHASCO Capsules DHA/d	Duration
Healthy Volunteers				
Agren (1996, 1997) Vidgren (1997)	55	Healthy male adults	1.68 g	15 wks
Theobald (2004, 2007)	38	Healthy men and women, aged 40-65 years	700 mg	90 days
Nelson (1997) Kelley (1998, 1999) Ferretti (1998)	11	Healthy males	6 g	90 days
Arterburn (2008)	32	Healthy adults	600 mg	14 days
Innis (1996)	32	Healthy males	600 mg 1700 mg 2900 mg	14 days
Conquer (1998)	19	Healthy Asian Indian	750 mg 1500 mg	42 days
Otto (2000a)	86	Healthy women	285 mg 570 mg	4 wks
[a]Arterburn (2007)	96	Healthy adults	200 mg 600 mg 1000 mg	4 wks

(Cont. on next page)

Table 15.A., cont. DHASCO (DHA) Exposure Table from Selected Studies.

Investigator/ First Author	Subjects	Condition	DHASCO Capsules DHA/d	Duration
Lindsay (2000)	10	Healthy adults	45 mg DHA/kg)	3 periods of 10 hours
Conquer (1997a)	24 8	Healthy vegetarians and omnivores	1620 mg	42 days
Benton (1998)	140	Healthy females	400 mg	50 days
Johnson (2008a, 2008b)	49	Females, aged 60 to 80 years	320 mg	4 mos
Patients				
Eye Disorders				
Berson (2004a, 2004b)	221	Retinitis pigmentosa	1200 mg	5 yrs
Wheaton (2003)	44	Retinitis pigmentosa	400 mg	4 yrs
Pregnancy and Reproduction				
Conquer (1997b)	1	Pregnant vegetarians	540 mg	9 mos
Innis (2008)	135	Pregnant women	400 mg	From 14-16 wks gestation until delivery
Otto (2000b)	24	Healthy, pregnant women in their second trimester	570 mg	4 wks
Jensen (1999, 2000, 2001)	147	Lactating women	200 mg	120 days
Makrides (1996) Gibson (1997)	52	Lactating women	200 mg 400 mg 900 mg 1300 mg	12 wks
Jensen (2000)	24	Lactating women	230 mg 170 mg 260 mg	6 wks
Fidler (2000)	10	Lactating women	200 mg	14 days

(Cont. on next page)

Table 15.A., cont. DHASCO (DHA) Exposure Table from Selected Studies.

Investigator/ First Author	Subjects	Condition	DHASCO Capsules DHA/d	Duration
Stark (2004)	32	Post-menopausal women +/- HRT	2.8 g	4 wks
Conquer (1997)	28	Astenozoo-spermic men (< 50% motility)	400 mg 800 mg	3 mos
Metabolism and Nutrition				
Kelley (2007, 2008)	34	Hyperlidipemia	3 g	3 mos
Engler (2004, 2005)	20	Hyperlipidemia children	1.2 g	6 wks
Davidson (1997)	27	Hyperlipidemia Hypertriglyceridemia	1250 mg 2500 mg	6 wks
Keller (2007)	40	Statin-treated, cardiac risk subjects	2 g	6 wks
Lloyd-Still (2001)	19	Cystic fibrosis	50 mg	6 mos
Brown (2001)	5	Cystic fibrosis	70 mg/kg	6 wks
Denkins (2002)	12	Overweight adults	1.8 g	12 wks
Harding (1999) Gillingham (1999)	14	Children w/ metabolic deficiencies	65 or 130 mg/kg	2-5 yrs
Psychiatric and Neurology				
Marangell (2006)	10	Bipolar disorder	2 g	1 yr
Mischoulon (2008)	35	Major depressive disorder	1 g 2 g 4 g	12 wks
Marangell (2003)	36	Major depressive disorder	2 g	6 wks
Martek-MIDAS (2009)	245	Memory complaints	900 mg	6 mo
NIA (2009)	200	Alzheimer's disease	2 g	18 mo

[a]Martek-sponsored study, monitored under GCP guidelines.

In both the published and unpublished studies, including IIS, subjects received moderate to high levels of DHA from DHASCO oil as capsules; safety outcomes were measured and any resulting adverse events were recorded. Over 500 healthy volunteers received up to 6 g of DHA for up to 90 days. Over 1,400 subjects with medical conditions received up to 12 g per day of DHA, and participated in studies for up to 5 years.

Regulations Applicable to SCOs

Novel foods and novel food ingredient regulations are established in the EU, Australia and New Zealand, Canada, and other parts of the world. In the U.S., the FDA regulates foods that would be considered novel foods in other parts of the world as food additives under existing laws. All novel food ingredients and food additives require pre-market approval by competent authorities before being introduced into the food supply. In the U.S., food ingredients that are considered GRAS are excluded from the food additive definition and are exempt from pre-market approval requirements. Novel food regulations in the EU, Australia and New Zealand, and Canada, as well as food additive and GRAS regulations in the U.S., will be described in more detail, as these regulations are applicable to SCOs.

United States-Food Ingredients

In response to public concerns about the increased use of chemicals in foods and food processing, Congress passed the 1958 Food Additives Amendment to the Federal Food, Drug, and Cosmetic Act (FDCA). The basic purpose of the amendment was to require that, before a new additive could be used in food, its manufacturer must demonstrate the safety of the additive to the FDA. The amendment defined the terms "food additive" [FDCA §201(s)] and "unsafe food additive" [FDCA §409(a)] and established a pre-market approval process for food additives [FDCA §409(b) through (h)]. When passing the amendment, Congress recognized that many substances intentionally added to food would not require a formal pre-market review by the FDA to assure their safety. For example, the safety of some substances could be established by a long history of use in food or by virtue of the nature of the substances, their customary or projected conditions of use, and the information generally available to scientists. Therefore, Congress enacted a two-step definition of "food additive" [FDCA §201(s)]. The first step broadly includes any substance, the intended use of which results, or may reasonably be expected to result, directly or indirectly, in its becoming a component or otherwise affecting the characteristics of food. The second step, however, excludes from the definition of "food additive" substances that are generally recognized, among experts qualified by scientific training and experience to evaluate their safety, as having been adequately shown through scientific procedures (or, in the case of substances used in food prior to January 1, 1958, through either scientific procedures or through experience based on common use in food) to be safe, under the conditions of their intended use. This exception to the food additive definition came to be known as the "GRAS Exemption". Many substances that are commonly used in foods today are legally marketed in the U.S. under the GRAS exemption.

One of the key elements of the GRAS exemption is that a GRAS substance may be lawfully marketed for a particular use without FDA review or approval. Nevertheless, many manufacturers have found it useful to have a statement from the FDA agreeing with the manufacturer's GRAS determination. Initially, the FDA issued informal "opinion letters" concerning the GRAS status of substances. The opinion letters, however, were issued only to the specific person requesting the letter and, therefore, did not provide industry-wide notification of the agency's GRAS decision. To address this and other concerns, the FDA adopted the GRAS affirmation petition process, a voluntary administrative process whereby manufacturers could petition the FDA to affirm that a substance was GRAS under certain conditions of use. If the FDA agreed with the petitioner's GRAS determination, a regulation was published in the Code of Federal Regulations, affirming the GRAS status of the substance. The GRAS affirmation petition process was intended to provide a mechanism for the official recognition of lawfully made GRAS determinations. The GRAS affirmation petition process not only facilitated awareness of FDA approval in the U.S. but also highlighted the approval in Europe and elsewhere for the international food industry.

However, the GRAS affirmation petition process turned out to be extremely resource-intensive, involving a comprehensive review of each petition and requiring these petitions to undergo a rulemaking process for each substance affirmed as GRAS. As a result, GRAS petitions languished at the agency for years, even decades, without the publication of a final regulation. As a result of the problems encountered with the GRAS petition process, the FDA proposed a "GRAS notification" procedure in 1997. This procedure was intended to replace the GRAS affirmation petition process. Under the GRAS notification procedure, the FDA evaluates whether a GRAS "notice," supplied by the manufacturer, provides a sufficient basis for a GRAS determination and whether information in the notice, or otherwise available to the FDA, raises issues that might lead the agency to question whether use of the substance is GRAS. Within ninety days of receipt of the notice, the FDA will respond in writing as to whether it has identified a problem with the notice. To provide the industry with information on prior GRAS notices, the FDA publishes a list of all submitted GRAS notices, along with the agency's response, on the FDA website. Although the GRAS notification regulation has never been finalized, the FDA has adopted the procedure as a replacement for the GRAS affirmation petition process.

If a potential new ingredient cannot be determined to be GRAS, a manufacturer must file a petition proposing the issuance of a regulation prescribing the conditions under which the proposed additive may be safely used. The manufacturer supplies the FDA with all pertinent data, especially safety data, and the agency then conducts a comprehensive review of all the safety data and determines if the ingredient is safe for its intended use.

Content of a GRAS Notification

The GRAS process is considered to be rigorous, flexible, credible, and transparent. To date, manufacturers of SCOs have followed a GRAS process to establish safety, as SCOs possess a variety of characteristics amenable to this process. SCOs are usually derived from novel sources or processes, and such diversity requires safety considerations that are clear, but not overly prescriptive, because of the multitude of issues attached to each type of oil and the organism from which it was derived.

Any person may notify the FDA of a claim that a particular use of a SCO is exempt from the statutory pre-market approval requirements based on the notifier's determination that such use is GRAS. Within thirty days of receipt of a notice, the FDA will acknowledge receipt of the notice by informing the notifier in writing, and, within ninety days of receipt of the notice, the FDA will respond to the notifier in writing. Copies of the GRAS exemption claim submitted to the agency, along with the letter issued to the notifier, acknowledging receipt of the notification, and subsequent letter(s) issued by the agency, regarding the notification, are accessible for public review.

The FDA has provided guidance on how to submit a GRAS notification. The content of the notification shall include the following information: (a) a claim, dated and signed by the notifier, that a particular use of a substance is exempt from the pre-market approval requirements of the FDCA, because the notifier has determined that such substance is GRAS; (b) detailed information about the identity of the substance, including methods of manufacturing (excluding any trade secrets and including, for substances of natural biological origin, source information, such as genus and species), characteristic properties, any content of potential human toxicants, and specifications for food-grade materials; (c) information on any self-limiting levels of use; and (d) a detailed summary of the basis for the notifier's determination that a particular use of the notified substance is exempt from the pre-market approval requirements of the FDCA. Such a determination may be based either on scientific procedures or on common use in food. For a GRAS determination based on scientific procedures, such a summary shall include: (a) a comprehensive discussion of, and citations to, generally available and accepted scientific data, information, methods, or principles that the notifier relies upon to establish safety; (b) a comprehensive discussion of any reports, investigations, or other information that may appear to be inconsistent with the GRAS determination; and (c) the basis for concluding that there is consensus among experts, qualified by scientific training and experience to evaluate the safety of substances added to food, that there is reasonable certainty that the substance is not harmful under the intended conditions of use.

Europe

Novel foods are foods, food ingredients, and food production methods that have not been used before 1997 for human consumption to a significant degree within the European Community. Regulation No. 258/97—Novel Food and Novel Food Ingredients—lays out detailed rules for the authorization of novel foods and novel food ingredients. Recent regulation 1829/2003 concerning genetically modified food and feed was established to separate genetically modified food and feed from novel foods and to set up an EU system to trace genetically modified organisms, introduce labeling of genetically modified feed, reinforce existing labeling rules for genetically modified food, and establish an authorization procedure for genetically modified organisms in food and feed and their deliberate release into the environment.

Foods commercialized in at least one Member State before the entry into force of the Regulation on Novel Foods are on the EU market under the "principle of mutual recognition." Novel foods must undergo a safety assessment before being placed on the EU market. Only those products considered to be safe for human consumption are authorized for marketing.

Companies that want to place a novel food on the EU market need to submit an application, in accordance with Commission Recommendation No. 97/618/EC, which concerns the scientific information and the safety assessment report required. Novel foods or novel food ingredients may follow a simplified procedure, only requiring notifications from the company, when they are considered by a national food assessment body as "substantially equivalent" to existing foods or food ingredients (with regard to their composition, nutritional value, metabolism, intended use, and the level of undesirable substances contained therein).

Novel Foods Full Application Process

Foods, food ingredients, and productions methods are determined to be novel according to guidelines, and usually in consultation with the competent authority, in the Member State where the application is submitted. If a product is determined to be a novel food, the applicant prepares a dossier for submission and presents it to a Member State. The Member State has ninety days to review the dossier and provide an "opinion." The ninety-day review process can take much longer, depending on questions raised by the Member State undertaking the review and responses provided by the applicants. Assuming a favorable opinion is generated by the Member State conducting the review, the dossier is then passed on to the European Commission and the other Member States, who have sixty days to raise "reasoned objections." If, during the course of the sixty-day Member State review process, reasoned objections are raised that cannot be resolved, the European Food Safety Authority (EFSA) may be enlisted for an opinion. The EFSA serves as an independent point of reference for scientific opinion and may be requested to provide an opinion not only by the Commission, but also by the European Parliament and the Member States. If there are no objections raised by Member States, the reasoned objections are satisfied by the applicant, or the EFSA offers a favorable opinion to counter the reasoned objections, then the product is approved, and a Commission Decision is passed by the Standing Committee on the Food Chain and Animal Health, and it is published in the Official Journal of the European Communities.

Commission Regulation 1852/2001 provides that the following information must be made public: (a) name and address of the applicant; (b) description allowing identification of the food or food ingredient; (c) intended use of the food or food ingredient; (d) summary of the dossier, except for those parts for which confidentiality has been requested; and (e) date of receipt of a complete request. The Regulation also provides that the Commission must make the initial assessment report available to the public, except for any information identified as confidential.

Regulation 1852/2001 also lays down rules for the protection of information provided by applicants when requesting authorization under the Novel Foods Regulation. Pursuant to Regulation 1852/2001, Member States may not divulge information identified as confidential, with the exception of the information that needs to be made public in order to protect human health. When submitting a novel food application, applicants may indicate what information, relating to the manufacturing process, should be kept confidential on the grounds that its disclosure might harm their competitive position. This information must be duly justified, and it is then up to the competent authority, in consultation with the applicant, to decide which information will be kept confidential.

The "Substantial Equivalence" Concept

Article 3 of the Novel Foods Regulation introduces the concept of "substantial equivalence" and a simplified notification procedure for the case of foods or food ingredients that, on the basis of an opinion delivered by competent bodies, are "substantially equivalent" to "existing foods or food ingredients as regards their composition nutritional value, metabolism, intended use, and the level of undesirable substances contained therein". The concept of substantial equivalence embodies the idea that existing organisms or products used as foods or food sources can serve as a basis for comparison when assessing the safety and nutritional value of a food or food ingredient that has been modified or is new. Article 5 of the Novel Foods Regulation states that, in the case that a food or food ingredient has been determined to be substantially equivalent (Article 3), the applicant shall notify the Commission when the food or food ingredient is placed on the market. Applicants may market substantially equivalent food or food ingredients immediately after notification of the Commission; they do not have to wait for approval. The Commission is required to forward a copy of the notification and relevant details to Member States, if requested. Member States may oppose the marketing of such a product on their territory if they have "detailed grounds" of considering that the use of a food or a food ingredient endangers human health or the environment (see Article 12 of the Regulation). The Commission publishes a summary of those notifications in the "C" series of the Official Journal of the European Communities.

European Food Safety Authority (EFSA)

Following a series of food scares in the 1990s that undermined consumer confidence in the safety of the food chain, the EU concluded that it needed to establish a new scientific body charged with providing independent and objective advice on food safety issues associated with the food chain. Its primary objective would be to "... contribute to a high level of consumer health protection in the area of food safety, through which consumer confidence can be restored and maintained." The result was the creation of the EFSA. In 2003, the five Scientific Committees supplying the Commission with scientific advice on food safety were transferred to the EFSA. The EFSA provides independent scientific advice on all matters linked to food and feed safety and communicates with the public in an open and transparent way on all matters within its remit.

Australia and New Zealand

Novel foods and novel food ingredients are regulated by Standard 1.5.1—Novel Foods—of the Australia New Zealand Food Standards Code. The Standard prohibits the sale of novel foods, unless they are listed in the Standard and comply with any special conditions noted. Foods deemed to be novel require an application to the authority, namely, Food Standards Australia New Zealand (FSANZ). FSANZ assesses the safety for human consumption of each novel food for which an application is made, prior to approval. The determination of to whether or not a food ingredient is novel is made in accordance with the definition of "non-traditional" and "novel," as defined in the Standard. As recently amended (Amendment No. 95 – 2007), non-traditional means (a) a food that does not have a history of human consumption in Australia or New Zealand; (b) a substance derived from a food, where that substance does not have a history of human consumption in Australia or New Zealand, other than as a component of the food; or (c) any other substance, where that substance, or the source from which it is derived, does not have a history of human consumption as a food in Australia or New Zealand. SCOs are considered novel foods (i.e., non-traditional foods), requiring assessments of public health and safety regarding (a) the potential for adverse effects in humans; (b) the composition or structure of this food; (c) the process by which this food has been prepared; (d) the source from which they are derived; (e) patterns and levels of consumption of this food; or (f) any other relevant matters.

Companies that want to place a novel SCO on the Australia New Zealand market need to submit an application to amend Standard 1.5.1—Novel Foods—and approve the use of the SCOs for the intended conditions of use. Novel SCOs are required to undergo a pre-market safety assessment. The objective of the assessment is to determine whether it is appropriate to amend the Foods Standards Code and permit the use of the SCO as a novel food. The process for amending the Australia New Zealand Food Standards Code is prescribed in the Food Standards Australia New Zealand Act of 1991. There are several stages involved in this process. Following an application and review by FSANZ, an initial assessment report is prepared, which outlines possible issues and options and identifies affected parties and questions. The initial assessment report is released for public consultation. Following public consultation, submissions are collated and analyzed, and a draft assessment report is prepared, using information provided by the applicant, stakeholders and other sources. A scientific risk assessment is prepared, risk analysis is completed, a risk management plan is developed, and a draft assessment is released for public consultation. Comments received on the draft assessment are analyzed; amendments are made, as appropriate; and a final assessment report is generated. The FSANZ Board approves or rejects the final assessment report, and the Ministerial Council is notified. Following Ministerial Council approval, an amendment is made to the Food Standards Code, which is published in the Commonwealth Gazette and the New Zealand Gazette, and use of the SCO is authorized.

Canada

The definition of "novel food," as provided in the Food and Drug Regulations published in the Canada Gazette Part II (October 27, 1999) is: (a) a substance, including a microorganism, that does not have a history of safe use as a food; (b) a food that has been manufactured, prepared, preserved, or packaged by a process that (i) has not been previously applied to that food and (ii) causes the food to undergo a major change; and (c) a food that is derived from a plant, animal, or microorganism that has been genetically modified such that (i) the plant, animal, or microorganism exhibits characteristics that were not previously observed, (ii) the plant, animal, or microorganism no longer exhibits characteristics that were previously observed, or (iii) one or more characteristics of the plant, animal or microorganism no longer fall within the anticipated range for that plant, animal, or microorganism. In Canada, SCOs fall with the definition of novel food, are regulated as such, and require pre-market approval.

Regulation requires petitioners to notify the Novel Foods Section of the Food Directorate of Health Canada prior to marketing or advertising a novel SCO in Canada. Pre-market notifications permit Health Canada to conduct a safety assessment of the novel food, to demonstrate that it is safe, before it is allowed on the Canadian market. The Novel Foods Section is responsible for communicating to petitioners and receiving novel foods notification/submission materials and initiating the review process. The Novel Foods Section distributes submission materials to relevant Food Directorates (e.g., Bureau of Chemical Safety, Bureau of Nutritional Sciences, and the Bureau of Microbial Hazards) for their respective reviews. Evaluators review the novel food notification package and determine whether or not the product is considered a novel food. If considered novel, a safety assessment is then conducted. Evaluators may make requests for additional information if relevant data has not been included in the original notification/submission package. At the completion of the safety assessment—only if all members of the evaluation team agree there are no health risks associated with the consumption of the novel food product in question—a proposal is drafted and presented to the Food Rulings Committee. If the Committee finds the proposal acceptable, the petitioner is notified, in writing, by the Director General that, based on an evaluation of the submitted data, Health Canada has no objection to the sale of the novel food product in Canada, as specified in the letter.

Current Regulatory Status of SCOs

The regulatory status of SCOs in select regulatory jurisdictions is summarized in Table 15.B.

Table 15.B. Status of SCOs in Select Regulatory Jurisdictions

Trade Name	Single Cell Organism	United States	European Union	Australia & New Zealand	Canada
DHASCO	*C. cohnii*	GRAS, 2001	History of Use	History of Use	Novel Food, 2002, 2006
DHASCO-S	*Schizochytrium* sp.	GRAS, 2004	Novel Food, 2003	Novel Food, 2002	Novel Food, 2006
DHA-45 Oil	*Ulkenia* sp.	GRAS, Withdrawn	Novel Food, 2003	Novel Food, 2005	Not Authorized
ARASCO	*M. alpina*	GRAS, 2001	History of Use	History of Use	Novel Food, 2002
SUNTGA40S	*M. alpina*	GRAS, 2006	Novel Food, 2008	Not Authorized	Novel Food, 2003

Footnotes
DHASCO, ARASCO, and SUNTGA40S GRAS for use in infant formula in U.S.
DHASCO-S GRAS for use in food and beverage applications in U.S.
DHASCO and ARASCO established history of use in Europe and Australia and New Zealand and are not considered novel foods
DHASCO-S novel food approval for use in foods in EU
DHA-45 oil notified as substantially equivalent to DHASCO-S in EU
SUNTGA40S novel food approval for use in infant formula in EU
DHASCO-S and DHA-45 oil novel food approval for use in food and beverage applications in Australia and New Zealand
DHASCO, ARASCO and SUNTGA40S novel food approval for use in infant formula in Canada
DHASCO-S and DHASCO novel food approval for use in food and beverage applications in Canada

United States

Several GRAS notifications for SCOs have been posted on the FDA website, and, at the time of writing, four have been successfully reviewed by the agency (http://www.cfsan.fda.gov/~rdb/opa-gras.html, accessed March 14, 2009).

In 2001, the FDA responded to Martek GRAS Notification GRN 000041 that DHASCO, derived from the microalgal species *C. cohnii*, and ARASCO, derived from the soil fungus *M. alpina*, as sources of DHA and ARA, are GRAS, through scientific procedures, when added to term infant formulas. Based on the information provided by Martek, as well as other information available to the FDA, the agency had no questions regarding the conclusion that ARASCO and DHASCO are GRAS sources of ARA and DHA under the intended conditions of use—i.e., when added to infant formulas for consumption by healthy term infants at a level of up to 1.25% each of total dietary fat and at a ratio of DHA to ARA of 1:1 to 1:2 (Food and Drug Administration, 2001a).

In response to GRAS Notice No. GRN 000080 (Food and Drug Administration, 2001b), based on information provided by Mead Johnson™ Nutrition, as well as other available information, the FDA had no questions regarding the conclusion that ARASCO is GRAS, under the intended conditions of use—i.e., when used in combination with DHASCO and at a 50% increase, relative to that proposed by Martek in GRN 000041.

In 2006, the FDA responded to a submission, on behalf of Ross Products Division, Abbott Laboratories (Ross), regarding the use of ARA-containing SCO from *Mortierella alpina* (referred to as AA-rich fungal oil by the FDA and as SUNTGA40S by Ross), with no objection (GRAS Notice No. GRN 000094). The intended conditions of use for this SCO were for addition to preterm infant formulas at target mean concentrations of 0.4% ARA and 0.25% DHA and for addition to term infant formula at target mean concentrations of 0.4% ARA and 0.15% DHA (Food and Drug Administration, 2006).

In the case of a new infant formula that contains a SCO (e.g., DHASCO, ARASCO, or SUNTGA40S), the manufacturer of the infant formula must make a submission to the FDA, providing the required assurances about its formula, at least 90 days before the formula is marketed under Section 412 of the FDCA. The infant formula manufacturer that intends to market an infant formula containing a new ingredient bears the responsibility for submission required by Section 412, not the manufacturer of the ingredient itself.

GRAS Notice No. GRN 000137 (Food and Drug Administration, 2004) was based on scientific procedures regarding the use of DHASCO-S (DHA oil from *Schizochytrium* sp.) as a direct food ingredient in specified food categories at specified use levels. Based on the information provided by Martek, as well as other information available to the FDA, the agency had no questions regarding the conclusion that this SCO is GRAS, under the intended conditions of use. DHASCO-S is allowed as a dietary supplement and an article of trade in the U.S. under a Dietary Supplement Health and Education Act (DSHEA) notification. Pursuant to this act and consistent with the final regulations published by the FDA, a new dietary ingredient submission was made to the FDA for DHASCO derived from *Schizochytrium* sp. The FDA acknowledged receipt of the new dietary ingredient notification and did not respond with comment. The submission was placed on public display at Dockets Management Branch (Food and Drug Administration, 1997b).

European Union

There are a number of SCOs currently approved for use within the EU. DHASCO and ARASCO, produced by Martek, were commercialized in at least one Member State before the entry into force of the Regulation on Novel Foods on May 15, 1997 and are on the EU market under the "principle of mutual recognition." The Netherlands' Ministry of Public Health, Welfare and Sport granted exemption for the addition of ARASCO and DHASCO to preterm and term infant formula. This exemption was published in the State Journal on March 8, 1995 (Netherlands State Journal, 1995).

Martek received approval for DHA-rich oil derived from *Schizochytrium* sp. (DHASCO-S). This SCO was the subject of novel foods approval and is currently being marketed for select food applications, as listed in the annex of the decision. This decision was published in the Official Journal of the European Communities (2003).

Nutrinova notified the Commission in 2003 of its intention to market a novel SCO product (DHA45-oil) obtained from the microalga *Ulkenia* sp., in accordance with Article 5 of the Novel Food Regulation (EC) 258/97. The German competent authority agreed with the company's claim that the product was "substantially equivalent" to the oil obtained from *Schizochytrium* sp. Following the determination of substantial equivalence, the European Commission was notified, and Nutrinova was able to market the oil as a food ingredient for the same food categories and use levels as DHA-rich oil from *Schizochytrium* sp. The Nutrinova process has been taken over by Lonza, Switzerland. In 2002, an applicant (Innovalg—S.A.R.L, Bouin, France) notified the Commission of their intention to market a marine microalga, *Odontella aurita*, as a novel food, referring to an opinion on the "substantial equivalence" from a French competent authority. *O. aurita* is stated to be rich in the polyunsaturated fatty acid EPA. The product, which consists of dried algae, is intended for use in a range of food products. Notifications for the microalga *O. aurita* and DHA45-oil from *Ulkenia* sp., made under Article 5 of the Novel Food Regulation (EC) 258/97, are listed on the Commission website (see Notifications Pursuant to Article 5 of Regulation (EC) No 258/97 of the European Parliament and of the Council, http:// www.europa.eu.int/comm/food/food/biotechnology/novelfood/notif_list_en.pdf accessed March 14, 2009).

Abbott Laboratories (now, Suntory Limited) made a request in 1999 to the competent authority of the Netherlands to place ARA-rich oil from *Mortierella alpina* on the market as a novel food. In 2005, the competent food assessment body of the Netherlands issued its initial assessment report. In that report, it came to the conclusion that ARA-rich oil from *M. alpina* can safely be used in formula for premature and term infants. The Commission forwarded this initial assessment report to all Member States, and, within a 60-day period, reasoned objections to the marketing of the product were raised. Therefore, the European Food Safety Authority (EFSA) was consulted. In 2008, the EFSA adopted an opinion related to the safety of "fungal oil from *Mortierella alpina*." In the opinion, the EFSA came to the conclusion that the fungal oil from *M. alpina* is a safe source of ARA to be used in infant formula and follow-on formula. On the basis of the scientific assessment, it was established that fungal oil from *M. alpina* could be placed on the community market as a novel food ingredient for use in infant and follow-on formulas (Official Journal of the European Union, 2008).

Australia and New Zealand

There are a number of SCOs currently approved for use in Australia and New Zealand. DHASCO (from *C. cohnii*) and ARASCO (from *M. alpina*), produced by Martek, were added to infant formulas in Australia and New Zealand beginning around 1998, before Standard 1.5.1—Novel Foods—of the Australia New Zealand Food Standards Code was implemented. Based on the fact that DHASCO and ARASCO were on the market prior to novel foods regulation and that considerable evidence existed for the safe use of infant formula containing DHASCO and ARASCO, FSANZ (concluded that sufficient knowledge existed to enable their safe use when added to formula; novel foods application was not required. Nevertheless, Martek submitted data and information to FSANZ, allowing safety review that culminated in a published safety report (Food Standards Australia New Zealand, 2003).

FSANZ approved the use of Martek DHASCO-S oil from *Schizochytrium* algae as novel food ingredients in food products for DHA enrichment. FSANZ concluded that the composition of the *Schizochytrium* sp. microalgae and the oil derived from *Schizochytrium* sp. are comparable to other traditional sources of DHA, and *Schizochytrium* sp. algae and oil provide an alternative source of omega-3 fatty acids in foods (Australia New Zealand Food Authority (ANZFA), 2002). In 2002, approval was published in the Commonwealth of Australia Gazette (2002).

Nutrinova (now, Lonza) received approval in 2005 to market a DHA-rich oil (DHA45-oil), obtained from the microalga *Ulkenia* sp., as a novel food (Food Standards Australia New Zealand, 2005). FSANZ concluded that the oil provides an alternative source of omega-3 fatty acids, when used as a food ingredient. In 2005, the approval was published in the Commonwealth of Australia Gazette (2005).

Canada

In Canada, DHASCO (from *C. cohnii*) and ARASCO (from *M. alpina*) were subjected to a novel food pre-market notification for use as sources of the nutrients DHA and ARA, respectively, in human milk substitutes (infant formula). Health Canada (2002) notified Martek that it had no objections to the use of DHASCO and ARASCO as sources of DHA and ARA in infant formulas.

A pre-market notification for the use of ARA oil from *M. alpina* was submitted to Health Canada by Suntory. Health Canada (2003) posted a novel food decision on their website.

DHASCO and DHASCO-S oil (from *C. cohnii* and *Schizochytrium* sp., respectively) were subject to novel food pre-market notifications for use in a number of food and beverage applications. Health Canada (2006) had no objection to the sale of these DHASCOs as food ingredients for the general population.

Summary

The safety of the SCOs that have been evaluated to date is based on several lines of evidence, including: (a) the inherent safety of fatty acids and other components of these oils, their presence in food (including human breast milk), the small quantities expected to be consumed, and knowledge of their metabolism; (b) the absence of reports of pathogenicity or toxigenicity of the source organisms used for their production; (c) published results of nonclinical safety studies in rodent and nonrodent species demonstrating no unexpected, treatment-related, dose-dependent, adverse toxicological effects; (d) human clinical studies in target populations monitoring safety outcomes and documenting no serious treatment-related adverse events; and (e) historical safe use of these products, including use in infant formulas (preterm and term), as dietary supplements, and as food ingredients.

Novel foods and novel food ingredient regulations are established in the EU, Australia and New Zealand, Canada, and other parts of the world. In general, SCOs having significant history of use prior to novel food regulations are considered traditional foods and are not regulated as novel foods. This precedent applies to DHASCO (from *C. cohnii*) and ARASCO (from *M. alpina*) in the EU and in Australia and New Zealand. SCOs not having significant history of use are regulated as novel foods and proceed through a formal pre-market review and approval process. Manufacturers wishing to obtain approval for the use of a new ingredient in the EU, Australia and New Zealand, and Canada traditionally approach the competent authority to consult on the appropriate approach (e.g., novel food application) to obtain pre-market clearance. Novel foods regulation in the EU provides two routes for authorization of novel foods: a full procedure and a simplified procedure, based on the concept of "substantial equivalence." Under Article 4 (the "full procedure"), an initial assessment from one Member State is made and circulated for formal review by the competent authorities in other Member States. If Member States are not unanimous at this stage, a decision is taken by majority vote. The European Food Safety Authority (EFSA) may be asked to offer an opinion on reasoned objections, related to the scientific risk assessment, before a vote is taken. Under Article 5 (the "simplified procedure"), an applicant can apply to a Member State for an opinion on "substantial equivalence," which is then forwarded to the European Commission, along with a notification of the applicant's intention to market the product.

The FDA has the primary responsibility for regulating new food ingredients, including SCOs used as food ingredients in the U.S. Manufacturers may propose the addition of new food ingredients by either filing a food additive petition with FDA, which requests a formal pre-market review, or making a "generally recognized as safe" (GRAS) determination. The GRAS notification process has become the route of choice not only for the introduction of new food ingredients, but also specifically for several new SCOs used in infant formulas and food applications.

Using DHASCO (DHA oil from *C. cohnii*), ARASCO (ARA oil from *M. alpina*), and SUNTGA40S (ARA oil from *M. alpina*) in infant formulas is considered GRAS in the U.S., following examination by qualified experts. These conclusions were reviewed by the FDA as a part of the GRAS notification process with no objections. The use of DHASCO-S (from *Schizochytrium* sp.) in food and beverage applications is GRAS and has been reviewed by the FDA as a part of the GRAS notification process. DHASCO and DHASCO-S are also marketed as dietary supplements. DHASCO-S oil was the subject of a New Dietary Pre-market Notification submitted to the FDA under the provisions of the Dietary Supplement Health and Education Act (DSHEA).

DHASCO (from *C. cohnii*) and ARASCO (from *M. alpina*) are not considered novel foods in the EU based on a significant history of use prior to 1997, the year novel foods regulations were implemented. DHASCO-S (from *Schizochytrium* sp.), DHA-45 oil (from *Ulkenia* sp.), and SUNTGA40S (from *M. alpina*) are considered novel foods within the EU. DHASCO-S and SUNTGA40S were evaluated under the full authorization procedure for use in foods and infant formula, respectively. DHA-45 oil was determined to be substantially equivalent to DHASCO-S oil, was notified under the simplified procedure, and was subsequently placed on the market for use in foods.

In Australia and New Zealand, DHASCO (from *C. cohnii*) and ARASCO (from *M. alpina*) are not considered novel foods, based on a history of use that dates back to 1998. DHASCO-S (from *Schizochytrium* sp.) and DHA-45 oil (from *Ulkenia* sp.) are considered novel foods in Australia and New Zealand and were reviewed and approved for use as ingredients in food and beverage applications.

Health Canada has evaluated DHASCO (from *C. cohnii*), ARASCO (from *M. alpina*), and SUNTGA40S (from *M. alpina*) as novel foods for use in infant formula and issued novel food decisions allowing such use. DHASCO and DHASCO-S (from *Schizochytrium* sp.) are also considered novel foods for use in food and beverage applications, with novel foods decisions issued by Health Canada.

Numerous clinical studies in preterm infants, full-term infants, children, pregnant/lactating women, and many other populations support the safety of SCOs. SCOs have been added to commercial infant formula since 1995, and their safety profile is well established. These formulas are now available in more than seventy-five countries worldwide. It is estimated that 50 million infants have received formulas containing SCOs with no known reported pattern of adverse events associated with the consumption of any of the formulas. Almost all preterm infants (a vulnerable population) and 90% of term infants in the U.S. are fed infant formulas with SCOs.

References

Agren, J.J.; O. Hanninen; A. Julkunen; L. Fogelholm; H. Vidgren; U. Schwab; O. Pynnonen; M. Uusitupa. Fish Diet, Fish Oil, and Docosahexaenoic Acid Rich Oil Lower Fasting and Postprandial Plasma Lipid Levels. *Eur. J. Clin. Nutr.* **1996**, *50*, 765–771.

Agren, J.J.; S. Vaisanen; O. Hanninen; A.D. Muller; G. Hornstra. Hemostatic Factors and Platelet Aggregation After a Fish-Enriched Diet of Fish Oil or Docosahexaenoic Acid Supplementation.

Prostaglandins Leukot. Essent. Fatty Acids **1997,** *57,* 419–421.

American Academy of Pediatrics, Committee on Nutrition. *Pediatric Nutrition Handbook,* 6th edn.; American Academy of Pediatrics: Elk Grove Village, IL, 2009; pp 17, 85.

ANZFA. *Final Assessment Report (Inquiry-Section 17) for Application A428: DHA-Rich Dried Marine Micro Algae (Schizochytrium sp.) and DHA-Rich Oil Derived from Schizochytrium sp. as Novel Food Ingredients.* Australia New Zealand Food Authority 09/028 May 8, 2002 .

Arterburn, L.M.; H.A. Oken; E. Bailey Hall; J. Hammersley; C.N. Kuratko; J.P. Hoffman. Algal-Oil Capsules and Cooked Salmon: Nutritionally Equivalent Sources of Docosahexaenoic Acid. *J. Am. Diet. Assoc.* **2008,** *108,* 1204–1209.

Arterburn, L.M.; H.A. Oken; J.P. Hoffman; E. Bailey-Hall; G. Chung; D. Rom; J. Hamersley; D. McCarthy. Bioequivalence of Docosahexaenoic Acid from Different Algal Oils in Capsules and in a DHA-Fortified Food. *Lipids* **2007,** *42,* 1011–1024.

Benton, D., Ed. *Fatty Acid Intake and Cognition in Healthy Volunteers. NIH Workshop on Omega-3 Essential Fatty Acids and Psychiatric Disorders,* Bethesda, MD; 1998.

Berson, E.L.; B. Rosner; M.A. Sandberg; C. Weigel-DiFranco; A. Moser; R.J. Brockhurst; K.C. Hayes; C.A. Johnson; E.J. Anderson; A.R. Gaudio; et al. Clinical Trial of Docosahexaenoic Acid in Patients with Retinitis Pigmentosa Receiving Vitamin A Treatment. *Arch Ophthalmol.* **2004a,** *122,* 1297–1305.

Berson, E.L.; B. Rosner; M.A. Sandberg; C. Weigel-DiFranco; A. Moser; R.J. Brockhurst; K.C. Hayes; C.A. Johnson; E.J. Anderson; A.R. Gaudio; et al. Further Evaluation of Docosahexaenoic Acid in Patients with Retinitis Pigmentosa Receiving Vitamin A Treatment: Subgroup Analyses. *Arch Ophthalmol.* **2004b,** *122,* 1306–1314.

Brown, N.E.; M.T. Clandinin; S.P. Man; A.B. Thomson; Y.K. Goh; J. Jumpsen. Docosahexaenoic Acid (DHA) Feeding in Cystic Fibrosis Patients. *Pediatr. Pulmonol.* **2001,** *22,* 492A.

Calder, P.C. Dietary Arachidonic Acid: Harmful, Harmless or Helpful? *Br. J. Nutr.* **2007,** *98,* 451–453.

Clandinin, M.T.; J.E. Van Aerde; K.L. Merkel; C.L. Harris; M.A. Springer; J.W. Hansen; D.A. Dersen-Schade. Growth and Development of Preterm Infants Fed Infant Formulas Containing Docosahexaenoic Acid and Arachidonic Acid. *J. Pediatr.* **2005,** *146,* 461-468.

Commonwealth of Australia. *Amendment No. 60 to the Food Standards Code. Australia New Zealand Food Authority.* Gazette No. FSC 2, June 20, 2002.

Commonwealth of Australia. *Amendment No. 78 to the Food Standards Code. Australia New Zealand Food Authority.* Gazette No. FSC 20, May 26, 2005.

Conquer, J.A.; B.J. Holub. Supplementation with an Algae Source of Docosahexaenoic Acid Increases (n-3) Fatty Acid Status and Alters Selected Risk Factors for Heart Disease in Vegetarian Subjects. *J. Nutr.* **1996,** *126,* 3032–3039.

Conquer, J.A.; B.J. Holub. Dietary Docosahexaenoic Acid as a Source of Eicosapentaenoic Acid in Vegetarians and Omnivores. *Lipids* **1997a,** *32,* 341–345.

Conquer, J.A.; B.J. Holub. Docosahexaenoic Acid (Omega-3) and Vegetarian Nutrition. *Veget. Nutr.* **1997b,** *1-2,* 42–49.

Conquer, J.A.; B.J. Holub. Effect of Supplementation with Different Doses of DHA on the Levels of Circulating DHA as Non-Esterified Fatty Acids in Subjects of Asian Indian Background. *J.*

Lipid Res. **1998,** *39,* 286–292.

Davidson, M.H.; K.C. Maki; J. Kalkowski; E.J. Schaefer; S.A. Torri; K.B. Drennan. Effect of Docosahexaenoic Acid on Serum Lipoproteins in Patients with Combined Hyperlipidemia: A Randomized, Double-blind, Placebo-controlled Trial. *J. Am. Coll. Nutr.* **1997,** *16,* 236–243.

Denkins, Y.M.; J.C. Lovejoy; S.R. Smith. Omega-3 PUFA Supplementation and Insulin Sensitivity. *FASEB* **2002,** *16,* A24.

Dodge, J.D. Dinoflagellate Taxonomy. *Dinoflagellates*; D.L. Spector, Ed.; Academic Press: Orlando, 1984; pp 17–42.

Domsch, K.H.; W. Gams; T. Anderson, Eds. Mortierella. *Compendium of Soil Fungi;* Academic Press: Orlando, 1980; pp 431–460.

Engler, M.M.; M.B. Engler; L.M. Arterburn; E. Bailey; E.Y. Chiu; M. Malloy; M.L. Mietus-Snyder. Docosahexaenoic Acid Supplementation Alters Plasma Phospholipid Fatty Acid Composition in Hyperlipidemic Children: Results from the Endothelial Assessment of Risk from Lipids in Youth (EARLY) Study. *Nutr. Res.* **2004,** *24,* 721–729.

Engler, M.M.; M.B. Engler; M.J. Malloy; S.M. Paul; K.R. Kulkarni; M.L. Mietus-Snyder. Effect of Docosahexaenoic Acid on Lipoprotein Subclasses in Hyperlipidemic Children (the EARLY Study). *Am. J. Cardiol.* **2005,** *95,* 869–871.

Ferretti, A.; G.J. Nelson; P.C. Schmidt; G. Bartolini; D.S. Kelley; V.P. Flanagan. Dietary Docosahexaenoic Acid Reduces the Thromboxane/Prostacyclin Synthetic Ratio in Humans. *J. Nutr.* **1998,** *9,* 88–92.

Fidler, N.; T. Sauerwald; A. Pohl; H. Demmelmair; B. Koletzko. Docosahexaenoic Acid Transfer into Human Milk after Dietary Supplementation: A Randomized Clinical Trial. *J. Lipid Res.* **2000,** *41,* 1376–1383.

Food and Drug Administration, Department of Health and Human Services. Substances Affirmed as Generally Recognized as Safe: Menhaden Oil. *Fed. Reg.* **1997a,** *62,* 30751–30752.

Food and Drug Administration. *Notification of a New Dietary Ingredient.* Submitted by Monsanto Company to Office of Special Nutritionals (HFS-450), Center for Food Safety and Applied Nutrition, U.S Food and Drug Administration, Washington, D.C., December 19, 1997b. Published online: http://www.fda.gov/ohrms /dockets/dockets/ 95s0316/rpt0017_01.pd, (accessed March 24, 2004).

Food and Drug Administration, Agency Response Letter. *GRAS Notice No. GRN 000041.* U.S. Food and Drug Administration. Department of Health and Human Services. May 17, 2001a.

Food and Drug Administration, Agency Response Letter. *GRAS Notice No. GRN 000080.* U.S. Food and Drug Administration. Department of Health and Human Services, 2001b.

Food and Drug Administration, Agency Response Letter. *GRAS Notice No. GRN 000137.* U.S. Food and Drug Administration. Department of Health and Human Services, 2004.

Food and Drug Administration, Agency Response Letter. *GRAS Notice No. GRN 000094.* U.S. Food and Drug Administration. Department of Health and Human Services. April 18, 2006.

Food Standards Australia New Zealand. *DHASCO and ARASCO Oils as Sources of Long-Chain Polyunsaturated Fatty Acids in Infant Formula: A Safety Assessment.* Technical Report Series No. 22, 2003. Published online: http://www.foodstandards.gov.au (accessed July 20, 2009).

Food Standards Australia New Zealand. *Final Assessment Report for Application A522: DHA-Rich*

Micro-Algal Oil from Ulkenia *sp. as a Novel Food*. Food Standards Australia New Zealand 02/05, March 23, 2000.

Gibson, R.A.; M.A. Neumann; M. Makrides. Effect of Increasing Breast Milk Docosahexaenoic Acid on Plasma and Erythrocyte Phospholipid Fatty Acids and Neural Indices Of Exclusively Breast Fed Infants. *Eur. J. Clin. Nutr.* **1997,** *51,* 578-584.

Gillingham, M.; S. van Calcar; D. Ney; J. Wolff; C. Harding Dietary Management of Long-Chain 3-Hydroxyacyl-CoA Dehydrogenase Deficiency (LCHADD) A Case Report and Survey. *J. Inherit. Metab. Dis.* **1999,** *22,* 123-131.

Harding, C.O.; M.B. Gillingham; S.C. van Calcar; J.A. Wolff; J.N. Verhoeve; M.D. Mills. Docosahexaenoic Acid and Retinal Function in Children with Long Chain 3-Hydroxyacyl-CoA Dehydrogenase Deficiency. *J. Inherit. Metab. Dis.* **1999,** *22,* 276-280.

Harris, W.S. Expert Opinion: Omega-3 Fatty Acids and Bleeding—Cause for Concern? *Am. J. Cardiol.* **2007,** *99,* S44-S46.

Health Canada. *Novel Food Decision: DHASCO and ARASCO Oils as Sources of Docosahexaenoic Acid (DHA) and Arachidonic Acid (ARA) in Human Milk Substitutes, 2002.* Published online: http://www.novelfoods.gc.ca (accessed July 20, 2009).

Health Canada. *Novel Food Decision: SUN-TGA40S as a Source of Arachidonic Acid in Infants Formulas, 2003.* Published online: http://www.novelfoods.gc.ca (accessed July 20, 2009).

Health Canada. *Novel Food Decision: DHASCO Oil as a Source of Docosahexaenoic Acid (DHA) in Foods, 2006.* Published online: http://www.novelfoods.gc.ca (accessed July 20, 2009).

Henriksen, C.; K. Haugholt; M. Lindgren; A.K. Aurvag; A. Ronnestad; M. Gronn; R. Solberg; A. Moen; B. Nakstad; R.K. Berge; et al. Improved Cognitive Development Among Preterm Infants Attributable to Early Supplementation of Human Milk with Docosahexaenoic Acid and Arachidonic Acid. *Pediatrics* **2008,** *121,* 1137-1145.

Innis, S.M.; J.W. Hansen. Plasma Fatty Acid Responses, Metabolic Effects, and Safety of Microalgal and Fungal Oils Rich in Arachidonic and Docosahexaenoic Acids in Healthy Adults. *Am. J. Clin. Nutr.* **1996,** *64,* 159-167.

Innis, S.M.; R.W. Friesen. Essential n-3 Fatty Acids in Pregnant Women and Early Visual Acuity Maturation in Term Infants. *Am. J. Clin. Nutr.* **2008,** *87,* 548-557.

Johnson, E.J.; H.Y. Chung; S.M. Caldarella; D.M. Snodderly. The Influence of Supplemental Lutein and Docosahexaenoic Acid on Serum Lipoproteins, and Macular Pigmentation. *Am. J. Clin. Nutr.* **2008a,** *87,* 1521-1529.

Johnson, E.J.; K. McDonald; S.M. Caldarella; H.Y. Chung; A.M. Troen; D.M. Snodderly. Cognitive Findings of an Exploratory Trial of Docosahexaenoic Acid and Lutein Supplementation in Older Women *Nutr. Neurosci.* **2008b,** *11,* 75-83.

Jensen, C.; A. Llorente; R. Voigt; T. Prager; J. Fraley; Y. Zou; M. Berretta, W. Heird. Effects of Maternal Docosahexaenoic Acid (DHA) Supplementation on Visual and Neurodevelopmental Function of Breast-fed Infants and Indices of Maternal Depression and Cognitive Interference. *Pediatr. Res.* **1999,** *45,* 284A.

Jensen, C.L.; M. Maude; R.E. Anderson; W.C. Heird. Effect of Docosahexaenoic Acid Supplementation of Lactating Women on the Fatty Acid Composition of Breast Milk Lipids and Maternal and Infant Plasma Phospholipids. *Am. J. Clin. Nutr.* **2000,** *71,* 292S-299S.

Jensen, C.L.; R.G. Voigt; T.C. Prager; Y.L. Zou; J.K. Fraley; J.C. Rozelle; M.R. Turcich; A.M. Llorente; R.E. Andersen; W.C. Heird. Effects of Maternal Docosahexaenoic Acid Intake on Visual Function and Neurodevelopment in Breast Fed Term Infants. *Am J. Clin. Nutr.* **2005**, *82*, 125-132.

Jensen, C.L.; R.G. Voigt; T.C. Prager; Y.L. Zou; J.K. Fraley; J. Rozelle; M. Tureich; A.M. Llorente; W.C. Heird. Effects of Maternal Docosahexaenoic Acid (DHA) Supplementation on Visual Function and Neurodevelopment of Breast-fed Infants. *Pediatr. Res.* **2001**, *49*, 448A.

Keller, D.D.; S. Jurgilas; B. Perry; J. Blum; B. Farino; J. Reynolds; L. Keilson. Docosahexaenoic Acid (DHA) Lowers Triglycerides and Improves Low Density Lipoprotein Particle Size in a Statin-treated Cardiac Risk Population. *J. Clin. Lipidol.* **2007**, *1*, 151A.

Kelley, D.S.; P.C. Taylor; G.J. Nelson; B.E. Mackey. Dietary Docosahexaenoic Acid and Immuno-competence in Young Healthy Men. *Lipids* **1998**, *33*, 559-566.

Kelley, D.S.; P.C. Taylor; G.J. Nelson; P.C. Schmidt; A. Ferretti; K.L. Erickson; R.Yu; R.K. Chandra; B.E. Mackey Docosahexaenoic Acid Ingestion Inhibits Natural Killer Cell Activity and Production of Inflammatory Mediators in Young Healthy Men. *Lipids* **1999**, *34*, 317-324.

Kelley, D.S.; D. Siegel; M. Vemuri; B.E. Mackey. Docosahexaenoic Acid Supplementation Improves Fasting and Postprandial Lipid Profiles in Hypertriglyceridemic Men. *Am. J. Clin. Nutr.* **2007**, *86*, 324-333.

Kelley, D.S.; D. Siegel; M. Vemuri; G.H. Chung; B.E. Mackey. Docosahexaenoic Acid Supplementation Decreases Remnant-like Particle-cholesterol and Increases the (n-3) Index in Hyper-triglyceridemic Men. *J. Nutr.* **2008**, *138*, 30-35.

Kiy, T.; M. Rusing; D. Fabritius. Production of Docoahexaenoic Acid by the Marine Microalga, *Ulkenia* sp. *Single Cell Oils*, 1st edn.; Z. Cohen; C. Ratledge, Eds.; AOCS Press: Champaign, IL, 2005, pp 99-106.

Koletzko, B.; I. Cetin; J.T. Brenna for the Perinatal Lipid Intake Working Group. Dietary Fat Intakes for Pregnant and Lactating Women. *Br. J. Nutr.* **2007**, *98*, 873-877.

Koletzko B.; E. Lien; C. Agostini; H. Bohles; C. Campoy; I. Cetin; T. Decsi; J. Dudenhausen; C. Dupont; S. Forsyth; et al. The Roles of Long-chain Polyunsaturated Fatty Acids in Pregnancy, Lactation and Infancy: Review of Current Knowledge and Consensus Recommendations. *J. Perinat. Med.* **2008**, *36*, 5-14.

Kris-Etherson, P.M.; S. Innis. Position of the American Dietetic Association and Dietitians of Canada: Dietary Fatty Acid . *J. Am. Diet. Assoc.* **2007**, *107*, 1599-1611.

Kroes, R.; E.J. Schaefer; R.A. Squire; G.M. Williams. A Review of the Safety of DHA45-Oil. *Food Chem. Toxicol.* **2003**, *41*, 1433-1446.

Kusumoto, A.; Y. Ishikura; H. Kawashima; Y. Kiso; S. Takai; M. Miyazaki. Effects of Arachidonate-Enriched Triacylglycerol Supplementation on Serum Fatty Acids and Platelet Aggregation in Healthy Male Subjects with a Fish Diet. *Br. J. Nutr.* **2007**, *98*, 626-635.

Lien, E.L. Toxicology and Safety of DHA. *Lipids* **2009**, in press.

Lindsay, C.; K. Boswell; C. Becker; H. Oken; D. Kyle; L. Arterburn. Kinetics of Absorption of DHA from DHASO Oil. *Inform* **2000**, *11*, S110-S111.

Lloyd-Still, J.D.; C.A. Powers; D.R. Hoffman; K. Boyd-Trull; L.M. Arterburn; D.C. Benisek; L.A. Lester. Blood and Tissue Essential Fatty Acids after Docosahexaenoic Acid Supplementation in

Cystic Fibrosis, *Pediatr. Pulmonol.* **2001**, *22*, 263A.

Makrides, M.; M.A. Neumann; R.A. Gibson. Effect of Maternal Docosahexaenoic Acid (DHA) Supplementation on Breast Milk Composition. *Eur. J. Clin. Nutr.* **1996**, *50*, 352-357.

Marangell, L.B.; J.M. Martinez; H.A. Zboyan B. Kertz; H.F.S. Kim; L.J. Puryear. A Double-blind, Placebo-controlled Study of the Omega-3 Fatty Acid Docosahexaenoic Acid in the Treatment of Major Depression. *Am. J. Psychiatry* **2003**, *160*, 996-998.

Marangell, L.B.; T. Suppes; T.A. Ketter; E.B. Dennehy; H. Zboyan; B. Kertz; A. Nierenberg; J. Calabrese; S.R. Wisniewski; G. Sachs. Ω-3 Fatty Acids in Bipolar Disorder: Clinical and Research Implications. *Prostaglandins Leukot. Essent. Fatty Acid* **2006**, *75*, 315–21.

Martek Biosciences Corporation. *Memory Improvement with Docosahexaenoic Acid Study (MIDAS).* Published online: http://clinicaltrials.gov/ct2/show/NCT00278135?term= Martek&rank=4 (accessed April 2009).

Mischoulon, D.; C. Best-Popescu; M. Laposata; W. Merens; J.L. Murakami; S.L. Wu; G.I. Papakostas; C.M. Dording; S.B. Sonawalla; A.A. Nierenberg, et al. A Double-blind Dose-finding Pilot Study of Docosahexaenoic Acid (DHA) for Major Depressive Disorder. *Eur. Neuropsychopharmacol.* **2008**, *18*, 639-645.

National Institute on Aging. *DHA (Docosahexaenoic Acid), an Omega-3 Fatty Acid, in Slowing the Progression of Alzheimer's Disease.* Published online http://clinicaltrials.gov/ ct2/show/ NCT00440050?term=Martel&rank=10 (accessed April 2009).

Nelson, G.J.; P.S. Schmidt; G.L. Bartolini; D.S. Kelley; D. Kyle. The Effect of Dietary Docosahexaenoic Acid on Platelet Function, Platelet Fatty Acid Composition, and Blood Coagulation in Humans. *Lipids* **1997a**, *32*, 1129-1136.

Nelson, G.J.; P.C. Schmidt; G. Bartolini; D.S. Kelley; S.D. Phinney; D. Kyle; S. Silbermann; E.J. Schaefer. The Effect of Dietary Arachidonic Acid on Plasma Lipoprotein Distributions, Blood Lipid Levels, and Tissue Fatty Acid Composition in Humans. *Lipids* **1997b**, *32*, 427-433. *Exemption of Novel Food.* March 8, 1995, page 4.

Office of Food Additive Safety. *Toxicological Principles for the Safety Assessment of Direct Food Additives and Color Additives Used in Food Redbook II-Draft.* Center for Food Safety and Applied Nutrition, Food and Drug Administration: Washington, D.C., 2001.

Office of Food Additive Safety. *Redbook 2000, Toxicological Principles for the Safety of Food Ingredients.* Online Center for Food Safety and Applied Nutrition, Food and Drug Administration. Published online: http://www.cfsan.fda.gov/~redbook/red-toca.htm (accessed March 24, 2004).

Official Journal of the European Communities. *Commission Decision of 5 June 2003 Authorising the Placing on the Market of Oil Rich in DHA (Docosahexaenoic Acid) from the Microalgae Schizochytrium sp. as a Novel Food Ingredient under Regulation (EC) No. 258/97 of the European Parliament and of the Council (2003/427/EC).* OJ L 144/13, 12.6.03, 2003.

Official Journal of the European Union. *Commission Decision of 12 December 2008 Authorising the Placing on the Market of Arachidonic Acid Rich Oil from Mortierella alpina as a Novel Food Ingredient under Regulation (EC) No. 258/97 of the European Parliament and of the Council (2008/968/ EC).* OJ L 344/123, 20.12.08, 2008.

Otto, S.J.; A.C. van Houwelingen; G. Hornstra. The Effect of Different Supplements Containing Docosahexaenoic Acid on Plasma and Erythrocyte Fatty Acids of Healthy Non-Pregnant Women. *Nutr. Res.* **2000a**, *20*, 917-927.

Otto, S.J.; A.C. van Houwelingen; G. Hornstra. The Effect of Supplementation with Docosa-hexaenoic and Arachidonic Acid Derived from Single Cell Oils on Plasma and Erythrocyte Fatty Acids of Pregnant Women in the Second Trimester. *Prostoglandins Leukot. Essent. Fatty Acids* **2000b,** *63,* 323-328.

Ryan, A.S.; M.A. Keske; J.P. Hoffman; E.B. Nelson. Clinical Overview of Algal Docosahexaenoic Acid: Effects on Triglyceride Levels and Other Cardiovascular Risk Factors. *Am. J. Ther,* **2009,** *16,* 183-192.

Ryan, A.S.; E.B. Nelson. Assessing the Effect of Docosahexaenoic Acid on Cognitive Functions in Healthy, Preschool Children: A Randomized, Placebo-controlled, Double-blind Study. *Clin. Pediatr.* **2008,** *47,* 355–362.

Sanders, T.A.B.; K. Gleason; B. Griffin; G.J. Miller. Influence of a Triacylglycerol Containing Docosahexaenoic Acid (22:6n-3) and Docosapentaenoic Acid (22:5n-6) on Cardiovascular Risk Factors in Healthy Men and Women. *Br. J. Nutr.* **2006,** *95,* 525-531.

Scholer, H.; E.N. Mueller; M. Schipper. Mucorales. *Fungi Pathogenic for Humans and Animals,* D. Howard, Ed.; Marcel Dekker, New York, 1983; 9.

Schwellenbach, L.J.; K.L. Olson; K.J. McConnell; R.S. Stolcpart; J.D. Nash; J.A. Merenich for the Clinical Pharmacy Cardiac Risk Services Study Group. The Triglyceride-lowering Effects of a Modest Dose of Docosahexaenoic Acid Alone Versus in Combination with Low Dose Eicosap-entaenoic Acid in Patients with Coronary Artery Disease and Elevated Triglycerides. *J. Am. Coll. Nutr.* **2006,** *25,* 480-485.

Seyberth, H.W.; O. Oelz T. Kennedy; B.J. Sweetman; A. Danon; J.C. Frolich; M. Heimberg; J.A. Oates. Increased Arachidonate in Lipids after Administration to Man: Effects on Prostaglandin Synthesis. *Clin. Pharmacol. Ther.* **1975,** *18,* 521-529.

Stark, K.D.; B.J. Holub. Differential Eicosapentaenoic Acid Elevations and Altered Cardiovascular Disease Risk Factor Responses after Supplementation with Docosahexaenoic Acid in Postmeno-pausal Women Receiving and Not Receiving Hormone Replacement Therapy. *Am. J. Clin. Nutr.* **2004,** *79,* 765-773.

Steidinger, K.A.; D.G. Baden. Toxic Marine Dinoflagellates. *Dinoflagellate;* D.L. Spector, Ed.; Academic Press; Orlando, 1984; p. 201-262.

Streekstra, H. On the Safety of *Mortierella alpina* for the Production of Food Ingredients, such as Arachidonic Acid. *J. Biotechnol.* **1997,** *56,* 153-165.

Theobald, H.E.; P.J. Chowiencyk; R. Whittall; S.E. Humphries; T.A. Sanders. LDL Cholesterol-Raising Effect of Low Dose Docosahexaenoic Acid in Middle-aged Men and Women. *Am. J. Clin. Nutr.* **2004,** *79,* 558-563.

Theobald, H.E.; A.H. Goodall; N. Sattar; D.C. Talbot; P.J. Chowienczyk; T.A. Sanders. Low-dose Docosahexaenoic Acid Lowers Diastolic Blood Pressure in Middle-aged Men and Women. *J. Nutr.* **2007,** *137,* 973-978.

The Threshold Working Group of the FDA. *Approaches to Establish Thresholds for Major Food Aller-gens and for Gluten in Food,* U.S. Food and Drug Administration, June 2005. Published online: http://www.cfsan.fda.gov/~dms/alrgn.html (accessed Jan 2009).

Van den Hoek, C.; D.G. Mann; H.M. Jahns. *Algae: An Introduction to Phycology,* Cambridge University Press: Cambridge, 1995.

Vidgren, H.M.; J.J. Agren; U. Schwab; T. Rissanen; O. Hanninen; M.I.J. Uusitupa. Incorporation of n-3 Fatty Acids into Plasma Lipid Fractions, and Erythrocyte Membranes, and Platelets During Dietary Supplementation with Fish Oil, and Docosahexaenoic Acid Rich Oil Among Healthy Young Men. *Lipids* 1997, *32*, 697-705.

Wesler, A.R.; C.E.H. Dirix; M.J. Bruins; G. Hornstra. Dietary Arachidonic Acid Dose-dependently Increases the Arachidonic Acid Concentration in Human Milk. *J. Nutr.* 2008, *138*, 2190-2197.

Wheaton, D.H.; D.R. Hoffman; K.G. Locke; R.B. Watkins; D.G. Birch. Biological Safety Assessment of Docosahexaenoic Acid Supplementation in a Randomized Clinical Trial for X Linked Retinitis Pigmentosa. *Arch. Ophthalmol.* 2003, *121*, 1269-1278.

Yuhas, R.; K. Pramuk; E.L. Lien. Human Milk Fatty Acid Composition from Nine Countries Varies Most in DHA. *Lipids* 2006, *41*, 851–858.

Zeller, S. Safety Evaluation of Single Cell Oils and the Regulatory Requirements as Food Ingredients. *Single Cell Oils*, Z. Cohen; C. Ratledge, Eds.; AOCS Press: Champaign, IL, 200 ; pp.161-181.

Appendix I: Clinical Studies of Infants Supplemented with DHA and ARA SCOs

Agostoni, C.; G.V. Zuccotti; G. Radaelli; R. Besana; A. Podestà; A. Sterpa; A. Rottoli; E. Riva; M. Giovannini. Docosahexaenoic Acid Supplementation and Time at Achievement of Gross Motor Milestones in Healthy Infants: A Randomized, Prospective, Double-blind, Placebo-controlled Trial. *Am. J. Clin. Nutr.* 2009, *89*, 64–70.

Birch, E.E.; D.R. Hoffman; R. Uauy; D.G. Birch; C. Prestidge. Visual Acuity and the Essentiality of Docosahexaenoic Acid and Arachidonic Acid in the Diet of Term Infants. *Pediatr. Res.* 1998, *44*, 201–209.

Birch, E.E.; S. Garfield; D.R. Hoffman; R. Uauy; D.G. Birch. A Randomized Controlled Trial of Early Dietary Supply of Long-Chain Polyunsaturated Fatty Acids and Mental Development in Term Infants. *Dev. Med. Child Neurol.* 2000, *42*, 174–181.

Birch, E.E.; D.R. Hoffman; Y.S. Castaneda; S.L. Fawcett; D.G. Birch; R.D. Uauy. A Randomized Controlled Trial of Long-Chain Polyunsaturated Fatty Acid Supplementation of Formula in Term Infants after Weaning at 6 wk of Age. *Am. J. Clin. Nutr.* 2002, *75*, 570–580.

Birch, E.E.; Y.S. Castaneda; D.H. Wheaton; D.G. Birch; R.D. Uauy; D.R. Hoffman. Visual Maturation of Term Infants Fed Long-Chain Polyunsaturated Fatty Acid-Supplemented or Control Formula for 12 mo. *Am. J. Clin. Nutr.* 2005, *81*, 871–879.

Burks, W.; S. M. Jones; C.L. Berseth; C. Harris; H.A. Sampson; D.M. Scalabrin. Hypoallergenicity and Effects on Growth and Tolerance of a New Amino Acid-based Formula with Docosahexaenoic Acid and Arachidonic Acid. *J. Pediatr.* 2008, *153*, 266–271.

Clandinin, M.T.; J.E. Van Aerde; A. Parrott; C.J. Field; A.R. Euler; E.L. Lien. Assessment of the Efficacious Dose of Arachidonic and Docosahexaenoic Acids in Preterm Infant Formulas: Fatty Acid Composition of Erythrocyte Membrane Lipids. *Pediatr. Res.* 1997, *42*, 819–825.

Clandinin, M.T.; J.E. Van Aerde; A. Parrott; C.J. Field; A.R. Euler; E. Lien. Assessment of Feeding Different Amounts of Arachidonic and Docosahexaenoic Acids in Preterm Infant Formulas on the Fatty Acid Content of Lipoprotein Lipids. *Acta Paediatr.* **1999**, *88,* 890–896.

Clandinin, M.T.; J.E. Van Aerde; K.L. Merkel; C.L. Harris; M.A. Springer; J.W. Hansen; D.A. Diersen-Schade. Growth and Development of Preterm Infants Fed Infant Formulas Containing Docosahexaenoic Acid and Arachidonic Acid. *J. Pediatr.* **2005**, *146,* 461–468.

Field, C.J.; C.A. Thomson; J.E. Van Aerde; A. Parrott; A. Euler; E. Lien; M.T. Clandinin. Lower Proportion of CD45R0+ Cells and Deficient Interleukin-10 Production by Formula-Fed Infants, Compared with Human-Fed, Is Corrected with Supplementation of Long-Chain Poly-unsaturated Fatty Acids. *J. Pediatr. Gastroenterol. Nutr.* **2009**, *31,* 291–299.

Florendo, K.N.; B. Bellflower; A. van Zwol; R.J. Cooke. Growth in Preterm Infants Fed Either a Partially Hydrolyzed Whey or an Intact Casein/Whey Preterm Infant Formula. *J. Perinatol.* **2009**, *29,* 106–111.

Foreman-van Drongelen, M.M.; A.C. van Houwelingen; A.D. Kester; C.E. Blanco; T.H. Hasaart; G. Hornstra. Influence of Feeding Artificial-Formula Milks Containing Docosahexaenoic and Arachidonic Acids on the Postnatal Long-Chain Polyunsaturated Fatty Acid Status of Healthy Preterm Infants. *Br. J. Nutr.* **1997**, *76,* 649–667.

Gibson, R.A.; D. Barclay; H. Marshall; J. Moulin; J.C. Maire; M. Makrides. Safety of Supple-menting Infant Formula with Long-Chain Polyunsaturated Fatty Acids and Bifidobacterium lactis in Term Infants: A Randomised Controlled Trial. *Br J Nutr.* **2009**, 1-8.

Henriksen, C.; K. Haugholt; M. Lindgren; A.K. Aurvag; A. Ronnestad; M. Gronn; R. Solberg; A. Moen; B. Nakstad; R.K. Berge; et al. Improved Cognitive Development among Preterm Infants Attributable to Early Supplementation of Human Milk with Docosahexaenoic Acid and Arachi-donic Acid. *Pediatrics* **2008**, *121,* 1137–1145.

Hoffman, D.R.; E.E. Birch; D.G. Birch; R. Uauy; Y.S. Castaneda; M.G. Lapus; D.H. Wheaton. Impact of Early Dietary Intake and Blood Lipid Composition of Long-Chain Polyunsaturated Fatty Acids on Later Visual Development. *J. Pediatr. Gastroenterol. Nutr.* **2000**, *31,* 540–553.

Hoffman, D.R.; E. Birch; Y.S. Castañeda; S.L. Fawcett; D.G. Birch, D.G; R. Uauy. Dietary Docosahexaenoic Acid (DHA) and Visual Maturation in the Post-weaning Term Infant. *Invest Ophthalmol Vis Sci.* **2001**, *42,* S122A.

Hoffman, D.; E. Ziegler; S.H. Mitmesser; C.L. Harris; D.A. Diersen-Schade. Soy-based Infant Formula Supplemented with DHA and ARA Supports Growth and Increases Circulating Levels of These Fatty Acids in Infants. *Lipids* **2008**, *43,* 29–35.

Innis, S.M.; D.H. Adamkin; R.T. Hall; S.C. Kalhan; C. Lair; M. Lim; D.C. Stevens; P.F. Twist; D.A. Diersen-Schade; C.L. Harris; et al. Docosahexaenoic Acid and Arachidonic Acid En-hance Growth with No Adverse Effects in Preterm Infants Fed Formula. *J. Pediatr.* **2002**, *140,* 547–554.

Makrides, M.; R.A. Gibson; A.J. McPhee; C.T. Collins; P.G. Davis; L.W. Doyle; K. Simmer; P.B. Colditz; S. Morris; L.G. Smithers; et al. Neurodevelopmental Outcomes of Preterm Infants Fed High-Dose Docosahexaenoic Acid: A Randomized Controlled Trial. *JAMA* **2009**, *301,* 175–182.

Morris, G.; J. Moorcraft; A. Mountjoy; J.C. Wells. A Novel Infant Formula Milk with Added Long-Chain Polyunsaturated Fatty Acids from Single-Cell Sources: A Study of Growth, Satisfac-tion and Health. *Eur. J. Clin. Nutr.* **2000**, *54,* 883–886.

Siahanidou, T.; A. Margeli; C. Lazaropoulou; E. Karavitakis; I. Papassotiriou; H. Mandyla. Circulating Adiponectin in Preterm Infants Fed Long-Chain Polyunsaturated Fatty Acids (LCPUFA)-supplemented Formula—A Randomized Controlled Study. *Pediatr Res.* **2008**, *63,* 428–432.

Smithers, L.G.; R.A. Gibson; A. McPhee; M. Makrides. Effect of Two Doses of Docosahexaenoic Acid (DHA) in the Diet of Preterm Infants on Infant Fatty Acid Status: Results from the DINO trial. *Prostaglandins Leukot. Essent. Fatty Acids* **2008**, *79,* 141–146. Epub Oct 23, 2008.

Smithers, L.G.; R.A. Gibson; A. McPhee; M. Makrides. Higher Dose of Docosahexaenoic Acid in the Neonatal Period Improves Visual Acuity of Preterm Infants: Results of a Randomized Controlled Trial. *Am. J. Clin. Nutr.* **2008**, *88,* 1049–1056.

Van Wezel-Meijler, G.; M.S. van der Knaap; J. Huisman; E.J. Jonkman; J. Valk; H.N. Lafeber. Dietary Supplementation of Long-Chain Polyunsaturated Fatty Acids in Preterm Infants: Effects on Cerebral Maturation. *Acta Paediatr.* **2002**, *91,* 942–950.

Vanderhoof, J.A. Hypoallergenicity and Effects on Growth and Tolerance of a New Amino Acid-based Formula with DHA and ARA. *J. Pediatr. Gastroenterol. Nutr.* **2008**, *47 Suppl 2,* S60–261.

Vanderhoof, J.; S. Gross; T. Hegyi. A Multicenter Long-Term Safety and Efficacy Trial of Preterm Formula Supplemented with Long-Chain Polyunsaturated Fatty Acids. *J. Pediatr. Gastroenterol. Nutr.* **2000**, *31,* 121–127.

Vanderhoof, J.; S. Gross; T. Hegyi; T. Clandinin; P. Porcelli; J. DeCristofaro; T. Rhodes; R. Tsang; K. Shattuck; R. Cowett; et al. Evaluation of a Long-Chain Polyunsaturated Fatty Acid Supplemented Formula on Growth, Tolerance, and Plasma Lipids in Preterm Infants up to 48 Weeks Postconceptional Age. *J. Pediatr. Gastroenterol. Nutr.* **1999**, *29,* 318–326.

16

Nutritional Aspects of Single Cell Oils: Applications of Arachidonic Acid and Docosahexaenoic Acid Oils

Andrew J. Sinclair[a] and Anura Jayasooriya[b]

[a]*School of Exercise & Nutrition Sciences, Deakin University, Burwood, Victoria;* [b]*Department of Food Science, RMIT University, Melbourne, Victoria, Australia;*

Introduction

One of the driving forces for the development of single cell oils (SCOs) containing long-chain polyunsaturated fatty acids (LCPUFA) was the presence in human milk of two particular LCPUFA: docosahexaenoic acid (DHA) and arachidonic acid (AA). Until recently, these polyunsaturated fatty acids (PUFA) had not been added to infant formulas. Once it was recognized that these two PUFA played an important role in brain function, attempts were made to provide these PUFA from natural sources, such as fish oils and egg phospholipids. It was relatively easy to obtain DHA from certain oils, such as tuna oil (Hawkes et al., 2002); however, providing AA was more difficult. When it was found that AA-containing oils were produced by certain species of soil fungi (Wynn & Ratledge, 2000), research soon established that it was possible to harvest this oil in commercial quantities. Similarly, a DHA-containing oil from a marine microalgae was used to produce commercial quantities of DHA (De Swaaf et al., 2003).

The brain has the second highest concentration of lipids in the body, after adipose tissue, with 36-60% of the nervous tissue being lipids (Svennerholm, 1968). The lipids in the brain are complex and include glycerophospholipids (GPL), sphingolipids (sphingomyelin and cerebrosides), gangliosides, cholesterol with little or no triglycerides, and cholesterol esters (Sastry, 1985). Brain GPL contain a high proportion of LCPUFA— mainly DHA, AA, and docosatetraenoic acid (C22:4n-6), with very small amounts of α-linolenic acid (ALA) and linoleic acid (LA). The proportion of DHA and AA in the GPL of brain grey matter is higher than that found in white matter (Svennerholm, 1968; Pullarkat & Reha, 1978), with phosphatidylethanolamine (PE) and phosphatidylserine (PS) containing the most

DHA of all GPL, while PE and PI contain the highest proportions of AA. The DHA plus AA content of the adult cerebral cortex is approximately 6% dry weight and 2% of the white matter (Svennerholm, 1968). The n-6 content (20:4n-6 plus 22:4n-6) of the cerebral cortex is similar to that of the DHA level, and, in white matter, there is a higher proportion of n-6 than of n-3 PUFA (Svennerholm, 1968). The highest proportion of DHA in membrane lipids is found in the disk membranes of the rod outer segments of photoreceptor cells in the retina (Fliesler & Anderson, 1983; Boesze-Battaglia & Albert, 1989). Carrie et al. (2000) showed that the proportion of DHA in eleven different regions of the rat brain varied from 7% GPL FA in the pituitary gland to 22% in the frontal cortex. The variation in the proportion of AA ranged from 5% in the pons medulla to 18% in the pituitary gland. This gland was the only region where the proportion of AA exceeded that of DHA.

DHA and AA are present in other tissues in the body, but in lower proportions. For example, in the guinea pig, the proportion of DHA in all tissues, except neural tissue, was < 0.5% total tissue FA, while it was 6-7% total fatty acids (TFA) in the whole brain (Fu et al., 2001). On a whole body basis, the brain contained approximately 22-25% of the total DHA found in the body, with approximately 50% of the DHA found in the carcass (muscle and adipose tissue). The same study showed that the AA was mostly distributed in the carcass (70%), with only 2% in the brain.

High levels of DHA and AA in the brain grey matter of over thirty different mammalian species (Sinclair, 1975) led to early speculations that these PUFA play a crucial role in the nervous system. In the 1970s, n-6 PUFA were regarded as essential for humans, while n-3 PUFA were only thought to be essential for fish and other marine species. The first clue to a physiological role of n-3 FA in mammals came when it was reported that dietary n-3 PUFA fed to rats led to nearly double the retinal response to visual stimulation of rats, compared with the responses of rats fed n-6 PUFA (Wheeler et al., 1975). Since then, intensive study of the role of DHA in the brain has revealed that DHA plays a vital role in many different parts of the brain, including (a) membrane-related events (membrane order can influence the function of membrane receptors, such as rhodopsin) (Litman et al., 2001; Feller et al., 2002); regulation of dopaminergic and serotoninergic neurotransmission (Zimmer et al., 2000); regulation of membrane-bound enzymes (Na/K-dependent ATPase) (Bowen & Clandinin, 2002); signal transduction via effects on inositol phosphates, diacylglycerol, and protein kinase C (Vaidyanathan et al., 1994); alteration of ion flux through voltage-gated K^+ and Na^+ channels (Seebungkert & Lynch, 2002; Leaf et al., 2002); (b) metabolic events (regulation of the synthesis of eicosanoids derived from AA) (Kurlack & Stephenson, 1999) as a precursor of docosatrienes and 17S resolvins (novel anti-inflammatory mediators) derived from DHA (Hong et al., 2003); (c) gene expression (regulation of the expression of many different genes in short- and long-term studies of the rat brain) (Urquiza et al., 2000; Puskas et al., 2003); and (d) cellular events, such as the regulation of phosphatidyl serine levels (Garcia et al., 1998),which appear to be involved in the protection of neural cells from apoptotic death (Akbar & Kim, 2002); stimulation of neurite outgrowth in PC-12 brain or neuron cells (Ikemoto et al., 1997; Martin, 1998); selective accumulation of DHA by synaptic growth cones during neuronal development (Ikemoto et al., 1997; Martin, 1998); regulation of neuron size (Ahmad et al., 2002); regulation of nerve growth factor (Ikemoto et al., 2000). In addition, DHA is a precursor of neuroprostanes which are DHA oxidation products (Roberts et al., 1998; Fam et al., 2002).

AA is the predominant n-6 PUFA in mammalian brain and neural tissue and, like DHA, is found in sn-2 position of the glycerol backbone of membrane GPL. AA, therefore, plays a key role in membrane function. The release of AA from membrane GPL is due to the receptor-mediated activation of phospholipase A_2, or the phospholipase C and diacylglycerol lipase (Jones et al., 1996). Once released, AA can become a substrate for oxidative enzymes, such as cyclooxygenases (COX-1, COX-2), lipoxygenase, or cytochrome P450 monoxygenases, which convert it to a number of bioactive eicosanoids like prostaglandins, prostacyclins, thromboxanes, and leukotrienes (Tapiero et al., 2002; Herschman, 1996). The functional role of AA appears to be mediated either by the FA itself or through the bioactive metabolites produced by oxidative reactions. For example, AA exerts diverse actions on acetylcholine receptors, such as a short-term depression by blocking the receptor and long-lasting potentiation by activating the protein kinase C pathway (Nishizaki et al., 1998). Furthermore, synaptic activation of glutamate receptors has been reported to release AA, which suggests AA has a role in synaptic transmission (Dumuis et al., 1990).

Abnormalities in AA metabolism have been linked to a number of brain disorders, such as bipolar disorder (Rapoport & Bosetti, 2002), Alzheimer's disease (Breitner et al., 1995), schizophrenia (Peet et al., 2002), and ischemia (Nogawa et al., 1997). COX-2, an enzyme that converts AA to eicosanoids, is highly expressed in different regions of the brain, such as the hippocampus, cortex, and amygdala (Yamagata et al., 1993). Age-dependent cognitive deficits and neuronal apoptosis have been reported in transgenic mice overexpressing COX-2, with a concurrent increase in prostaglandin levels in the brain (Andreasson et al., 2001). This finding suggests that neuronal COX-2 may contribute to the pathophysiology of age-related diseases. The reduced skin flushing response to niacin in schizophrenic subjects has been known for years; since the primary mechanism of action is conversion of AA to prostaglandin D2, this reduced flushing also suggests an abnormal AA metabolism in these subjects (Skosnik & Yao, 2003). Treatment of schizophrenic patients with a 2 g dose of ethyl-eicosapentaenoic acid (EPA, 20:5n-3) leads to a significant improvement of the condition, with an elevated level of AA in erythrocyte FA (Peet et al., 2002); it is speculated this improvement may result from the inhibition of phospholipase A2 by EPA (Peet et al., 2002).

There is a rapid increase in the weight of the human brain post-natally, until the infant reaches about two years of age. Associated with this increase, there is a rapid accretion of DHA and AA in the infant brain during the first postnatal year

(Martinez, 1992). It is thought that the DHA and AA for brain growth largely derive from mothers' milk. Breast-feeding provides at least 49 mg of DHA and 93 mg of AA to the infant each day, depending on the PUFA level in the mother's milk (Mitoulas et al., 2003). It is known that milk LCPUFA levels can be influenced by diet; usually LCPUFA are in the range from 0.2 to 1.0% of total milk FA (Makrides et al., 1996).

The fetal brain is believed to be able to produce a limited amount of DHA from ALA, and the liver may also be able to produce some DHA (Salem et al., 1996); however, this amount is believed to be insufficient for optimal development (Horrocks & Yeo, 1999). It has been argued that, based on the rate of accretion of DHA into the human brain, there is a need to supply DHA via breast milk or infant formula for at least the first six months of life (Cunnane et al., 2000). There has been little discussion about the capacity of newborn infants to synthesize sufficient AA for brain growth. This limited capacity to synthesize LCPUFA has been a driving force in developing infant formulas containing these FA.

Another rationale as to why it is necessary to add LCPUFA to infant formulas is based on the decline in blood LCPUFA levels after birth. Blood levels of DHA and AA in infants fed standard formulas are lower than those of breast-fed infants (Sanders & Naismith, 1979; Innis et al., 2002). Formulas containing LCPUFA can increase blood LCPUFA levels so they more closely resemble those found in breast-fed infants (Makrides et al., 1995; Auestad et al., 2001). Two studies have examined brain PUFA levels in formula-fed and breast-fed infants and both found that there was a significantly lower level of DHA, but not AA, in the brain tissue of children who had mainly been fed on infant formulas (those lacking DHA and AA) (Gibson et al., 1994; Farquharson et al., 1995).

These studies underpin a discussion that has been conducted throughout the world on the importance of adding DHA and AA to infant formulas. Before such action was undertaken, many studies were conducted in animals and primates.

SCO Studies in Animals: PUFA Levels in Tissues and Functional Studies

Many studies conducted on SCO in animals have been concerned with the efficiency of these oils in supplying tissues with PUFA, especially in relation to brain PUFA and brain function. This section will discuss several examples from relevant literature, whichillustrate that LCPUFA from SCO sources are bioavailable and can influence physiological function. Other studies have been concerned with safety of these oils; we deal with this topic in the following section.

Ward et al. (1998) studied the effects of adding LCPUFA from SCO on the brain and red blood cell FA composition. The rat pups were reared artificially from day five to day eighteen, post-natally, using a gastromy tube. The study compared three levels of DHA and three levels of AA in a factorial design. The basal diet contained LA and ALA, but no LCPUFA. The results showed that supplementing the formula with AA or DHA during the period of brain development increased deposition of these PUFA in the brain and red blood cells. Furthermore, it was found that increasing levels of each PUFA affected the levels of other PUFA (the highest dietary AA decreased tissue DHA levels and vice versa).

Abedin et al. (1999) compared the efficiency of ALA versus DHA in contributing DHA to various tissues (liver, heart, retina, and brain) in guinea pigs. In this study, LCPUFA from SCO were fed to guinea pigs fed from 3 weeks until 15 weeks of age. The LA content in the diets was constant (17% TFA) with ALA content varying from 0.05% (diet S) to 1% (diet A) and 7% (diet C). Diet A had an LA:ALA ratio of 17.5:1 and was structured to closely replicate the principal LCPUFA found in human breast-milk (0.9% AA and 0.6% DHA). In retina and brain phospholipids, the high ALA diet (diet C) or dietary DHA supplementation produced moderate increases in DHA levels when compared with diet S (low ALA diet). There was no change in retinal or brain AA following dietary AA supplementation. This stasis contrasted with the levels of AA in the liver and heart, where dietary DHA and AA supplements led to large increases (up to ten-fold) in tissue levels of these PUFA. The data confirmed that dietary ALA was less effective than dietary DHA supplementation (on a g/g basis) in increasing tissue DHA levels and that tissues vary greatly in their response to exogenous AA and DHA; the levels of these long-chain metabolites are most resistant to change in the retina and brain, compared with the liver and heart.

A recent study in rhesus monkey neonates examined the effect on neuromotor development of including DHA and AA in the rearing infant formula (Champoux et al., 2002). In this study, twenty-eight nursery-reared rhesus macaque infants were divided into two groups, one of which was fed a formula with DHA (1% of the fat) and AA (1% of the fat) from SCO sources; the other group was fed the standard formula, devoid of LCPUFA. Neurobehavioral tests were conducted weekly from the seventh day of life until day 30. Plasma DHA and AA concentrations in the supplemented group were significantly higher than the control group at 4 weeks of age. Monkeys fed the supplemented formula showed stronger orientation and motor skills than those fed the standard formula; the most pronounced differences were at day 7 and 14. Supplementing the formula with LCPUFA had no influence on temperament. These data support the inclusion of LCPUFA in infant formulas for optimal development.

Some studies have compared the effects of diets with DHA alone versus those with both DHA and AA. Auestad et al. (2003) conducted one such study, in which they measured the auditory brainstem-evoked response (ABR). In previous studies, juvenile offspring of rats fed high-DHA diets through gestation and lactation had a longer ABR that was associated with higher DHA and lower AA proportion in the brain (Stockard et al., 2000). In the Auestad et al. (2003) study, the ABR was assessed in juvenile rats fed high-DHA diets postnatally and compared with diets containing both DHA and AA. They found that the DHA and AA levels in the brain increased with supplementation. In contrast to the earlier study by Champoux et al. (2002),

Auestad et al. found that ABR was shorter in the high-DHA group than in the DHA plus AA group and did not differ from the unsupplemented or dam-reared suckling group. Clearly, further studies are needed to understand the relationship between dietary DHA and the development of the auditory system over a range of DHA intakes and discrete periods of development.

Blanaru et al. (2004) examined the effect of an increasing dose of AA (0.3 to 0.75% fat) with a constant level of DHA (0.1% fat) from SCO sources on piglet bone massin a study starting at day 5 and finishing at day 20 after birth. They decided to perform this study due to a surge of interest in the effect of PUFA on bone biology (Watkins and Seifert, 2001). Bone modeling was unaffected by the different treatments; however, the whole body bone mineral content was elevated in piglets fed 0.6 and 0.75% AA. The effect of altering the dietary intake of DHA in this model is not known.

Some studies have compared different sources of LCPUFA for various outcomes in animals. For example, Mathews et al. (2002) compared SCO sources of DHA and AA with egg phospholipids, as a source of these PUFA, on overall animal health and safety. In this study, piglets consumed a skim milk formula from day 1 until day 16 after birth. The formulas with LCPUFA provided 0.3 and 0.6 g/100 g TFA as DHA and AA, respectively. The control group received no DHA or AA. There was no difference in gross liver histology between the groups; the apparent dry matter digestibility was 10% greater in the SCO and control groups than in the group fed egg phospholipid PUFA. The plasma DHA proportion was higher in the SCO group than in the egg phospholipids group, while the plasma AA proportion was higher in the SCO group than in the control group. In summary, these studies revealed that LCPUFA from SCO sources were bioavailable and that they had the capacity to alter physiological function in small animals and primates.

Safety Aspects of SCO

The main concern about SCO derived from algae and fungi has been that they are new food ingredients without a history of safe use in infant feeding anywhere in the world. This concern has meant that these oils have had to undergo extensive toxicological testing in various animal species. The results have been favorable (Boswell et al., 1996; Merritt et al., 2003), and authorities in several countries have approved the use of these SCO in infant formulas. In the United States, the Food and Drug Administration (FDA) has given safe status to SCO, thus permitting their use in infant formulas (US FDA, 2001) (see also, Chapters 15 and 17).

SCO Studies in Infants

Following successful trials on bioavailability and the safety of SCO in animals and primates, there have been a number of trials in term and pre-term infants that have included LCPUFA from various sources in infant formulas. The authors have identified twelve randomized clinical trials (RCT) designed to test the efficacy and safety of adding either n-3 LCPUFA (DHA and EPA) or a combination of DHA and AA to formulas for term infants, which have been published in full. Four of these studies used SCO (Auestad et al., 2001; Birch et al., 1998; Gibson et al., 1998; Morris et al., 2000), while the remainder used other lipid sources (Agostini et al., 1995; Auestad et al., 1997; Carlson et al., 1996a; Horby et al., 1998; Lucas et al., 1999; Makrides et al., 1999, 2000; Willatts et al., 1998). It is doubtful whether the source of LCPUFA has much effect on the LCPUFA status of the infant (Sala-Vila et al., 2004).

Trial Design and Treatments

The four trials involved healthy-term infants fed formulas from near birth, and all but one had a breast-fed reference group. Most trials appeared to have adequate randomization and masking procedures, and most presented power calculations for their primary outcome measurements. Therefore, the trials involving term infants are generally of good methodological quality.

The levels of n-3 LCPUFA used in the trials ranged from 0.1 to 1% total fat, while the AA ranged from 0.4 to 0.7% TFA. Four trials assessed the effect of supplementation with n-3 LCPUFA, including no AA (Makrides et al., 1995; Auestad et al., 1997; Makrides et al., 1999; Horby et al., 1998).

Outcomes

The benefits to visual acuity of adding LCPUFA to formulas, assessed by both visual-evoked potential or Teller cards, have been reported in some studies (Birch et al., 1998), while other studies have shown no difference between LCPUFA-supplemented and LCPUFA-unsupplemented infants (Auestad et al., 1997). A systematic review of three trials on visual acuity in term infants indicated an improvement in card acuity with LCPUFA treatment only at 2 months of age (Sangiovanni et al., 2000a). There is also mixed evidence for supporting the effect of dietary LCPUFA on more global measures of development (Bayley Scales of Infant Development). Birch et al. (2000) have reported thesome benefits of dietary LCPUFA; however, the larger studies conducted by Scott et al. (1998) and Auestad et al. (2001) showed no effect of LCPUFA supplementation on the Bayley Scales of Infant Development. Possible interpretations of these data include a small individual effect—that only a proportion of infants will benefit—or the presence of confounding variables. Further studies are needed to elucidate this issue.

There have been no negative findings regarding LCPUFA supplementation of infant formulas in relation to growth in term infants, despite the fact that four trials have supplemented formulas with DHA alone, without added AA, for periods of up to 1 year, resulting in the AA status of infants being depleted. Therefore, there is no evidence that n-3 LCPUFA supplementation of term infant formulas results in perturbations of growth.

Trials Involving LCPUFA in Preterm Infants

The last trimester of pregnancy is the time when DHA accretion in the brain and nervous system is at its greatest velocity. Therefore, many preterm infants, especially those born before 30 weeks, are born with negligible body stores of DHA; they are subsequently fed milk that contains no DHA or levels that are much lower than what these infants would have received if they were still in utero. Preterm infants are more at risk than term infants of disturbed DHA accumulation and, thus, may have the most to gain from DHA supplementation.

The authors are aware of thirteen RCT (random clinical trials) reported in at least twenty separate papers. Four of these studies used SCO (Clandinin et al., 1997; Innis et al., 2002; O'Connor et al., 2001; Vanderhoof et al., 2000), while the remainder used lipids from other sources (Birch et al., 1992; Bougle et al., 1999; Carlson et al., 1993, 1996b, 1996c; Faldella et al., 1996; Fewtrell et al., 2002, 2004; Lapillonne et al., 2000; Ryan et al., 1999; Werkman & Carlson, 1996; Uauy et al., 1994). These trials were designed to test the efficacy and safety of varying levels of DHA, EPA, and AA in the diets of preterm infants.

Trial Design and Treatments

All the trials reported adequate concealment of allocation, and, in general, their methodological quality is more robust than that found in earlier trials.

Outcomes

The original trials that showed a benefit on electroretinographic responses and visual acuity all supplemented with fish oil. Of the SCO trials, only that of O'Connor et al. (2001) assessed visual function and showed a benefit in improving visual evoked potential (VEP) acuity but not Teller card acuity. Of the three trials that assessed global development, one reported an advantage of LCPUFA supplementation on psychomotor development in infants born at less than 1250 g (O'Connor et al., 2001). The data from O'Connor et al. (2001) suggest that the more immature and sick infants may have the most to gain from LCPUFA supplementation; this finding highlights the fact that some subgroups may be more sensitive to the effects of LCPUFA. Smithers et al. (2008) assessed visual responses of preterm infants fed human milk (HM) and formula with a DHA concentration estimated to match the intrauterine accretion rate (high-DHA group), compared with infants fed HM and formula containing DHA at current concentrations. A double-blind, randomized, controlled trial studied preterm infants born at < 33 wks gestation and fed HM, formula containing 1% DHA (high-DHA group), or formula containing approximately 0.3% DHA (current practice; control group) until these infants reached their estimated due date (EDD). Both groups received the same concentration of arachidonic acid. It was reported that by 4 months (corrected age), the high-DHA group exhibited an acuity that was 1.4 cpd higher than the control group (high-DHA: 9.6 +/- 3.7 cpd, n = 44; control: 8.2

+/- 1.8 cpd; n = 51; P = 0.025). Therefore, the DHA requirement of preterm infants may be higher than that currently provided by preterm formula or HM of Australian women. Further research is required to best maximize the potential benefits of DHA on early childhood development.

Although most trials show that there is no effect of LCPUFA supplementation on growth, a recent trial has suggested an enhancement of growth (Innis et al., 2002). Two separate systematic reviews and meta-analyses, combining growth data from all published RCT, showed no difference in any growth parameter between supplemented and unsupplemented infants (Simmer et al., 2008). Although the outcome data from the individual LCPUFA intervention trials with visual outcomes consistently indicated the beneficial effects of LCPUFA supplementation, the two available systematic reviews/meta-analyses of these data have not been completed (Sangiovanni et al., 2000b; Simmer et al., 2008). One review could not combine the visual outcome data because of differing assessments, methodologies, and assessment times (Simmer et al., 2008). The other review included data from three randomized trials and one non-randomized study and concluded that there was a beneficial effect of LCPUFA treatment on visual acuity at 2 and 4 months, corrected age (Sangiovanni et al., 2000b). Thus, despite the promising beneficial effect of DHA supplementation on the neural outcomes of preterm infants, trials with standard methodologies and follow-up of infants beyond 12 months, corrected age, are necessary to more precisely assess the beneficial extent of LCPUFA supplementation.

SCO Studies in Adults

There have been relatively few studies of SCO in adults, apart from those of Nelson and his colleagues who conducted two separate studies that involved feeding a small group of volunteers either DHA or AA derived from SCO. These studies led to a number of papers by the group, published during the period 1997–1999 (Emken et al., 1998; Kelly et al., 1998a, 1998b, 1999; Nelson et al., 1997a-d).

The aim of the DHA study was to examine the effects of feeding DHA-rich triacylglycerol (TAG) on FA composition, eicosanoid production, select activities of human peripheral blood mononuclear cells, plasma lipoprotein concentration, and FA composition of plasma lipids and adipose tissue. The 120 day study with eleven healthy men was conducted at the metabolic research unit of Western Human Nutrition Research Center. Four subjects (control group) were fed the stabilization diet or basal diet (15, 30, and 55% energy from protein, fat, and carbohydrates, respectively) throughout the study; the remaining seven subjects were fed the basal diet for the first 30 days, followed by 6 g DHA/d for the next 90 days. DHA replaced an equivalent amount of LA; the two diets were comparable in their total fat and all other nutrients. The ratio of saturated plus *trans* FA to monounsaturated FA to PUFA in the diets was 10:10:10. Both diets were supplemented with 20 mg D α-tocopherol acetate/d.

The white blood cell FA composition, eicosanoid production, immune cell functions, plasma lipoprotein concentrations, and the plasma and adipose tissue FA composition were examined on days 30 and 120. There was an increase in white blood cell DHA from 2.3 to 7.4% and a decrease in the proportion of AA from 19.8 to 10.7%. The researchers also saw that the production of prostaglandin E_2 (PGE_2) and leukotriene B_4 (LTB_4) fell by 60-75% in response to lipopolysaccharide. Natural killer cell activity, in vitro secretion of interleukin-1β, and tumor necrosis factor α were significantly reduced by DHA feeding (Kelly et al., 1998a, 1999). The concentrations of plasma cholesterol, low-density lipoprotein, and apolipoproteins [A_1, B, and lipoprotein (a)] were unchanged after 90 days, but TAG levels were significantly reduced, while high-density lipoprotein-C and apolipoprotein E levels increased significantly. The proportion of plasma DHA rose from 1.8 to 8.1% after 90 days on the high-DHA diet. Interestingly, the plasma EPA levels rose from 0.4 to 3.4% in subjects on the high-DHA diet, despite the diet being devoid of EPA. The DHA proportion in adipose tissue rose significantly from 0.1 to 0.3%, but the amount of EPA did not change (Nelson et al., 1997a). The effects of the high DHA diet on platelet aggregation were also studied in this experiment; however, no significant effects were found in blood coagulation parameters, platelet function, or thrombotic tendency (Nelson et al., 1997c).

The authors also studied the effect of the DHA-rich diet on the conversion of deuterium-labelled 18:2n-6 and 18:3n-3 to long-chain PUFA. The labelled compounds were administered as TAG at the end of the DHA-feeding period and blood samples were taken over the following 72 h. DHA supplementation significantly reduced the concentrations of most deuterium-labelled n-6 and n-3 LCPUFA metabolites in plasma lipids. For example, the accumulation of deuterium-labelled 20:5n-3 and 22:6n-3 was depressed by 76 and 88%, respectively. The accumulation of deuterium-labelled 20:3n-6 and 20:4n-6 also decreased by 72% for both PUFA (Emken et al., 1999). The authors calculated that the accumulation of n-3 LCPUFA metabolites synthesized from 18:3n-3 would be reduced from about 120 mg/d to 30 mg/d by supplementation with 6.5 g DHA/d. It was calculated that the accumulation of n-6 LCPUFA metabolites, synthesized from 18:2n-6, would be reduced from about 800 mg/d to 180 mg/d. The authors suggested that the health benefits associated with this level of DHA supplementation would result from reduced accretion of n-6 long-chain PUFA and increased levels of n-3 LCPUFA in tissue lipids.

A study on AA was conducted by the same group, in which ten healthy men lived at a metabolic research unit for 130 d (Nelson et al 1997d). All subjects were fed a basal diet containing 27 energy percentage (en%) fat, 57 en% carbohydrate, 16 en% protein, and 200 mg AA/d for the first and last 15 d of the study. Additional AA (1.5 g/d) was incorporated into the diet of six men from day 16 to day 65, while the remaining four subjects continued to eat the basal diet. The ratio of saturated plus *trans* FA to monounsaturated FA to PUFA in the diets was 7:10:7. The diets of the two groups were crossed-over from day 66 to day 115. Dietary AA had no significant effect on the blood cholesterol levels, lipoprotein distribution, or apoprotein levels.

The plasma TFA composition was markedly enriched in AA after 50 d ($P <$ 0.005). The AA proportion in plasma PL increased from 10.3 on the basal diet to 19.0% after the AA-enriched diet. There was a significant rise in the proportion of AA in the red blood cells that mainly replaced LA. The adipose tissue FA composition was not influenced by the AA-enriched diet (Nelson et al., 1997d). Platelet aggregation in the platelet-rich plasma was determined using ADP, collagen, and AA. There were no significant differences in platelet aggregation before and after subjects consumed the AA-enriched diet (Nelson et al., 1997b). There were no significant changes in any of the indices of blood clotting (prothrombin time, partial thromboplastin time, antithrombin III levels, and in vivo bleeding times) between diet groups. Surprisingly, there was only a small change in platelet AA proportion during the AA feeding period. The in vitro secretion of LTB_4 and PGE_2, from in vitro stimulated white blood cells, significantly increased after the AA-enriched diet, but this diet did not alter the secretion of tumor necrosis factor α; interleukins-1 β, -2, -6; and the receptor for interleukin-2 (Kelly et al., 1998a). At the end of each diet phase, each subject was dosed with about 3.5 g of deuterium-labelled 18:2n-6 as TAG. The concentrations of deuterium-labeled 20:3n-6 and 20:4n-6 were both 48% lower ($P < 0.05$) in plasma lipids from the AA-enriched group, compared with the low AA diet group (Emken et al., 1998).

Conclusion

The commercial development of SCO and their FDA approval has allowed the addition of LCPUFA from SCO sources to infant formulas. This advance has enabled the composition of infant formulas to approach that of human milk, a goal whose achievement is sought by both formula companies and parents. Future research on SCO should examine the nutritional benefits of SCO in adults.

Acknowledgments

Anura Jayasooriya was a postgraduate student, funded through RMIT University, at the time this manuscript was prepared and is currently working at the University of Peradeniya, Sri Lanka. The assistance of Ms Gunveen Kaur, Deakin University, and Anupama Pasam, Roshan Wimalasinghe, and T. Chandrasiri in the manuscript preparation is gratefully acknowledged.

References

Abedin, L.; E.L. Lien; A.J. Vingrys; A.J. Sinclair. The Effects of Dietary Alpha-Linolenic Acid Compared with Docosahexaenoic Acid on Brain, Retina, Liver, and Heart in the Guinea Pig. *Lipids* **1999,** *34,* 475–482.

Agostoni, C.; S. Trojan; R. Bellu; E. Riva; M. Giovannini. Neurodevelopmental Quotient of Healthy Term Infants at 4 Months and Feeding Practice: The Role of Long-Chain Polyunsaturated Fatty Acids. *Pediatr. Res.* **1995,** *38,* 262–266.

Ahmad, A.; T. Moriguchi; N. Salem. Decrease in Neuron Size in Docosahexaenoic Acid-Deficient Brain. *Pediatr. Neurol.* **2002,** *26,* 210–218.

Akbar, M.; H.Y .Kim. Protective Effects of Docosahexaenoic Acid in Staurosporine-Induced Apoptosis: Involvement of Phosphatidyl-3-kinase Pathway. *J. Neurochem.* **2002,** *82,* 655–665.

Andreasson, K.I.; A. Savonenko; S. Vidensky; J.J. Goellner; Y.A. Zhang; W.E. Shaffer Kaufmann; P.F. Worley; P. Isakson; A.L. Markowska. Age-Dependent Cognitive Deficits and Neuronal Apoptosis in Cyclooxygenase-2 Transgenic Mice. *J. Neuroscii* **2001,** *21,* 8198–8209.

Auestad, N.; J. Stockard-Sullivan; S.M. Innis; R. Korsak; J. Edmond. Auditory Brainstem Evoked Response in Juvenile Rats Fed Rat Milk Formulas with High Docosahexaenoic Acid. *Nutr. Neuroscii* **2003,** *6,* 335–341.

Auestad, N.; M.B. Montalto; R.T. Hall; K.M. Fitzgerald; R.E. Wheeler; W.E. Connor; M. Neuringer; S.L. Connor; J.A. Taylor; E.E. Hartmann. Visual Acuity, Erythrocyte Fatty Acid Composition, and Growth in Term Infants Fed Formulas with Long Chain Polyunsaturated Fatty Acids for One Year. Ross Pediatric Lipid Study. *Pediatr. Res* **1997,** *41,* 1–10.

Auestad, N.; R. Halter; R.T. Hall; M. Blatter; M.L. Bogle; W. Burks; J.R. Erickson; K.M. Fitzgerald; V.Dobson; S.M. Innis; et al. Growth and Development in Term Infants Fed Long-chain Polyunsaturated Fatty Acids: A Double-masked, Randomized, Parallel, Prospective, Multivariate Study. *Pediatrics* **2001,** *108,* 372–381.

Birch, D.G.; E.E. Birch; D.R. Hoffman; R.D. Uauy. Retinal Development in Very-low-birthweight Infants Fed Diets Differing in Omega-3 Fatty Acids. *Invest. Ophthalmol. Vis. Sci.* **1992,** *33,* 2365–2376.

Birch, E.E.; D.R. Hoffman; R. Uauy; D.G. Birch; C. Prestidge. Visual Acuity and the Essentiality of Docosahexaenoic Acid and Arachidonic Acid in the Diet of Term Infants. *Pediatr. Res.* **1998,** *44,* 201–209.

Birch, E.E.; S. Garfield; D.R. Hoffman; R. Uauy; D.G. Birch. A Randomized Controlled Trial of Early Dietary Supply of Long-Chain Polyunsaturated Fatty Acids and Mental Development in Term Infants. *Dev. Med. Child Neurol.* **2000,** *42,* 174–181.

Blanaru, J.L.; J.R. Kohut; S.C. Fitzpatrick-Wong; H.A. Weiler. Dose Response of Bone Mass to Dietary Arachidonic Acid in Piglets Fed Cow Milk-Based Formula. *Am. J. Clin. Nutr.* **2004,** *79,* 139–147.

Boesze-Battaglia, K; A.D. Albert. Fatty Acid Composition of Bovine Rod Outer Segment Plasma Membrane. *Exp. Eye Res.* **1989,** *49,* 699–701.

Boswell, K.; E.K. Koskelo; L. Carl; S. Glaza; D.J. Hensen; K.D. Williams; D.J. Kyle. Preclinical Evaluation of Single-Cell Oils that Are Highly Enriched with Arachidonic Acid and Docosahexaenoic Acid. *Food Chem. Toxicol.* **1996,** *34,* 585–593.

Bougle, D.; P. Denise; F. Vimard; A. Nouvelot; M.J. Penneillo; B. Guillois. Early Neurological and Neuropsychological Development of the Preterm Infant and Polyunsaturated Fatty Acids Supply. *Clin. Neurophysiol.* **1999,** *110,* 1363–1370.

Bowen, R.A.; M.T. Clandinin. Dietary Low Linolenic Acid Compared with Docosahexaenoic Acid Alter Synaptic Plasma Membrane Phospholipids Fatty Acid Composition and Sodium-Potassium Atpase Kinetics in Developing Rats. *J. Neurochem.* **2002,** *83,* 764–774.

Breitner, J.C.; K.A. Welsh; M.J. Helms; P.C. Gaskell; B.A. Gau; A.D. Roses, M.A. Pericak-Vance;

A.M. Saunders. Delayed Onset of Alzheimer's Disease with Nonsteroidal Anti-inflammatory and Histamine h2 Blocking Drugs. *Neurobiol. Aging* **1995**, *16*, 523–530.

Carlson, S.; A. Ford; S. Werkman; J. Peeples; W. Koo. Visual Acuity and Fatty Acid Status of Term Infants Fed Human Milk and Formulas with and without Docosahexaenoate and Arachidonate from Egg Yolk Lecithin. *Pediatr. Res.* **1996**, *39*, 882–888.

Carlson, S.E; S.H. Werkman. A Randomized Trial of Visual Attention of Preterm Infants Fed Docosahexaenoic Acid until Two Months. *Lipids* **1996a**, *31*, 85–90.

Carlson, S.E.; S.H. Werkman; E.A. Tolley. Effect of Long-Chain n-3 Fatty Acid Supplementation on Visual Acuity and Growth of Preterm Infants with and without Bronchopulmonary Dysplasia. *Am. J. Clin. Nutr.* **1996b**, *63*, 687–697.

Carlson, S.E.; S.H. Werkman; P.G. Rhodes; E.A. Tolley. Visual-Acuity Development in Healthy Preterm Infants: Effect of Marine-Oil Supplementation. *Am. J. Clin. Nutr.* **1993**, *58*, 35–42.

Carrie, I.; M. Clement; D. de Javel; H. Frances ; J.M. Bourre. Specific Phospholipid Fatty Acid Composition of Brain Regions in Mice. Effects of n-3 Polyunsaturated Fatty Acid Deficiency and Phospholipid Supplementation. *J. Lipid Res.* **2000**, *41*, 465–472.

Champoux, M.; J.R. Hibbeln; C. Shannon; S. Majchrzak; S.J. Suomi; N. Salem; J.D. Higley. Fatty Acid Formula Supplementation and Neuromotor Development in Rhesus Monkey Neonates. *Pediatr Res.* **2002**, *51*, 273–281.

Clandinin, M.T.; J.E. Van Aerde ; A. Parrott; C.J. Field; A.R. Euler; E.L. Lien. Assessment of the Efficacious Dose of Arachidonic and Docosahexaenoic Acids in Preterm Infant Formulas: Fatty Acid Composition of Erythrocyte Membrane Lipids. *Pediatr. Res.* **1997**, *42*, 819–825.

Cunnane, S.C.; V. Francescutti; J.T. Brenna; M.A. Crawford. Breast-Fed Infants Achieve a Higher Rate of Brain and Whole Body Docosahexaenoate Accumulation than Formula-Fed Infants Not Consuming Dietary Docosahexaenoate. *Lipids* **2000**, *35*, 105–111.

De Swaaf, M.E; L. Sijtsma; J.T. Pronk. High-Cell-Density Fed-Batch Cultivation of the Docosahexaenoic Acid Producing Marine Alga *Cryptecodinium cohnii. Biotechnol. Bioeng.* **2003**, *81*, 666–672.

Dumuis, A.; J.P. Pin; K. Oomagari; M. Sebben; J. Bockaert. Arachidonic Acid Released from Striatal Neurons by Joint Stimulation of Ionotropic and Metabotropic Quisqualate Receptors. *Nature* **1990**, *347*, 182–184.

Emken, E.A.; R.O. Adlof; S.M. Duval; G.J. Nelson. Effect of Dietary Arachidonic Acid on Metabolism of Deuterated Linoleic Acid by Adult Male Subjects. *Lipids* **1998**, *33*, 471–480.

Emken, E.A.; R.O. Adlof; S.M. Duval; G.J. Nelson. Effect of Dietary Docosahexaenoic Acid on Desaturation and Uptake *in vivo* of Isotope-labeled Oleic, Linoleic, and Linolenic Acids by Male Subjects. *Lipids* **1999**, *34*, 785–791.

Faldella, G.; M.R. Govoni Alessandroni; E. Marchiani; G.P. Salvioli; P.L. Biagi; C. Spano. Visual Evoked Potentials and Dietary Long Chain Polyunsaturated Fatty Acids in Preterm Infants. *Arch. Dis. Child. Fetal. Neonatal Ed.* **1996**, *75*, F 108-112.

Fam, S.S.; L.J. Murphey; E.S. Terry; W.E. Zackert; Y. Chen; L. Gao; S. Pandalai; G.L. Milne; L.J. Roberts; N.A. Porter; T.J. Montine; J.D. Morrow. Formation of Highly Reactive A-Ring and J-Ring Isoprostane-like Compounds (a4/j4-neuroprostanes) *in vivo* from Docosahexaenoic Acid. *J. Biol. Chem.* **2002**, *277*, 36076–36084.

Farquharson, J.; E.C. Jamieson; K.A. Abbasi; W.J. Patrick; R.W. Logan; F. Cockburn. Effect of Diet on the Fatty Acid Composition of the Major Phospholipids of Infant Cerebral Cortex. *Arch. Dis. Child.* **1995**, *72*, 198–203.

Feller, S.E.; K. Gawrisch; A.D. Mackerell. Polyunsaturated Fatty Acids in Lipid Bilayers: Intrinsic and Environmental Contributions to Their Unique Physical Properties. *J. Am. Chem. Soc.* **2002**, *124*, 318–326.

Fewtrell, M.S.; R. Morley; R.A. Abbott; A. Singhal; E.B. Isaacs; T. Stephenson; U. MacFadyen; A. Lucas. Double-blind, Randomized Rrial of Long-Chain Polyunsaturated Fatty Acid Supplementation in Formula Fed to Pre-term Infants. *Pediatrics* **2002**, *110*, 73–82.

Fewtrell, M.S.; R.A. Abbott; K. Kennedy; A. Singhal; R. Morley; E. Caine; C. Jamieson; F. Cockburn; A. Lucas. Randomized, Double-blind Trial of Long-Chain Polyunsaturated Fatty Acid Supplementation with Fish Oil and Borage Oil in Preterm Infants. *J. Pediatr.* **2004**, *144*, 471–479.

Fliesler, S.J., R.E. Anderson. Chemistry and Metabolism of Lipids in the Vertebrate Retina. *Prog. Lipid Res.* **1983**, *22*, 79–131.

Fu, Z.; N.M. Attar-Bashi; A.J. Sinclair. 1-14c-Linoleic Acid Distribution in Various Tissue Lipids of Guinea Pigs Following an Oral Dose. *Lipids* **2001**, *36*, 255–260.

Garcia M. C.; G. Ward; Y.C. Ma; N. Salem; H.Y. Kim. Effect of Docosahexaenoic Acid on the Synthesis of Phosphatidylserine in Rat Brain in Microsomes and c6 Glioma Cells. *J. Neurochem.* **1998**, *70*, 24–30.

Gibson, R.A.; M. Makrides; M.A. Neumann; K. Simmer; E. Mantzioris; M.J. James. Ratios of Linoleic Acid to Alpha-Linolenic Acid in Formulas for Term Infants. *J. Pediatr.* **1994**, *125*, S48–55.

Gibson, R.A.; M. Makrides; J.S. Hawkes; M.A. Neumannand; A.R. Euler. A Randomized Trial of Arachidonic Acid Dose in Formulas Containing Docosahexaenoic Acid in Term Infants. *Essential Fatty Acids and Eicosanoids: Invited Papers from the Fourth International Congress*; R.A. Riemersma, R. Wilson, Eds.; American Oil Chemists' Society: Champaign, IL, **1998**; pp 147–153.

Hawkes, J.S.; D.L. Bryan; M. Makrides; M.A. Neumann; R.A. Gibson. A Randomized Trial of Supplementation with Docosahexaenoic Acid-Rich Tuna Oil and Its Effects on the Human Milk Cytokines Interleukin 1 Beta, Interleukin 6, and Tumor Necrosis Factor Alpha. *Am. J. Clin. Nutr.* **2002**, *75*, 754–760.

Herschman, H.R. Prostaglandin Synthase 2. *Biochim. Biophys. Acta* **1996**, *1299*, 125–140.

Hong, S.; K. Gronert; P.R. Devchand; R.L. Moussignac; C.N. Serhan. Novel Docosatrienes and 17s-Resolvins Generated from Docosahexaenoic Acid in Murine Brain, Human Blood, and Glial Cells. Autocoids in Inflammation. *J. Biol. Chem.* **2003**, *278*, 14677–14687.

Horby Jorgensen, M; G. Holmer; P. Lund; O. Hernell; K.F. Michaelsen. Effect of Formula Supplemented with Docosahexaenoic Acid and Gamma-Linolenic Acid on Fatty Acid Status and Visual Acuity in Term Infants. *J. Pediatr. Gastroenterol. Nutr.* **1998**, *26*, 412–421.

Horrocks, L.A.; Y.K. Yeo. Health Benefits of Docosahexaenoic Acid (DHA). *Pharmacol. Res.* **1999**, *40*, 211–225.

Ikemoto, A.; T. Kobayashi; S. Watanabe; H. Okuyama. Membrane Fatty Acid Modifications of pc12 Cells by Arachidonate or Docosahexaenoate Affect Neurite Outgrowth but Not Norepinephrine Release. *Neurochem. Res.* **1997,** *22,* 671–678.

Ikemoto, A.; A. Nitta; S. Furukawa; M. Ohishi; A. Nakamura; Y. Fujii; H. Okuyama. Dietary n-3 Fatty Acid Deficiency Decreases Nerve Growth Factor Content in Rat Hippocampus. *Neurosci. Lett.* **2000,** *285,* 99–102.

Innis, S.; D.H. Adamkin; R.T. Hall; S.C. Kalhan; C. Lair; M. Lim; D.C. Stevens; P.F. Twist; D.A. Diersen-Schade; C.L. Harris; et al. Docosahexaenoic Acid and Arachidonic Acid Enhance Growth with No Adverse Effects in Preterm Infants Fed Formula. *J. Pediatr.* **2002,** *140,* 547–554.

Jones, C.R.; T. Arai; J.M. Bell; S.I. Rapoport. Preferential *in vivo* Incorporation of [3H]Arachidonic Acid from Blood in Rat Brain Synaptosomal Fractions Before and After Cholinergic Stimulation. *J. Neurochem.* **1996,** *67,* 822–829.

Kelley, D.; P.C. Taylor; G.J. Nelson; B.E. Mackey. Arachidonic Acid Supplementation Enhances Synthesis of Eicosanoids without Suppressing Immune Functions in Young Healthy Men. *Lipids* **1998a,** *33,* 125–130.

Kelley, D.S.; P.C. Taylor; G.J. Nelson; B.E. Mackey. Dietary Docosahexaenoic Acid and Immunocompetence in Young Healthy Men. *Lipids* **1998b,** *33,* 559–566.

Kelley, D.S.; P.C. Taylor; G.J. Nelson; P.C. Schmidt; A. Ferretti; K.L. Yu; R. Erickson; R.K. Chandra; B.E. Mackey. Docosahexaenoic Acid Ingestion Inhibits Natural Killer Cell Activity and Production of Inflammatory Mediators in Young Healthy Men. *Lipids* **1999,** *34,* 317–324.

Kurlack, L.O.; T.J. Stephenson. Plausible Explanations for Effects of Long Chain Polyunsaturated Fatty Acids on Neonates. *Arch. Dis. Child Fetal Neonatal Ed.* **1999,** *80,* 148–154.

Lapillonne, A.; J.C. Picaud; V. Chirouze; J. Goudable; B. Reygrobellet; O. Claris; B.L. Salle. The Use of Low-EPA Fish Oil for Long-Chain Polyunsaturated Fatty Acid Supplementation of Preterm Infants. *Pediatr. Res.* **2000,** *48,* 835–841.

Leaf, A.; Y.F. Xiao; J.X. Kang. Interactions of n-3 Fatty Acids with Ion Channels in Excitable Tissues. *Prostaglandins Leukot. Essent. Fatty Acids* **2002,** *67,* 113–120.

Litman, B.J.; S.L. Niu; A. Polozova; D.C. Mitchell. The Role of Docosahexaenoic Acid Containing Phospholipids in Modulating g Protein-Coupled Signalling Pathways: Visual Transduction. *J. Mol. Neurosci.* **2001,** *16,* 237–242.

Lucas, A.; M. Stafford; R. Morley; R. Abbott; T. Stephenson; U. MacFadyen; A. Elias-Jones; H. Clements. Efficacy and Safety of Long-Chain Polyunsaturated Fatty Acid Supplementation of Infant-Formula Milk: A Randomised Trial. *Lancet* **1999,** *354,* 1948–1954.

Makrides, M.; M. Neumann; K. Simmer; J. Pater; R. Gibson. Are Long-Chain Polyunsaturated Fatty Acids Essential Nutrients in Infancy? *Lancet* **1995,** *345,* 1463–1468.

Makrides, M.; M.A. Neumann; R.A. Gibson. Effect of Maternal Docosahexaenoic Acid (DHA) Supplementation on Breast Milk Composition. *Eur. J. Clin. Nutr.* **1996,** *50,* 352–357.

Makrides, M.; M.A. Neumann; K. Simmer; R.A. Gibson. Dietary Long-Chain Polyunsaturated Fatty Acids Do Not Influence Growth of Term Infants: A Randomized Clinical Trial. *Pediatrics* **1999,** *104,* 468–475.

Makrides, M.; M.A. Neumann; K.Simmer; R.A. Gibson. A Critical Appraisal of the Role of Dietary Long-Chain Polyunsaturated Fatty Acids on Neural Indices of Term Infants: A Randomized, Controlled Trial. *Pediatrics* **2000**, *105*, 32–38.

Martin, R.E. Docosahexaenoic Acid Decreases Phospholipase A2 Activity in the Neurites/Nerve Growth Cones of PC12 Cells. *J. Neurosci. Res.* **1998**, *54*. 805–813.

Martinez, M. Tissue Levels of Polyunsaturated Fatty Acids during Early Human Development. *J. Pediatr* **1992**, *120*, S129–138.

Mathews, S.; W.T. Oliver; O.T. Phillips ; J. Odle; D.A. Diersen-Schade; R.J. Harrell. Comparison of Triglycerides and Phospholipids as Supplemental Sources of Dietary Long-Chain Polyunsaturated Fatty Acids in Piglets. *J. Nutr.* **2002**, *132*, 3081–3089.

Merritt, R.; N. Auestad; C. Kruger; S. Buchanan. Safety Evaluation of Sources of Docosahexaenoic Acid and Arachidonic Acid for Use in Infant Formulas in Newborn Piglets. *Food Chem. Toxicol.* **2003**, *41*, 897–904.

Mitoulas, L.R.; L.C. Gurrin; D.A. Doherty; J.L. Sherriff; P.E. Hartmann. Infant Intake of Fatty Acids from Human Milk over the First Year of Lactation. *Br. J. Nutr.* **2003**, *90*, 979–986.

Morris, G.; J. Moorcraft; A. Mountjoy; J.C. Wells. A Novel Infant Formula Milk with Added Long-Chain Polyunsaturated Fatty Acids from Single-Cell Sources: A Study of Growth, Satisfaction and Health. *Eur. J. Clin. Nutr.* **2000**, *54*, 883–886.

Nelson, G.J.; P.C. Schmidt; G.L. Bartolini; D.S. Kelley; D. Kyle. The Effect of Dietary Docosahexaenoic Acid on Plasma Lipoproteins and Tissue Fatty Acid Composition in Humans. *Lipids* **1997a**, *32*, 1137–1146.

Nelson, G.J.; P.C. Schmidt; G. Bartolini; D.S. Kelley; D. Kyle. The Effect of Dietary Arachidonic Acid on Platelet Function, Platelet Fatty Acid Composition, and Blood Coagulation in Humans. *Lipids* **1997b**, *32*, 421–425.

Nelson, G.J.; P.S. Schmidt; G.L. Bartolini; D.S. Kelley; D. Kyle. The Effect of Dietary Docosahexaenoic Acid on Platelet Function, Platelet Fatty Acid Composition, and Blood Coagulation in Humans. *Lipids* **1997c**, *32*, 1129–1136.

Nelson, G.J.; P.C. Schmidt; G. Bartolini; D.S. Kelley; S.D. Phinney; D. Kyle; S. Silbermann; E.J. Schaefer. The Effect of Dietary Arachidonic Acid on Plasma Lipoprotein Distributions, Apoproteins, Blood Lipid Levels, and Tissue Fatty Acid Composition in Humans. *Lipids* **1997d**, *32*, 427–433.

Nishizaki, T.; T. Matsuoka; T. Nomura; K. Sumikawa. Modulation of Ach Receptor Currents by Arachidonic Acid. *Brain Res. Mol. Brain Res.* **1998**, *57*, 173–179.

Nogawa, S.; F. Zhang; M.E. Ross; C. Iadecola. Cyclo-oxygenase-2 Gene Expression in Neurons Contributes to Ischemic Brain Damage. *J. Neurosci.* **1997**, *17*, 2746–2755.

O'Connor, D.; R. Hall; D. Adamkin; N. Auestad; M. Castillo; S.L. Connor; W.E. Connor; K.Fitzgerald; S. Groh-Wargo; E.E. Hartmann; et al. Ross Preterm Lipid Study. Growth and Development in Preterm Infants Fed Long-Chain Polyunsaturated Fatty Acids: A Prospective, Randomized Controlled Trial. *Pediatrics* **2001**, *108*, 359–371.

Peet, M.; D.F. Horrobin; Group, E.-E. Multicentre Study Group. A Dose-Ranging Exploratory Study of the Effects of Ethyl-Eicosapentaenoate in Patients with Persistent Schizophrenic Symptoms. *J. Psychiatr. Res.* **2002**, *36*, 7–18.

Pullarkat, R.K; H. Reha. Acyl and Alk-1'-enyl Group Composition of Ethanolamine Phosphoglycerides of Human Brain. *J. Neurochem.* **1978**, *31*, 707–711.

Puskas, L.G.; K. Kitajka; C. Nyakas; G. Barcelo-Coblijn; T. Farkas. Short-Term Administration of Omega 3 Fatty Acids from Fish Oil Results in Increased Transthyretin Transcription in Old Rat Hippocampus. *Proc. Natl. Acad. Sci.* **2003**, *100*, 1580–1585.

Rapoport, S.I.; F. Bosetti. Do Lithium and Anticonvulsants Target the Brain Arachidonic Acid Cascade in Bipolar Disorder? *Arch. Gen. Psychiatry* **2002**, *59*, 592–596.

Roberts, L.J.; T.J. Montine; W.R. Markesbery; A.R. Tapper; P. Hardy; S. Chemtob; W.D. Dettbarn; J.D. Morrow. Formation of Isoprostane-like Compounds (Neuroprostanes) *in vivo* from Docosahexaenoic Acid. *J. Biol. Chem.* **1998**, *273*, 13605–13612.

Ryan, A.S.; M.B. Montalto; S. Groh-Wargo; F. Mimouni; J. Sentipal-Walerius; J. Doyle; J.S. Siegman; A.J. Thomas. Effect of DHA-Containing Formula on Growth of Preterm Infants to 59 Weeks Postmenstrual Age. *Am. J. Human Biol.* **1999**, *11*, 457–467.

Sala-Vila, A.; A.I. Castellote; C. Campoy; M. Rivero; M. Rodriguez-Palmero; M.C. Lopez-Sabater. The Source of Long-Chain PUFA in Formula Supplements Does Not Affect the Fatty Acid Composition of Plasma Lipids in Full-Term Infants. *J. Nutr.* **2004**, *134*, 868–873.

Salem, N.; B. Wegher; P. Mena; R. Uauy. Arachidonic and Docosahexaenoic Acids are Biosynthesized from Their 18-Carbon Precursors in Human Infants. *Proc. Natl. Acad. Sci.* **1996**, *93*, 49–54.

Sanders, T.A.; D.J. Naismith. A Comparison of the Influence of Breast-Feeding and Bottle-Feeding on the Fatty Acid Composition of the Erythrocytes. *Br. J. Nutr.* **1979**, *41*, 619–623.

San Giovanni, J.P.; C.S. Berkey; J.T. Dwyer; G.A. Colditz. Dietary Essential Fatty Acids, Long-Chain Polyunsaturated Fatty Acids, and Visual Resolution Acuity in Healthy Full-Term Infants: A Systematic Review. *Early Hum. Dev.* **2000a**, *57*, 165–188.

San Giovanni, J.P.; S. Parra-Cabrera; G.A. Colditz; C.S. Berkey; J.T. Dwyer. Meta-Analysis of Dietary Essential Fatty Acids and Long-Chain Polyunsaturated Fatty Acids as They Relate to Visual Resolution Acuity in Healthy Preterm Infants. *Pediatrics* **2000b**, *105*, 1292–1298.

Sastry, R.S. Lipids of Nervous Tissue: Composition and Metabolism. *Prog. Lipid Res.* **1985**, *24*, 69–176.

Scott, D.; J.S. Janowsky; R.E. Carroll; J.A. Taylor; N. Auestad; M.B. Montalto. Formula Supplementation with Long-Chain Polyunsaturated Fatty Acids: Are There Developmental Benefits? *Pediatrics* **1998**, *102*, E59 .

Seebungkert, B.; J.W. Lynch. Effects of Polyunsaturated Fatty Acids on Voltage-Gated K$^+$ and Na$^+$ Channels in Rat Olfactory Receptor Neurons. *Eur. J. Neurosci.* **2002**, *16*, 2085–2094.

Simmer, K.; S.M. Schulzke; S. Patole. Long Chain Polyunsaturated Fatty Acid Supplementation in Preterm Infants. *Cochrane Database Syst. Rev* **2008**, Jan 23. CD000375.

Sinclair, A.J. Long-Chain Polyunsaturated Fatty Acids in the Mammalian Brain. *Proc. Nutr. Soc.* **1975**, *34*, 287–291.

Skosnik, P.; J.K. Yao. From Membrane Phospholipid Defects to Altered Neurotransmission: Is Arachidonic Acid a Nexus in the Pathophysiology of Schizophrenia? *Prostaglandins Leukot. Essent. Fatty Acids* **2003**, *69*, 367–384.

Smithers, L.; R. Gibson; A. McPhee; M. Makrides. Higher Dose of Docosahexaenoic Acid in Neonatal Period Improves Visual Acuity of Preterm Infants: Results of a Randomized Controlled Trial. *Am. J. Clin. Nutr.* **2008**, *88*, 1049–1056.

Stockard, J.E.; M.D. Saste; V.J. Benford; L. Barness; N. Auestad; J.D. Carver. Effect of Doco-sahexaenoic Acid Content of Maternal Diet on Auditory Brainstem Conduction Times in Rat Pups. *Dev. Neurosci.* **2000**, *22*, 494–499.

Svennerholm, L. Distribution and Fatty Acid Composition of Phosphoglycerides in Normal Human Brain. *J. Lipid Res.* **1968**, *9*, 570–579.

Tapiero, H.; G.N. Ba; P. Couvreur; K.D. Tew. Polyunsaturated Fatty Acids (PUFA) and Eico-sanoids in Human Health and Pathologies. *Biomed. Pharmacother.* **2002**, *56*, 215–222.

U.S. Food and Drug Administration. Agency Response Letter, GRAS Notice No. Grn 000041, Addressed to Henry Linsert Jr, Martek Biosciences Corporation International, May 21, 2001.

Uauy, R.; D.R. Hoffman; E.E. Birch; D.G. Birch; D.M. Jameson; J. Tyson. Safety and Efficacy of Omega-3 Fatty Acids in the Nutrition of Very Low Birth Weight Infants: Soy Oil and Marine Oil Supplementation of Formula. *J. Pediatr.* **1994**, *124*, 612–620.

Urquiza, A.M.D.; S. Liu; M. Sjoberg; R.H. Zetterstrom; W. Griffiths; J. Sjovall; T. Perlmann. Docosahexaenoic Acid, A Ligand for the Retinoid X Receptor in Mouse Brain. *Science* **2000**, *290*, 2140–2144.

Vaidyanathan, V.V.; K.V. Rao; P.S. Sastry. Regulation of Diacylglycerol Kinase in Rat Brain Membranes by Docosahexaenoic Acid. *Neurosci. Lett.* **1994**, *179*, 171–174.

Vanderhoof, J.; S. Gross; T. Hegyi. A Multicenter Long-Term Safety and Efficacy Trial of Preterm Formula Supplemented with Long-Chain Polyunsaturated Fatty Acids. *J. Pediatr. Gastroenterol. Nutr.* **2000**, *31*, 121–127.

Ward, G.R.; Y.S. Huang; E. Bobik; H.C. Xing; L. Mutsaers; N. Auestad; M. Montalto; P. Wainwright. Long-Chain Polyunsaturated Fatty Acid Levels in Formulae Influence Deposition of Docosahexaenoic Acid and Arachidonic Acid in Brain and Red Blood Cells of Artificially Reared Neonatal Rats. *J. Nutr.* **1998**, *128*, 2473–2487.

Watkins, B.; Y. Li; M.F. Seifert. Nutraceutical Fatty Acids as Biochemical and Molecular Modulators of Skeletal Biology. *J. Am. Coll. Nutr.* **2001**, *20*, 410S–416S.

Werkman, S.H.; S.E. Carlson. A Randomized Trial of Visual Attention of Preterm Infants Fed Docosahexaenoic Acid until Nine Months. *Lipids* **1996**, *31*, 91–97.

Wheeler, T.G.; R.M. Benolken; R.E. Anderson. Visual Membranes: Specificity of Fatty Acid Precursors for the Electrical Response to Illumination. *Science* **1975**, *188*, 1312–1314.

Willatts, P.; J.S. Forsyth; M.K. DiModugno; S. Varma; M. Colvin. Effect of Long-Chain Polyunsaturated Fatty Acids in Infant Formula on Problem Solving at 10 Months of Age. *Lancet* **1998**, *352*, 688–691.

Wynn, J.; C. Ratledge. Evidence that the Rate-Limiting Step for the Biosynthesis of Arachidonic Acid in *Mortierella alpina* Is at the Level of the 18:3 to 20:3 Elongase. *Microbiology* **2000**, *146*, 2325–2331.

Yamagata, K.; K.I. Andreasson; W.E. Kaufmann; C.A. Barnes; P.F. Worley. Expression of a Mitogen-Inducible Cyclooxygenase in Brain Neurons: Regulation by Synaptic Activity and Glucocorticoids. *Neuron* **1993**, *11*, 371–386.

Zimmer, L.; S. Delion-Vancassel; G. Durand; D. Guilloteau; S. Bodard; J.C. Besnard; S. Chalon. Modification of Dopamine Neurotransmission in the Nucleus Accumbens of Rats Deficient in n-3 Polyunsaturated Fatty Acids. *J. Lipid Res.* **2000**, *41*, 32–40.

17

Recent Developments in the Human Nutrition of Polyunsaturated Fatty Acids from Single Cell Oils

Connye N. Kuratko, James P. Hoffman, Mary E. Van Elswyk, and
Norman Salem, Jr

Martek Biosciences Corporation, 6480 Dobbin Road, Columbia, MD 21045

Introduction

The literature that reports on the health benefits of single cell oils (SCOs) is growing. As an important source of docosahexaenoic acid (DHA; C22:6n-3), algal oil is used in research designs to define the nutrient-specific benefits to infant development and to cardiovascular and cognitive health in adults. As a result of the publication of several authoritative documents linking DHA and health (Food and Nutrition Board 2005; Koletzko et al., 2007a, 2008), the public and healthcare providers are becoming more aware of the importance of this nutrient in the human diet.

SCOs also are important sources of arachidonic acid (ARA, 20:4n-6). As will be discussed in the infant section (see below), ARA is a necessary addition to DHA-containing infant formula. Many international guidelines that regulate infant feeding formulations mandate the addition of ARA to DHA-containing formula on at least a one-to-one ratio (Codex Alimentarius Commission, 2007; European Commission, 2006).

In this chapter, an overview of clinical research using SCOs since 2005 will be presented. In most of these studies, algal oil was selected as a source of DHA, and data were evaluated according to specific health benefits, related to DHA intake. Algal oil was often selected by the investigator as the source of DHA due to its defined fatty acid composition, its proven safety profile, and its sustainability for production. The clinical studies included in this review report the use of DHASCO® [docosahexaenoic acid (DHA) oil derived from *Crypthecodinium cohnii*] and ARASCO® [arachidonic acid (ARA) oil derived from *Mortierella alpina*], particularly in infant formulas. Other oils included are DHA oils derived from *Schizochytrium* sp. (DHASCO-S) and *Ulkenia* sp. (Ulkenia DHA).

Use of SCO in Infant Formula

The period from late gestation to four years of age is one in which the central nervous system sees critical and rapid growth, especially of the brain and eye. During this time, DHA accretion is the most rapid and contributes to both structural and functional development (Dobbing & Sands, 1973), with brain mass increasing approximately three-fold. DHA is passed to the infant from the mother via the placenta during gestation and via breast milk during nursing (Clandinin et al., 1980; Dutta-Roy, 2000; Putnam et al., 1982; Koletzko et al., 2007b). The long-chain polyunsaturated fatty acids (LCPUFA)—docosahexaenoic acid (DHA) and arachidonic acid (ARA)—are always supplied in human milk, but LCPUFA levels, particularly DHA levels, are dependent on maternal intake (Yuhas et al., 2006). Therefore, LCPUFA intake is important for both mother and infant throughout the perinatal period.

Since the introduction of a combination of DHA and ARA supplements for infant formulas in 1998, an estimated 45 million newborns in over seventy countries have consumed SCO-supplemented formulas, with a highly favorable safety profile (Data on file: Martek Biosciences Corp. Columbia, MD, USA). Clinical studies have established that such use is not only safe but, especially for preterm infants, helpful in optimizing growth and neurocognitive functions. Cohorts of infants initially participating in trials of DHA and ARA formula supplementation are now in early childhood, and, for these children, the benefits of supplementation continue to be demonstrated in many instances.

Below is a review of the clinical literature published from 2005 until January 2009. It is clear that the standard of care now includes the combined addition of DHA and ARA in studies of both term and preterm infant formulas. Not only are these recent studies consistent in design but they also have avoided the growth-associated problems detected in earlier studies. Some of those early studies demonstrated a reduction in ARA status in infants supplemented with DHA alone, and, in some preterm studies, this lower level of ARA resulted in growth reduction (Carlson et al., 1996; Ryan et al.,1999). In all of the following studies with combined DHA and ARA supplementation, growth was not compromised, and, in preterm infants, the studies often demonstrated growth and other benefits of the combined supplementation. Furthermore, early studies had inconsistent outcomes, most likely related to differences in study design. Some early studies, for example, provided very low levels of LCPUFA in the supplemented formula, utilized supplemented formulas for a period of time that was too brief to demonstrate any clinical effect, or selected endpoints or measurement times that lacked sensitivity. As the following review highlights, current studies with few exceptions were designed to include somewhat higher levels of DHA and ARA in the formula and to evaluate supplementation for longer periods of time—in some cases, throughout the first year of life.

SCO Supplementation in Preterm Infants

Clandinin et al. (2005) reported the results of a robust clinical comparison of formula supplemented with DHA and ARA (from DHASCO and ARASCO), DHA and ARA (from fish oil and ARASCO), or control formula in a double-blind, randomized, multi-site, prospective, two-phase, controlled trial. The study was designed to evaluate the safety and benefits of feeding 361 preterm infants formulas containing DHA and ARA from a variety of sources until 92 weeks postmenstrual age (PMA), with follow-up until 118 weeks PMA. Term infants (n = 105) who were exclusively breast-fed for ≥ 4 months served as the Reference Group. The weight of the algal-DHA group was significantly greater than the fish oil-DHA group from 66 to 118 weeks PMA; the fish-DHA group at 118 weeks PMA did not differ from term infants at 118 weeks PMA. The algal-DHA group was significantly longer than the control group at 48, 79, and 92 weeks PMA and the fish-DHA group at 57, 79, and 92 weeks PMA, but it did not differ from term infants from 79 to 118 weeks PMA. Supplemented groups had higher Bayley mental and psychomotor development scores at 118 weeks PMA than did the control group. Supplementation did not increase morbidity or adverse events. The authors concluded: "Feeding formulas with DHA and AA from algal and fungal oils resulted in enhanced growth. Both supplemented formulas provided better developmental outcomes than unsupplemented formulas" (Clandinin et al., 2005).

The DINO [Docosahexaenoic Acid (DHA) for the Improvement of Neurodevelopmental Outcome] study—a randomized, double-blind, placebo-controlled trial in Australia—examined the effect of increasing DHA concentration in human milk and formula on the circulating fatty acids (Smithers et al., 2008a), neurodevelopment (Makrides et al., 2009), and visual response (Smithers et al., 2008b) of preterm infants. Mothers of infants receiving breast milk were given either DHA supplements (900 mg per day) or a matching placebo. The breast milk DHA level in the placebo-treated mothers was approximately 0.25% (of total fatty acids), while that of the DHA-treated mothers was approximately 0.85%. Infants receiving formula were given either high concentration DHA-supplemented infant formula (DHA = approximately 1.1% of total fatty acids, from DHASCO) or standard DHA preterm formula (DHA = approximately 0.42% of total fatty acids, from DHASCO). Both infant formulas also contained ARASCO. Enhanced or standard DHA supplementation was utilized until the expected full-term delivery date (approximately nine weeks after birth). During this period, infants received either breast milk exclusively, a combination of breast milk and formula, or formula alone. At 18 months of age, the 657 infants were evaluated using standard development tests. A preplanned analysis by gender found that premature girls on the high DHA diet achieved a mean score approximately five points higher on the mental development test than girls receiving the standard DHA diet, resulting in, approximately, a 55% reduction in mild mental delay and, approximately, an 80% reduction in significant mental delay, a statistically significant improvement versus the control group. However, the same benefit was not found in the male infants, which resulted in no statistically significant differences overall between all infants on high-dose DHA versus the control. The reason for the gender difference is unclear. The authors theorize that male infants may need a higher level of DHA than female infants in order to see benefits and note that "given the lack of an alternative therapy for cognitive delay in this group of infants and the apparent safety of the current dose of DHA, further studies are warranted." When the same investigators compared the fatty acid status of 143 preterm infants receiving high 1% DHA versus the standard 0.2-0.3% DHA human milk or formula, they found that high DHA human milk or formula raised, but did not saturate, erythrocyte phospholipids with DHA and suggest that "milk exceeding 1% DHA may be required to increase DHA status to levels seen in term infants." At 2 months corrected age (CA), visual acuity of the high-DHA group (n = 54) did not differ from the control group (n = 61), but by 4 months CA, the high-DHA group exhibited an acuity that was 1.4 cycles per degree higher than the control group. VEP latencies and anthropometric measurements did not differ between the groups. The authors suggest that the amount of DHA generally fed to preterm infants in human milk or in formula may not be sufficient for optimal visual development.

Henriksen et al. (2008), in a randomized, double-blind, placebo-controlled study in Norway, evaluated the effect of supplementation with DHASCO and ARASCO on cognitive development at 6 months in 141 preterm infants receiving human milk. The SCOs were sonicated into human milk and given by gavage. There was no difference in adverse events or growth between the groups. At a 6-month follow-up, the supplemented group performed better on the problem solving subscore. The authors concluded that supplementation with DHA/ARA for preterm infants fed human milk in the early neonatal period was associated with better recognition memory and higher problem solving scores at 6 months.

Siahanidou et al. (2008) compared the effect of DHASCO- and ARASCO-supplemented formula with unsupplemented formula on oxidative stress in 104 healthy preterm infants and concluded that supplementation of infant formulas with LCPUFA does not affect lipid peroxidation or increase oxidative stress in healthy preterm infants. This same study group also compared effects on serum adiponectin and lipid concentrations of a DHA/ARA SCO-supplemented formula with those of unsupplemented formula in 60 preterm infants and reported that circulating adiponectin concentrations were higher with the LCPUFA formula, which correlated with a favorable lipidemic profile.

SCO Supplementation in Term Infants

Agostoni et al. (2009) conducted a multicenter, randomized, prospective, double-blind, placebo-controlled trial of DHA supplementation on the achievement of four gross motor milestones (sitting without support, hands-and-knees crawling, standing alone, and walking alone) in healthy infants who received supplementation with either 20 mg liquid DHA (n = 580) or a placebo (n = 580) once daily, orally, throughout the first year of life. Regression analysis showed improvements in several developmental milestones as the result of DHA supplementation. The supplemented infants exhibited a significantly shorter time for being able to sit without support ($P <$ 0.0001). "Despite the 1-wk advance in sitting without support associated with DHA supplementation, no demonstrable persistent effects of DHA supplementation on later [gross] motor development milestones were found. Thus, the long-term clinical significance of the 1-wk change in sitting without support, if any, remains unknown." There were significant improvements in several secondary endpoints,: including the abilities to reach out and touch objects ($P < 0.0001$), bring a toy to the mouth ($P <$ 0.0001), and say their first comprehensible word ($P < 0.0001$). Supplemental DHA exerted a beneficial effect on these secondary endpoints, particularly language.

Birch et al. (2007) studied visual acuity via sweep visual evoked potential in a double-masked, randomized, controlled clinical trial in 103 term infants randomly assigned to either a control formula or formula supplemented with DHASCO and ARASCO (0.36% and 0.72%, respectively, of total fatty acids). Stereoacuity in the LCPUFA-supplemented group was significantly better at 17 weeks of age, but not at 39 and 52 weeks of age. By 17 and 39 weeks, the red blood cell DHA concentration in the LCPUFA-supplemented group was more than double and more than triple, respectively, that found in the control group. The growth of infants fed LCPUFA-supplemented and control formulas did not differ significantly; both diets were well tolerated. The authors concluded that LCPUFA-supplementation of term infant formula during the first year of life yields clear differences in visual function and in total red blood cell lipid composition.

Burks et al. (2008) conducted two studies in healthy term infants. The first study compared the effect on growth in 165 healthy infants of a new amino acid-based formula (AAF) versus an extensively hydrolyzed formula (EHF), which was used as the control, with both DHA and ARA at levels similar to those in human milk worldwide. The second study evaluated the hypoallergenicity of the new AAF in 32 infants and children with a confirmed cow's milk allergy. The investigators concluded that the new AAF with DHASCO- and ARASCO-supplemented formula was hypoallergenic, safe, and supported growth in healthy term infants.

Similarly, Hoffman et al. (2008) compared a control, unsupplemented, soy-based formula with a DHA+ARA-supplemented (from DHASCO and ARASCO), soy-based formula in a double-blind, parallel group trial to evaluate safety, benefits, and growth from 14 to 120 days of age in 244 healthy term infants. The authors concluded that "feeding healthy term infants [a] soy-based formula supplemented with DHA and ARA from single cell oil sources at concentrations similar to human milk significantly increased circulating levels of DHA and ARA when compared with the control group. Both formulas supported normal growth and were well tolerated." Likewise, Vanderhoof (2008) examined the hypoallergenicity and the effects on growth and tolerance of a new amino acid-based formula with DHASCO and ARASCO in a randomized study of healthy term infants and observed no allergic reactions in a double-blind, placebo-controlled food challenge in subjects receiving either Nutramigen AA or placebo (Neocate). Vanderhoof concluded that Nutramigen AA sustains growth and is well tolerated in babies exhibiting a cow's milk allergy.

Gibson et al. (2008) conducted a single-center, randomized, double-blind, controlled, parallel-group trial to examine the safety of supplementing infant formula with long-chain polyunsaturated fatty acids and *Bifidobacterium lactis* in 142 term infants, who were fed either supplemented or unsupplemented formula for 7 months. There was no significant difference in growth between the two groups. There were no significant differences in mean length and head circumference. No influence of the supplements on the response to vaccines was observed. Thus, the authors concluded that the growth characteristics of term infants fed the starter formula, which contained a probiotic plus LCPUFA, were similar to those exhibited by standard, formula-fed infants.

SCO Maternal Supplementation during Lactation

Breast-feeding is the preferable method for infant feeding, and breast milk always contains both DHA and ARA. The amount of DHA in human milk is highly variable, however, and is directly related to maternal diet. Therefore, tests to determine the effect of maternal DHA supplementation on health outcomes in breast-fed infant have interested researchers. One such study used DHASCO as the DHA supplement provided to lactating mothers. In the study, Jensen et al. (2005) determined the effect of DHASCO (200 mg/d x 4 mo) supplementation of breast-feeding mothers on the neurodevelopment and visual status of their term infants (n = 115) in a double-blind, placebo-controlled trial, and reported a higher ($P < 0.01$) Bayley Psychomotor Development Index (eye-hand coordination) at 30 months of age.

After reviewing these studies as a group, it is clear that DHASCO and ARASCO supplementation in infants continues to be well studied; their effects on health and developmental outcomes, within specific populations of infants, are beneficial. Of particular interest is the cohort of infants, established several years ago by Birch et al. (2007), who were fed DHA- and ARA-supplemented formula for the first 4 to 12 months of their lives. These infants have been followed into childhood and are still showing the benefits of early supplementation for their visual and cognitive development. In addition, recent studies in premature infants show the importance of DHA and ARA for development in this special group; they indicate that the levels of DHA and ARA currently used in premature formulations may not be sufficient to allow developmental outcomes equal to those of breast-fed infants.

Based on these and earlier findings, expert groups currently recommend that, if LCPUFA supplementation is provided, DHA be included at a level of at least 0.2% of total fatty acids, ARA be included at the same level or greater, and the level of eicosapentaenoic acid (EPA, 20:5n-3) be no higher than the level of DHA (Codex Alimentarius 2007; Koletzko et al., 2007b; Commission of the European Communities, 2006). Recommendations are even stronger for premature infants, with expert groups calling for the requirement of LCPUFA supplementation for premature infants, both during hospitalization and after discharge (ESPGHAN, 2007). Expert groups also acknowledge the importance of DHA in the maternal diet and recommend that pregnant and lactating women consume at least 200 mg DHA/d (Koletzko et al., 2007a).

The Use of SCO for Cardiovascular Health

The role of fish and fish oil in the reduction of all-cause mortality, myocardial infarction, and cardiac and sudden death is well established for those with cardiovascular disease (Wang, 2006). Evidence also continues to accumulate in support of the role of fish consumption and fish oil supplementation in the primary prevention of cardiovascular disease (Wang et al., 2006). The effective bioactives in fish and fish oils are predominately the long-chain omega-3 fatty acids, EPA and DHA. Oils derived from single-cell microalgae provide sustainable alternatives to fish and fish oils for the intake of these important nutrients. Two commercially available SCOs, derived from *Crypthecodinium cohnii* (DHASCO) and *Schizochytruim sp.* (DHASCO-S), are rich in DHA (40% by weight) and have been demonstrated to be bioequivalent sources of DHA when compared with fish (Arterburn et al., 2008). Since fish and fish oils provide a combination of EPA and DHA, but DHASCO is EPA-free and DHASCO-S contains only 1.5%, by weight, as EPA, a review of the cardiovascular benefits resulting from supplementation with these and other SCOs is warranted. The following discussion will review the clinical literature published since 2005 regarding the role of n-3 and n-6 LCPUFA from SCOs in cardiovascular outcomes.

The Effects of SCOs on Lipoproteins

It is well documented that the consumption of fish and fish oil has no effect on total cholesterol but results in significant changes in the amount and distribution of cholesterol carried by LDL and HDL particles. Specifically, an average 6% increase in LDL cholesterol has been reported in response to fish oil supplementation, often accompanied by an average 1.6% increase in HDL (Balk et al., 2006). It appears that LDL cholesterol increases are driven by reductions in hepatic lipogenesis. The combination of reduced triglyceride production, VLDL assembly and secretion from the liver, and rapid conversion of secreted VLDL to LDL promotes LDL and is consistent with the effect of other potent triglyceride lowering agents, such as fibrate drugs (Jacobson, 2008). Furthermore, changes in LDL are often, but not always (see review by Hartweg et al., 2007; Lu et al., 1999), accompanied by a redistribution of LDL away from small, dense, atherogenic LDL (sdLDL) toward large, buoyant, less atherogenic particles (Minihane et al., 2000; Griffin et al., 2006).

Studies of DHA-rich SCOs published since 2005 also confirm the ability of these oils to promote HDL and favorably impact LDL particle subclasses, despite providing DHA alone or DHA with minimal EPA. Four randomized, controlled trials (RCT) have investigated DHA-rich SCOs for the modification of lipoproteins among mildly hyperlipidemic individuals (Engler et al., 2005; Maki et al., 2005; Schwellenbach et al., 2006; Kelley et al., 2007). Maki et al. studied healthy, middle-aged adults (n = 57) with HDL levels below the average for their gender (≤44 mg/dl for men; ≤ 54 mg/dl for women) during a 6-week, double-blind, placebo controlled (olive oil) trial. Subjects were supplemented with 1.52 g DHA/d from DHASCO-S, while they continued following their typical diet. Subjects taking statins were excluded, but those stabilized on blood pressure-lowering drugs were allowed to participate. LDL cholesterol increased significantly (+12%, P < 0.001), but the cholesterol carried by sdLDL decreased significantly (-10 %; P < 0.025) . No change in HDL was observed. Similarly, Kelley et al. (2007) reported increased LDL cholesterol and LDL particle size redistribution but no change in HDL among mildly hypertriglyceridemic (150-400 mg/dl) adults. Adult men (n = 34) were supplemented with 3 g/d of DHA from DHASCO during a 90-day RCT. Subjects taking lipid-lowering medications were excluded. By day 45 of the study, a 12.6% increase in LDL (P < 0.05) was observed, with no further elevations noted at the end of the trial. Large, buoyant LDL particles increased by 120% (P < 0.007), intermediate LDL particles were reduced by about one half (P < 0.02), and sdLDL particles were reduced (NS) by an average of 312 nmol/l. In addition, Kelley et al. (2008) reported a 21% reduction in remnant-like particle-cholesterol (RLP-C). Engler et al. (2005) reported the effect of 1.2 g DHA/d from DHASCO on LDL cholesterol levels after 6 weeks among children (n = 20) aged 9–19 years with hyperlipidemia (LDL ≥ 130 mg/dl), who participated in a 6-month, cross-over study. Children were stable on a National Cholesterol Education Program II diet and were not taking lipid-lowering medications. DHASCO had no effect on LDL cholesterol, likely due to the LDL cholesterol-lowering effect of the background diet, but still favorably influenced LDL particle distribution. Specifically, sdLDL particles decreased 48% (P = 0.002), and large, buoyant LDL increased 91% (P = 0.004), as compared with a corn/soy oil placebo. HDL cholesterol was unchanged, but the large, buoyant subclass of HDL (HDL$_2$) increased 14% compared to placebo (P = 0.01).

Schwellenbach et al. (2006) compared DHASCO-S (1 g DHA/d) with fish oil (1.2 g DHA and EPA/d) for the modification of blood lipids among mildly hypertriglyceridemic (>200 but <750 mg/dl) adults (n = 116) with documented coronary artery disease. Subjects were excluded if they exhibited poorly controlled diabetes or were taking fish oil supplements. Otherwise, concomitant use of medications was acceptable, with about 50% of the population using statin drugs. Likely due to the concomitant use of statins, no increase in LDL was observed in either group. Interestingly, however, DHASCO-S resulted in a 5.5% increase in HDL from baseline (P < 0.5) which was significantly greater (P < 0.4) than that observed for fish oil.

Lipid modification studies are most commonly undertaken among subjects with lipid abnormalities. However, two RCTs have been published since 2005 that report the effects of SCOs on lipoproteins among normolipidemic adults (Sanders et al. , 2006; Oe et al., 2008). Sanders et al. supplemented normolipidemic adults (n = 79) aged an average of 30 years with 1.5 g DHA/d from DHASCO-S during a 4-week RCT. Subjects were excluded if they were taking lipid- or blood pressure-lowering medications and were allowed to consume their usual diet, with the exception of oily fish, throughout the study. DHASCO-S resulted in a 10% increase (P < 0.01) in LDL, which was largely compensated for by a 9% increase in HDL (P < 0.001). No significant effect on LDL particle distribution was noted. Finally, Oe et al. (2008) studied the combined effects of 240 mg DHA with 240 mg ARA and 1 mg astaxanthin per day versus an olive oil placebo among healthy, elderly (n = 28) Japanese subjects during a 3-month RCT. Bioactives were derived from unspecified SCOs. Subjects were excluded if they exhibited uncontrolled hypertension, diabetes, or obesity. Subjects were allowed to maintain their normal diet. Not surprisingly, given the small supplemental dose of DHA and the high background contribution of DHA and EPA from the typical Japanese diet (Iso, 2006), no effects on blood lipids were observed.

Collectively, these limited data suggest that SCOs promote HDL among normolipidemic and statin-treated hyperlipidemic adults and may do so to a greater extent (average = 3.6%) than that reported previously for fish oils (1.6%). These data also indicate that SCOs are similar to fish oils with regard to their potential to elevate LDL cholesterol. It is not possible, however, due to the limited number of studies and inconsistency regarding LDL particle outcomes reported (i.e., size, particle number, and cholesterol content), to adequately compare SCOs with fish oil, in regards to LDL particle outcomes. However, the current literature suggests that DHA-rich SCOs profoundly lower the concentration of sdLDL. Changes in LDL cholesterol in response to SCOs are likely driven by effects on liver lipid metabolism, which ultimately result in concomitant triglyceride reductions.

The Effects of SCOs on Triglyceride Levels

Decrease of triglyceride concentrations in hypertriglyceridemic individuals appear to be one of the most profound lipid effects of EPA and DHA from fish oil, averaging 15% reductions from baseline (Balk, 2006). In keeping with this finding, the American Heart Association recommends 2-4 g/d of EPA and DHA for triglyceride reduction (Kris-Etherton et al., 2002). Ryan et al. (2008) recently reviewed studies of SCOs from either DHASCO or DHASCO-S for triglyceride reduction among normal and hyperlipidemic individuals. Sixteen studies were reviewed; eleven considered DHASCO, and five assessed DHASCO-S. Triglycerides were decreased an average of 21% among dyslipidemic adults in response to an average dose of 2 g DHA/d from either SCO. Reductions (−17%) were also observed among normolipidemic adults, with the greatest reductions among those with baseline triglyceride levels close to 150 mg/dl. From the studies reviewed, one conducted by Schwellenbach et al. (2006) is of particular interest, as it included a head-to-head comparison of DHASCO-S and fish oil. The DHASCO-S supplement in this study provided 1 g/d DHA, while the fish oil supplement provided 1 g DHA and an additional 252 mg of EPA. Despite providing 25% more n-3 LCPUFA per day, triglyceride reductions were not significantly different between the groups, and a greater percentage of subjects in the DHASCO-S group, compared with the fish oil group (24.6% and 10.2%, respectively), achieved the triglyceride goal of < 150 mg/dl. In addition, based on the studies reviewed by Ryan et al. (2008), it would appear that the tolerability of DHASCO and DHASCO-S is quite good. In fact, Schwellenbach et al. (2006) reported that a significantly greater number of subjects in the fish oil group (n = 33), compared with the DHASCO-S group (n = 12), complained of a fishy flavor. This observation is important as it has been reported that the primary barrier to compliance with fish oil supplementation for lipid-lowering is fishy off-flavors (Bays, 2008). Taken together, these data suggest that DHA-rich SCOs are an effective substitute for fish oil to promote triglyceride reduction.

The Effects of SCOs on Cardiac and Vascular Function

In the United States, about 73 million adults over the age of twenty have hypertension (American Heart Association, 2007). Blood pressure levels higher than 140/90 mm Hg are associated with 69% of people who have a first heart attack, 77% who have a first stroke, and 74% who have congestive heart failure (American Heart Association, 2008). Even those with high–normal blood pressure (130-139 mm Hg systolic/85-89 mm Hg diastolic) are at increased risk, with studies suggesting a greater risk (1.5-2.5 times) of suffering a heart attack, stroke, or heart failure within 10 years, as compared with those individuals exhibiting the optimal blood pressure (Vasan et al., 2001). The role of fish oils in the modest reduction of blood pressure was confirmed in a recent meta-analysis of RCTs (n = 8) among mild and moderately hypertensive adults, suggesting a –2.3 mm Hg reduction in systolic blood pressure and a –2.1 mm Hg in diastolic blood pressure, with an average dose of 4 g/d (Dickinson et al., 2006). Most recently, four studies have reported on the effects of DHA-rich SCOs on heart rate and blood pressure among normotensive adults. Kelley et al. (2007), discussed above, reported that both systolic and diastolic blood pressure were significantly (P < 0.05) reduced by 7 and 3 mm Hg, respectively, at the mid-point of their study but were no longer significant at the end, despite the continued reduction of systolic blood pressure of –2.8 mmHg in response to a 3 g/d supplement of DHASCO. Additionally, heart rate decreased –5.8 beats per minute (bpm) (P < 0.05) by the mid-point of their study and –3.4 bpm (P < 0.07) by the end of 90 days.

In contrast, Theobald et al. (2007) studied 40 healthy, normotensive, male (126/81 mm Hg) and female (117/77 mm Hg) subjects, ages 45-65, randomized in a cross-over design to either a DHASCO supplement (700 mg DHA/d) or a refined olive oil placebo. A consistent, significant (P < 0.01) reduction of –3.3 mm Hg in diastolic blood pressure was reported after the DHA supplement intervention. These results are 1 mm Hg greater than has been previously reported for hypertensive subjects on much higher doses of fish oil. Theobald et al. also reported a non-significant –2.1 bpm

(P = 0.15) reduction in heart rate after DHA treatment, as compared to the placebo group. Interestingly, this study also measured endothelial function in response to both salbutanol and glyceryl trinitrate but found no effect on vasodilation or arterial stiffness, thus providing little evidence to explain the mechanism responsible for the observed blood pressure reduction.

Two studies of normolipidemic adults found no significant effect of DHASCO (Sanders et al., 2006) or low-dose unspecified SCOs (Oe et al., 2008) on blood pressure or heart rate, although a –3.1 mm Hg average reduction (NS) in response to DHASCO-S was reported in the first study. While the very low dose of DHA (240 mg/d) likely accounts for the lack of effect on blood pressure in the Oe et al. (2008) study, in both of these studies, baseline blood pressure was already considered optimal. For example, diastolic blood pressures averaged 74 mm Hg in these studies, as compared to Kelley et al. (2007) and Theobald et al. (2007), where baseline blood pressures averaged ≥79 mm Hg. Since optimal is defined as blood pressure <120/80 mmHg (Kannel et al., 2008), it is possible that blood pressure and heart rate reductions in response to SCOs may only be consistently observed among those with typical or slightly elevated blood pressures,; studies among those with optimal blood pressure may require greater statistical power to detect significant reductions.

Finally, Oe et al. (2008) investigated the effects of low dose SCO-DHA, -ARA, and astaxanthin (see above) on coronary flow velocity reserve (CFVR). CFVR, accomplished using transthoracic Doppler echocardiography, is considered a useful index of coronary circulation, with impaired flow suggestive of subclinical coronary atherosclerosis. While, as noted above, the low dose dietary supplement failed to influence blood lipids or blood pressure and heart rate, the authors observed a significant increase in CFVR among these healthy, elderly Japanese subjects. Specifically, CVFR significantly (P < 0.01) increased 15%, as compared to the placebo, after 3 months of supplementation. These results are surprising given the high intake of EPA and DHA in the typical Japanese diet (Iso, 2008), and it is unclear what the implications are for Western populations where the intake of EPA and DHA is typically low.

The Effects of SCOs on Atherosclerotic Risk Factors

Inflammation is believed to be an important contributor to the development of atherosclerotic plaques. Numerous inflammatory markers have been identified as useful in predicting cardiovascular risk, although none are generally accepted as surrogate markers for cardiovascular disease. Three studies published since 2005 have investigated the effect of DHA-rich SCOs on C-reactive protein (CRP) and other inflammatory markers, such as E-selectin and interleukin-6. In the population studied by Kelley et al. (2009), as described above, 3 g/d of DHA from DHASCO resulted in a significant (P < 0.05) 15% reduction in CRP, as compared to the placebo group, among mildly hypertriglyceridemic men after 90 days. The authors suggest that the observed CRP reductions were comparable to those observed among patients using statins and speculate that this reduction would have decreased further with continued

DHA use. Kelley et al (2009) also reported a significant reduction in circulating neutrophils (−11.7%; P < 0.05), interleukin-6 (−23%), and granulocyte monocyte colony stimulating factor (−21%) but no effect on E-selectin; there was, however, a 7% increase in anti-inflammatory matrix metalloproteinase-2.

In contrast, Theobald et al. (2007) found no effect of DHA-rich SCOs on any of the inflammatory markers they investigated. Specifically, no effect in response to 700 mg DHA/d from DHASCO was reported for CRP or IL-6. Consistent with the results of Kelley et al. (2009), there was no effect of DHASCO on E-selectin. Similarly, Sanders et al. (2006) reported no effect of 1.5 g DHA/d from DHASCO-S on CRP among healthy volunteers. One likely conclusion is that, given that both null studies provided one quarter to one half the dose supplied by Kelley et al. (2009), higher doses are required to impact inflammatory markers.

Excessive platelet activation and aggregation is a key contributor to atherothrombosis, making anti-platelet therapies a mainstay in the prevention of atherosclerotic heart disease (Krotz, 2008). Among polyunsaturated fatty acids, n-3 LCPUFA are generally believed to have weak anti-aggregatory properties, while n-6 fatty acids are traditionally assumed to be pro-aggregatory. Kusumoto et al. (2007) examined the effect of ARASCO supplementation on platelet aggregation among healthy, non-smoking, male Japanese subject (n = 24). The study consisted of a 1-week run-in period, followed by a 4-week supplementation with 838 mg ARA/d versus an olive oil placebo. Platelet aggregation in platelet-rich plasma was measured at the end of 4 weeks using ADP, collagen, and ARA stimulation. ARA supplementation did not impact platelet aggregation in response to any agonist tested. The authors speculate that these results would be broadly applicable, as the baseline ARA levels in their subjects were similar to those previously reported for American and European populations. Importantly, however, the authors also report an average intake of EPA and DHA of 882 mg/d among their subjects, as determined by a 7-day weighed intake; the typical American diet is estimated to contain about one-fifth this amount of n-3 LCPUFA per day (Food and Nutrition Board, 2005). It is possible that a failure to find pro-aggregatory effects of ARA in this population may have resulted from the provision of a nearly identical amount of n-3 LCPUFA, thus maintaining platelet homeostasis. Replication of these results in populations with lower background n-3 LCPUFA intake is needed.

Emerging evidence suggests the role of higher dose, DHA-rich SCOs in the reduction of a variety of inflammatory markers that may promote atherosclerosis. Recent studies also suggest the beneficial role of DHA-rich SCOs in the promotion of HDL, the reduction of triglycerides, and the redistribution of LDL toward less atherogenic particles. Current studies further suggest that slightly lower doses of DHA-rich SCOs than those required for fish oils may be needed to achieve desired lipid outcomes and lower blood pressure. Despite the provision of little or no EPA, current research consistently supports the role of DHA-rich SCOs in the promotion of cardiovascular health.

SCO Supplementation and Cognition

The cognitive benefits of dietary DHA do not end after infancy. Although clinical trials involving toddlers and children are limited, benefits of DHA supplementation were seen in a recent randomized, placebo- controlled double-blind study. In the study, healthy 4-year-old children received 400 mg/d docosahexaenoic acid from algal oil or a matching placebo for 4 months. In regression analysis, there was a statistically significant, positive association between the blood level of DHA and higher scores on the Peabody Picture Vocabulary Test, a test of listening comprehension and vocabulary acquisition (Ryan and Nelson, 2008).

Brain health, including memory, is a high-ranking priority and often listed as one of the top health-related concerns of aging populations in the U.S. and other countries. A certain degree of memory loss and decline in cognitive function may be considered a normal part of aging; however, there is a growing awareness that certain aspects of healthy lifestyles may prevent or delay the expression of negative symptoms. DHA is an integral component of all mammalian membranes. It is the major structural and functional n-3 LCPUFA in the brain and retina throughout life. As a consequence, the presence of DHA at adequate levels in neural tissue is essential for optimal brain function during aging, as well as during development.

A body of literature—comprised of both observational studies and randomized, controlled trials using fish oil—indicates the importance of DHA, specifically, for brain health and the maintenance of cognitive function in many, if not all, domains. In an extensive, evidence-based review and meta-analysis conducted by the Agency for Healthcare Research Quality (AHRQ), DHA was recognized as the specific omega-3 fatty acid associated with a lower risk of dementia and Alzheimer's disease. The report specifically concludes, "Total omega-3 FA consumption and consumption of DHA (but not ALA or EPA) were associated with a significant reduction in the incidence of Alzheimer's" (MacLean et al., 2005).

The association of DHA with a decreased risk of dementia is also supported by a recent study from the Framingham cohort. In this study, Schaefer et al. (2006) demonstrated that patients with the top quartile of plasma DHA showed a 47% reduction in all types of dementia , but, in contrast, there was no association between EPA levels and a diminution of dementia. In the accompanying editorial on this study, Morris estimated that 180 mg DHA/day is an adequate protective dose (Schaefer et al., 2006).

Johnson et al. (2008) published the results of a small, randomized, double-blind, controlled study that tested the effect of DHA (from DHASCO)+lutein supplement on cognitive function and eye health. In the study, 49 women (aged 60–80 years) were randomized to receive 800 mg DHA/d, 12 mg lutein/d, a combination of DHA and lutein, or a placebo. After 4 months, verbal fluency scores improved significantly in the DHA, lutein, and combined treatment groups (P < 0.03). Memory scores and the rate of learning improved significantly in the combined treatment group (P < 0.03). Those subjects who were supplemented with the combination also displayed a trend toward more efficient learning (P = 0.07). Measures of mental processing speed, accuracy, and mood were not affected by supplementation.

In addition, two recently completed major, randomized, double-blind, placebo-controlled, parallel assignment interventional studies were designed to determine the effect of higher levels of DHA on neurocognitive function in the elderly. Both of these studies were designed using SCO as a source of DHA. The first of these upcoming trials determined the effects of DHA on improving cognitive function in normal elderly with memory complaints but with a normal Mini Mental Exam score. The study supplemented subjects with 900 mg of algal DHA per day and assessed working memory, memory retention, attention, and executive function in order to determine the potential for DHA to improve the symptoms of cognitive decline associated with aging (MIDAS, clinicaltrials.gov). The results indicated that algal DHA 900mg/day for 6 months was associated with a significant improvement compared to placebo in paired association learning, an episodic memory test. Additionally a significant decrease in heart rate was noted (Yurko-Mauro 2009). The second trial sponsored by the National Institutes of Health, supplemented Alzheimer's Disease patients with 2 g algal DHA/d in order to determine the potential for DHA to delay the progression of Alzheimer's disease. Patients were allowed to continue conventional Alzheimer's Disease therapy. Specifically, the study included 400 individuals with mild to moderate Alzheimer's disease and sought to measure changes in rate of cognitive and functional decline by assessing cognitive, behavioral, and functional status, as well as the global severity of dementia (DHA, National Inst. Aging, clinical trials.gov). Preliminary results were presented at the International Conference on Alzheimer's Disease, July 2009. As reported, the results did not demonstrate any benefit of DHA on the primary end points. However, on a planned secondary analysis, patients with an APOE4 negative status showed a significant improvement with algal DHA therapy compared to placebo (alz.org press release 2009). Further research in this area is needed to elucidate the full benefit of DHA in cognitive decline and the importance of early supplementation.

Conclusions

The studies reported in this review highlight the contribution of SCOs to a growing body of data associating health benefits with both DHA and ARA, for fetal and infant development, and DHA, for adults. In addition, current evidence supports the importance of DHA supplementation not only for pregnant and nursing women, but also for all adults, regardless of age. Because food sources of DHA are limited, the primary source being predatory/fatty fish, fortified foods become increasingly important as we try to optimize DHA intake. While many people cannot or will not consume adequate amounts of fish to satisfy their requirement for DHA, others are voicing valid concerns regarding the sustainability of worldwide fish supplies. SCOs provide a renewable source for this essential fatty acid that is free of oceanic contamination and makes a significant contribution to human health.

References

Aggett, P.J.; C. Agostoni; I. Axelsson; M. De Curtis; O. Goulet; O. Hernell; B. Koletzko; H.N. Lafeber; K.F. Michaelsen; et al; ESPGHAN Committee on Nutrition. Feeding Preterm Infants after Hospital Discharge: A Commentary by the ESPGHAN Committee on Nutrition. *J. Pediatr. Gastroenterol. Nutr.* **2006,** *42,* 596–603.

Agostoni, C.; G.V. Zuccotti; G. Radaelli; R. Besana; A. Podestà; A. Sterpa; A. Rottoli; E. Riva; M. Giovannini. Docosahexaenoic Acid Supplementation and Time at Achievement of Gross Motor Milestones in Healthy Infants: A Randomized, Prospective, Double-blind, Placebo-controlled Trial. *Am. J. Clin. Nutr.* **2009,** *89,* 64–70.

Alz.org Press Release. Results from Trials of DHA in Alzheimer's Disease and Age-Related Cognitive Decline. Vienna, Austria. International Conference on Alzheimer's Disease (ICAD), July 12, 2009. http://www.alz.org/icad/2010 _release_071209

American Heart Association. Common Misconceptions About High Blood Pressure. Published online: http://www.americanheart.org/presenter.jhtml;jsessionid=CPAUCL00ZTCWSCQFCXP SCZQ?identifier=3008517 (Accessed Apr 15, 2009).

American Heart Association and American Stroke Association. Know the Facts, Get the Stats, 2007. Published online: http://www.americanheart.org/downloadable/heart/116861545709855-1041%20KnowTheFactsStats07_loRes.pdf (Accessed Apr 15, 2009).

Arterburn, L.M.; H.A. Oken; E. Bailey Hall; J. Hamersley; C.N. Kuratko; J.P. Hoffman. Algal-Oil Capsules and Cooked Salmon: Nutritionally Equivalent Sources of Docosahexaenoic Acid. *J. Am. Diet. Assoc.,* **2008,** *108,* 1204–1209.

Balk, E.M.; A.H. Lichtenstein; M. Chung; B. Kupelnic; P. Chew; J. Lau. Effects of Omega-3 Fatty Acids on Serum Markers of Cardiovascular Disease Risk: A Systematic Review. *Atherosclerosis* **2006,** *18,* 19–30.

Bays, H. Rationale for Prescription Omega-3-acid Ethyl Ester Therapy for Hypertriglyceridemia: A Primer for Clinicians. *Drugs Today (Barc).* **2008,** *44,* 205–246.

Birch, E.E.; Y.S. Castaneda; D.H. Wheaton; D.G. Birch; R.D. Uauy; D.R. Hoffman. Visual Maturation of Term Infants Fed Long-Chain Polyunsaturated Fatty Acid-Supplemented or Control Formula for 12 Mo. *Am. J. Clin. Nutr.* **2005,** *81,* 871–879.

Burks, W.; S.M. Jones; C.L. Berseth; C. Harris; H.A. Sampson; D.M. Scalabrin. Hypoallergenicity and Effects on Growth and Tolerance of a New Amino Acid-Based Formula with Docosahexaenoic Acid and Arachidonic Acid. *J. Pediatr.* **2008,** *153,* 266–271.

Carlson, S.E.; S.H. Werkman; E.A. Tolley. Effect of Long-Chain n-3 Fatty Acid Supplementation on Visual Acuity and Growth of Preterm Infants with and without Bronchopulmonary Dysplasia. *Am. J. Clin. Nutr.* **1996,** *63,* 687–697.

Clandinin, M.T.; J.E. Chappell; S. Leong; T. Heim; P.R. Swyer; G.W. Chance. Intrauterine Fatty Acid Accretion Rates in Human Brain: Implications for Fatty Acid Requirements. *Early Hum. Dev.* **1980,** *4,* 121–129.

Clandinin, M.T.; J.E. Van Aerde; K.L. Merkel; C.L. Harris; M.A. Springer; J.W. Hansen; D.A. Diersen-Schade. Growth and Development of Preterm Infants Fed Infant Formulas Containing Docosahexaenoic Acid and Arachidonic Acid. *J Pediatr.* **2005,** *146,* 461–468.

Codex Alimentarius Commission. *Report of the 28th Session of the CODEX Committee on Nutrition and Foods For Special Dietary Uses.* Codex Alimentarius Commission, Revised July 2007. Published online: http://www.codexalimentarius.net/download/standards/288/CXS_072e.pdf (accessed Apr 2, 2009).

The Commission of the European Communities. Commission Directive 2006/141/EC of 22 December 2006 on Infant Formulae and Amending Directive 1999/21/EC. *Official Journal of the European Union.* 30.12.2006:L401/1401/33. Published online: http://eur-lex.europa.eu/LexUriServ/LexUriServ.do?uri=OJ:L:2006:401:0001:0033:EN:PDF (accessed Apr 2, 2009).

DHA (Docosahexaenoic Acid), an Omega 3 Fatty Acid, in Slowing the Progression of Alzheimer's Disease. National Institute on Aging, Martek Biosciences Corp. Published online: http://clinicaltrials.gov/ct2/show/NCT00440050?term=dha+AND+alzheimers&rank=1 (accessed Apr 2, 2009).

Dickinson, H.A.; J.M. Mason; D.J. Nicolson; F. Campbell; F. Beyer; J. Cook; B. Williams; G.A. Ford. Lifestyle Interventions to Reduce Raised Blood Pressure: A Systematic Review of Randomized Controlled Trials. *J. Hypertens.* **2006,** *24,* 215–233.

Dobbing, J.; J. Sands. Quantitative Growth and Development of Human Brain. *Arch. Dis. Child.* **1973,** *48,* 757–767.

Dutta-Roy, A.K. Transport Mechanisms for Long-Chain Polyunsaturated Fatty Acids in the Human Placenta. *Am. J. Clin. Nutr.* **2000,** *71*(1Suppl), 315S–22S.

Engler, M.M.; M.B. Engler; M.J. Malloy; S.M. Paul; K.R. Kulkarni; M.L. Mietus-Snyder. Effect of Docosahexaenoic Acid on Lipoprotein Subclasses in Hyperlipidemic Children (The EARLY Study). *Am. J. Cardiol.* **2005,** *95,* 869–871.

Food and Nutrition Board, Institute of Medicine of the National Academies. *Dietary Reference Intakes for Energy, Carbohydrate, Fiber, Fat, Fatty Acids, Cholesterol, Protein, and Amino Acids.* Washington, DC: National Academies Press; 2005.

Gibson, R.A.; D. Barclay; H. Marshall; J. Moulin; J.C. Maire; M. Makrides. Safety of Supplementing Infant Formula with Long-Chain Polyunsaturated Fatty Acids and *Bifidobacterium lactis* in Term Infants: A Randomised Controlled Trial. *Br. J. Nutr.* **2009,** *Jan 12,* 1–8. [Epub ahead of print]

Griffin, M.D.; T.A. Sanders; I.G. Davis; L.M. Morgan; D.J. Millward; F. Lewis; S. Slaughter; J.A. Cooper; G.J. Miller; B. A. Griffin. Effects of Altering the Ratio of Dietary n-6 to n-3 Fatty Acids on Insulin Sensitivity; Lipoprotein Size; and Postprandial Lipemia in Men and Postmenopausal Women Aged 45-70 y: The OPTILIP Study. *Am. J. Clin. Nutr.* **2006,** *84,* 1290–1298.

Hartweg, J.; A.J. Farmer; R. Perera; R.R. Holman; H.A. Neil. Meta-analysis of the Effects on n-3 Polyunsaturated Fatty Acids on Lipoproteins and Other Emerging Lipid Cardiovascular Risk Markers in Patients with Type 2 Diabetes. *Diabetologia.* **2007,** *50,* 1593–1602.

Henriksen, C.; K. Haugholt; M. Lindgren; A.K. Aurvag; A. Ronnestad; M. Gronn; R. Solberg; A. Moen; B. Nakstad; R.K. Berge; et al. Improved Cognitive Development among Preterm Infants Attributable to Early Supplementation of Human Milk with Docosahexaenoic Acid and Arachidonic Acid. *Pediatrics.* **2008,** *121,* 1137–1145.

Hoffman, D.; E. Ziegler; S.H. Mitmesser; C.L. Harris; D.A. Diersen-Schade. Soy-based Infant Formula Supplemented with DHA and ARA Supports Growth and Increases Circulating Levels of These Fatty Acids in Infants. *Lipids.* **2008,** *43,* 29–35.

Innis, S.M.; R.W. Friesen. Essential n-3 Fatty Acids in Pregnant Women and Early Visual Acuity Maturation in Term Infants. *Am. J. Clin. Nutr.* **2008,** *87,* 548–557.

Iso, H. Changes in Coronary Heart Disease Risk among Japanese. *Circulation.* **2008,** *118,* 2725–2729.

Jacobson, T.A. Role of n-3 Fatty Acids in the Treatment of Hypertriglyceridemia and Cardiovascular Disease. *Am. J. Clin. Nutr.* **2008,** *87,* 1981S–1990S.

Jensen, C.L.; R.G. Voigt; T.C. Prager; Y.L. Zou; J.K. Fraley; J.C. Rozelle; M.R. Turcich; A.M. Llorente; R.E. Anderson; W.C. Heird. Effects of Maternal Docosahexaenoic Acid Intake on Visual Function and Neurodevelopment in Breastfed Term Infants. *Am. J. Clin. Nutr.* **2005,** *82,* 125–132.

Kannel, W.B.; P.A. Wolf; J. Verter; P.M. McNamara. Framingham Study Insights on Hazards of Elevated Blood Pressure. *JAMA* **2008,** *300,* 2545–2547.

Kelley, D.S.; D. Siegel; M. Vemuri; B.E. Mackey. Docosahexaenoic Acid Supplementation Improves Fasting and Postprandial Lipid Profiles in Hypertriglyceridemic Men. *Am. J. Clin. Nutr.* **2007,** *86,* 324–333.

Kelley, D.S.; D. Siegel; M. Vemuri; G.H. Chung; B.E. Mackey. Docosahexaenoic Acid Supplementation Decreases Remnant-like Particle-Cholesterol and Increases the (n-3) Index in Hypertriglyceridemic Men. *J. Nutr.* **2008,** *138,* 30–35.

Kelley, D.S.; D. Siegel; D.M. Fedor; Y. Adkins; B.E. Mackey. DHA Supplementation Decreases Serum C-Reactive Protein and Other Markers of Inflammation in Hypertriglyceridemic Men. *J. Nutr.* **2009** [Epub ahead of print]

Koletzko, B.; I. Cetin; J.T. Brenna; Perinatal Lipid Intake Working Group; Child Health Foundation; Diabetic Pregnancy Study Group; European Association of Perinatal Medicine; European Association of Perinatal Medicine; European Society for Clinical Nutrition and Metabolism; European Society for Paediatric Gastroenterology; Hepatology and Nutrition; Committee on Nutrition; International Federation of Placenta Associations; International Society for the Study of Fatty Acids and Lipids. Dietary Fat Intakes for Pregnant and Lactating Women. *Br. J. Nutr.* **2007a,** *98,* 873–877.

Koletzko, B.; E. Larqué; H. Demmelmair. Placental Transfer of Long-Chain Polyunsaturated Fatty Acids (LC-PUFA). *J. Perinat. Med.* **2007b,** *35,* S5–S11.

Koletzko, B.; E. Lien; C. Agostoni; H. Böhles; C. Campoy; I. Cetin; T. Decsi; J.W. Dudenhausen; C. Dupont; S. Forsyth; et al; World Association of Perinatal Medicine Dietary Guidelines Working Group. The Roles of Long-Chain Polyunsaturated Fatty Acids in Pregnancy; Lactation and Infancy: Review of Current Knowledge and Consensus Recommendations. *J. Perinat. Med.* **2008,** *36,* 5–14.

Kris-Etherton, P.M.; W.S. Harris; L.J. Appel; American Heart Association, Nutrition Committee. Fish Consumption; Fish Oil; Omega-3 Fatty Acids; and Cardiovascular Disease. *Circulation* **2002,** *106,* 2747–2757.

Krotz, F.; H.Y. Sohn; V. Klauss. Antiplatelet Drugs in Cardiological Pactice: Established Strategies and New Developments. *Vasc. Health Risk Manage.* **2008,** *4,* 637–645.

Kusumoto, A.; Y. Ishikura; H. Kawashima; Y. Kiso; S. Takai. Effects of Arachidonate-Enriched Triacylglycerol Supplementation on Serum Fatty Acids and Platelet Aggregation in Healthy Male Subjects with a Fish Diet. *Br. J. Nutr.* **2007,** *98,* 626–635.

Lu, G.; S.L. Windsor; W.S. Harris. Omega-3 Fatty Acids Alter Lipoprotein Subfraction Distributions and the in vitro Conversion of Very Low Density Lipoproteins to Low Density Lipoproteins. *J. Nutr. Biochem.* **1999**, *10,* 151–158.

Maclean, C.H.; A.M. Issa; S.J. Newberry; W.A. Mojica; S.C. Morton; R.H. Garland; L.G. Hilton; S.B. Traina; P.G. Shekelle. Effects of Omega-3 Fatty Acids on Cognitive Function with Aging; Dementia; and Neurological Diseases. *Evid. Rep. Technol. Assess. (Summ.)* **2005**, *114,* 1 –3.

Maki, K.C.; M.E. Van Elswyk; D. McCarthy; S.P. Hess; P.E. Veith; M. Bell; P. Subbaiah; M.H. Davidson. Lipid Responses to a Dietary Docosahexaenoic Acid Supplement in Men and Women with Below Average Levels of High Density Lipoprotein Cholesterol. *J. Am. Coll. Nutr.* **2005,** *24,* 189–199.

Makrides, M.; R.A. Gibson; A.J. McPhee; C.T. Collins CT; P.G. Davis; L.W. Doyle; K. Simmer; P.B. Colditz; S. Morris; L.G. Smithers; et al. Neurodevelopmental Outcomes of Preterm Infants Fed High-Dose Docosahexaenoic Acid: A Randomized Controlled Trial. *JAMA.* **2009,** *301,* 175–182.

Memory Improvement with Docosahexaenoic Acid Study (MIDAS). Martek Biosciences Corp. Published online: http://clinicaltrials.gov/ct2/show/NCT00278135?term=midas&rank=2 (accessed Apr 2, 2009).

Minihane, A.M.; S. Kahn; E.C. Leigh-Firbank; P.Talmud; J.W. Wright; M.C. Murphy; B.A. Griffin; C.M. Williams. ApoE Polymorphism and Fish Oil Supplementation in Subjects with an Atherogenic Lipoprotein Phenotype. *Arterioscler. Thromb. Vasc. Biol.* **2000,** *20,* 1990–1997.

Oe, H.; T. Hozumi; E. Murata; H. Matsuura; K. Negishi; Y. Matsumura; S. Iwata; K. Ogawa; K. Sugioka; Y. Takemoto; et al. Arachidonic Acid and Docosahexaenoic Acid Supplementation Increases Coronary Flow Velocity Reserve in Japanese Elderly Subjects. *Heart.* **2008,** *94,* 316–321.

Putnam, J.; Carlson S.; DeVoe P.; Barness L. The Effect of Variations in Dietary Fatty Acids on the Fatty Acid Composition of Erythrocyte Phosphatidylcholine and Phosphatidylethanolamine in Human Infants. *Am J Clin Nutr.* **1982,** *36,* 106–114.

Ryan, A.S.; M.B. Montalto; S. Groh-Wargo; F. Mimouni; J. Sentipal-Walerius; J. Doyle; J.S. Siegman; A.J. Thomas. Effect of DHA-Containing Formula on Growth of Preterm Infants to 59 Weeks Postmenstrual Age. *Am. J. Hum. Biol.* **1999,** *11,* 457–467.

Ryan, A.S. and E.B. Nelson. Assessing the effect of docosahexaenoic acid on cognitive functions in healthy, preschool children: a randomized, placebo-controlled, double-blind study. *Clin Pedi,* **2008,** 47, 355-362.

Ryan, A.S.; M.A. Keske; J.P. Hoffman; E.B. Nelson. Clinical Overview of Algal-Docosahexaenoic Acid: Effects on Triglyceride Levels and Other Cardiovascular Risk Factors. *Am. J. Ther.* **2009,** *16,* 183–192.

Sanders, T.A.; K. Gleason; B. Griffin; G.J. Miller. Influence of an Algal Triacylglycerol Containing Docosahexaenoic acid (22 : 6n-3) and Docosapentaenoic Acid (22 : 5n-6) on Cardiovascular Risk Factors in Healthy Men and Women. *Br. J. Nutr.* **2006,** *95,* 525–531.

Schaefer, E.J.; V. Bongard; A.S. Beiser; S. Lamon-Fava; S.J. Robins; R. Au; K.L. Tucker; D.J. Kyle; P.W. Wilson; P.A. Wolf. Plasma Phosphatidylcholine Docosahexaenoic Acid Content and Risk of Dementia and Alzheimer disease: The Framingham Heart Study. *Arch. Neurol.* **2006,** *63,* 1545–1550.

Schwellenbach, L.J.; K.L. Olson; K.J. McConnell; R.S. Stolcpart; J.D. Nash; J.A. Merenich. The Triglyceride-Lowering Effects of a Modest Dose of Docosahexaenoic Acid Alone versus in Combination with Low Dose Eicosapentaenoic Acid in Patients with Coronary Artery Disease and Elevated Triglycerides. *J. Am. Coll. Nutr.* **2006,** *25,* 480–485.

Siahanidou, T.; C. Lazaropoulou; K. Michalakakou; I. Papassotiriou; C. Bacoula; H. Mandyla. Oxidative Stress in Preterm Infants Fed a Formula Containing Long-Chain Polyunsaturated Fatty Acids (LCPUFA). *Am. J. Perinatol.* **2007,** *24,* 475–479.

Siahanidou, T.; A. Margeli; C. Lazaropoulou; E. Karavitakis; I. Papassotiriou; H. Mandyla. Circulating Adiponectin in Preterm Infants Fed Long-Chain Polyunsaturated Fatty Acids (LCPUFA)-Supplemented Formula—A Randomized Controlled Study. *Pediatr. Res.* **2008,** *63,* 428–432.

Smithers, L.G.; R.A. Gibson; A. McPhee; M. Makrides. Effect of Two Doses of Docosahexaenoic Acid (DHA) in the Diet of Preterm Infants on Infant Fatty Acid Status: Results from the DINO trial. *Prostaglandins Leukot. Essent. Fatty Acids.* **2008a,** *79,* 141–146.

Smithers, L.G.; R.A. Gibson; A. McPhee; M. Makrides. Higher Dose of Docosahexaenoic Acid in the Neonatal Period Improves Visual Acuity of Preterm Infants: Results of a Randomized Controlled Trial. *Am. J. Clin. Nutr.* **2008b,** *88(4),* 1049–1056.

Theobald, H.E.; A.H. Goodall; N. Sattar; D.C. Talbot; P.J. Chowienczyk; T.A. Sanders. Low-Dose Docosahexaenoic Acid Lowers Diastolic Blood Pressure in Middle-aged Men and Women. *J. Nutr.* **2007,** *137,* 973–978.

Vanderhoof, J.A. Hypoallergenicity and Effects on Growth and Tolerance of a New Amino Acid-based Formula with DHA and ARA. *J. Pediatr. Gastroenterol. Nutr.* **2008,** *47,* S60–S61.

Vasan, R.S.; M.G. Larson; E.P. Leip; J.C. Evans; C.J. O'Donnell; W.B. Kannel; D. Levy. Impact of High-Normal Blood Pressure on the Risk of Cardiovascular Disease. *N Engl J Med.* **2001,** *345,* 1291–1297.

Wang, C.; W.S. Harris; M. Chung; A.H. Lichtenstein; E.M. Balk; B. Kupelnick; H.S. Jordan; J. Lau. N-3 Fatty Acids from Fish or Fish-Oil Supplements; But Not   -Linolenic Acid; Benefit Cardiovascular Disease Outcomes in Primary- and Secondary-Prevention Studies: A Systematic Review. *Am. J. Clin. Nutr.* **2006,** *84,* 5–17.

Yuhas, R.; K. Pramuk; E. L. Lien. Human Milk Fatty Acid Composition from Nine Countries Varies Most in DHA. *Lipids* **2006,** *41,* 851–858.

Yurko-Mauro K. et al. Results of the MIDAS Trial: Effects of Docosahexaenoic Acid on Physiological and Safety Parameters in Age-related Cognitive Decline. Abstract 09-A-2137-ALZ presented 7/12/09 at International Conference on Alzheimer's Disease (ICAD 2009) Vienna. http://www. alz.org/icad/documents/abstracts/abstracts_dha_ICAD09.pdf

18

Applications of Single Cell Oils for Animal Nutrition

Jesus R. Abril, Todd Wills, and Flint Harding

Martek Biosciences, 4909 Nautilus Court North, Suite 208, Boulder Colorado, USA. 80301

Introduction

The importance of the foods we eat in maintaining a healthy lifestyle has become evident as we move into the twenty-first century. A well-balanced diet and moderate exercise, in most cases, meet this need. From the perspective of health-conscious consumers, one of the most commonly monitored nutrients is fat. In the past, we have routinely seen diet fads come and go, but the one nutrient that seems to occupy everyone's minds is fat. The food industry has responded quickly to consumer concerns regarding fat, and there have been many trends, including the move from "partially hydrogenated fat" to "fully hydrogenated, no *trans* fat." In addition to these changes, the scientific community has gained a more thorough understanding of the importance of polyunsaturated fats or fatty acids (PUFAs) and their contribution to a healthy diet. Recently, there has been a shift in focus from PUFAs, in general, to, more specifically, n-3 fats (omega-3). Furthermore, particular interest in the health benefits of these fatty acids has led to differentiation between short- and long-chain PUFAs, and we are now gaining an even better understanding of the differences between long-chain fatty acids, EPA (eicosapentaenoic acid, 20:5 n-3), and DHA (docosahexaenoic acid, 22:6 n-3).

The nutritional importance of EPA and DHA is now well defined, but consumption levels generally fall far below recommended levels. Since current eating habits in the Western world will not allow consumers to reach recommended intakes of n-3 fatty acids with fish consumption alone, the remaining amounts of these fatty acids need to be supplied by other means. Such alternatives could come from animal products enriched with high concentrations of n-3 fatty acids, direct fortification of foods with n-3 fatty acids, or dietary supplementation with fish or algal oil capsules.

Although EPA and DHA are native components of cell membranes in livestock, the n-3 fat composition of livestock has generally decreased over the past 150 years. Such decreases are primarily due to domestication and changes in animal feed rations. Modern agricultural practices have shifted livestock diets from a wild and natural feed environment (which offers a variety of plant sources containing short-chain n-3 fatty acid precursors and LCPUFAs) to confined feed operations, which are designed to improve production and yields. Such confined feed operations predominantly utilize cereal grains and seed oils as feed sources. This move from a more natural, plant-based diet to a cereal-based diet has resulted in an undesirable escalation in the omega-6 fatty acid content of livestock tissues, at the expense of the n-3 fatty acids. Meat, milk, and eggs from animals found in the wild will generally have more n-3 fatty acids when compared with similar products from domesticated animals.

The gross composition of domestic animal muscle tissue varies considerably due to differences in species, breed, sex, age, activity, and nutrition. Variation in composition is primarily influenced by the accumulation of lipids from the diet. Lipids are a diverse group of biological substances made up of both polar and non-polar compounds. Neutral (non-polar) lipids are comprised of fatty acids esterified to either glycerol (triglycerides) or a fatty alcohol and are commonly found in cells as storage fats and oils and as intermuscular deposits in adipose tissue. Phospholipids (polar lipids) are the primary lipids associated with biological membranes. The polar nature of phospholipids arises from their phosphate-containing headgroup, which is linked to the third carbon of the glycerol molecule. This group leads to the greatest variation in the physical properties of the phospholipids (Sigfusson, 2000). Furthermore, phospholipids play an important role in governing the quality of meat during both raw processing and cooking, and they are important flavor precursors because of their high content of long-chain PUFAs. The high PUFA content of phospholipids [20–50% of total fatty acids (TFA) in the phospholipids] is primarily represented by long-chain fatty acids with 18, 20, and 22 carbons with two to six double bonds. In muscle tissue, the phospholipid content is relatively constant and less influenced by species, breed, nutrition, and age taking into consideration that phospholipids are constituents of cell membranes. Differences do, however, exist between fatty acid compositions of different muscle fiber types. Red muscle tissue has a higher concentration of phospholipids than white muscle tissue, and, therefore, has a higher percentage of PUFA. A change in the fatty acid profile of cell membranes alters membrane properties and other physiological functions. For this reason, the PUFA proportion of phospholipids is strictly controlled by a complex enzymatic system, consisting of desaturases and elongases, and is responsible for the conversion of both linoleic acid (LA) and linolenic acid (ALA) to their long-chain metabolites. These enzymes act on both n-6 and n-3 fatty acids but have a preference for the n-3 (Brenner, 1989). In addition, there is competition between n-6 and n-3 fatty acids for incorporation into the phospholipids (Raes et al., 2004).

Numerous feeding trials have been carried out using different species and breeds of livestock, aimed at bringing the polyunsaturated fatty acid/saturated fatty acid (P:S) ratio of meat closer to the recommended value (>0.7), as well as decreasing the n-6/n-3 ratio (<5) (Raes et al., 2004). Such work has focused on modifying the fatty acid composition of animal tissue through nutritional means by including specific oils or oilseeds, marine products, or types of forage in the animal's diet. This research has shown that the P:S ratio is primarily determined by dietary fatty acids, species of animal, and, to a lesser extent, breed of animal.

In the past, meat and other animal products were estimated to have a much higher ratios of n-3 to n-6 fatty acids (1:2 to 1:3), whereas today the ratios have dropped to 1:10, 1:15, or even lower Such a profound shift comes with detrimental effects to the human population that customarily consumes these animal products. Since some meats naturally have a P:S ratio of around 0.1, they have been implicated in causing an imbalanced fatty acid intake in today's consumers (Wood et al., 2003). As we have previously discussed, this imbalance is a result of feeding elevated levels of 18:2 (linoleic acid), which is directly passed from cereal based diets, thereby creating the n-6/n-3 imbalance.

Much of the original work in the area of PUFA enrichment in domestic animals has focused both on fish-based long-chain n-3 ingredients (fish oil and fish meal) and on the readily available short-chain ALA (alpha linolenic acid), as these were the only ingredients available at the time. In the late 1980s, biotech companies started developing alternative sources to fish oils, which they accomplished through the large-scale production of single cell oils (SCOs) that contained significant quantities of long-chain PUFAs. Two of the pioneers in this area were Martek Biosciences, producing DHA via the single cell organism, *Crypthecodinium cohnii*, and OmegaTech, Inc., producing DHA via the single cell organism, *Schizochytrium* sp. In 2002, these two companies merged and maintained the Martek name. Since that time, Martek has remained focused on producing SCOs for infant formula companies and mainstream food companies, as well as the animal feed industry.

The food industry has taken advantage of the nutritional benefits of n-3 fatty acids by introducing new functional foods that are directly fortified with n-3 rich oils. The increasing demand for n-3 PUFA enriched foods has also directed awareness to the potential of products from domestic animals as a means for delivering n-3 fatty acids to consumers. This chapter deals with the different approaches, challenges, and results of enriching eggs, milk, and meat products via the direct feeding of n-3 rich single-cell organisms to domestic animals. SCOs may represent the best alternative to achieve this goal because of their efficiency of transfer and ease of use in the well-known feeding operations that are currently used in animal husbandry. This chapter will also touch on the health and performance benefits of feeding long-chain n-3 PUFAs to animals and will discuss the nutritional benefits for companion animals and horses.

Enrichment of Animal Products with Long-Chain Omega-3 PUFAs

It is evident that concentrations of long-chain n-3 PUFAs—mainly, EPA and DHA—are normally found at very low levels in animal food products (e.g., eggs, meat, milk, etc.). Increasing evidence of the benefits associated with the consumption of these healthy fats

Table 18.A. Comparison of the Primary Product Benefits of the Major Sources of ω3 Fatty Acids, for Use in Feeds.

Product Criteria	n-3 Fatty Acid Feed Ingredient			
	Fish Oil	Fish Meal	Flaxseed	*Schizochytrium*
Limits to Use	90–120 mg DHA/egg	60–80 mg DHA/egg	80 mg DHA/egg	>200 mg DHA/egg
Taste/Odor Problems	High	High	Low	None
Proprietary Technology	No	No	No	Yes
Price	Low	Low	Low	High
Efficiency of DHA Incorporation (%)	40	40	3–6	55–65
Stability	Low	Medium	High	High
Ease of Use	Difficult	Easy	Easy	Easy
Health Issues	Fatty liver syndrome	None	Increase in double yolks	None
Environmental Contaminants	Trace PCB and dioxin	None	None	None

has prompted the animal feed industry to find practical and efficient ways to boost or reintroduce these "good fats" into the food chain. Animal enrichment with LCPUFAs provides a set of challenges for the industry that can be generalized into the following categories: feed type, animal species, and source of the n-3 fat. The most common ingredients used in feed applications as sources of n-3s are flax seeds, fish oils, and single-cell organisms. Table 18.A illustrates a general comparison of these ingredients, as applied to poultry feeding. The efficiency of n-3 fatty acid deposition varies tremendously a0nd is highly dependent on animal species, source of n-3 being used in the feed, and carbon chain length. Bourre (2005) indicated that feeding extracts of flax or rapeseed resulted in a two-, six-, or ten-fold increase in the levels of n-3 fatty acids found in beef, pork, and chicken, respectively, whereas, when feeding fish extracts or algal oils, DHA levels increased two, six, seven, and twenty times for beef, eggs, chickens, and fish, respectively.

Efficiency of deposition and transfer decreases in larger animals and becomes even more limiting in the case of ruminants. With this information in mind, the source of n-3 and its form (e.g., seed or free oil) will have a definitive impact on the efficiency, percent transferred, and type of fatty acids deposited. Whole-cell dried algae presents a set of unique characteristics, which enables it to overcome many issues seen in competing sources. Such superiority is directly evident in the case of ruminants where lipid hydrogenation is of great concern. Single-cell organisms (e.g., whole-cell dried algae) offer a "fat by-pass effect" for the case of ruminants; therefore, they directly protect LCPUFAs from the ruminal effects of micro-flora and enzymes. As in the case of simple gastric systems, single-cell organisms offer a more stable form of delivery, as well as a more user-friendly ingredient, in terms of handling. These advantages are primarily due to the fact that lipid bodies are an integral part of the whole cell; thus, they minimize surface oil and gain valuable oxidative protection from native cell walls.

Poultry Enrichment

The rate and efficiency of PUFA uptake and incorporation has received considerable attention. Generally, it is agreed that the most efficient animal at transferring and depositing LCPUFAs are laying hens, followed by broiler chickens. In laying hens, the rate of transfer can be between 50 and 60%, depending on feed composition, whereas, in the case of broiler chickens, efficiency is closer to 25% (Abril & Barclay, 1998).

Layers

Eggs are considered to be one of the most complete foods, from a nutritional point of view. Despite their nutritional value and low price, eggs have received a bad reputation due to their relatively high cholesterol content (ca. 250 mg contained in yolk) and the perception that cholesterol-rich foods lead to coronary heart disease (CHD) and atherosclerosis. Eggs and poultry meat that contain high levels of n-3 fatty acids can be produced by using feed ingredients rich in ALA, EPA, and DHA. Many studies have focused on altering the fatty acid profile of eggs by introducing the n-3 fatty acids in layer rations.

Recently, Kralik et al. (2008) supplemented feed for laying hens ("layer feeds") with combinations of rapeseed and fish oil, thereby replacing a portion of soybean oil. These researchers were specifically interested in measuring laying hen performance and the fatty acid profiles of their eggs, primarily focusing on the n-3 PUFAs. Layer feeds were supplemented with either 3.5 or 1.5% fish oil or rapeseeds. Their findings indicated significant differences in laying hen performance, as well as differences in the egg yolk fat composition. Increases in the content of ALA, EPA, and DHA were reported, with a lower deposition of ARA (arachidonic acid, 20:4n6). The results also show a greater increase of ALA at the lower level of rapeseed supplementation, which can be explained by the higher level of fish oil in the diet. These findings may contradict similar feeding studies where sources of ALA were fed (Van Elswyk, 1997). Scheideler & Froning (1996) also reported a linear response in ALA deposition in egg yolk fats when feeding flax seeds at 5, 10, and 15% of the diet, with a 2.31, 4.18, and 6.83% increase in ALA, respectively. An increase in DHA was noted in eggs derived from the feed group that contained supplemental fish oil—2.87% as compared to 0.52% for the control egg. EPA was not included in the report, suggesting that EPA does not deposit to any great extent in eggs, even when feeding fish oil containing high levels of EPA (Van Elswyk, 1997; Abril & Barclay, 1998; Hargis et al., 1991; Damiani et al., 1994; Herber & Van Elswyk, 1996).

A study by Basmacioglu et al. (2003), where laying hens were fed combinations of fish oil and flax seed, clearly demonstrated the competitive effect of incorporating ALA as opposed to DHA. Control eggs, containing 0.65% DHA (% of TFA), increased to 3.29% when the diet contained 1.5% fish oil. A substantial reduction was evident (decrease to 2.9%) when the fish oil was complemented with 4.32% flax seeds. Doubling the amount of flax seeds, from 4.32 to 8.64%, did not show a significant difference in the amount of DHA deposited. Also, it is noteworthy that cholesterol levels were not affected by any of the diets, a finding that has been demonstrated in many other studies.

Ethoxyquin-stabilized menhaden oil (1.5% of diet) has been compared to the single-cell marine alga *Schizochytrium* sp. (2.4 and 4.8% of diet) as a source of the long-chain PUFAs EPA and DHA (Herber & Van Elswyk, 1996). Results indicated that, at the end of 4 weeks, the egg yolk fat contained 8.1, 8.8, and 11.5 mg/g of DHA, respectively, with the 2.4% marine algae performing very similarly to the 1.5% menhaden oil treatment. Also, it was noted that EPA did not deposit more than 12 mg/yolk, accounting for less than 5% of the diet's total EPA. The authors hypothesized that the natural carotenoids in marine algae may have provided extra antioxidant protection for the feed, as well as for the resulting animal products. Fredriksson et al. (2006) experimented with the feeding of single-cell microalga, specifically, *Nanochloropsis oculata*, to laying hens. The goal of this work was to enrich eggs by feeding algae in up to 20% of the hens' ration (dry matter basis), combined with either rapeseed oil or corn oil. Hens were supplemented for four weeks, allowing the fatty acids to equilibrate. The microalgae contained 3.6% lipids on a dry matter basis, with 37.1% of EPA (mostly found in the polar fraction), no DHA, and 6.7% of ALA. When added to the feed, the single-cell organism provided 1.5% EPA to the 10% algae feed, 3.1% EPA to the 20% algae feed as TFA on a dry matter (DM) basis, and 4.7 to 8.4% ALA, depending on feed composition (described in Table 18.B). Egg yolk lipids were analyzed as the neutral and polar fractions, triacylglycerides and phospholipids, respectively. Results indicated that EPA was elongated to DHA, to some extent. In the case of rapeseed oil fed alone, DHA also increased, possibly indicating an elongation-desaturation process. Unfortunately, the authors did not detail the basal

Table 18.B. Selected Fatty Acids Found in Egg Yolks from Layers Fed Different Levels of Rapeseed Oil, Corn Oil, and Algae.

Phospholipids	% (TFA)				
	RO	RO+	RO+	RO/CO	RO/CO+
18:2 n-6	13.7[a]	13.7[a]	13.2[a]	17.0[b]	16.8[b]
18:3 n-3	0.5[a]	0.4[a]	0.4[a]	0.3[b]	0.2[b]
20:5 n-3	0.1[a]	0.3[b]	0.5[c]	n/d	0.2[d]
22:6 n-3	6.5[a]	6.9[a]	7.4[c]	5.1[b]	5.8[d]
Triacylglycerols					
18:2 n-6	14.5[a]	14.9[a]	14.6[a]	20.1[b]	19.9[b]
18:3 n-3	2.1[a]	2.4[b]	2.5[b]	1.4[c]	1.4[c]
20:5 n-3	0.1[a]	0.1[a]	0.1[a]	0.02[b]	0.06[ab]
22:6 n-3	0.3	0.3	0.3	0.2	0.2

From: Fredriksson et al. (2006)
RO=Rapeseed oil, CO=Corn oil, (+) = Plus 10% algae,
a-d: Different letters differ significantly (p < 0.05)
n/d = Not detected

Table 18.C. Typical Fatty Acid Content (mg/egg) of Eggs from Hens Fed DHA GOLD™ (Providing 165 mg DHA/hen/day) and 2.5% Flaxseed, Compared with Eggs from Control Hens.

Fatty Acid	Control Egg	DHA Gold™ Egg
C14:0 (myristic)	19.2	21.2
C14:1 (myristoleic)	0.0	8.0
C16:0 (palmitic)	1240.1	1103.0
C16:1 (pamitoleic)	129.4	194.9
C18:0 (stearic)	430.9	314.6
C18:1 (oleic)	1863.7	1414.7
C18:2 (linoleic)	688.9	510.3
C18:3 (linolenic)	15.0	94.9
C20:2 (eicosadienoic)	42.6	29.2
C20:4 (arachidonic)	86.8	53.3
C20:5 (EPA)	0.0	10.2
C22:0 (behenic)	0.0	11.7
C22:5 (DPA, ω6)	24.2	13.1
C22:5 (DPA, ω3)	0.0	7.3
C22:6 (DHA, ω3)	27.6	135.1
Total ω3/egg	43.0	248.0
Long-chain ω3	28.0	153.0
Short-chain ω3	15.0	95.0
ω6/ω3 ratio	19:1	2:1

From: Abril & Barclay, 1998.
Data are from a dozen eggs pooled for each treatment.

levels of DHA and EPA present in the eggs derived from control layers who were not fed any preformed long-chain fatty acids or other precursors. Table 18.C illustrates the fatty acid composition of eggs produced by feeding the microalga *Schizochytrium* sp., as compared with eggs that were produced by excluding, as much as possible, the precursor ALA and excluding completely the preformed EPA and DHA fatty acids. Table 18.D concurs with the previous study, indicating that most of the long-chain PUFAs do end up in the phospholipid fraction.

Broilers

Another area of interest, with respect to poultry, is in the area of meat production—specifically, in this case, broiler chickens. Moving from egg to muscle enrichment is accompanied by a decrease in the efficiency of transfer and deposition of these LCPUFAs. One interesting observation and consideration of broiler enrichment is the relatively short life cycle of a broiler chicken versus a laying hen. Broilers have a life span measured in just a few weeks, compared with laying who are kept in production for several years. This difference creates a new variable that needs to be taken into consideration in order to attain a maximum rate of transfer of LCPUFAs. In any case

Table 18.D. Fatty Acid Distribution in Egg Phospholipid and Triacylglycerol Fractions from DHA-Enriched Eggs as the Percentage of Total Fatty Acids (as Fatty Acids Methyl Esters).

Fatty Acid	Phospholipid	Triacylglycerol
C14:0	0.00	0.68
C16:0	33.70	26.75
C16:1	0.00	3.47
C18:0	18.10	7.61
C18:1	24.20	46.43
C18:2	10.70	12.13
C18.3	0.00	1.73
C20:1	0.00	0.12
C20:2	0.10	0.77
C20:4	3.00	0.02
C20:5ω3	0.20	0.00
C22:5ω6	0.80	0.00
C22:5ω3	0.20	0.00
C22:6ω3	8.80	0.39

From: Abril & Barclay, 1998.

of animal enrichment with DHA, there is a lag time before the animal products reach a level of DHA that can be considered commercially viable or nutritionally enhanced. In the case of eggs, it takes between 7 and 10 days of feeding before laying hens reach this level, whereas broilers accomplish this goal in about 10 to 14 days. Enrichment of broiler meat is more economically feasible as the animal gets closer to its commercial weight, and enrichment can be limited to as little as 14 days, pre-slaughter. One such example was demonstrated when 3.6 g of DHA was supplemented to broiler chickens, representing approximately 18 g of the single-cell algae (DHA GOLD®) fed over the production cycle (in this case, 14 days prior to slaughter). Supplementation resulted in DHA uptake in white meat and increased from 14 mg/100 g in the control meat to 88 mg/100 g in the DHA-fed group. In the case of dark meat, the amounts were increased from 24 to 54 mg/100 g. Differences in absolute amounts present in white and dark meat are, most likely, due to the phospholipid content of different muscle types. One interesting aspect of DHA enrichment was the substantial shift observed in the n-6/n-3 ratio—from 14:1 to 4:1 in white meat and 37:1 to 11:4 in the dark meat (control and DHA-enriched, respectively) (Abril & Barclay, 1998). In another study, broiler chickens were fed ground flax or whole flax seeds, and researchers monitored the fatty acid uptake and composition of the meat produced. As expected, the whole flax seeds had a minimum impact on ALA increases and even less of an effect on the elongation/desaturation path to EPA and DHA. Ground flax seeds had greater influence, but enrichment was still on the low side. The greatest impact was seen on ALA concentration: up to 11 mol % for the 7.5% ground flax seed in the diet versus 3.2 mol % for the same level of whole flaxseed. DHA from elongation of the ALA levels was not statistically significant for any of the treatments, remaining at 0.2 to 0.3

mol% (Roth-Maier et al., 1998). The authors of this study seem to have missed the fact that broiler chickens need coarse stones included in their diets to help grind their feed. As a result, the whole flax seeds may have just passed through the intestinal tract without being digested. This oversight could partially explain why low levels of ALA were observed in the treatment containing whole seeds.

In a study done by Manilla et al. (1999), they investigated broiler chicken diets supplemented with 4% linseed oil or 4% cod liver oil. They observed a significant increase of ALA in breast muscle for the linseed oil diet, but not as much of an increase of DHA (10% TFA and 2.6% TFA, as compared with the control of 0.6% TFA and 3.4% TFA, respectively). On the other hand, the cod liver oil-supplemented diet showed a significant increase in DHA levels, as compared with the control diet (9.9% TFA and 3.4% TFA, respectively). Therefore, it is clear that the efficiency of conversion of ALA to DHA in broilers is low, whereas the transfer of pre-formed long-chain PUFAs proves much more efficient. This finding also applies in the case of marine microalgae, with regard to DHA efficiency of transfer from diet to muscle (Manilla et al., 1999). The enrichment of broiler muscle by the utilization of fish oils is limited to about 1 or 2% inclusion in the diet (Lopez-Ferrer et al., 2001). This limitation is primarily due to a deteriorating sensory quality of the meat above these levels. In this case, feeding 2% and 4% fish oil, combined with 6% and 4% tallow, respectively, showed a linear response in DHA enrichment of thigh muscle—from 1.03% to 2.42% (% TFA as methyl esters)—and a substantial increase from the 0.1% found in control meat. One and two weeks before slaughter (a 38-day period), the fish oil diets were replaced by a mixture of fish oil, linseed oil, and tallow at 1.3% and 4.0%, respectively. As expected, DHA levels decreased from 2.42% to 1.72% and 1.42% for the 1- and 2-week periods; ALA increased from 2.98% to 7.28% and 10.28% for the same time period. The reason the authors used this type of approach was to improve the sensory quality of the meat, since the 2% and 4% fish oil diets would yield unacceptable sensory results.

Feeding the single-cell algae *Schizochytrium* at 2.8% and 5.5%, as compared to 2.1% menhaden fish oil, produced a substantial increase in DHA levels (109, 196, and 106 mg/100 g of breast meat, respectively), with the highest level of marine algae providing the highest levels of enrichment. Although the composition of fish oil naturally contains higher levels of EPA (135 g/kg), as compared to DHA (91 mg/kg), levels of EPA in the muscle were only reported at 42.7 mg/100 g. The study also demonstrated that feeding 2% fish oil might be the upper limit before sensory quality deteriorates in broiler meat. Also, the researchers noted that the 2.8% marine algae diet was the only one that maintained acceptable sensory scores throughout the 82 days of storage (Mooney et al., 1998).

Essentially, it is possible to modify the n-3 fatty acid profile of broiler meat by feeding fish oil, linseed oil, and marine microalgae. As noted above, however, limitations apply in the case of fish oil because of the unacceptable off-flavors associated with enriched meat. In the case of linseed oil, the use of whole seeds creates a problem of inefficient digestion, which limits the subsequent transfer and uptake of ALA. This issue is often overcome, however, as linseed oil can be added to the feed, yielding a substantial increase in tissue ALA, but with limited elongation and desaturation of the long-chain PUFAs. Therefore, feeding a single-cell microorganism rich in long-chain PUFAs, mainly DHA, provides the best solution for LCPUFA meat enrichment. Such a strategy creates a meat product enriched in important LCPUFA n-3 fatty acids, while minimizing or eliminating potential sensory issues stemming from the use of fish oil.

Swine

Large animal species, such as swine, present an opportunity to improve their meat quality through diet by feeding them sources of n-3 fatty acids. In the case of pork meat, one must deal with larger feed intakes, as well as higher fat content, which will present added challenges for this type of enrichment.

Native levels of n-3 PUFAs are fairly low or non-existent and are rarely reported in swine (Irie & Sakimoto, 1992). In general, PUFAs are primarily located in fat, with smaller amounts deposited in the lean muscle. After feeding elevated levels of fish oil (2, 4, and 6%) from sardines, Irie and Sakimoto removed biopsy samples of subcutaneous fat. These authors detected increases in EPA and DHA levels during the first week of feeding the fish oil diets, also noting that the rate of increase was greater in the first two weeks than during the last two. EPA and DHA increased from 0.07% and 0.25% to 0.92% and 1.10%, respectively, for the 6% fish oil diet. They also reported that most of the fatty acid changes occurred within 4 to 5 weeks after diet introduction.

In another study, feeding 5 and 20 g of fish oil in a basal pig diet resulted in a rapid increase in concentrations of EPA, DPA (docosapentaenoic acid, 22:5n-3), and DHA in fat tissue, but these acids increased to a much lesser extent in muscle. Additionally, it was noted that more than two weeks were needed to influence the n-6/n-3 ratio (Taugbol, 1993). The rate of incorporation for n-3 fatty acids was less for muscle than for fat tissue, as shown in the diet containing 20 g fish oil in Table 18.E. The largest increases observed in this study were for the longer-chain fatty acids, DHA and DPA, and, to a lesser extent, EPA.

Feeding the precursors to the long-chain n-3 fatty acids, DHA and EPA, has been primarily accomplished by the inclusion of linseed (flax), and, to a lesser extent, Chia seeds. Feeding 4.0 g/kg of ALA, as a test diet for boars and gilts, showed no statistical difference between the sexes in muscle tissue enrichment for omega-3 fatty

Table 18.E. Effect of Feeding 20 g Fish Oil in the Diet on Long-Chain PUFAs in Fat and Muscle Tissue of Swine.

	Values Reported as % Area		
	DHA	EPA	DPA
Fat	15.41	7.21	10.30
Muscle	7.76	6.20	6.62

Data from: Taugbol, 1993

Table 18.F. Omega-3 Fatty Acid Composition of Muscle (*L. lumborum*) from Pigs Fed Test Diet.

	Values Reported as mg/100 g Muscle			
	ALA	EPA	DPA	DHA
Boar	16.9	8.3	13.0	5.6
Gilt	15.7	8.3	12.1	5.3

Data from: Enser et al., 2000.

acids (the control contained 1.9 g/kg of ALA), as seen in Table 18.F. Control diets already had a significant amount of DHA—4.4 and 3.7 mg/100 g for gilt and boar, respectively. Therefore, elongation to DHA after feeding higher amounts of ALA did not account for a significant amount of this fatty acid, although it was statistically different from the control (Enser et al., 2000).

An interesting study was conducted by Haak et al. (2008), where they investigated the effect of duration and time of feeding n-3 PUFA sources on the fatty acid composition of the logissimus thoracis muscle in swine. This study was carried out by feeding the swine fish oil or crushed linseed for approximately 17 weeks (until they reach 100 kg slaughter weight). A barley/wheat diet was supplemented with either 3% or 6% of crushed linseed (L) or fish oil (F), respectively, providing 15.9 g/100 g of ALA or 2.31 and 3.53% of EPA plus DHA for the two diets. Additionally, the fish oil diet contributed 3.79 g/100 of ALA, which was similar to the basal (B) diet (3.41 g/100 and no detectable EPA or DHA). The pigs were separated into seven dietary treatment groups, and the feed regime was further divided into two fattening phases (Phase 1: 8 weeks on test diet; Phase 2: 6–9 weeks on test diet, until 100 kg slaughter weight was achieved). Three dietary groups were supplied with the same fatty acid source during both fattening phases. For the other four dietary groups, the fatty acid source was switched after the first phase (i.e., BB, LL, FF, BL, BF, LF, FL; the first letter indicating feed during Phase 1 and the second letter, feed during Phase 2). Results indicated that the muscle tissue (L. thoracis) from animals fed linseed oil for the 17 weeks (throughout both phases) exhibited a significant increase in amounts of ALA and EPA; however, the DHA proportion did not differ from the basal diet, indicating an absence of the elongation pathway to this fatty acid. The fish oil diet (FF) resulted in a six-fold increase in both EPA and DHA over the basal diet, as compared to a three- and five-fold increase of EPA and DHA, respectively, in the linseed feeding. This result demonstrates that some ALA elongated to EPA, but not to the longer-chain DHA. As expected, feeding fish oil continuously, as compared with feeding fish oil only in Phase 1, resulted in greater EPA and DHA deposition. When fish oil was fed either during Phase 1 or 2, only the proportion of DHA was influenced; it was greater when fish oil was administered during the second phase. This finding further supports the importance of feeding pre-formed DHA throughout a maintenance diet, or in the later stages of a finishing diet, to achieve and maintain meaningful levels of this fatty acid. This work further demonstrates that the only way to increase the DHA content of muscle is to supply DHA directly to the animal in its diet.

Marriott et al. (2002b) investigated uptake of DHA in various target tissues when feeding swine a DHA algal meal supplement. These researchers fed pigs either 125 or 250 g over the course of 42 days. Additionally, they employed a feeding regimen where 85% of the DHA was fed during days 0–21 (dose period), and the remaining 15% was provided during days 22–42 (maintenance period). Results indicated that DHA supplementation had no effect on feed consumption, feed efficiency, or average daily gain. The percent of fatty acids was determined for hams, loins, shoulders, and bellies (cured). As expected, DHA enrichment had no apparent effect on ALA across both treatment and muscle type. Dietary inclusion of 125 g DHA did not increase EPA in muscles; however, the upper dose (250 g) resulted in a two- to three-fold increase in EPA. Since negligible levels of this fatty acid were present in the feed, one can assume that such an increase was a direct result of DHA retro-conversion back to the shorter-chain EPA. Tissue concentrations of DHA followed dose response closely and displayed a linear increase in DHA across all muscle samples (Fig. 18.1). Enrichment with algal DHA resulted in a four- to ten-fold increase in DHA when pigs were fed 125 and 250 g of algae biomeal, respectively.

Marriott et al. (2002a) investigated feeding the single-cell microorganism *Schizochytrium* sp. to pigs. The whole-cell alga was mixed with corn-soybean feed at two different levels, and the dose was distributed over 42 days, pre-slaughter. They reported that algal supplementation resulted in significant DHA enrichment of muscle tissue without affecting the sensory qualities of pork cuts. The single-cell algae, rich in DHA (DHA GOLD®), was fed at 0.625 kg and 1.25 kg, equivalent to 0.125 kg and 0.250 kg of pure DHA. throughout the 42-day period. Also, the algae doses were distributed over two periods, 0 to 21 days and 22 to 42 days. In the first period, 85% of the dose was fed, and the remaining 15% was fed in the second period, before slaughter. The authors measured carcass characteristics, sensory traits of muscles, rancidity development, and bacon characteristics. It was concluded that feeding DHA at the above rates caused no adverse effects on the measured variables and led to the creation

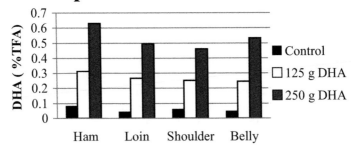

Fig. 18.1. DHA Deposition in Tissue (% TFA) from Pork Supplemented 125 and 250 g DHA. (Data from: Marriott et al., 2002b.)

of healthier pork meat (Marriott et al., 2002a). Previously, Leskanich et al. (1993) and Rhee et al. (1990) had shown that pigs fed with diets containing n-3 fatty acids could enrich the tissue with these fatty acids without affecting carcass characteristics, lean color, or sensory traits.

A safety assessment of the above single-cell microorganism for pigs was conducted by feeding increasing levels of DHA, either over a 120-day life cycle or 42 days, pre-slaughter, following the same 85% and 15% dose distribution, except for the 120th day, where the dose was distributed evenly. DHA was fed at 250, 750, and 1250 g for 42 days or 785 g over the 120 days. The study concluded that this single-cell organism could be safely fed at up to five times the recommended commercial level (Abril et al., 2003).

In summary, effective LCPUFA enrichment of pork meat is possible through inclusion of the single-cell microorganism *Schizochytrium* sp., trade name DHA GOLD®. Studies reviewed here indicate PUFA enrichment can effectively occur without causing any adverse effects on the quality characteristics or sensory traits of different meat cuts. This practice results in a healthier fatty acid composition cut, with a higher ratio of n-3 to n-6 fatty acids, more specifically, with the longer-chain n-3 fatty acid, DHA.

Ruminants

In recent years there has been an increased interest in ways to manipulate the fatty acid composition of ruminant animals. Givens et al. (2001) reviewed various studies that had measured the EPA and DHA content of foods. These researchers noted that dietary sources of EPA and DHA are almost entirely confined to fish products. However, lean meat (beef, lamb, and pork) did contain reasonable amounts of EPA and DHA (a total of 3.5, 5.0, and 7.0 g/kg in beef, lamb, and pork, respectively). This result compares with values of 25, 16, 18, and 16 g/kg for Atlantic mackerel, Atlantic herring, farmed salmon, and bluefin tuna, respectively (Rymer et al., 2003). Due to the fact that per capita fish consumption in the Western world is low and concerns surrounding the sustainability of our fisheries exist, alternative sources of dietary EPA and DHA are needed. Since the consumption of meat and milk products from ruminants accounts for a very large percentage of fat intake in the human diet and animal sources of food contribute 74% of dietary saturated fat (Mir et al., 2003), these products lend themselves as a great "vehicle" for nutrient delivery. Unfortunately, the fats from these sources are primarily saturated and do not contain appreciable amounts of EPA and DHA. For this reason, altering the fatty acid composition of meat and milk is of great importance.

Ruminal Metabolism of PUFAs

Manipulating the diet of ruminants has proven to be an effective way to increase the PUFA content of muscle and milk. This objective is generally accomplished by feeding grains or forages high in C18 fatty acids, fish products, or single-celled marine algae.

Unfortunately, the ruminal digestive system makes PUFA uptake a very difficult goal to achieve, primarily because dietary lipids are extensively and rapidly hydrolyzed by microbial enzymes in the rumen. The rumen environment is the site of intense microbial lipid metabolism. Lipolysis of dietary glycolipids, phospholipids, and triglycerides leads to free fatty acids that are hydrogenated by microbes to become more saturated end products (AbuGhazaleh et al., 2003). Although hydrogenation occurs at a lower rate than lipolysis, only a few polyunsaturated fatty acids may be found in the rumen. Most biohydrogenation (>80%) occurs in association with the fine food particles, a process that has been attributed to extracellular enzymes of bacteria, either associated with the feed or free in suspension (Harfoot & Hazlewood, 1988). The major substrates are linoleic and linolenic acids, and the rate of rumen biohydrogenation of fatty acids is typically faster with increasing unsaturation. For most diets, linoleic acid (LA) and linolenic acid (ALA) are hydrogenated to the extent of 70–95% and 85–100%, respectively (Doreau & Ferlay, 1994). This extensive level of hydrogenation is reduced when diets high in concentrates are fed, which can be attributed to the inhibition of lipolysis at the low pH that is typically observed with these diets (Van Nevel & Demeyer, 1995, 1996). Hydrogenation is also adversely affected when excessive unprotected lipids are present in the diet. How fat interferes with microbial fermentation is not clear, but it is believed to result from either the coating of feed particles or a direct toxic effect on the microorganisms of the rumen.

The rate and extent of LA and ALA hydrogenation is well-known. However, while C18 fatty acids are certainly hydrogenated, there has been much controversy regarding the fate of longer-chain fatty acids. Ashes et al. (1992) observed that biohydrogenation of EPA and DHA did not occur in vitro, to any great extent. This finding was further supported by in vitro observations by Gulati et al. (1999) and Dohme et al. (2003). Ashes et al. (1992) and Offer et al. (1999, 2001) found that a substantial proportion of EPA and DHA must escape biohydrogenation, as their inclusion in the diet of sheep or cattle resulted in significant increases in the concentration of these PUFAs in the plasma or serum. Ashes et al. (2000) also indicated that unless the lipid is protected from rumen metabolism, the efficiency of transfer of dietary EPA and DHA to the duodenum is very limited. Contrary to results published by these researchers, extensive biohydrogenation of long-chain PUFAs, particularly, EPA and DHA, has been reported in vivo by Doreau and Chilliard (1997), Chilliard et al. (2000), and Kitessa et al. (2001b). Doreau and Chilliard (1997) infused fish oil directly into the rumen of dairy cows and measured the fatty acid composition of the resultant duodenal digesta. A substantial reduction in the concentration of EPA and DHA in the fatty acid fraction in the duodenal digesta was noted and compared with the infusate. This reduction was comparable to an apparent hydrogenation of approximately 0.96 and 0.83 for EPA and DHA, respectively. The actual values may be slightly lower, due to a possible dilution of the duodenal digesta fatty acid fraction with microbial lipids.

Overall, it would seem reasonable to conclude that dietary EPA and DHA, which escape ruminal metabolism, are absorbed from the small intestine in the form of lipoproteins. Their long-chain length may inhibit their absorption, to some extent, although it is possible that their high degree of unsaturation will facilitate uptake. Once absorbed, they are likely to be carried in the lymph and drained into the circulation system via the thoracic duct, through which these fatty acids are transported to the various target tissues and organs where storage and/or further metabolism occur (Rymer et al., 2003).

PUFA Impact on Growth Performance

Since we know long-chain PUFAs are not completely hydrogenated and broken down during ruminal metabolism and are found in plasma, a closer look is necessary to understand their impact on growth performance and the resulting fatty acid composition in the tissue of n-3 supplemented animals. If PUFA enrichment is to be a viable means of altering the fatty acid profile of animal products, there must be minimal impact on animal growth, while attaining an appreciable uptake of EPA and DHA.

Questions have been raised concerning the impact of n-3 supplementation and daily feed intake of ruminants. Wistuba et al. (2006) reported that feed intake decreased for a diet containing 3% fish oil. Similar findings were also reported by Kook et al. (2001) when a fish oil supplement was included at 5%. A linear effect in decreased dry matter intake (DMI) was also reported by Mandell et al. (1997) for both fish oil inclusion rate and time on feed. Despite the deleterious effect of fish meal on DMI, feed efficiency was not affected by either level or duration. However, Wonsil et al. (1994) reported that a diet containing 3% added oil (1.5% fish oil and 1.5% stearic acid) and 7.1% total fat had no effect on the apparent total tract digestibility of dry matter (DM), acid digestible fiber (ADF), organic matter (OM), or nitrogen (N). Additionally, Scollan et al. (2001a) reported that cattle consuming a diet consisting of 6% total fat (2% from fish oil) had no apparent negative effect on feed intake. High concentrations of unprotected polyunsaturated oils are seldom used in ruminant diets, as high levels of dietary fat disturb the rumen environment and rumen processes (Harford & Hazlewood, 1988). For this reason, the level of lipids in ruminant diets should be limited to avoid adverse effects on ruminal fermentation. Decreases in feed intake are most likely due to several factors, such as palatability, energy density of the diet, and/or total fat (Wistuba et al., 2006).

Dietary Manipulation: Grain vs. Forage

Pasturing animals and feeding grains high in ALA (linseed) have been shown to have a positive effect on the levels of beneficial fatty acids in ruminants. In a study involving British lambs, fed grass, and Spanish lambs, fed milk and concentrates, analogous differences in fatty acid composition were observed. For example, the grass-fed lambs had a higher muscle concentration of n-3 PUFAs, and the milk concentrate-fed lambs had higher concentrations of the n-6 PUFA (Sanudo et al., 2000). In the study

Table 18.G. Fatty Acid Composition (g/100g TFA) of Muscle (maximus longissimus) Lipid from Cattle Fed on Grass or Concentrates.

Fatty Acid	Grass	Concentrates
18: 2n-6 (linoleic)	2·50	8·28
20: 3n-6	0·26	0·53
20: 4n-6 (arachidonic)	0·84	2·32
18: 3n-3 (α-linolenic)	1·23	0·52
20:4n-3	0·19	0·03
20:5n-3 (eicosapentaenoic)	0·51	0·20
22: 5n-3 (docosapentaenoic)	0·76	0·48
22: 6n-3 (docosahexaenoic)	0·07	0·05

From: Wood & Enser, 1997.

described in Table 18.G, steers finished on grass had higher concentrations of ALA and all other n-3 fatty acids than young bulls given a barley-soybean concentrate diet, which had higher concentrations of linoleic acid and all other n-6 fatty acids. This result is partly explained by the fact that ALA is the major fatty acid in grass lipids, whereas the cereals and oil seeds used in concentrate diets are major sources of linoleic acid (Wood & Enser, 1997). Scollan et al. (2001a) conducted a trial by feeding steers a lightly bruised linseed diet for 120 days, with grass silage used as forage in a ratio with concentrate of 60/40. Results indicated that the linseed treatment resulted in an increase of total intramuscular n-3 fatty acids, ALA, EPA, and DPA, but no difference could be observed in the DHA content. French et al. (2000) looked at the effects of lowering the concentrate proportion in a concentrated, –grass-based diet, using continental crossbred steers. The intramuscular fat of the *longissimus* linearly increased in n-3 fatty acid content when the amount of concentrate in the diet decreased, while no effect was observed on the n-6 content. Despite the large difference in the dietary LA and ALA supplies between the treatment groups, these authors only observed an effect of the diet on the intramuscular ALA content. No effect of the diet was detected on EPA, LA, or ARA, whereas DPA and DHA were not measured. Results such as these further underscore the inefficient process of ALA elongation to more highly unsaturated end products. One can assume that, based on negligible levels of EPA found in the tissue, the same finding would hold true for the longer-chain DHA. When ruminants are fed on diets that are predominantly grain, there is a change in rumen microbial populations, and the hydrogenation of PUFAs is reduced, leading to the absorption and deposition of a lower proportion of saturated fatty acids. On the other hand, although grass (forage) diets increase the deposition of n-3 PUFAs, the production of saturated fatty acids is also increased, and the P:S value in tissue becomes less favorable (Wood et al., 1999). Overall, the use of forage, vegetable oils, and whole grains results in a very limited increase of EPA and DHA in muscle tissue. Fish-based products and marine algae seem to be the only food sources that stimulate deposition of these LCPUFAs to a high extent. It can, therefore, be concluded that DHA formation is strictly metabolically regulated and cannot be substantially influenced by diet for ruminants, at least when it is not supplied by means of fish or algal-based ingredients.

Increasing PUFA Content in Muscle Tissue

The supplementation of ruminant diets, throughout the finishing phase, with fish meal/oil has previously been shown to result in an increased intramuscular concentration of n-3 fatty acids (Mandell et al., 1997; Kook et al., 2001; Ponnampalam et al., 2001, 2002; Kitessa et al., 2001c; Scollan et al., 2001a; Choi et al., 2000). Mandell et al. (1997) investigated uptake of long-chain PUFA into the beef *longissimus* muscle when feeding steers fish meal (containing 5.87 g/kg EPA and 9.84 g/kg DHA) at 5 or 10% (dietary dry matter) for up to 168 days. Fish meal supplementation resulted in increased intramuscular deposition of EPA and DHA. The amount and duration of supplementation seemed to play a significant role in uptake, as higher levels were achieved when feeding increased amounts over longer periods. These researchers noted a five- to seven-fold increase in EPA concentration over the course of the 168 days. Additionally, they observed a twenty-six- to twenty-eight-fold increase in DHA concentration. On a mg/100 g tissue basis, fish meal supplementation increased DHA concentrations up to 500%, as compared with steers fed a control diet. Kook et al. (2001) looked at *longissimus* fatty acid composition of Korean cattle that resulted when a fish oil supplement was included in the diet at 5%. On a total fatty acid basis, this inclusion rate corresponds to 2.0% and 4.8% of EPA and DHA, respectively. Fish oil supplementation resulted in a two-fold increase in EPA and an eighteen-fold increase in the DHA composition of muscle. Recently, lambs were fed a fish meal (168 g DM per day) or barley (179 g DM per day)/fish meal (84 g DM per day) diet (Ponnampalam et al., 2002) and a fish oil or fish oil/sunflower meal diet (Ponnampalam et al., 2001). These four treatments resulted in a two- to four-fold increase in the deposition of EPA and DHA in the *longissimus*, while no effect on ALA deposition was observed, compared with the alfalfa and oat-based control diet (Table 18.H). Although dietary intake of DHA and EPA was higher when fish oil was fed, compared with the fish meal diet, no uptake differences were observed between the two in the resulting intramuscular n-3 content. Increased EPA and DHA concentrations in intramuscular fat were also observed by Kitessa et al. (2001c) who fed protected tuna oil to lambs, compared with a tallow diet (Table 18.H). Similar results were obtained when fish meal (Mandell et al., 1997) or fish oil (Choi et al., 2000; Scollan et al., 2001b) was fed to steers (Table 18.H). From these studies, it is reasonable to conclude that supplementing diets with preformed sources of EPA and DHA markedly increases the total n-3 content of intramuscular fat. Such increases are seen as a direct result of increased intramuscular concentrations of EPA and DHA, which requires that considerable amounts of these fatty acids reach the small intestine and thus escape ruminal hydrogenation (Raes et al., 2004). Appreciable

Table 18.H. Effect of Feeding Fish Oil or Meal to Ruminants on the Long-Chain PUFA of *Longissimus* (g/100g of TFA) and Its Effect on the P/S and n-6/n-3 Ratio.

Lamb	LA	ARA	ALA	EPA	DPA	DHA	P/S	n-6/n-3	Reference
Control Diet[a]	4.58	1.39	0.79	0.61	NA	0.44	0.13	NA	Kitessa et al. (2001)
Protected Tuna Oil[a]	8.27	1.64	1.06	1.81	NA	1.01	0.23	NA	Kitessa et al. (2001)
Basal Diet[b]	2.23	0.70	0.67	0.27	0.34	0.10	0.17	1.80	Ponnampalam et al. (2001)
1.5% Fish Oil[b]	2.25	0.77	0.68	0.90	0.56	0.47	0.16	1.40	Ponnampalam et al. (2001)
1.5% Fish Oil +9% Sunflower Meal[b]	2.44	0.77	0.57	0.77	0.51	0.44	0.19	1.30	Ponnampalam et al. (2001)
Basal Diet[b]	3.72	1.19	0.91	0.47	0.50	0.20	NA	2.45	Ponnampalam et al. (2002)
168 g DM Fish Meal per Day[b]	2.97	0.72	0.83	0.93	0.65	0.47	NA	1.29	Ponnampalam et al. (2002)
179 g DM Barley per Day + 84 g DM Fish Meal per Day[b]	3.56	0.78	0.89	0.90	0.63	0.48	NA	1.50	Ponnampalam et al. (2002)
Beef									
Control Diet[c]	3.02	0.97	0.68	0.35	0.61	0.07	0.08	2.27	Choi et al. (2000)
30 g Fish Oil/kg DM +88 g Linseed/kg DM[c]	2.35	0.73	0.86	0.43	0.60	0.16	0.08	1.77	Choi et al. (2000)
Control Diet[d]	3.30	0.70	0.40	0.10	0.30	0.01	0.08	5.63	Mandell et al. (1997)
5% Fish Meal during 168 d[d]	3.30	0.50	0.40	0.50	0.40	0.28	0.09	4.09	Mandell et al. (1997)
10% Fish Meal during 168 d[d]	2.40	0.30	0.40	0.70	0.40	0.26	0.06	1.72	Mandell et al. (1997)
Control Diet[c]	2.32	0.65	0.58	0.30	0.57	0.07	0.06	2.13	Scollan et al. (2001)
30 g Fish Oil/kg DM + 118 g Linseed/kg DM[c]	1.81	0.50	0.83	0.47	0.59	0.16	0.06	1.20	Scollan et al. (2001)
59.6 g Fish Oil/kg DM[c]	1.43	0.32	0.61	0.55	0.55	0.12	0.04	1.04	Scollan et al. (2001)

From: Raes et al., 2004.

NA: data not available; ARA: arachidonic acid (C20:4n-6); DHA: docosahexanoic acid (C22:6n-3); DPA: docosapentaenoic acid (C22:5n-3); EPA: eicosapentaenoic acid (C20:5n-3); LA: linoleic acid (C18:2n-6).

[a] The control diet was tallow-based.

[b] The basal diet contained mainly alfalfa and oats and served also as the control diet.

[c] The control diet was based on Megalac (palm-oil-based, high in C16:0).

[d] The control diet was mainly composed of high moisture corn and alfalfa hay.

alterations in the intramuscular content of n-3 fatty acids in ruminants can be effectively achieved by dietary manipulation with fish/algal oils.

Increasing PUFA in Milk

The natural levels of DHA and EPA in bovine or ovine milk are extremely low, which indicates that negligible chain elongation and desaturation of the C18 and C20 polyenoic fatty acids occurs in the mammary gland of ruminants. The DHA and EPA content of milk can be increased by dietary supplementation with fish oils or marine algae (Abril et al., 2009; Papadoupoulos et al., 2002; Chilliard et al., 2000; Franklin et al., 1999; Gulati et al., 1999). Unfortunately, EPA and DHA are not transferred into milk fat with high efficiency, even when protected sources of marine oils and algae are fed. Feeding protected sources of the n-3 long-chain PUFA does significantly increase the EPA and DHA content of total milk fatty acids, although absolute levels are generally below 10 g/kg of fatty acids. Even at these low concentrations, milk can supply a significant proportion of the EPA and DHA required to redress the imbalanced n-6:n-3 ratio of PUFA in current human diets (Rymer et al., 2003). Franklin et al. (1999) initiated algae feeding trials in milk cows, comparing a protected and non-protected full fat algae supplement. Cows were fed algae, which were protected with a xylose coating, which contributed 6.07% DHA of total fat to the diet. Results indicated overall performance characteristics were not affected, although the DMI decreased. Neither the actual amount of milk produced nor the energy corrected milk (adjusted for fat and protein content) were affected by feeding marine algae; however, percentage and yield of fat decreased. The proportion of DHA in milk was also greater in cows fed the xylose-protected algae than in the unprotected counterpart. Transfer efficiency was two-fold higher in protected feed versus unprotected (16.7% and 8.4%, respectively). It is not clear from the literature if the increase in EPA and DHA in milk fat after feeding protected marine oils is in TAG or in phospholipids of the milk fat globule and secreted cell membranes, a question that needs to be investigated, since mechanisms for increasing membrane phospholipids in milk could be important for increasing EPA and DHA concentrations in milk (Rymer et al., 2003).

Papadoupoulos et al. (2002) initiated a study feeding ewes (small ruminant) diets containing low, medium, and high concentrations of the marine alga *Schizochytrium* sp., where 23.5, 47, and 94 g were incorporated into a concentrated diet, respectively. The algae contained 147 g/kg of DHA and 38.5 % fat, on a dry matter basis. Milk fat was enriched from a non-detected level of DHA to 4.3, 6.9, and 12.4 g/kg of total fatty acids for the low, medium, and high diets. On the other hand, EPA levels were increased to a lower degree from non-detected to 0.4, 1.2, and 2.4 g/kg fatty acids, and ALA levels were not significantly affected, staying around 3.0 g/kg.

Milk fat is a rich natural dietary source of conjugated linoleic acid (CLA), which has been the focus of a considerable number of research efforts in recent years. Data have revealed severalpotential health-promoting properties associated with particular isomers of CLA, primarily, *cis*-9, *trans*-11 CLA, and *trans*-10, *cis*-12 CLA.

Conjugated linoleic acids have been shown to affect atherosclerosis, diabetes, the immune system, bone mineralization, body fat accretion, and nutrient partitioning (McGuire & McGuire, 2000). Additionally, there is an growing body of evidence that *cis*-9 *trans*-11 CLA suppresses chemically induced tumor development in animal models (Parodi, 1999; Kritchevsky, 2000). Such reported health benefits of CLA make it desirable to increase the concentrations and yields of CLA in dairy products.

Including EPA and DHA in the diet of ruminants has been shown to increase the concentrations of CLA, EPA, and DHA in milk (Scollan et al., 2001a; Donovan et al., 2000). The uptake of n-3 fatty acids is fairly straightforward in that once the preformed fats are consumed, they escape rumen metabolism and are deposited throughout the body and milk. Increased concentrations of CLA appear to be related to an effect on the biohydrogenating activity and/or microbial community structure in the rumen. Fish oil is known to inhibit the reduction of *trans*-18:1 fatty acids to 18:0 in vivo (Scollan et al., 2001b). Supplementing the diet with 200–300 g fish oil/day has been shown to inhibit rumen bio-hydrogenation in dairy cows and steers, increasing the CLA content of milk and meat (Shingfield et al., 2003). Similar studies have been conducted regarding the feeding of DHA algal meal (Or-Rashid et al., 2008). The objective of this work was to evaluate the ruminal fatty acid changes of CLA, VA (vaccenic acid), and 18:1 *trans*-10, as a result of algal meal supplementation. The inclusion of algal biomeal (9.6 g DHA/d) decreased 18:0 by 80%, while increasing total *trans*-18:1 from 19 to 43% of total fatty acids. Such a drastic change is attributed to an inhibition of the reduction step of *trans*-18:1 to 18:0 by the presence of the long-chain PUFA DHA. The total percentage of *cis*-18:1 increased by algal meal supplementation, with largest contributor being the 18:1 *cis*-9 isomer (46.4% greater than the control diet). A greater proportion of this isomer in the ruminal content of animals fed the algal rations may indicate that the algae partially protected it from biohydrogenation. One additional health benefit realized when supplementing diets with EPA and DHA is the alteration of C:18 fatty acids. Recent work conducted by Wistuba et al. (2006) demonstrated that feeding fish oil at 3% for 70 days resulted in a increase in the proportion of oleic acid (C18:1). These results are consistent with the findings of Ashes et al. (1992). Such an outcome is highly desired, as increasing the proportion of C18:0 and C18:1 would be beneficial to the beef industry, because these fatty acids are hypocholesteremic in humans.

Health Benefits Passed to Ruminants

One important, but often overlooked, aspect of n-3 supplementation in animal diets is the potential health benefits passed to the animal. Limited research has been conducted in this area, but the potential of economic importance exists. One such area is the high rate of embryonic mortality in the cattle industry. It has been estimated that 20 to 30% of all bovine embryos die within the first 30 days of gestation (Dunne et al., 2000), costing the U.S. beef industry millions of dollars annually in lost meat and milk production. The addition of fish meal to the diet has been reported to increase fertility in lactating dairy and beef cattle (Burke et al., 1997; Burns et al., 2002). Previous studies have shown that fish meal supplementation increases endometrial n-3 fatty acids in beef cows. Such an increase in endometrial n-3 fatty acids may suppress uterine prostaglandin $F_{2\alpha}$ synthesis during the period of maternal recognition of pregnancy and, potentially, decrease embryonic death loss (Burns et al., 2003). Additional areas of n-3 importance for animal health surely exist. One area that could provide economic benefit is the organic, or all-natural, sector of livestock production. Producers in this sector have limited means for treating sick animals and face a difficult decision when animals become ill: administer antibiotics and lose the market value associated with a "natural" animal or forego treatment and risk death loss. Although research has not been conducted in this area, n-3 fatty acids are known to promote health and well-being throughout all stages of life.

Applications for SCOs in Companion and Performance Animals

LCPUFA Requirements in Cats and Dogs

Feline animals, both big and domestic cats, are obligate carnivores and have an absolute requirement for LCPUFAs (both n-3 and n-6) because of a deficiency in certain desaturase enzymes ($\Delta 5$- and $\Delta 6$-desaturases), which are required for LCPUFA synthesis from their 18-C precursors, linoleic and linolenic acids (Pawlosky et al., 1994). Consequently, cats require a dietary source of preformed DHA and ARA, and most cat foods add both acids in the form of fish or animal organ meats (e.g., liver). Canine animals—including coyotes, wolves, foxes, jackals, and domesticated dogs—are partly considered omnivores, and, although they do possess the enzymes required to synthesize DHA and ARA from linolenic and linoleic acids, respectively, these enzymes are not well developed. Dogs get most of their required DHA and ARA from diets of meat, fish, and/or animal organs. Both cats and dogs exhibit a particularly high requirement for dietary DHA and ARA during the perinatal period, especially, in the adult female during gestation and lactation, as well as in the growing infant (kitten or puppy) during the early stages of neurological growth and development.

In the case of dogs, a number of studies have demonstrated improvements in the visual acuity and learning ability of puppies when they receive diets supplemented with DHA (Heinemann et al., 2005; Heinemann & Bauer, 2006). This finding is, perhaps, not surprising since DHA is a fundamental building block of brain and retinal tissues in dogs and cats, just as it is in human infants (Crawford, 1993). Canine milk contains DHA and is a natural source of this ingredient for the suckling puppy, but the level of DHA in the milk is dependent on the mother's diet. In an attempt to elevate the DHA level in canine milk, Bauer et al. (2004) have shown that, if the lactating female receives a diet supplemented with preformed DHA, her milk will also be enriched in DHA (Wright-Rodgers et al., 2005). However, if ALA (a precursor for DHA, found primarily in flax oil) was provided in the diet of the mother, there was

Table 18.I. National Research Council (NRC) Recommendations for EPA and DHA in Canine and Feline Diets.

	NRC Recommendations for EPA and DHA Omega-3 Fatty Acids			
	Adequate Intake (mg/kg BW)	Recommended Allowance (mg/kg BW)	EPA Limit	Safe Upper Limit (mg/kg BW)
Bitches, Late Gestation and Peak Lactation	60	60	60%	1400
Puppies, after Weaning	36	36	60%	770
Adult Dogs	30	30	60%	360
Queens, Late Gestation & Peak Lactation	6.7	4.4	60%	-
Kittens, after Weaning	5	5	60%	-
Adult Cats	2.5	2.5	20%	-

From: NRC Publication, 2006. Nutrient Requirements for Dogs and Cats.

no elevation in the DHA level of the milk and, therefore, no increased DHA flowing to the puppy (Bauer et al., 2004). This result parallels a number of studies and observations with humans (Doughman et al., 2007). In order to provide DHA and ARA to growing puppies and kittens, preformed fatty acids need to be provided either indirectly, to the young via their mother's milk, or directly, in the diet after weaning.

Scientific bodies and industry groups understand the importance of including LCPUFAs in the diets of cats and dogs. The National Research Council (NRC) publication "Nutrient Requirements for Dogs and Cats 2006" provides recommended EPA/DHA dosage, based on animal body weight, through the various stages of life (Table 18.I). The maintenance dose for adult dogs is recommended to be 50–60% EPA and 40–50% DHA. It is important to note that the Association of American Feed Control Officials (AAFCO) does not currently recognize EPA or DHA as essential fatty acids.

Alternative Sources of DHA and ARA

With the recent recognition of the value of DHA in puppy nutrition, many manufacturers are now adding preformed DHA directly to puppy food, either from fish oil/meal or from a SCO source. Fish meal is a source of DHA as a consequence of the fact that 5–10% by weight of the fish meal is fish oil, which comes along with commercially prepared fish meals. Because the manufacturing process, from the harvesting of fish (i.e., the reduction fishery) through the processing and separating of oil from the meal, results in a significant amount of oxidation, DHA levels are not very consistent in fish meal, and the quality of fish meal, as a companion animal feed, varies considerably. In many cases, the oxidative damage is minimized by including the antioxidant ethoxyquin in bulk fish meal. Fish oil, on the other hand, can be cleaned up using distillation and/or deodorization, enabling the product to become more acceptable for these animals. Furthermore, alternative antioxidants can be used to avoid the presence of ethoxyquin in the pet food; however, most fish oils contain EPA at the same levels of DHA. When fish oil was first tested as a potential source for DHA in human infant milk, it was rejected because the EPA in the fish oil suppressed the baby's ARA levels and resulted in an inhibition of growth.

In human neonatal nutrition, the SCO from microalgae is being used worldwide, because it is safe, free of contaminants often found in fish oil (e.g., dioxin, PCBs, methyl mercury, etc.), and because it contains little or no EPA (for a review of this reasoning, see Chapters 15 and 20). It seems logical, therefore, that the major producers of pet food might also consider providing this form of DHA to their customers. In fact, a recent consumer panel study by Advanced BioNutrition Corp (Columbia, MD), comparing puppy kibbles produced with a fish oil coating to kibbles containing algal DHA at the same levels, showed that SCO-containing kibbles were significantly preferred over the fish oil kibbles in both fresh and aged kibbles, due to a significant odor problem associated with the fish oil kibbles (Kyle, personal communication). Technical analysis of the kibbles also showed that the peroxide value of the kibbles produced with the fish oil increased rapidly with storage time, whereas it remained relatively stable and low in the SCO-produced kibbles. A number of pet food manufacturers have now chosen to use the SCO-containing algal DHA source (DHA Gold®, Martek Biosciences Corp, Columbia, MD) in high-end dog foods and treats. Although ARA has not yet been added to pet foods (dog or cat) from a SCO source, it has been added to nearly all human infant formulas in the United States, and it is only a matter of time before it will also be added to companion animal feeds—particularly for the neonatal period.

LCPUFA Requirements in Horses

Horses are herbivores by nature and, as such, have no requirement for preformed DHA or ARA because they have the ability to convert the linoleic and linolenic acids in their diets into the more critical LCPUFAs. However, there may be certain animals, particularly equestrian athletes, that endure stress and physical demands on their bodies, which could require dietary supplements to achieve optimal nutritional status. Furthermore, during the neonatal period, it may also be critical for these animals to receive specific dietary enrichment of these LCPUFAs. Although there is not a large collection of data on the impact of dietary supplementation of DHA on horses, there is a growing body of information on its effect in mammals, in general. For example, DHA enrichment has been shown to improve vascular compliance and reduce the fragility of red cells (Nestel et al., 2002), which represents critical parameter for equestrian athletes in a race, because of the enormous volumes of blood flow through the lungs at maximum exertion. Indeed, this volume commonly leads to Exercise-Induced Pulmonary Hemorrhage (EIPH) as noted by blood flecks in the lungs or around the nostrils following exertion. Improved vascular compliance and an increased flexibility of red cell membranes should reduce the amount of exercise-induced damage to the lungs. There have also been some reports that DHA-supplementation significantly reduces the recovery time for exercise-induced injuries (Clayton & Rutter, 2004). Farriers have reported more rapid recover from injury when animals were provided supplemental DHA in the form of an algal SCO product. Finally, the neonatal period for horses, as is the case for other mammals, is a period where elevated DHA levels in the mother and the offspring may be of significant advantage. For an equestrian athlete, more rapid neurological development at a young age could be an important attribute.

Perhaps the most intriguing observation related to DHA in the equestrian field comes from a number of publications noting the significant improvement of sperm motility and survivability in stallions supplemented with DHA (Brinsko et al., 2005) or of sperm stored in a medium enriched in DHA (Kaeoket et al., 2008). The association of DHA with male fertility has been known in the agriculture industry (Gliozzi et al., 2009), and there have also been several reports on the correlation of infertility with low sperm DHA content in humans (Aksoy et al., 2006; Tavilani et al., 2006). Again, this finding is not surprising when considered with the knowledge that sperm contain a relatively high level of DHA (Lenzi et al., 2000).

Alternative Sources of DHA for Horses

Feed for horses will contain the same sources of DHA that are used in the case of companion animals, and are limited to fish meal/oil or SCO sources of DHA. Horses, however, are generally more affected by taste and odor issues than cats and dogs. As a result, fish meal/oil sources are particularly problematic when provided to horses. Even if fish oil is further processed to have little odor or taste and reduced levels of contaminants, such as PCBs or dioxin, the oil itself still contains a significant quantity of EPA. In mammals, including humans, EPA is well-known to increase bleeding time (Mark & Sanders, 1994). In situations of extreme exertion resulting in EIPH (Exercise Induced Pulmonary Haemorrhage), it is not useful to have a supplement that provides the advantages of improved vascular compliance and red cell flexibility if it is also coupled with the disadvantage of longer bleeding times. Clearly, a source of DHA without EPA would be preferred. An algal SCO-containing product is already on the market that is particularly tailored for horses and contains predominantly DHA, with little or no EPA. The product Magnitude® has been launched for stallions that are in the transition from performer to producer. Based on studies from several universities, the product has been shown to increase sperm concentration by 78% and sperm daily output by 46%, as well as increasing sperm motility and the percentage of morphologically normal sperm (Brinsko et al., 2005; Harris et al., 2005; Squires, 2005). With results such as these, coupled with the high value attributed to the performance of equestrian athletes, there is potential for a significant increase in the use of algal SCO-containing products with horses in the future.

Conclusions

LCPUFA manipulation in animal products is of great importance to producers, due to the limited means of altering the fatty acid profile through selective breeding or through feed modification (other than LCPUFA supplementation). Throughout this chapter, we have demonstrated that animal products can be successfully enriched with DHA via the inclusion of LCPUFA-rich single-cell microorganisms in animal rations. Additionally, SCO supplementation in companion animals has been shown to improve growth and performance in many species. These sources of LCPUFAs have inherent advantages when used as a feed supplement, as opposed to utilizing LCPUFAs from animal by-products, such as fish oil or fish meal, which have many drawbacks. The quality and stability of fish-based sources are of great concern and have been shown to compromise the sensory quality of enriched muscle tissue. Additionally, the inclusion rate often becomes self-limiting due to palatability issues. Plant-based sources of omega-3 fatty acids offer limited functionality, since ALA is not elongated to EPA or DHA to any great extent in vivo. SCOs provide unique functionality in that the dried cells offer superior oxidative stability, ease of handling, efficiency of transfer, and natural protection, in the case of ruminants. For these reasons, DHA-rich SCO offers a unique solution for the feed industry.

References

Abril, R.; W. Barclay. Production of Docosahexaenoic Acid-Enriched Poultry Eggs and Meat Using and Algae-Based Feed Ingredient. *The Return of ω3 Fatty Acids into the Food Supply: I. Land-Based Animal Food Products and Their Health Effects*; A.P. Simopoulos, Ed.; World Review of Nutrition and Dietetics; S. Karger: Switzerland, **1998**; *83*, pp 77–88.

Abril, R.; J. Garrett; S.G. Zeller; W. J. Sander; R.W. Mast. Safety Assessment of DHA-Rich Microalgae from *Schizochytrium* sp. Part V: Target Animal Safety/Toxicity Study in Growing Swine. *Reg. Toxicol. Pharmacol.* **2003,** *37*, 73–82.

Abril, J.R.; W.R. Barclay; A. Mordenti; M. Tassinari; A. Zotti. U.S. Patent 7,504,121B2, **2009.**

AbuGhazaleh, A.; D. Schingoethe; A. Hippen; K. Kalscheur. Conjugated Linoleic Acid and Vaccenic Acid in Rumen, Plasma, and Milk of Cows Fed Fish Oil and Fats Differing in Saturation of 18 Carbon Fatty Acids. *J. Dairy Sci.* **2003,** *86*, 3648–3660.

Aksoy, Y.; H. Aksoy; K. Altinkaynak; H.R. Aydin; A. Ozkan. Sperm Fatty Acid Composition in Subfertile Men. *Prostaglandins Leukot. Essent. Fatty Acids* **2006,** *75*, 75–79.

Ashes, J.R.; B.D. Siebert; S.K. Gulati; A.Z. Cuthbertson; T.W. Scott. Incorporation of n-3 Fatty Acids of Fish Oil into Tissue and Serum Lipids of Ruminants. *Lipids* **1992**, *27*, 629–631.

Ashes, J.R.; S.K. Gulati; S.M. Kitessa; E. Fleck; T.W. Scott. Utilization of Rumen Protected n-3 Fatty Acids by Ruminants. *Recent Advances in Animal Nutrition*; P.C. Garnsworthy, J. Wiseman, Eds.; Nottingham University Press: Nottingham, UK, **2000**; pp 128–140.

Basmacıoğlu, H.; M. Çabuk; K. Ünal; K. Özkan; S. Akkan; H. Yalçın. Effects of Dietary Fish oil and Flax Seed on Cholesterol and Fatty Acid Composition of Egg Yolk and Blood Parameters of Laying Hens. *S. African J. Anim. Sci.* **2003,** *33*, 266–273.

Bauer, J.E.; K.M. Heinemann; K.E. Bigley; G.E. Lees; M.K. Waldron. Maternal Diet Alpha-Linolenic Acid during Gestation and Lactation Does Not Increase Docosahexaenoic Acid in Canine Milk. *J. Nutr.* **2004,** *134,* 2035S–2038S.

Bauman, D.E.; J.M. Grinari. Nutritional Regulation of Milk Fat Synthesis. *Annu. Rev. Nutr.* **2003,** *23,* 203–227.

Bourre, J.-M. Where to Find Omega-3 Fatty Acids and How Feeding Animals with Diet Enriched in Omega-3 Fatty Acids to Increase Nutritional Value of Derived Products for Human: What is Actually Useful? *J. Nutr. Health & Aging* **2005,** *9,* 232–242.

Brenner, R.R. Factors Influencing Fatty Acid Chain Elongation and Desaturation. *The Role of Fats in Human Nutrition*; A.J. Vergroesen, M. Crawford, Eds.; Academic Press: San Francisco, California, 1989; pp 45–80.

Brinsko, S.P.; D.D. Varner; C.C. Love; T.L. Blanchard; B.C. Day; M.E. Wilson. Effect of Feeding a DHA-Enriched Nutriceutical on the Quality of Fresh, Cooled and Frozen Stallion Semen. *Theriogenology,* **2005,** *63,* 1519–1527.

Burke, J.M.; C.R. Staples; C.A. Risco; R.L. de la Sota; W.W. Thatcher. Effect of Ruminant Grade Menhaden Fish Meal on Reproductive and Productive Performance of Lactating Dairy Cows. *J. Dairy Sci.* **1997,** *80,* 3386–3398.

Burns, P.D.; T.R. Bonnette; T.E. Engle; J.C. Whittier. Case Study: Effects of Fishmeal Supplementation on Fertility and Plasma Omega-3 Fatty Acid Profiles in Primiparous, Lactating Beef Cows. *Prof. Anim. Sci.* **2002,** *18,* 373–379.

Chilliard, Y.; A. Ferlay; R.M. Mansbridge; M. Doreau. Ruminant Milk Fat Plasticity: Nutritional Control of Saturated, Polyunsaturated, Trans and Conjugated Fatty Acids. *Ann. Zootech.* **2000,** *49,* 181–205.

Choi, N.J.; M. Enser; J.D. Wood; N.D. Scollan. Effect of Breed on the Deposition in Beef Muscle and Adipose Tissue of Dietary n-3 Polyunsaturated Fatty Acids. *Anim. Sci.* **2000,** *71,* 509–519.

Clayton, D.; R. Rutter. Inflammatory Disease Treatment. International Patent Application No. PCT/GB2004/002707; WO04/112776, 2004.

Crawford, M.A. The Role of Essential Fatty Acids in Neural Development: Implications for Perinatal Nutrition. *Am. J. Clin. Nutr.* **1993,** *57,* 703S–710S.

Damiani, P.; L. Cossignani; M.S. Simonetti; F. Santinelli; M. Castellini; F. Valfre. Incorporation of n-3 PUFA into Hen Egg Yolk Lipids. I: Effect of Fish Oil on the Lipid Fractions of Egg Yolk and Hen Plasma. *Ital. J. Food Sci.* **1994,** *6,* 275–292.

Dohme, F.; V.I. Fievez; K. Raes; D.I. Demeyer. Increasing Levels of Two Different Fish Oils Lower Ruminal Biohydrogenation of Eicosapentaenoic and Docosahexaenoic Acid in vitro. *Anim. Res.* **2003,** *52,* 309–320.

Donovan, D.C.; D.J. Schingoethe; R.J. Baer; J. Ryali; A.R. Hippen; S.T. Franklin. Influence of Dietary Fish Oil on Conjugated Linoleic Acid and Other Fatty Acids in Milk Fat from Lactating Dairy Cows. *J. Dairy Sci.* **2000,** *83,* 2620–2628.

Doreau, M.; Y. Chilliard. Effects of Ruminal or Post-ruminal Fish Oil Supplementation on Intake and Digestion in Dairy Cows. *Reprod. Nutr. Dev.* **1997,** *37,* 113–124.

Doreau, M.; A. Ferlay. Digestion and Utilization of Fatty Acids by Ruminants. *Anim. Feed Sci. Tech.* **1994,** *45,* 379–396.

Doughman, S.D.; S. Krupanidhi; C.B. Sanjeevi. Omega-3 Fatty Acids for Nutrition and Medicine: Considering Microalgae Oil as a Vegetarian Source of EPA and DHA. *Curr. Diabetes Rev.* **2007,** *3,* 198–203.

Dunne, L.D.; M.G. Disken; J.M. Sreenan. Embryo and Foetal Loss in Beef Heifers between Day 14 of Gestation and Full Term. *Anim. Reprod. Sci.* **2000,** *58,* 39–44.

Enser, M.; R. Richardson; J. Wood; B. Gill; P. Sheard. Feeding Linseed to Increase the n-3 PUFA of Pork: Fatty Acid Composition of Muscle, Adipose Tissue, Liver and Sausages. *Meat Sci.* **2000,** *55,* 201–212.

Franklin, S.; K. Martin; R. Baer; D. Schingoethe; A. Hippen. Dietary Marine Algae (*Schizochytrium* sp.) Increases Concentrations of Conjugated Linoleic, Docosahexaenoic and Transvaccenic Acids in Milk of Dairy Cows. *J. Nutr.* **1999,** *129,* 2048–2054.

Fredriksson, S.; K. Elwinger; J. Pickova. Fatty Acid and Carotenoid Composition of Egg Yolk as an Effect of Microalgae Addition to Feed formula for Laying Hens. *Food Chem.* **2006,** *99,* 530–537.

French, P.; C. Stanton; E.G. Lawless; E.G. O'Riordan; F.J. Monahan; P.J. Caffrey; A.P. Moloney. Fatty Acid Composition, Including Conjugated Linoleic Acid of Intramuscular Fat from Steers Offered Grazed Grass, Grass Silage or Concentrate-based Diets. *J. Anim. Sci.* **2000,** *78,* 2849–2855.

Givens, D.I.; B.R. Cottrill; M. Davies; P.A. Lee; R.J. Mansbridge; A.R. Moss. Sources of n-3 Polyunsaturated Fatty Acids Additional to Fish Oil from Livestock Diets, A Review (Revised Version). *Nutr. Abs. Rev., Series B, Livestock Feeds and Feeding* **2001,** *71,* 53R–83R.

Gliozzi, T.M.; L. Zaniboni; A. Maldjian; F. Luzi; L. Maertens; S. Cerolini. Quality and Lipid Composition of Spermatozoa in Rabbits Fed DHA and Vitamin E Rich Diets. *Theriogenology* **2009,** *71,* 910–919.

Gulati, S.K.; J.R. Ashes; T.W. Scott. Hydrogenation of Eicosapentaenoic and Docosahexaenoic Acids and Their Incorporation into Milk Fat. *Anim. Feed Sci. Tech.* **1999,** *79,* 57–64.

Haak, L.; S. De Smet; D. Fremaut; K. Walleghem; K. Raes. Fatty Acid Profile and Oxidative Stability of Pork as Influenced by Duration and Time of Dietary Linseed or Fish Oil Supplementation. *J. Anim. Sci.* **2008,** *86,* 1418–1425.

Harfoot, C.G.; G.P. Hazlewood. Lipid Metabolism in the Rumen. The Rumen Microbial Ecosystem; P.N. Hobson, Ed.; Elsevier: New York, USA, 1988; pp 382–426.

Hargis, P.S.; M.E. Van Elswyk; B.M. Hargis. Dietary Modification of Yolk Lipid with Menhaden Oil. *Poultry Sci.* **1991,** *70,* 874–883.

Harris, M.A.; L.H. Baumgard; M.J. Arns; S.K. Webel. Stallion Spermatozoa Membrane Phospholipids Dynamics Following Dietary n-3 Supplementation. *An. Reprod. Sci.* **2005,** *89,* 275.

Heinemann, K.M.; J.E. Bauer. Docosahexaenoic Acid and Neurologic Development in Animals. *J. Am. Vet Med. Assoc.* **2006,** *228,* 700–705.

Heinemann, K.M.; M.K. Waldron; K.E. Bigley; G.E. Lees; J.E. Bauer. Long-Chain (n-3) Polyunsaturated Fatty Acids Are More Efficient than Alpha-Linolenic Acid in Improving Electroretinogram Responses of Puppies Exposed during Gestation, Lactation, and Weaning. *J. Nutr.* **2005,** *135,* 1960–1966.

Herber, S.M.; M.E. Van Elswyk. Dietary Marine Algae Promotes Efficient Deposition of n-3 Fatty Acids for the Production of Enriched Shell Eggs. *Poultry Sci.* **1996,** *75,* 1501–1507.

Irie, M.; M. Sakimoto. Fat Characteristics of Pigs Fed Fish Oil Containing Eicosapentaenoic and Docosahexaenoic Acids. *J. Anim. Sci.* **1992,** *70,* 470–477.

Kaeoket, K.; P. Sang-Urai; A. Thamniyom; P. Chanapiwat; M. Techakumphu. Effect of Docosahexaenoic Acid on Quality of Cryopreserved Boar Semen in Different Breeds. *Reprod. Dom. Anim.* doi: 10.1111/j.1439-0531.2008.01239.x

Keady, T.W.J.; C.S. Mayne; D.A. Fitzpatrick. Effects of Supplementation of Dairy Cattle with Fish Oil on Silage Intake, Milk Yield and Milk Composition. *J. Dairy Res.* **2000,** *67,* 137–153.

Kitessa, S.M.; S.K. Gulati; J.R. Ashes; E. Fleck; T.W. Scott; P.D. Nichols. Utilisation of Fish Oil in Ruminants. II. Transfer of Fish Oil Fatty Acids into Goats' Milk. *Anim. Feed. Sci. Technol.* **2001a,** *89,* 201–208.

Kitessa, S.M.; S.K. Gulati; J.R. Ashes; E. Fleck; T.W. Scott; P.D. Nichols. Utilisation of Fish Oil in Ruminants: Fish Oil Metabolism in Sheep. *Anim. Feed Sci. Tech.* **2001b,** *89,* 189–199.

Kitessa, S.M.; S.K. Gulati; J.R. Ashes; T.W. Scott; E. Fleck. Effect of Feeding Tuna Oil Supplement Protected Against Hydrogenation in the Rumen on Growth and *n-3* Fatty Acid Content of Lamb Fat and Muscle. *Austr. J. Agric. Res.* **2001c,** *52,* 433–437.

Kook, K.; B.H. Choi; S. Sun; F. Garcia; K. Myung. Effect of Fish Oil Supplement on Growth Performance, Ruminal Metabolism, and Fatty Acid Composition of Longissimus Muscle in Korean Cattle. *Asian-Aust. J. Anim. Sci.* **2002,** *15,* 66–71.

Kralik, G.; Z. Gajcevic; Z. Skrtic. The Effect of Different Oil Supplementations on Laying Performance and Fatty Acid Composition on Egg Yolk. *Ital. J. Anim. Sci.* **2008,** *7,* 173–183.

Kritchevsky, D. Antimutagenic and Some Other Effects of Conjugated Linoleic Acid. *Br. J. Nutr.* **2000,** *83,* 459–465.

Lenzi, A.; L. Gandini; V. Maresca; R. Rago; P. Sgro; F. Dondero; M. Picardo. Fatty Acid Composition of Spermatozoa and Immature Germ Cells. *Mol. Hum. Reprod.* **2000,** *6,* 226–231.

Leskanich, C.O.; K.R. Matthews; C.C. Warkup; R.C. Noble; M. Hazzledine. The Effect of Dietary Oil Containing (n-3) Fatty Acids on the Fatty Acid, Physicochemical, and Organoleptic Characteristics of Pig Meat and Fat. *J. Anim. Sci.* **1997,** *75,* 673–683.

Lopez-Ferrer, S.; M. Baucells; A. Barroeta; M.A. Grashorn. N-3 Enrichment of Chicken Meat. 1. Use of Very Long-Chain Fatty Acids in Chicken Diets and Their Influence on Meat Quality: Fish Oil. *Poultry Sci.* **2001,** *80,* 741–752.

Mandell, I.; J. Buchanan-Smith; B. Holub; C. Campbell. Effects of Fish Meal in Beef Cattle Diets on Growth Performance, Carcass Characteristics, and Fatty Acid Composition of Longissimus Muscle. *J. Anim. Sci.* **1997,** *75,* 910–919.

Manilla, H.; F. Husveth. *N-3* Fatty Acid Enrichment and Oxidative Stability of Broiler Chicken: A Review. *Acta Alimentaria* **1999,** *28,* 235–249.

Mark, G.; T.A. Sanders. The Influence of Different Amounts of n-3 Polyunsaturated Fatty Acids on Bleeding Time and in vivo Vascular Reactivity. *Br. J. Nutr.* **1994,** *71,* 43–52.

Marriott, N.G.; J.E. Garret; M.D. Sims; J.R. Abril. Characteristics of Pork with Docosahexaenoic Acid Supplemented Diet. *J. Muscle Foods* **2002a,** *13,* 253–263.

Marriott, N.G.; J.E. Garrett; M.D. Sims; J.R. Abril. Performance Characteristics and Fatty Acid Composition of Pigs Fed a Diet with Docosahexaenoic Acid. *J. Muscle Foods* **2002b,** *13,* 265–277.

McGuire, M.A.; M.K. McGuire. Conjugated Linoleic Acid (CLA): A Ruminant Fatty Acid with Beneficial Effects on Human Health. *J. Anim. Sci.* **2000**, *77*, 1–8.

Mir, P.; M. Ivan; M. He; B. Pink; E. Okine; L. Goonewardene; T.A. McAllister; R. Weselake; Z. Mir. Dietary Manipulation to Increase Conjugated Linoleic Acids and Other Desirable Fatty Acids in Beef: A Review. *Can. J. Anim. Sci.* **2003**, *83*, 673–685.

Mooney, J.W.; E.M. Hirschler; A.K. Kennedy; A.R. Sams; M.E. Van Elswyk. Lipid and Flavor Quality of Stored Breast Meat from Broilers Fed Marine Algae. *J. Sci. Food Agric.* **1998**, *78*, 134–140.

Nestel, P.; H. Shige; S. Pomeroy; M. Cehun; M. Abbey; D. Raederstorff. The n-3 Fatty Acids Eicosapentaenoic Acid and Docosahexaenoic Acid Increase Systemic Arterial Compliance in Humans. *Am. J. Clin. Nutr.* **2002**, *76*, 326–330.

Offer, N.W.; M. Marsden; J. Dixon; B.K. Speake; F.E. Thacker. Effect of Dietary Fat Supplements on Levels of n-3 Poly-unsaturated Fatty Acids, Trans Acids and Conjugated Linoleic Acid in Bovine Milk. *Anim. Sci.* **1999**, *69*, 613–625.

Offer, N.W.; B.K. Speake; J. Dixon; M. Marsden. Effect of Fish Oil Supplementation on Levels of (*n-3*) Poly-unsaturated Fatty Acids in the Lipoprotein Fractions of Bovine Plasma. *Anim. Sci.* **2001**, *73*, 523–531.

Or-Rashid, M.M.; J. Kramer; M. Wood; B. McBride. Supplemental Algal Meal Alters the Ruminal Trans-18:1 Fatty Acid and Conjugated Linoleic Acid Composition in Cattle. *J. Anim. Sci.* **2008**, *86*, 187–196.

Papadoupoulos, G.; C. Goulas; E. Apostolaki; R. Abril. Effects of Dietary Supplements of Algae, Containing Polyunsaturated Fatty Acids, on Milk Yield and Composition of Milk Products in Dairy Ewes. *J. Dairy Res.* **2002**, *69*, 357–365

Parodi, P.W. Conjugated Linoleic Acid and Other Anticarcinogenic agents of Bovine Milk Fat. *J. Dairy Sci.* **1999**, *82*, 1339–1349.

Pawlosky, R.; A. Barnes; N. Salem Jr. Essential Fatty Acid Metabolism in the Feline: Relationship between Liver and Brain Production of Long-Chain Polyunsaturated Fatty Acids. *J. Lipid Res.* **1994**, *35*, 2032–2040.

Ponnampalam, E.N.; A.J. Sinclair; A.R. Egan; S.J. Blakeley; D. Li; B.J. Leury. Effect of Dietary Modification of Muscle Long Chain n-3 Fatty Acid on Plasma Insulin and Lipid Metabolites, Carcass Traits and Fat Deposition in Lambs. *J. Anim. Sci.* **2001**, *79*, 895–903.

Ponnampalam, E.N.; A.J. Sinclair; B.J. Hosking; A.R. Egan. Effects of Dietary Lipid Type on Muscle Fatty Acid Composition, Carcass Leanness and Meat Toughness in Lambs. *J. Anim. Sci.* **2002**, *80*, 628–636.

Raes, K.; S. De Smet; D. Demeyer. Effect of Dietary Fatty Acids on Incorporation of Long Chain Polyunsaturated Fatty Acids and Conjugated Linoleic Acid in Lamb, Beef and Pork Meat: A Review. *Anim. Feed Sci. Tech.* **2004**, *113*, 199–221.

Rhee, K.S.; T.L. Davidson; H.R. Cross; Y.A. Ziprin. Characteristics of Pork Products from Swine Fed a High Monosaturated Fat Diet: Part I—Whole Muscle Products. *Meat Sci.* **1990**, *27*, 329–341.

Roth-Maier, D.; K. Eder; M. Kirchgessner. Live Performance and Fatty Acid Composition of Meat in Broiler Chickens Fed Diets with Various Amounts of Ground or Whole Flaxseed. *J. Anim. Physio. Anim. Nutr.* **1998**, *79*, 260–268.

Rymer, C.; D. Givens; K. Whale. Dietary Strategies for Increasing Docosahexaenoic Acid (DHA) and Eicosapentaenoic Acid (EPA) Concentration in Bovine Milk: A Review. *Nutr. Abs. Rev.* **2003,** *7(4),* 9R–25R.

Sanudo, C.; M. Enser; M.M. Campo; G.R. Nute; G. Maria; I. Sierra; J.D. Wood. Fatty Acid Composition and Fatty Acid Characteristics of Lamb Carcasses from Britain and Spain. *Meat Sci.* **2000,** *54,* 339–346.

Scheideler, S.E.; G.W. Froning. The Combined Influence of Dietary Flaxseed Variety, Level, Form, and Storage Conditions on Egg Production and Composition among Vitamin E-Supplemented Hens. *Poultry Sci.* **1996,** *75,* 1221–1226.

Scollan, N.D.; N.J. Choi; E. Kurt; A.V. Fisher; M. Enser; J.D. Wood. Manipulating the Fatty Acid Composition of Muscle and Adipose Tissue in Beef Cattle. *Br. J. Nutr.* **2001a,** *85,* 115–124.

Scollan, N.D.; M.S. Dhanoa; N.J. Choi; W.J. Maeng; M. Enser; J.D. Wood. Biohydrogenation and Digestion of Long Chain Fatty Acids in Steers Fed on Different Sources of Lipid. *J. Agric. Sci.* **2001b,** *136,* 345–355.

Shingfield, K.J.; S. Ahvenjarvi; V. Toivonen; A. Arola; K.V.V. Nurmela; P. Huhtanen; J.M. Griinaari. Effect of Dietary Fish Oil on Biohydrogenation of Fatty Acids and Milk Fatty Acid Content in Cows. *Anim. Sci.* **2003,** *77,* 165–179.

Sigfusson, H. Partitioning of an Exogenous Lipid-Soluble Antioxidant between the Neutral and Polar Lipids of Minced Muscle. Ph.D. Dissertation, University of Massachusetts, Amherst, **2000;** p 3–8.

Simopoulos, A.P. Overview of Evolutionary Aspects of ω3 Fatty Acids in the Diet. *The Return of ω3 Fatty Acids into the Food Supply: I. Land-Based Animal Food Products and Their Health Effects*; A.P. Simopoulos, Ed.; World Review of Nutrition and Dietetics; S. Karger: Switzerland, **1998;** *83,* p 1–11.

Squires, E.L. Stallion Semen Characteristics Following Dietary Supplementation with Magnitude™. Colorado State University Research Report. **2005.**

Taugbøl, O. Omega 3 Fatty Acid Incorporation in Fat and Muscle Tissues of Growing Pigs Fed Supplements of Fish Oil. *J. Vet. Med.* **1993,** *40,* 93–101.

Tavilani, H.; M. Doosti; K. Abdi; A. Vaisiraygani; H.R. Joshaghani. Decreased Polyunsaturated and Increased Saturated Fatty Acid Concentration in Spermatozoa from Asthenozoospermic Males as Compared with Normozoospermic Males. *Andrologia* **2006,** *38,* 173–178.

Van Elswyk, M. Nutritional and Physiological Effects of Flax Seed in Diets for Laying Fowl. *World Poultry Sci. J.* **1997,** *53,* 253–264.

Van Nevel, C. J.; D. I. Demeyer. Lipolysis and Biohydrogenation of Soybean Oil in the Rumen in vitro: Inhibition by Antimicrobials. *J. Dairy Sci.* **1995,** *78,* 2797–2806.

Van Nevel, C. J.; D.I. Demeyer. Influence of pH on Lipolysis and Biohydrogenation of Soybean Oil by Rumen Contents in vitro. *Reprod. Nutr. Dev.* **1996,** *36,* 53–63.

Wistuba, T.; E. Kegley; J. Apple. Influence of Fish Oil in Finishing Diets on Growth Performance, Carcass Characteristics, and Sensory Evaluation of Cattle. *J. Anim. Sci.* **2006,** *84,* 902–909.

Wood, J.; M. Enser. Factors Influencing Fatty Acids in Meat and the Role of Antioxidants in Improving Meat Quality. *Brit. J. Nutr.* **1997,** *78,* S49–S60.

Wood, J.D.; M. Enser; A.V. Fisher; G.R. Nute; R.I. Richardson; P.R. Sheard. Manipulating Meat Quality and Composition. *Proc. Nutr. Soc.* **1999,** *58,* 363–370.

Wood, J.; R. Richardson; G. Nute; A. Fisher; M. Campo; E. Kasapidou; P. Sheard; M. Enser. Effects of Fatty Acids on Meat Quality: A Review. *Meat Sci.* **2003,** *66,* 21–32.

Wonsil, B.J.; J.H. Herbein; B.A. Watkins. Dietary and Ruminally Derived Trans-18:1 Fatty Acids Alter Bovine Milk Lipids. *J. Nutr.* **1994,** *124,* 556–565.

Wright-Rodgers, A.S.; M.K. Waldron; K.E. Bigley; G.E. Lees; J.E. Bauer. Dietary Fatty Acids Alter Plasma Lipids and Lipoprotein Distributions in Dogs during Gestation, Lactation, and the Perinatal Period. *J. Nutr.* **2005,** *135,* 2230–2235.

19

Applications of Single Cell Oils for Aquaculture

Mario Velasco-Escudero[a] and Hui Gong[b]

[a]Maricultura Negocios y Tecnología, S.L., C/. Pere Terre i Domenech 11, 08027 Barcelona, Spain; [b]College of Natural and Applied Science, University of Guam, Mangilao, Guam

Introduction

Aquaculture is the fastest growing food-producing sector in the world, with an average rate of increase of 8.8% per year since 1970, compared with only 1.2% for capture fisheries and 2.8% for terrestrial, farmed meat production systems over the same period (FAO, 2009). Farmed fish accounts for approximately 50% of all fish consumed globally, and this percentage is expected to continue to increase as a result of dwindling catches from capture fisheries and ever-increasing total and per capita sea food consumption (Turchini et al., 2009).

Such a steady increase in aquaculture production mainly results from the adoption of intensive farming practices that rely on formulated feeds. Fish meal and fish oil are the two most important ingredients in conventional aquafeeds because of their unique nutritional value. Fish meal and fish oil not only are good sources of high quality animal protein and essential fatty acids, but they also contain vitamins, digestible energy, and macro and trace minerals. It is also believed that the marine origin ingredients contain some unknown feeding stimulants or attractants for most aquaculture species. It was estimated that almost one third of aquaculture production came from "feeding species" that consumed almost 12 million tons of aquafeeds, containing 2.3 and 0.7 million tons of fish meal and fish oil, respectively (Tacon, 1999). As for fish oil, known for its high levels of ω-3 long-chain, highly unsaturated fatty acids (LC-HUFAs), 87% of the global supply of this oil is used for aquaculture feed.

On one hand, because of the high demand for fish meal and fish oil, their use in aquaculture profoundly impacts and will impact finite natural resources. Emerging environmental and safety issues associated with fish meal/fish oil production from natural fish stocks have instigated public concern over the sustainability of this natural resource. On the other hand, the general trend of global fish meal production remains relatively static or in a slight decline, due to overfishing. Climate changes, such as the *El Niño* phenomenon, periodically cause sharp declines in fish meal and fish oil supply. Given the fluctuating supply and variable availability of fish meal/fish oil, along with the growing demand for it in aquaculture and other animal production industries, prices will continue to increase, perhaps even dramatically, as demand for this growing industry outstrips supply. In the long run, a more sustainable approach is needed, and the search for possible replacements is under way. In this chapter, we will focus on advances in the use of single cell oils (SCOs) to replace fish oil in aquafeeds, providing a safe and sustainable alternative that will allow the continued advancement of the aquaculture industry.

LC-HUFAs, such as docosahexaenoic acid (DHA; C22:6 ω-3), eicosapentaenoic acid (EPA; C20:5 ω-3), and arachidonic acid (ARA; C20:4 ω-6), derived from microalgae, have been shown to be essential for growth and development of various fish larvae (Langdon & Waldock, 1981; Sargent et al., 1997; Brown, 2002; Harel et al., 2002). In a recent review, Brown (2002) summarized the proportion of these important LC-HUFAs in forty-six strains of microalgae (data from Volkman et al., 1989, 1991, 1993; Dunstan et al., 1994). The fatty acid content of these microalgae showed systematic differences, according to taxonomic group, and even significant differences between microalgae from the same class. Several microalgal species have moderate to high percentages of EPA (7 to 34% of total lipid content). Prymnesiophytes [e.g., *Pavlova* spp. and *Isochrysis* sp. (T.ISO)] and cryptomonads are relatively rich in DHA (0.2 to 11% of total lipid content) (Brown, 2002), whereas the highest content of AA (over 20% of dry weight; 32% of fatty acids) was found in the green alga *Parietochloris incisa* (see Chapter 10). However, most Chlorophytes (*Dunaliella* spp. and *Chlorella* spp.) have been shown to exhibit a lower nutritional value because of their deficiency in both C20 and C22 HUFAs, even though some species contain small amounts of EPA (up to 3.2% of total lipid content) (Brown et al., 1997; Brown, 2002). Prasinophytes contain significant proportions of C20 (*Tetraselmis* spp.) or C22 (*Micromonas* spp.) HUFAs, but rarely both. Although there is a preponderance of data related to the fatty acid profile of microalgae, the absolute amount of LC-HUFAs produced also depends on whether or not that species also produces oil (e.g., triacylglycerols), rather than carbohydrates (e.g., starch), as a storage product. In many cases, what originally looks like a viable LC-HUFA producer later turns out to be unusable if it cannot produce more than 20% of its mass in total lipid of the same composition.

SCO derived from unicellular algae seems to be a viable source of high quality ω-3 LC-HUFA to replace fish oil in aquaculture feed. Microalgae have a simpler fatty acid profile (much like plants), which facilitates purification (Medina et al., 1998). More importantly, microalgae are an unlimited and renewable resource that can be produced in industrial quantities under environmentally safe conditions (Harel et al., 2002). Through careful species selection and cultivation control, one can formulate particular fatty acid compositions (designer oils) to meet particular nutritional requirements (Medina et al., 1998; Atalah et al., 2007). In a recent review by Turchini et al. (2009), the use of DHA-rich oils derived from single-cell microalgae, such as *Schizochytrium* spp., *Crypthecodinium cohnii* (Biecheler, 1938), or *Phaeodactylum tricornutum* (Bohlin, 1897), was reported to have been successfully tested on gilthead sea bream (*Sparus aurata*; Linnaeus, 1758). Similarly, *Schizochytrium* spp. oil has also been successfully tested on Atlantic salmon (Miller et al., 2007). However, reducing the production cost of SCO remains as a challenge for such a promising innovation to become viable, commercially and globally.

Essential Fatty Acid Requirements in Aquatic Animals

Marine versus Terrestrial Environment

Because of their aquatic surroundings, fish and shellfish have unique characteristics, compared with terrestrial livestock. For example, fish and shellfish are poikilothermic; they maintain neutral buoyancy, operate in a three-dimensional world, excrete ammonia, and exhibit indeterminate growth. These features generally result in less energy allocated for locomotion and/or maintaining a constant body temperature, while requiring more calories to degrade amino acids into the end product of ammonia, instead of urea and/or uric acid, as in terrestrial animals. The metabolic rate of aquatic animals, as well as some physiological and nutritional functions, is also affected by fluctuations of salinity and water temperature. The nominal physiological and biochemical functioning of these animals ultimately depends on cellular membrane fluidities, which, to a large extent, determine the dietary fatty acids requirements in aquatic organisms.

Fatty acids are major components of various lipid classes (e.g., fats, phospholipids, sterols, and wax) and are involved in various metabolic and structural functions. Essential fatty acids (EFAs) are those that cannot be synthesized de novo, but are required to sustain an organism's growth and development.

Homeothermic, land-dwelling animals, such as mammals and birds, have two EFAs, namely, linoleic acid (LA, C18:2 ω-6) and α-linolenic acid (ALA, C18:3 ω-3). However, the dietary requirement for LA is generally about ten-fold higher than ALA, among most animals, including man. Consequently, the symptoms of EFA deficiency with LA are generally first to appear and are more severe than those seen for ALA deficiency. In both cases, the symptoms of EFA deficiency can be overcome by supplying dietary LC-HUFAs, which indicates that it is not the LA and ALA themselves that are required for nominal growth and development, but the elongation and desaturation products of these two fatty acids that are truly essential. The conversion of oleic acid (C18:1, ω-9) to LA and the conversion of LA to ALA would require the existence of a Δ12 desaturase. This enzyme has not been shown to be present in mammals, which explains the metabolic block in de novo formation of LA and ALA that results in their dietary essentiality. Because of the wealth of data on EFAs in mammals, it has been generally accepted that these are the EFA for all vertebrates, including fish.

The elongation and desaturation of fatty acids to form the LC-HUFAs occurs in a series of enzymatic steps, outlined in Fig. 19.1. Fish also appear to be unable to synthesize the ω-6 and ω-3 LC-HUFAs , unless precursor EFAs (LA and ALA) are provided from dietary sources. However, fish, like mammals, are able to synthesize saturated and monounsaturated fatty acids from acetate de novo, as is shown in Fig. 19.1. For example, fish can convert palmitic acid (C16:0) to the monoene palmitoleic acid (C16:1, ω-7) and stearic acid (C18:0) to the monoene oleic acid, using a Δ9 desaturase.

Almost all fish, like all other vertebrates, have the ability to convert ALA (C18:3 ω-3) into LC-HUFA (e.g., EPA and DHA) in vivo through an alternating succession of de-saturation and elongation (Sargent et al., 2002; Nakamura & Nara, 2004, Turchini et al., 2009). EPA can be produced by a process of elongation/Δ6 desaturation or Δ6 desaturation/elongation and then Δ5 desaturation from ALA. The eventual DHA production requires the combination of two further elongation steps, a Δ6 desaturation and, finally, one cycle of β-oxidation (a chain-shortening reaction), commonly referred to as the Sprecher Shunt (Sprecher et al., 1995; Turchini et al., 2009).

There is a competitive inhibition of the desaturation of ω-3 series fatty acids by those of the ω-6 series, and vice versa, because they both compete as substrates for the binding sites on the same enzymes, namely, the Δ5 and Δ6 desaturases. As in mammals, the ability to elongate and desaturate EFAs to LC-HUFAs is not the same in all species of fish. For example, the turbot is able to desaturate and elongate only 3-15% of LA or ALA to LC-HUFAs when given the ^{14}C-labelled fatty acid; whereas 70% of the label (^{14}C) from ALA was found in DHA in rainbow trout (Halver, 1980). This conversion rate is still very high, relative to that found in terrestrial mammals, which ranges from zero, in cats (obligate carnivores), to about 1%, in man (omnivores), and 10%, in mice (herbivores).

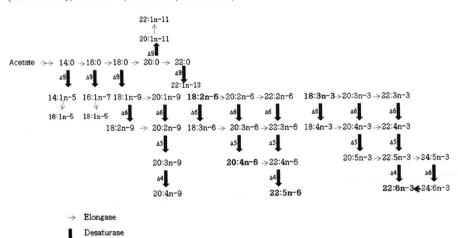

Fig. 19.1. Flow Diagram for fatty acid biosynthesis in fish (adapted from Izquierdo, 2005 and Halver, 1980).

It appears that high levels of C18 ω-6 or ω-3 fatty acids inhibit the synthesis and metabolism of C18:1 ω-9 in fish (Halver, 1980). It is interesting to note that the channel catfish, which also exhibits a negative growth response to dietary LA or ALA, incorporates very high levels of C18:1 into its body lipids. The inclusion of either LA or ALA in the diet reduces the levels of C18:1 fatty acids in body lipids. A similar reduction has also been observed in red sea bream liver phospholipid when either LA or ALA was added to the diet. The competitive inhibition desaturation of members from one series of fatty acids for members of another series is well established, with ω-3 > ω-6 > ω-9 being the usual order of potency for inhibition (Halver, 1980).

Compared to the fatty acid composition of lipids from major terrestrial livestock, some fish tissues contain relatively high concentrations of LC-HUFAs with a high content of EPA and DHA. LC-HUFAs are important components of all cell membranes, but they also make fish tissue highly vulnerable to lipid peroxidation (Lall, 2000). The difference between the fatty acid compositions of marine and freshwater fish is also distinguishable. The ω-6/ω-3 ratios are 0.37 and 0.16 for freshwater and marine fish, respectively. There is also a marked effect of salinity in the diet on the fatty acid composition of juveniles in certain salmonid species that favors a higher ω-6/ω-3 ratio in lower salinities (Tocher et al., 1995). Fatty acid composition and requirements vary considerably from species to species but are also affected by other factors, such as fish sex, size, developmental stage, previous dietary history, source and levels of dietary lipid, composition and nutrient content of experimental diets, and feeding duration and practices, as well as other environmental conditions (e.g., temperature).

The EFA requirements of freshwater fish can be met by supplying dietary LA and ALA, but meat from those species will remain notoriously low in health-supporting ω-3 LC-HUFAs. The EFA requirements of marine fish can only be met by supplying LC-HUFAs, such as EPA and DHA (typically from fish oil), and meat from these species is very rich in health-supporting ω-3 LC-HUFAs. Freshwater fish are able to elongate and desaturate ALA to DHA, albeit quite poorly, whereas marine fish, which lack or have even exhibit less activity of Δ5-desaturase, require the dietary supplementation of LC-HUFAs, such as ARA, EPA, and DHA. The quantitative requirements and signs of EFA deficiency in several freshwater and marine fish have been documented (NRC, 1993). Emphasis has been placed on the essentiality of ARA (C20:4 ω-6) for marine fish nutrition (Sargent et al., 1997; Izquierdo, 2005), indicating that a balanced ratio of dietary ARA, EPA, and DHA is prefereable.

Although ω-6 fatty acids are essential to growth and development, fish oils generally have low levels of these fatty acids and higher levels of ω-3 type fatty acids. There is evidence that ω-3 LC-HUFAs, which are present in relatively large concentrations in fish oil, play the equivalent role of EFAs for fish.

Currently there are over 140 fish and shellfish species in aquaculture; we will focus on several ones with major economic importance.

Tilapia

Tilapia mainly requires the dietary supplementation of ω-6 polyunsaturated fatty acids. Lim & Aksoy (2009) recently reviewed the fatty acid requirement of Tilapia, and this species has been reported to have a dietary requirement for LA. The optimum dietary levels of ω-6 LA or ARA have been estimated to be 0.5 and 1% for redbelly tilapia (*Tilapia zillii*) and Nile tilapia (*Oreochromis niloticus*), respectively. This information has not been determined for other species, but it has been suggested that blue tilapia (*O. aureus*) have an even higher requirement for ω-6 fatty acids. Tilapia may also have a requirement for ALA, but the optimum dietary requirement level for ω-3 fatty acids has not been determined. The presence of high levels of either ω-6 or ω-3 fatty acids may spare the requirement of the other, although ω-6 a ppears to have better growth-promoting effects than ω-3. Tilapia appear to possess the ability to desaturate and chain elongate LA to ARA (ω-6 pathway) and ALA to EPA and DHA (ω-3 pathway), but information on the nutritional value of fish oil for tilapia is inconsistent. Some studies have shown that the nutritional value of fish oil is no different from that of plant oils, while other studies reported the poor performance of fish oil-containing diets. Fish oil provided good spawning performance for tilapia broodstock reared in saline water, while, in freshwater, good reproductive performance was obtained with plant oil (soybean oil).

Rainbow Trout

Rainbow trout, a cold-water fish, requires n-3 fatty acids as EFA in its diet, and this EFA requirement can be met by supplementing the diet with 1.0% ALA (Halver, 1980). Although the dietary inclusion of LA (ω-6) in the diet may result in some improvement in growth and feed conversion, compared with EFA deficient diets, it will not prevent some EFA deficiency symptoms, such as "shock syndrome" (Halver, 1980). Furthermore, EFA deficiency also adversely affects internal haemostasis and immunocompetence in rainbow trout. EFA influence thrombocytes by affecting the aggregation capacity of these in rainbow trout, and thrombocytes from fish deficient in dietary EFA were weaker in their aggregatory response (Kfoury et al., 2006). Antibody production and in vitro killing of bacteria by macrophages were both compromised in rainbow trout by a dietary EFA deficiency (Kiron et al., 1995). Fish fed polyunsaturated fatty acids were stronger when resisting pathogens, but excess levels of n-3 HUFA may not be effective (Kiron et al., 1995). Dietary LA or ALA could be readily converted to C-20 and C-22 LC-HUFA of the same series; ALA exhibits higher a EFA value than LA, while EPA or DHA is superior to ALA, in terms of the EFA value in rainbow trout. Combining applications of EPA and DHA yields a better performance than applications of either EPA or DHA alone. The superior nutritional value of C-20 and C-22 carbon LC-HUFA is further supported by the excellent growth-promoting effects of dietary fish oils, such as pollock liver oil and salmon oil, for rainbow trout (Halver, 1980).

Salmon

Salmonids require mainly ω-3 polyunsaturated fatty acids in their diet. The dietary requirement of 1% of ALA or, perhaps, 0.5-1.0% HUFA is suggested for salmon. (Sargent et al., 2002). Ruyter et al. (2001) ran a 4 month feeding trial using semi-synthetic diets, which contained fatty acid methyl esters of either LA, ALA, or a mixture of equal amounts of EPA and DHA, on Atlantic salmon fry (4 g). In this study, fatty acids esters were added in a range of 0 to 2% (by dry weight). Results showed that increasing the levels of dietary ω-3 fatty acids, up to 1%, yielded faster growth rates in salmon fry, and the mixture of EPA and DHA was more beneficial in promoting fish growth than ALA. No significant effect on growth rate was seen when the dietary level of LA was increased. Dietary inclusion of ω-3 fatty acids reduced the mortality of salmon, while dietary LA had no such beneficial effects. These dietary treatments caused substantial changes in the fatty acid composition of blood and liver phospholipids (PL), whereas the total lipid fraction of the carcass was less affected. Increasing doses of LA in the diet resulted in an increased percentage of ARA in liver and blood PLs, while the percentage of mead acid (C20:3 ω-9; an indicator of EFA deficiency) decreased. The percentage of LA also increased in the liver, blood, and carcass. Dietary ALA resulted in increased percentages of ALA and EPA in liver PLs, while the percentage of Mead acid decreased. Consistent with mammalian studies, the addition of ALA did not significantly increase the percentage of DHA. Dietary ALA produced no significant changes in the composition of blood fatty acids, but it increased the percentage of ALA in the carcass. The dietary combination of the ω-3 fatty acids EPA and DHA resulted in an increased percentage of DHA in blood and liver lipids and a decreased percentage of mead acid; there were no changes in the percentage of EPA.

Marine Fish

Marine fish require ω-3 LC-HUFA, such as EPA and DHA, as EFA for their normal growth. Requirements range from 0.5 to 2.0%, varying in correspondence with the growth stage of newly hatched larvae receiving live foods to broodstock for reproduction (Watanabe, 1993). In general, concentrations of ω-3 LC-HUFA are high in fish oil and fish meal, and no EFA-related deficiencies will manifest themselves in fish offered these ingredients as sources of protein and lipids. EFA requirements are higher in the larval stage, compared to the juvenile stage, for each species. For example, red sea bream (*Pagrus major*) require 0.5% LC-HUFA in juvenile diets and 3.0% for larvae; yellow tail (*Seriola quinqueradiata*) require 2.0% for juveniles and <3.9% for larvae; striped jack (*Pseudocaranx dentex*) require 1.7% for juveniles and <3.0% for larvae; turbot (*Scophthalmus maximus*) require 0.8% for juveniles and 1.2-3.2% for larvae; 1.0% LC-HUFA are required by striped knifejaw (*Oplegnathus fasciatus*) and flounder (*Paralichthys olivaceus*) juveniles, in comparison to 3.0-4.0% and <3.5%, respectively, for their larvae (Watanabe, 1993). Because of the limited conversion from EPA to DHA, DHA confers a higher EFA value than EPA for the above larval marine fish (Watanabe, 1993).

Shrimp

The quantitative requirements for essential fatty acids have been reported for penaeid shrimp. Kanazawa et al. (1979) suggested that a dietary provision of 1.0% ω-3 LC-HUFA could be considered a minimal value for postlarval penaeids.

For *Penaeus monodon*, Chen & Tsai (1986) indicated a requirement for ω-3 LC-HUFA at 0.5-1.0% of the diet, while Rees et al. (1994) observed that post-larvae can grow well on an *Artemia* sp. diet, with ω-3 LC-HUFA ranging from 12 to 22 mg/g dry weight. Xu et al. (1994) suggested that, for *Farfantepenaeus chinensis*, the requirement for ALA may be between 0.7 and 1.0% of the diet. Once DHA was adequately provided in the diet (around 1.0%), growth, molt frequency, and survival were significantly greater than in animals fed a diet with 1.0% ALA. They concluded that, although ω-6 LA and ARA have beneficial effects on growth and survival, ω-3 LC-HUFAs, especially DHA, are the most potent EFA for this species. Lim et al. (1997) evaluated the growth response and fatty acid composition of juvenile shrimp (*Litopenaeus vannamei*) fed different dietary lipids. They found that menhaden oil, rich in ω-3 LC-HUFAs, was most nutritious for this species, and, among plant oils, those rich in ALA had a higher nutritional value than those rich in LA. They concluded that both ω-6 and ω-3 fatty acids appear to be essential in the diet, although ω-3 LC-HUFAs were required for maximum growth, feed efficiency, and survival. González-Félix et al. (2002a, 2002b, 2003b) evaluated the nutritional value and dietary requirement of LA and ALA, including different ratios of LA/ALA for juvenile *L. vannamei*, and compared the nutritional value of these fatty acids to that of ω-3 LC-HUFAs, in combination, which showed a higher nutritional value by producing significantly ($P < 0.05$) higher final weight and weight gain. Neither LA nor ALA, alone or in combination, improved growth significantly when compared with shrimp fed the basal diet of only palmitic and stearic acids. Dietary requirements for LA and ALA were not demonstrated under their experimental conditions. The same authors later re-evaluated the nutritional value of dietary LA and ALA, in comparison with the LC-HUFAs ARA, EPA, and DHA, alone or in combination with each other, at a level of 0.5% of diet (González-Félix et al., 2003a). All LC-HUFAs showed a higher nutritional value than LA and ALA and produced significantly ($P < 0.05$) higher final weight, weight gain, and total lipid content in shrimp muscle. Under their experimental conditions, *L. vannamei* appeared to be able to satisfy its EFA requirements with LC-HUFAs from either ω-3 (e.g., EPA and DHA) or ω-6 (e.g., ARA) families, fatty acids that exhibited higher nutritive values than LA or ALA. Moreover, the requirement for LA and ALA was, once again, not demonstrated. In spite of some reports, in which fatty acids of the ω-3 family have greater essential values than those of the ω-6 family for marine shrimp, *L. vannamei* might equally utilize ω-3 and ω-6 LC-HUFAs as the result of metabolic adaptations related to specific environmental conditions or feeding habits of this species and, possibly, phylogenetic differences. In addition, there is evidence that eicosanoids derived from ARA might be involved in the molting process of crustaceans and, thus, influence growth (Koskela et al., 1992). Consequently, this study suggests that the EFA value for this species may be determined by chain length and degree of unsaturation, with LC-HUFA having greater essential value than shorter-chain fatty acids, regardless of family.

L. *vannamei* appeared to be able to satisfy its ω-3 LC-HUFA requirements when these fatty acids were supplied at 0.5% of diet, but this level may be even lowered. Depressed growth was observed in shrimp fed diets with ω-3 LC-HUFA supplemented at 2.0%. In addition, this study showed that increasing the dietary lipid level had an effect on the lipid composition of shrimp by increasing lipid deposition in hepatopancreas and muscle tissue, but without a significant effect on growth.

Lytle et al. (1990) indicated there might be a delicate balance between ω-3 and ω-6 fatty acids and proposed a high ω-3/ω-6 ratio in maturation diets. In addition to high EPA and DHA levels, a moderate level of ARA was also recommended for inclusion in the maturation diet. An ω-3 to ω-6 ratio of approximately 2 to 1 was reported in mature ovaries of *P. semisulcatus* (Ravid et al., 1999) and *L. vannamei* (Wouters et al., 1999) spawners, while in nauplii of *L. vannamei*, this ratio increased to 3 to 1 (Wouters et al., 2001).

Sources of Essential Fatty Acids in Aquaculture

Fish Oil and Fish Meal

Fish oil has been traditionally the main source of essential LC-HUFAs in the aquaculture of marine species and, even more so, of marine stenohaline (which are not able to withstand a wide variation in water salinity) species. The fact that fish oil is embedded in fish meal (from 3 to 17% of dry matter, depending on the species) also makes this ingredient an important source of LC-HUFA (Hertrampf & Piedad-Pascual, 2000). Fish meal and fish oil manufacture follows this basic flow:

Fish oil and fish meal essential fatty acid composition varies, depending on the fish species used, catching season, geographical location, fishing ground, processing, and degree of oxidation (Hertrampf & Piedad-Pascual, 2000). Because high levels of unsaturated fatty acids facilitate the rapid oxidation of fish oil and fish meal, these products generally require the addition of antioxidants, such as ethoxyquin, to extend their shelf life.

Plant Oils

No plant oils contain LC-HUFAs (ARA, EPA, or DHA), although some plant oils, such as linseed and canola oils, may contain a high percentage of ALA (Turchini et al., 2009). When plant oils are used in aquafeeds, as they are for mammals, the conversion of ALA to DHA by the organism is minimal or none, and the muscular tissue fatty acid profile reflects the feed fatty acid composition (Hardy, 2006; Mourente et al., 2005; Mourente & Bell, 2006). Also, DHA, and not ALA, represents the lipid building blocks of neurological tissues (Masuda, 2003), making it a required nutrient during fish ontogeny and early developmental stages. Plant oils can be used, as long as the EFA requirements of the cultured species are met (Turchini et al., 2009).

Algal SCO Products

Unicellular algae are the only primary producers of DHA in the marine environment, and, as such, they are an excellent alternative to replace fish oil in aquafeeds. DHA-rich algal oils and meals have been tested as a fish oil alternative on gilthead seabream (Atalah et al., 2007; Ganuza et al., 2008), Atlantic salmon (Miller et al., 2007), Pacific white shrimp (Browdy et al., 2006; Patnaik et al., 2006) and Black tiger shrimp (Chandrasekar, 2008). In all cases, they have been successful in the partial or total replacement of fish oil and fish meal in aquafeeds. These SCO-containing algal meals have also been successfully used for the enrichment of rotifers and *Artemia* in fish larval culture (Harel et al., 2002; Copeman et al., 2002; Harel & Place, 2003), and this practice is now well established in commercial hatcheries (Lai, 2008). DHA SCO-rich algal products have also been used in finishing feeds to increase levels of DHA in the muscle tissue of cultured organisms, such as Tilapia (Laurin et al., 2006), thereby significantly improving their nutritional value for consumers.

New GMO Alternatives

Research is being conducted with genetically engineered yeast and plants to obtain oils containing LC-HUFA (Robert, 2006). Although some positive results have been obtained, the levels of DHA and EPA are still very low. Moreover, the negative perception consumers have of genetically modified organisms may delay the use of these products in aquafeeds.

Fish Meal and Fish Oil Dilemma

Sustainability in the Aquaculture Industry

Traditional capture fisheries are the last great manifestation of the hunter/gatherer development phase of mankind. The technology and resources of fisheries has advanced to the point that many wild populations with commercial value (e.g., cod, swordfish, tuna, anchoveta, pollock, etc.) are already caught at their maximum sustainable yields and/or overexploited. This problem is reflected by the fact that the production of global capture fisheries has been nearly static, at about 90 million MT, for the last 15 years (FAO, 2009). Aquaculture is an agribusiness dedicated to the managed production of aquatic organisms. During the last decade, aquaculture has been, and still is, the fastest growing food-producing sector, currently accounting for over 50% of the total world's fisheries (capture and aquaculture) (FAO, 2009). The absolute increase in output products from aquaculture is able to supply products to meet the increased demand from a growing world population, with changing nutritional habits, that recognizes the nutritional benefits of fish and seafood consumption.

However, our dependency on capture fish stocks for the inclusion of fish meal and oil in aquafeeds is raising concerns about the best use of available resources and the sustainability of aquaculture practices (FAO, 2007). This dependency on wild fish stocks may limit further growth of some aquacultured species, mainly, carnivorous species, in the future and represents a great challenge for us to overcome.

Captures In/Aquaculture Out (CIAO) index

The requirement in aquafeeds for meals and oils derived from wild captures (fish, squid, krill, etc.) to produce other aquatic organisms sets a dilemma regarding the best use of resources. Based on the amount of meals and oils derived from wild captures necessary to produce a unit of an aquacultured organism, species may be classified as net consumers or net producers of resources. An index can be calculated, based on inputs from wild captures and outputs from aquaculture production, to determine the sustainability of a species and a strategy for its production. Values above 1.0 would indicate the species and/or production strategy is a net consumer and values below 1.0 would indicate the species and/or production strategy is a net producer. Thus, as shown in Table 19.A, if a species and/or production strategy has an index for fish oil of 1.88, it will fall under the net consumer category, as over 1.8 MT of wild captured fish would be required to produce 1.0 MT of the cultured species. However, if algal biomass containing a high level of SCO is included within the nutritional production strategy, and fish oil and fish meal levels in the feed are reduced without affecting production results, then that same species will fall into the net producer category (Table 19.A).

Table 19.A. Example of CIAO Index Calculation without and with Use of Algal SCOs in Aquafeeds.

Captures In/Aquaculture Out (CIAO) Index				
Aquafeed Formulation		FCR[1]	CIAO Index	Raw Fish[2] Used kg / MT Produced
Fish meal used % (dw)	20	1.5	Fish meal **1.36**	1,363.6
Fish oil used % (dw)	10		Fish oil **1.88**	1,875.0
Captures In/Aquaculture Out Index Calculation—with SCO Use				
Aquafeed Formulation		FCR[1]	CIAO Index	Raw Fish[2] Used kg / MT Produced
Fish meal used % (dw)	10	1.5	Fish meal **0.68**	681.8
Fish oil used % (dw)	3		Fish oil **0.56**	562.5

[1]FCR: feed conversion ratio
[2]Raw fish yield: 22% fish meal, 8% fish oil

Recently completed commercial trials in Thailand with Pacific white shrimp under intensive culture conditions have demonstrated that reducing fish meal by 50% and eliminating fish oil from feeds containing an algal SCO product did not affect production results. With this approach, it was possible to swing this species from the net consumer category to net producer, resulting in an environmentally responsible and sustainable culture (Wilson, 2009).

Future of SCOs for Aquaculture

Consumer and Aquaculture Industry Perception of Sustainability

Both the consumer and the aquaculture industry have a positive perception of sustainability as an objective and end result that warrants the production and consumption of wholesome aquatic food in the future. Sustainability is viewed as having three components of responsibility: environmental, economic, and social. Continuous adaptation of the aquaculture industry towards sustainability, in regards to these three components, is necessary and eagerly pursued by all stakeholders that wish to be active participants in this growing global industry. Change to improve production practices is always desirable and aiming at sustainability, in its broad sense, is commendable for all stakeholders.

A Bright Future for SCO in Aquaculture

SCOs and meals have a bright future in aquaculture, particularly in their role as a renewable source of EFA, such as DHA and ARA. The importance of DHA as a functional nutrient is highlighted in this book, and more scientific evidence will surely be published. As more aquaculture nutritionists, feed mill formulators, and production managers become aware of the advantages of algal SCOs, their use will become common practice, driving their cost down and making their use even more widespread. It is imperative that algal SCO development and utilization materializes as an option to further facilitate the expansion of the aquaculture industry in an environmentally responsible fashion.

References

Atalah, E.; C.M. Hernandez-Cruz; M.S. Izquierdo; G. Roselund; M.J. Caballero; A. Valencia. Two Microalgae *Crypthecodinium cohnii* and *Phaeodactylum tricornutum* as Alternative Sources of Essential Fatty Acids in Starter Feeds for Seabream (*Sparus aurata*). *Aquaculture* **2007**, *270*, 178–185.

Brown, M.R. Nutritional Value of Microalgae for Aquaculture. *Avances en Nutrición Acuícola VI.* Memorias del VI Simposium Internacional de Nutrición Acuícola, Cancún, Quintana Roo, México, Sept 3–6, 2002; L.E. Cruz-Suárez, D. Ricque-Marie, M. Tapia-Salazar, M.G. Gaxiola-Cortés, N. Simoes, Eds.; 2002.

Brown, M.R.; S.W. Jeffrey; J.K. Volkman; G.A. Dunstan. Nutritional Properties of Microalgae for Mariculture. *Aquaculture* **1997**, *151*, 315–331.

Browdy, C.; G. Seaborn; H. Atwood; D.A. Davis; R.A. Bullis; T.M. Samocha; E. Wirth; J.W. Leffler. Comparison of Pond Production Efficiency, Fatty Acid Profiles, and Contaminants in *Litopenaeus vannamei* Fed Organic Plant-based and Fish-meal-based Diets. *J. World Aquaculture Society* **2006,** *37,* 437–451.

Chen, H.Y.; R.H. Tsai, The Dietary Effectiveness of *Artemia* nauplii and Microencapsulated Food for Postlarval *Penaeus monodon. Research and Development of Aquatic Animal Feed in Taiwan*; J.L. Chuang, S.Y. Shiau, Eds.; Fisheries Society of Taiwan: Taipei, 1986; Vol. I, Ser. 5, pp 73–79.

Copeman, L.A.; C.C. Parrish; J.A. Brown; M. Harel. Effects of Docosahexaenoic, Eicosapentaenoic, and Arachidonich acids on the Early Growth, Survival, Lipid Composition and Pigmentation of Yellowtail Flounder (Limanda ferruginea): A Live Food Enrichment Experiment. *Aquaculture* **2002,** *210,* 285–304.

Dunstan, G.A.; J.K. Volkman; S.M. Barrett; J.M. Leroi; S.W. Jeffrey. Essential Polyunsaturated Fatty Acids from Fourteen Species of Diatom (Bacillariophyceae). *Phytochemistry* **1994,** *35,* 155–161.

Food and Agriculture Organization of the United Nations (FAO). *Use of Wild Fish and/or Other Aquatic Species to Feed Cultured Fish and Its Implications to Food Security and Poverty Alleviation.* FAO Expert Workshop. FAO Fisheries and Aquaculture Department: Kochi, 2007.

Food and Agriculture Organization of the United Nations (FAO). *The State of World Fisheries and Aquaculture 2008.* FAO Fisheries and Aquaculture Department: Rome, 2009.

Ganuza, R.; T. Benitez-Santana; E. Atalah; O. Vega-Orellana; R. Ganga; M.S. Izquierdo. *Crypthecodinium cohnii* and *Schizochytrium* sp. as Potential Substitutes to Fisheries-derived Oils in Seabream (*Sparus aurata*) Microdiets. *Aquaculture* **2008,** *277,* 109–116.

González-Félix, M.L.; A.L. Lawrence; D.M. Gatlin III; M. Perez-Velazquez. Growth, Survival and Fatty Acid Composition of Juvenile *Litopenaeus vannamei* Fed Different Oils in the Presence and Absence of Phospholipids. *Aquaculture* **2002a,** *205,* 325–343.

González-Félix, M.L.; D.M. Gatlin III; A.L. Lawrence; M. Perez-Velazquez. Effect of Phospholipid on Essential Fatty Acid Requirements and Tissue Lipid Composition of *Litopenaeus vannamei* Juveniles. *Aquaculture* **2002b,** *207,* 151–167.

González-Félix, M.L.; D.M. Gatlin III; A.L. Lawrence; M. Perez-Velazquez. Nutritional Evaluation of Fatty Acids for the Open Thelycum Shrimp, *Litopenaeus vannamei*: II. Effect of Dietary n-3 and n-6 Polyunsaturated and Highly Unsaturated Fatty Acids on Juvenile Shrimp Growth, Survival and Fatty Acid Composition. *Aquacult. Nut.* **2003a,** *9,* 115–122.

González-Félix, M.L.; A.L. Lawrence; D.M. Gatlin III; M. Perez-Velazquez. Nutritional Evaluation of Fatty Acids for the Open Thelycum Shrimp, *Litopenaeus vannamei*: I. Effect of Dietary Linoleic and Linolenic Acids at Different Concentrations and Ratios on Juvenile Shrimp Growth, Survival and Fatty Acid Composition. *Aquacult. Nut.* **2003b,** *9,* 105–113.

Halver, J.E. Lipids and Fatty Acids. *Fish Feed Technology*; Lectures Presented at the FAO/UNDP Training Course in Fish Feed Technology, College of Fisheries, University of Washington, Seattle, Washington, USA, Oct 9–Dec 15 1978, 1980 . United Nations Development Programme, Food and Agriculture Organization of the United Nations: Rome, 1980.

Hardy, R.W. Fish Oil Replacement Shift Fatty Acid Content of Farmed Fish. *Global Aquacult. Adv.* **2006,** *6,* 48–49.

Harel, M.; A.R. Place. Tissue Essential Fatty Acid Composition and Competitive Response to Dietary Manipulations in White Bass (*Morone chrysops*), Striped Bass (*M. saxatilis*) and Hybrid

Striped Bass (*M. chrysops* X *M. saxatilis*). *Comp. Biochem. Physiol. B* **2003**, *135*, 83–94.

Harel, M.; W. Koven; I. Lein; Y. Bar; P. Behrens; J. Stubblefield; Y. Zohar; A.R. Place. Advanced DHA, EPA and ArA Enrichment Materials for Marine Aquaculture Using Single Cell Heterotrophs. *Aquaculture* **2002**, *213*, 347–362.

Hertrampf, J.W.; F. Piedad-Pascual. *Handbook on Ingredients for Aquaculture Feeds*; Kluwer Academic Publishers: Dordrecht, **2000**.

Izquierdo, M.S. Essential Fatty Acid Requirements in Mediterranean Fish Species. *Cah. Opt. Med.* **2005**, *63*, 91–102.

Kanazawa, A.; S. Teshima; S. Tokiwa. Biosynthesis of Fatty Acids from Palmitic Acid in the Prawn, *Penaeus japonicus*. *Memoirs of the Faculty of Fisheries, Kagoshima University* **1979**, *28*, 17–20.

Kfoury, J.R.; V. Kiron; N. Okamoto. Influence of Essential Fatty Acid Deprivation on Thrombocyte Aggregation in Rainbow Trout. *Braz. J. Vet. Res. Anim. Sci., Sao Paulo* **2006**, *43*, 74–80.

Kiron, V.; H. Fukuda; T. Takeuchi; T. Watanabe. Essential Fatty Acid Nutrition and Defense Mechanisms in Rainbow Trout *Oncorhynchus mykiss*. *Comp. Biochem. Physiol. A* **1995**, *111*, 361–367.

Koskela, R.W.; J.G. Greenwoog; P.C. Rothlisberg. The Influence of Prostaglandin E2 and the Steroid Hormones, 17 alpha-hydroxyprogesterone and 17 beta-estradiol on Moulting and Ovarian Development in the Tiger Prawn, *Penaeus esculentus*, Hastwell, 1879 (Crustacea: Decapoda). *Comp. Biochem. Physiol. A* **1992**, *101*, 295–299.

Lall, S.P. Nutrition and Health of Fish. *Avances en Nutrición Acuícola V. Memorias del V Simposium Internacional de Nutrición Acuícola*; L.E. Cruz-Suárez, D. Ricque-Marie, M. Tapia-Salazar, M.A. Olvera-Novoa, R. Civera-Cerecedo, Eds.; Nov 19-22, 2000. Mérida, Yucatán, México.

Langdon, C.J.; M.J. Waldock. The Effect of Algal and Artificial Diets on the Growth and Fatty Acid Composition of *Crassostrea gigas* Spat. *Journal of the Marine Biological Assoc. UK.* **1981**, *61*, 431–448.

Laurin, E.; B. Carpenter; M.P. Schreibman; J. Polle; R.A. Bullis. Tilapia Finishing Feed for ω-3 DHA Enrichment of the Fillets Using Single Cell Biomass from *Schizochytrium* sp. *Int. Aquafeed* **2006**, *9*, 28–29.

Lim, C.; H. Ako; C.L. Brown; K. Hahn. Growth Response and Fatty Acid Composition of Juvenile *Penaeus vannamei* Fed Different Sources of Dietary Lipid. *Aquaculture* **1997**, *151*, 143-153.

Lim, C.E.; M. Aksoy. Lipid and Fatty Acid Requirements of Tilapia. *Aquaculture America 2009*; World Aquaculture Society, February 15–18, 2009, Seattle, Washington, USA; p 194.

Lytle, J.S.; T.F. Lytle; J. Ogle. Polyunsaturated Fatty Acid Profiles as a Comparative Tool in Assessing Maturation Diets of *Penaeus vannamei*. *Aquaculture* **1990**, *89*, 287–299.

Masuda, R. The Critical Role of Docosahexaenoic Acid in Marine and Terrestrial Ecosystems: From Bacteria to Human Behavior. *The Big Fish Bang. Proceedings of the 26th Annual Larval Fish Conference*; H.I. Browman, A.B. Skiftesvik, Eds.; Institute of Marine Research: Bergen, 2003; pp 249–256.

Medina, A.R.; M.E. Grima; G.A. Jiménez; M.J. González. Downstream Processing of Algal Polyunsaturated Fatty Acid. *Biotechnol. Advan.* **1998**, *16*, 517–580.

Miller, M.R.; P.D. Nichols; C.G. Carter. Replacement of Fish Oil with Thraustochytrid *Schizochytrium* sp. L Oil in Atlantic Salmon Parr (*Salmo salar* L) Diets. *Comp. Biochem. Physiol. A* **2007**, *148*, 382–392.

Mourente, G.; J.G. Bell. Partial Replacement of Dietary Fish Oil with Blends of Vegetable Oils (Rapeseed, Linseed and Palm Oils) in Diets for European Sea Bass (*Dicentrarchus labrax* L.) over a Long Term Growth Study: Effects on Muscle and Liver Fatty Acid Composition and Effectiveness of a Fish Oil Finishing Diet. *Comparative Biochemistry and Physiology B* **2006**, *145*, 389–399.

Mourente, G.; J.E. Good; J.G. Bell. Partial Substitution of Fish Oil with Rapeseed, Linseed and Olive Oils in Diets for European Sea Bass (*Dicentrarchus labrax* L.): Effects on Flesh Fatty Acid Composition, Plasma Prostaglandins E2 and F2, Immune Function and Effectiveness of a Fish Oil Finishing Diet. *Aquacult. Nut.* **2005**, *11*, 25–40.

Nakamura M.T.; T.Y. Nara. Structure, Function, and Dietary Regulation of Delta-6, Delta-5, and Delta-9 Desaturases. *Ann. Rev. Nut.* **2004**, *24*, 345–376.

National Research Council. *Nutrient Requirements of Fish*. National Research Council. National Academy Press: Washington, DC, 1993.

Patnaik, S.; T.M. Samocha; D.A. Davis; R.A. Bullis; C.L. Browdy. The Use of PUFA-Rich Algal Meals in Diets for *Litopenaeus vannamei*. *Aquaculture Nutrition* **2006**, *12*, 395–401.

Ravid, T.; A. Tietz; M. Khayat; E. Boehm; R. Michelis; E. Lubzens. Lipid Accumulation in the Ovaries of a Marine Shrimp *Penaeus semisulcatus* (De Haan). *J. Exp. Biol.* **1999**, *202*, 1819–1829.

Rees, J. F.; K. Curé; S. Piyatiratitivorakul; P. Sorgeloos; P. Menasveta. Highly Unsaturated Fatty Acid Requirements of *Penaeus monodon* Postlarvae: An Experimental Approach Based on *Artemia* Enrichment. *Aquaculture* **1994**, *122*, 193–207.

Robert, S.S. Production of Eicosapentaenoic and Docosahexaenoic Acid-Containing Oils in Transgenic Land Plants for Human and Aquaculture Nutrition. *Marine Biotechnology* **2006**, *8*, 103–109.

Ruyter, R.; E. Ruyter; T. Ruyter. Essential Fatty Acids in Atlantic Salmon: Time Course of Changes in Fatty Acid Composition of Liver, Blood and Carcass Induced by a Diet Deficient in n-3 and n-6 Fatty Acids. *Aquacult. Nut.* **2001**, *6*, **109–117.**

Sargent, J. R.; L.A. McEvoy; J.G. Bell. Requirements, Presentation and Sources of Polyunsaturated Fatty Acids in Marine Fish Larval Feeds. *Aquaculture* **1997**, *155*, 117–127.

Sargent J.R.; D.R. Tocher; J.G. Bell. The Lipids. *Fish Nutrition*; J.E. Halver, R.W. Hardy, Eds.; Academic Press, Elsevier : San Diego, 2002; pp. 81–257.

Sprecher H.; D.L. Luthria; B.S. Mohammed; S.P. Baykousheva. Reevaluation of the Pathways for the Biosynthesis of Polyunsaturated Fatty Acids. *J. Lipid Res.* **1995**, *36*, 2471–2477.

Tacon, A.G.S. Trends in Global Aquaculture and Aquafeed Production: 1984-1996 Highlights. *Feed Manufacturing in the Mediterranean Region. Recent Advances in Research and Technology, CIHEAM/IAMZ, Zaragoza, Spain* . J. Brufau, A. Tacon, Eds.; 1999; Vol. 37, pp 107–122.

Tocher, D.R.; J.D. Castell; J.R. Dick; J. Sargent. Effects of Salinity on the Fatty Acid Composition of Total Lipid and Individual Glycerophospholipid Classes of Atlantic Salmon (*Salmo salar*) and Turbot (*Scophthalmus maximus*) Cells in Culture. *Fish Physiol. Biochem.* **1995**, *14*, 125–137.

Turchini, G.M.; B.E. Torstesen; W.-K. Ng. Fish Oil Replacement in Finfish Nutrition. *Rev. Aquacult.* **2009**, *1*, 10–57.

Volkman J.K.; S.W. Jeffrey; P.D. Nichols; G.I. Rogers; C.D. Garland. Fatty Acid and Lipid Composition of 10 Species of Microalgae Used in Mariculture. *J. Exp. Mar. Biol. Ecol.* **1989**, *128*, 219–240.

Volkman, J.K.; G.A. Dunstan; S.W. Jeffrey; P.S. Kearney. Fatty Acids from Microalgae of the Genus *Pavlova*. *Phytochemistry* **1991**, *30,* 1855–1859.

Volkman, J.K.; M. R. Brown; G.A. Dunstan; S.W. Jeffrey. The Biochemical Composition of Marine Microalgae from the Class Eustigmatophyceae. *J. Phycol.* **1993**, *29,* 69–78.

Watanabe, T. Importance of Docosahexaenoic Acid in Marine Larval Fish. *J. World Aquacult. Soc.* **1993**, *24,* 152–161.

Wouters, R.; L. Gómez; P. Lavens; J. Calderón. Feeding Enriched *Artemia* Biomass to *Penaeus vannamei* Broodstock: Its Effect on Reproductive Performance and Larval Quality. *J. Shellfish Res.* **1999**, *18,* 651–656.

Wouters, R.; P. Lavens; J. Nieto; P. Sorgeloos. Penaeid Shrimp Broodstock Nutrition: An Updated Review on Research and Development. *Aquaculture* **2001**, *201,* 1–21.

Xu, X.; J. Wenjuan; J.D. Castell; R. O'Dor. Essential Fatty Acid Requirement of the Chinese Prawn *Penaeus chinensis*. *Aquaculture* **1994**, *127,* 29–40.

PART 6

Synopsis and Overview

20

Future Development of Single Cell Oils

David J. Kyle

Advanced BioNutrition Corp., 6430 Dobbin Rd., Columbia, MD 21045

Introduction

The potential of microbial systems to produce oils and fats on a large scale has been recognized for over 100 years, but only within the past 25 years has this potential been realized using successful commercial processes. During the intense political and economic turmoil of World Wars I and II, research and development in microbial oil production was driven by strategic defense initiatives. In peacetime, strategic defense drives were replaced by a commercial focus to develop microbial oil production systems that yielded lower-cost equivalents of high value oils. Such applications included the production of cocoa butter substitutes from oleaginous yeast (Smit et al., 1992) and evening primrose oil substitutes from oleaginous fungi (Nakahara et al., 1992). The reduced cost associated with the production of these substitutes led researchers to rapidly develop an understanding of the capabilities of growing oleaginous microbes in deep tank culture; however, improving efficiencies of agricultural production and global distribution dropped the worldwide plant oil prices to levels where the microbial production was simply not economically competitive.

The recent commercial success of specialty single cell oils (SCOs) enriched in docosahexaenoic acid (DHA) and arachidonic acid (ARA) can be attributed to three major factors: (1) the accumulated knowledge acquired from earlier attempts to scale up other SCO processes; (2) a critical need of the infant formula industry for a "designer oil" with specific characteristics; and (3) no plant or animal sources of oils that could provide the required characteristics. As a result of the commercial success of the DHASCO® and ARASCO® processes, our understanding of how to cultivate oleaginous microorganisms in deep tanks—along with an understanding of how to harvest and process the oils contained therein—has expanded at an extremely rapid pace. However, if the needs of the infant formula industry change, or if an alternative plant or animal source is identified that can fulfill its requirements, the SCO process will again be challenged by another means of production. Furthermore, if new opportunities for SCO expansion are to be realized, new applications need to be identified that will give the designer SCO process unique advantages, in a manner similar to those of DHASCO and ARASCO in infant formulas.

A Short History of Commercial Production of SCOs

To better understand the future direction of SCO technology and applications, it is important to learn lessons from past successes and failures. The early development of SCO production has been well reviewed by Ratledge (1992; see also Chapter 1) who noted that, prior to the 1960s, interest in SCO was primarily based on academic curiosity, tempered by strong prospects for commercial production, should the need arise. In the last 30 years, full-scale production became a reality for several processes, but only in a few cases did this production also become an economic reality, generating a truly sustainable commercial process. Although many SCO sources and applications have been proposed over the past 30 years, only six have been scaled up, and only two have reached a significant commercial realization.

Cocoa Butter Equivalent (CBE) from Apiotrichum curvatum

Based on earlier work of Hammond et al. (1981), a team from the New Zealand Department of Scientific and Industrial Research attempted to commercialize this yeast-based SCO process for CBE production in 1988 (Davies, 1988). The group successfully scaled the process to 250 m³ and demonstrated the production of a SCO using an oleagenous yeast that yielded a palm oil equivalent. Although production economics were based on the use of a discounted whey feedstock (a byproduct of cheese processing), the projected operating costs were still over $700/T in 1988, and no commercial production of the SCO-CBE took place. The 2008/2009 commodity price for palm oil was about $550/ton, so the process has clearly not become economically viable for a palm oil equivalent yet. However, since the 2008/2009 commodity price for cocoa butter was about $2,300/T, this process could still become economically viable if the equivalent CBE could be produced again.

GLA Oil from Mortierella isabellina

Originally using *Mortierella isabellina* and, later, *Mortierella ramanniana,* the Idemizu Corporation (Japan) produced a SCO rich in γ-linolenic acid (GLA) for the domestic Japanese market in 1988 (Nakahara et al., 1992). Originally, the oil was used as a specialty food additive, but this SCO source is no longer on the market. It is still unclear how much of the oil was ever produced or why it was taken off the market, but it is likely that it was still uncompetitive with newly developed, plant-based GLA oils, such as evening primrose oil (EPO), borage oil (BO), and black current seed oil (BCO).

GLA Oil from Mucor circinelloides

We have a much clearer picture of the development of a GLA oil from *Mucor circinelloides* (formerly, *Mucor javanicus*) by Ratledge (2006) and the ultimate scale-up by

J. & E. Sturge Ltd. (UK). Commercial production began in 1985 with 220 m³ stirred tank fermentors that produced about 2 tons of oil per batch. The SCO product was marketed under the trade name of *Oil of Javanicus*. Although the process was not closed down until the company was sold in 1990, few batches were produced because the market for a specialty supplement or food additive containing GLA remained small and was well served by plant-based sources of GLA.

DHA Oil from Crypthecodinium cohnii

In the late 1980s, Kyle's group at Martek Biosciences was also developing a series of SCOs; group members quickly found an application for a unique SCO that was highly enriched in docosahexaenoic acid (DHASCO) and was produced by the marine dinoflagellate algae *Crypthecodinium cohnii* (Kyle, 1996). The application for this oil depended on a requirement by the infant formula industry for an oil rich in DHA, but devoid of eicosapentanoic acid (EPA) (Kyle et al., 1995). These prerequisites could not be met by any known plant or fish oils, and, as a result, commercial production of DHASCO in stirred tank fermentors at the 200 m³ scale is still continued today by Martek Biosciences Corp (USA). A detailed account of this process is given in Chapter 6.

ARA Oil from Mortierella alpina

In the late 1980s, Kyle's group also developed an oil rich in arachidonic acid (ARA) for infant formula applications, which was produced from the fungi *Mortierella alpina* (Kyle, 1997a). Like DHASCO, there is still no plant-based alternative, and animal sources (e.g., egg yolk) are more costly, on an ARA basis, and contain a wider array of fatty acyl components than the designer oil ARASCO (Kyle, 1997b). Martek also scaled up this process, discovering a way to enrich the ARA content of the oil so that it would be greater than 50% (Kyle, 2003); then, the company entered into an agreement with Gist-brocades (now DSM) in The Netherlands for production. The ARASCO product is now being produced in the USA in 200 m³ fermentors,; the background for this process has been described in detail in Chapter 5. A comparable oil is also being produced using a similar process by companies in other countries (e.g., Suntory Corp in Japan, and Cargill in China).

DHA Oil from Schizochytrium sp.

In the early 1990s, Kelco Corporation scaled up production of a second DHA-rich SCO using the marine microbe *Schizochytrium* sp., which was first described by Barclay (1992). The process was acquired by Martek Biosciences Corp in 2001, and although this process was very efficient and low cost, DHASCO had already been chosen by the infant formula industry by this time. The application of DHA oil from *Schizochytrium* was, therefore, destined to compete with fish oil for a position in the food additive industry as an oil, or in the animal feed

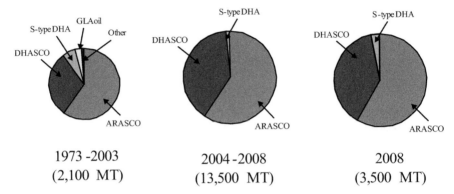

Fig. 20.1. Relative Proportion of Commercial SCO Production in 2008, Relative to the 30 Year Period from 1973–2003 and the 5 Year Period from 2004–2008. Production is attributed to various oils types: ARASCO from *Mortierella alpina*; DHASCO from *Crypthecodinium cohnii*; DHA S-type from *Schizochytrium* sp.; and GLA oil from *Mucor circinelloides*.

business as a DHA-rich whole cell biomass (see Chapter 18). In addition to the gel-cap nutraceutical market, this oil, provided by Martek, has now been included in over 100 commercial food products worldwide, ranging from fortified milks and yogurts to nutritional bars and fruit drinks. An account of the development of this process and its current status in both the food and feed business is reviewed in Chapters 4, 16–18.

The overall production of SCO in 2008 is estimated to be about 3,500 tons, of which about 60% was ARASCO and 30% was DHASCO (Fig. 20.1). The expansion of the production of, and market for, DHASCO and ARASCO has been so rapid that the production of SCOs other than the DHA oil from *Schizochytrium* has been dwarfed by that of these two products. Since their commercial introduction in 2000, it is estimated that approximately 9,000 MT of ARASCO from the *Mortierella* process (Kyle, 1997b), 6,000 MT of DHASCO from the *Crypthecodinium* process (Kyle et al., 1995), and 200 MT of S-Type DHASCO from the *Schizochytrium* process (Barclay, 1992) have been produced to date, realizing a commercial value of over $1.5 billion. The annual production rate of these three oils is probably close to 3,500 MT/yr and growing.

Future Sources

Since we have seen two successful examples of the commercialization of SCOs, the hunt is on to find alternative commercial sources of the same oils. They may develop out of the classical screening of new microbes or the genetic modification of lower-cost, plant-based systems to produce oils substantially equivalent to the microbial oils DHASCO and ARASCO.

New Sources of SCO

Since the discovery and application of SCO from *Crypthecodinium cohnii*, many groups have searched for other dinoflagellates related to *C. cohnii* that could be used as an alternative source of DHASCO. Thus far, no potential alternatives have emerged from this group of algae. Other algae have been identified as potential DHA and EPA producers, but these are generally photosynthetic and pose a greater scale-up challenge than those produced in fermentors (Cohen et al., 1995). Nevertheless, one company, Eau+, launched an algal SCO in 2007 that contained a combined 23% EPA + DHA. Searches for chytrids other than *Schizochytrium* have also provided potential alternatives for a DHA-containing SCO. Phylogenetically similar organisms have been reported, including *Ulkenia* (Tanaka et al., 2003) and *Labyranthula* (Yokochi et al., 2002), but the oil composition of these strains appears to be virtually identical to that originally reported for *Schizochytrium*, which makes their discrimination quite difficult. Because all chytrids appear to have a relatively high content of omega-6 docosapentaenoic acid (22:5 omega-6, not to be confused with the omega-3 docosapentaenoic acid precursor of DHA found in many fish oils), they are all easily distinguished from the DHASCO produced by *C. cohnii*.

Certain marine fungi represent possible candidates for the production of ARA-containing SCOs. However, none have yet been reported to have the production efficiency of *Mortierella alpina*, and most are producers of ARA-rich phospholipids, not ARA-rich triglycerides (Gandhi & Weete, 1991). Perhaps the closest alternative to the ARASCO product is an oil produced by the green alga *Parietochloris incisa*. Although it has not yet been commercialized, this algae produces over 50% of its weight in triacylglycerol, and its fatty acid composition is almost identical to that of ARASCO (Bigogno et al., 2002; Kyle, 2008).

The search for new producers of SCO will continue and there is no doubt that new species and strains will be identified with unique characteristics that may lead to their use as replacements for existing SCOs or their use in completely new applications. Some of the recent developments that have taken place with mutant strains of *Mortierella alpina* are detailed in the review given in Chapter 2.

Genetic Engineering of Microbes

The mutagenesis and selection of a yeast, using classical techniques to produce an oil with a composition similar to cocoa butter (i.e., high stearic acid content), has already been successfully demonstrated (Smit et al., 1992). Since yeasts grow very well in fermentors, they have been the focus of intensive efforts over the past few years to produce ARA, EPA, and DHA using recombinant DNA technology with considerable success (see Chapters 2 and 3). Key regulator genes involved in this process code for the $\Delta 5$ and $\Delta 6$ desaturases and PUFA elongases. Attempts to isolate these genes from *Mortierella* have been successful, and they have been transferred to yeast, primarily as a model system to identify the genes, rather than as a production system for EPA and ARA (Knutzon et al., 1998; Parker-Barnes et al., 2000; Pereira et al., 2004b). Using the oleaginous yeast *Yarrowia lipolytica*, Damude and colleagues from the DuPont Company have been able to produce genetically engineered yeast lines that produce 10% ARA in the oil (Damude et al., 2006a), 25% EPA in the oil (Damude et al., 2006b), and 5.6% DHA (Damude et al., 2006c) in the oil (see also Chapter 3). Although it is not considered oleaginous, common baker's yeast, *Saccharomyces cerevisiae*, was used by Patel and Rajyashri of the Avestha Grengraine PTV LTD (India) to demonstrate the production of DHA in recombinant yeast strains (Patel & Rajyashri, 2008). Both of these genetically modified sources should soon be commercialized to yield a product that could be competitive with the non-recombinant, natural algal sources presently being used to produce DHASCO.

An approach of genetically engineering microbes involves identifying photosynthetic algae with a high potential to produce SCOs and then converting this algae into a heterotrophic production mode. One characteristic that distinguishes algae capable of growing in the dark in the presence of glucose (i.e., heterotrophically) from photoautotrophic algae (99% of all algal species) is the presence of an active glucose transporter in the outer cell membrane; this was confirmed by converting the photoautotrophic algae *Phaeodactylum tricornutum* into a heterotroph by transforming the algae with a gene for glucose transport (Zaslavskaia et al., 2001; Apt et al., 2008). Such new technology opens up the potential to convert other photosynthetic algae, like *Parietochloris,* into economic, heterotrophic producers of SCO.

Genetic Engineering of Plants

Once genes coding for the various enzymes involved in the biosysnthesis of ARA, EPA, and DHA were isolated, and oils were produced in model microbial systems such as those discussed above, there was a concerted effort to transfer the same genes to agronomic plants for the large-scale, low-cost production of oils enriched in ARA, EPA, and/or ARA in field crops. If successful, this process could eventually replace SCO production for these products. However, the task has proven to be more daunting than originally anticipated. Initial attempts involved the use of genes from a DHA-producing bacteria (*Photobacterium profundum*) isolated from the intestines of certain deep water fish (Allen & Bartlett, 2002). It was quickly realized, however, that this organism did not produce DHA using the conventional fatty acid biosynthetic pathway involving $\Delta 5$ and $\Delta 6$ desaturases and the PUFA elongases found in plants. Rather, DHA was produced using a unique, polyketide-based pathway (Allen & Bartlett, 2002). An alternative source of genetic material was *Schizochytrium*, but, once again, its biosynthesis proved to involve a polyketide pathway, even though this was a eukaryotic organism (Metz et al., 2001). These authors have also recently shown that the PKS pathway produces free fatty acids (FFAs), not thioesters, so additional genes will be needed to ensure the FFAs will not be cytotoxic (Metz et al., 2009).

Isolation of genes from the fungus *Mortierella* has been more productive, and these genes have been used to transform canola (*Brassica napus*) for the production of EPA and ARA (Knutzon et al., 1998). More recently, Kajikawa et al. (2008) isolated

$\Delta 5$ and $\Delta 6$ desaturases and elongases from a liverwort (*Marchantia polymorpha*) and co-expressed them in tobacco and soybean. Tobacco transformants had ARA and EPA levels of 15.5% and 4.9% of the total fatty acids in the leaves, respectively, and the soybean transformants expressed ARA + EPA at levels of up to 19.5% of the total fatty acids in the seeds.

The biosynthetic pathway for DHA in mammals involves the "Sprecher Shunt" in which EPA is further elongated and desaturated to form 24:6 (omega-3), which is then cycled through one stage of β-oxidation to produce DHA (Sprecher, 1999). A simpler pathway from EPA to DHA would be the elongation of EPA to C22:5 (omega-3) and then its direct conversion to DHA using a $\Delta 4$-desaturase. Although the existence of such an enzyme is still hotly disputed in vertebrate systems, there is clear evidence that it does exist in certain marine algae (Tanon et al., 2003; Pereira et al., 2004b; Zhou et al., 2007).

Although some plants have now been genetically modified to produce ARA and EPA (but not DHA, to any significant extent), it may not be possible to produce an oil so highly enriched (up to 50%) in these fatty acids as is found in the SCOs without a significant disruption of the biochemistry and physiology of other plant tissues, including the roots, leaves, stems, and flowers, which could weaken the plant, as a whole. As a result, this high level of enrichment may remain a key differentiator between a SCO and a plant-based oil. Even if a plant could produce an oil with substantial ARA, EPA, or DHA enrichment, the oil would have to be harvested, processed, and stored very carefully to prevent oxidation. This oxidation problem has proven to be very difficult to overcome, even for flax seed oil, which contains high levels of 18:3 (omega-3). Consequently, the processing of these new, highly unsaturated plant oils may require the co-development of new technology for their transportation, extraction, and processing. Oils that have a high propensity for oxidation may also need to be treated with powerful antioxidants, such as ethoxyquin, prior to transportation or use in food or feed products, to prevent spoilage. In addition to the requirement for new process technology, there remains a strong political and social reluctance to use genetically modified organisms (GMOs) in foods. In the European Union, for example, no GMO materials are presently allowed in infant formulas. Thus, the design, development, and production of oils from these new GMO plants that produce ARA, EPA or DHA may actually be the lowest of the many hurdles to the commercialization of such a product.

Future Uses of SCO

It has been well established that SCOs can be produced in large quantities for commercial applications. From experience gained over the past 30 years, however, it is also clear that the future commercial use for SCOs will depend on their differentiation from existing or future alternatives or the identification of specific market niches where the higher cost of the SCO is offset by the value of the new application.

Infant Formula

The use of DHASCO and ARASCO has been well established in the infant formula industry, with nearly every major infant formula company worldwide using these SCOs for enrichment. Due to the well-established safety profile for these SCOs and their sensitivity, it is unlikely that the dominance of these products will change in the near future. The high concentration of DHA and ARA in these SCOs minimizes the quantity of their addition, and public concern over the use of GMO products, particularly in infant formulas, suggests that, even if lower-cost GMO plant sources were available, they would not be used in the foreseeable future. Thus, the production quantities for DHASCO and ARASCO should continue to rise as their market penetration increases, but will likely level off at a production quantity of about 5,000 tons per annum within 2–4 years.

New Food Applications

The use of SCOs in new food applications will depend on the application, commodity pricing of alternatives, and delivery form of the SCO. With respect to LC-PUFAs, the primary competition will be fish oil. Fish oil has three major concerns: (1) oxidative instability in food products due to high levels of unsaturated fatty acids, which lead to organoleptic problems; (2) the potential for carrying contaminants, such as dioxin, PCBs, methyl mercury, and other pollutants that concentrate in fish oils, and (3) the lack of ecological sustainability as fisheries collapse due to environmentally irresponsible overfishing. To overcome the first two problems, fish oils must be purified and microencapsulated to be usable in conventional, food-based systems. There is less need for additional stabilization of a designer SCO, such as DHASCO, since it is highly enriched only in DHA, with no other extraneous PUFAs that may act as focal points for oxidation. Furthermore, the oils directly extracted from these microbial sources will generally co-extract with the natural antioxidant package found within the cells themselves, which provides the stability required for the natural cell to survive.

The release of oils in the stomach, as is the case for gel-caps of fish oils or many microencapsulated fish oil products incorporated in certain food applications, results in premature oxidation of the oils and odiferous eructation ("fishy burps"). Thus, to fully exploit the use of SCOs rich in DHA (or fish oil, for that matter) in food applications, there is still a need to develop new delivery systems that will allow gastric protection and post-gastric release of the oils. Such delivery systems also need to allow the use of the encapsulated oils in a food matrix of high water activity (e.g., yogurt, spoonable dressings, or drinks), as well as in a matrix of low water activity (e.g., powdered formulations, food bars, breads). Such delivery systems are presently being developed, and they should greatly expand the use of SCOs in mainstream food products in the future. By 2010, such applications could increase SCO production by an additional 2,000–4,000 tons/annum.

Animal Feed Applications

When considering SCO applications, we should also consider other parts of the microbial biomass and its applications. Use of the whole SCO biomass, without the additional extraction costs and losses, may offer an economic opportunity for functional nutrition in the animal field. A number of patents have been filed on the use of SCO microbial biomass in animal feeds to improve the animal's DHA content or a part of the animal that is to be consumed (Barclay, 1996, 2006; Kyle, 2006). Examples include feeding *Schizochytrium* biomass to chickens in order to produce high DHA eggs (Herber & Van Elswyk, 1996) or feeding cows with a protected fish oil product to produce DHA-enriched milk (Wright et al., 2003). Such concepts are limited by the extent of the bioconversion of feed DHA into the DHA content of the commercial food product, without incurring substantial losses in general metabolism, conversion into unused body parts, or losses in the feces. In competition with this approach is a cracked-egg product, which contains highly deodorized and purified fish oil at a much lower cost, and a milk product where highly deodorized fish oil is added during the UHT (ultra-high temperature) packaging process at a lower price than providing a DHA-enriched feed to the cow.

With the rapid development of freshwater, aquacultured species, such as Tilapia, there will be a growing need for the DHA enrichment of filets offered to the consumer. Tilapia is a freshwater fish that can be grown with feed essentially devoid of any fish meal or fish oil. The general movement away from the use of fishmeal and fish oil in the aquaculture industry is primarily for cost savings and ecological sustainability reasons, but the net consequence is that the nutritional value of the final product is significantly diminished. For example, the EPA and DHA levels of commercially grown tilapia are less than one tenth of those found in salmon, and the nutritional value of commercially grown tilapia (rated as the omega-6 to omega-3 ratio) has been reported to be not much better than bacon (Weaver et al., 2008). Ironically, this consequence creates an interesting opportunity for the enrichment DHA levels in certain fish, such as tilapia, by using feeds that are enriched with a sustainable algal biomeal rich in DHA.

Although there may be niche markets for providing animals with SCO biomass containing certain fatty acids to enrich food product made by those animals (i.e., eggs or meat), there may be a much larger market in the provision of SCO biomass for the health benefit of the animal itself. The neurological growth and development of piglets, for example, is commonly used as a model for the human infant, because the stage of brain development at birth and the extent of postnatal brain development is almost identical in both species. Consequently, the DHA and ARA content of sow's milk is high (compared to cow's milk), putting the former more in line with human breast milk. Both of these facts indicate that growing piglets require adequate amounts of dietary DHA, both during gestation and while suckling. As the swine production industry continues to push the productivity of sows well beyond their natural limit, it would not surprise the author if the sows exhibited certain nutritional deficiencies (e.g., DHA deficiency) that might result in lower fecundity or poorer growth and development of the piglets. There is now a growing awareness that improving the DHA status of the sow during gestation and lactation can result in increased litter sizes and improved growth and survival of piglets (Rooke et al., 2001). This is an example where the addition of dietary DHA could result in a significant return on investment for the producer and would provide a long-term, sustainable product flow, much like that of human infant formula.

The key component in fish meal and fish oil, which is used so extensively in the aquaculture industry and cannot be replaced by plant-sourced materials, is, in fact, DHA. Fish meal/fish oil alternatives made from SCO biomass will ensure the future sustainability and safety of farmed aquaculture species, such as salmon and shrimp. However, such an environmentally responsible "Clean & Green" approach does not come without a cost, and a 10–15% premium on the price of end products (like the premium on DHA/ARA supplemented infant formula) will need to be accepted by most consumers. If we are correct, a major expansion in the production and use of SCO biomass may be the key to the future sustainability of the aquaculture industry.

Final Words

The commercial development of SCOs has gone through a series of ups and downs since their potential was first recognized over 100 years ago. Initially, SCOs were academic curiosities. Then, there were several attempts to commercialize them, but many failed due to lower cost alternatives in the market. The success story of DHASCO and ARASCO in infant formulas seems to be due to their emergence in the right place at the right time with no commercial alternative available for the industry. Since this industry shows reluctance to change, it is unlikely that alternatives will replace DHASCO or ARASCO in the near future, unless there is a substantial reason for the change. The scale-up and commercialization of these two oils was encouraged by previous work on other SCOs and has now validated the theory that designer SCOs could be commercially viable. Without the groundwork and knowledge base that was built so extensively by previous SCO work, the commercial success of the DHASCO and ARASCO processes would never have been realized. It is with this background that we look forward to a bright future for SCOs and the SCO biomass in solving many industry and socio-economic problems worldwide.

References

Allen, E.E.; D.H. Bartlett. Structure and Regulation of the Omega-3 Polyunsaturated Fatty Acid Synthase Genes from the Deep-sea Bacterium *Photobacterium profundum* Strain SS9. *Microbiology* **2002,** *148,* 1903–1913.

Apt, K.; F. Allnutt; D.J. Kyle; C. Lippmeyier. Trophic Conversion of Obligate Phototrophic Algae Through Metabolic Engineering. Martek Biosciences Corp. U.S. Patent 6,027,900, 2008.

Barclay, W. Process for the Heterotrophic Production of Microbial Products with High Concentrations of Omega-3 Highly Unsaturated Fatty Acids. Martek Biosciences Corp. U.S. Patent 5,130,242, 1992.

Barclay, W. Microfloral Biomass Having Omega-3 Highly Unsaturated Fatty Acids. Martek Biosciences Corp. U.S. Patent 5,518,918, 1996.

Barclay, W. Feeding Thraustochytriales to Poultry for Increasing Omega-3 Highly Unsaturated Fatty Acids in Eggs. Martek Biosciences Corp. U.S. Patent 7,033,584, 2006.

Bigogno, C.; I. Khozin-Goldberg; D. Adlerstein; Z. Cohen. Biosynthesis of Arachidonic Acid in the Oleaginous Microalga *Parietochloris incisa* (Chlorophyceae): Radiolabeling Studies. *Lipids* **2002,** *37,* 209–216.

Cohen, Z.; H.A. Norman; Y.M. Heimer. Microalgae as a Source of Omega 3 Fatty Acids. *World Rev. Nutr. Diet* **1995,** *77,* 1–31.

Damude, H.; P. Gillies; D.J. Macool; S.K. Picataggio; D.M.W. Pollak; J.J. Ragghianti; Z. Xue; N.S. Yadav; H. Zhang; Q.Q. Zhu. Arachidonic Acid Producing Strains of *Yarrowia lipolytica*. E. I. DuPont de Nemours & Co. U.S. Patent Application US2006/0094092, 2006a.

Damude, H.; P. Gillies; D.J. Macool; S.K. Picataggio; D.M.W. Pollak; J.J. Ragghianti; Z. Xue; N.S. Yadav; H. Zhang; Q.Q. Zhu. Eicosapentaenoic Acid Producing Strains of *Yarrowia lipolytica*. E. I. DuPont de Nemours & Co. U.S. Patent Application US2006/0115881, 2006b.

Damude, H.; P. Gillies; D.J. Macool; S.K. Picataggio; D.M.W. Pollak; J.J. Ragghianti; Z. Xue; N.S. Yadav; H. Zhang; Q.Q. Zhu. Docosahexaenoic Acid Producing Strains of *Yarrowia lipolytica*. E. I. DuPont de Nemours & Co. U.S. Patent Application US2006/0110806, 2006c.

Davies, R. Yeast Oil from Cheese Whey—Process Development. *Single Cell Oil;* R. Morton, Ed.; John Wiley & Sons Inc.: New York, NY, 1988; pp 99–146.

Gandhi, S.R.; J.D. Weete. Production of the Polyunsaturated Fatty Acids Arachidonic Acid and Eicosapentaenoic Acid by the Fungus *Pythium ultimum*. *J. Gen. Microbiol.* **1991,** *137,* 1825–1830.

Hammond, E.; B. Glatz; Y, Choi; M.T. Teasdale. *New Sources of Fats and Oils;* E. Pryde, L. Pricen, K. Mukhergee, Eds.; American Oil Chemists' Society Press: Champaign, IL, 1981; pp 171–187.

Herber, S.M.; M.E. Van Elswyk. Dietary Marine Algae Promotes Efficient Deposition of n-3 Fatty Acids for the Production of Enriched Shell Eggs. *Poult. Sci.* **1996,** *75,* 1501–1507.

Kajikawa, M.; K. Matsui; M. Ochiai; Y. Tanaka; Y. Kita; M. Ishimoto; Y. Kohzu; S. Shoji; K.T. Yamato; K. Ohyama; et al. Production of Arachidonic and Eicosapentaenoic Acids in Plants Using Bryophyte Fatty Acid Delta-6 Desaturase, Delta-6 Elongase, and Delta5-Desaturease Genes. *Biosci. Biotechnol. Biochem.* **2008,** *72,* 435–444.

Knutzon, D.S.; J.M. Thurmond; Y.S. Huang; S. Chaudhary; E.G. Bobik; G.M. Chan; S.J. Kirchner; P. Mukerji. Identification of Delta5-Desaturase from *Mortierella alpina* by Heterologous Expression in Bakers' Yeast and Canola. *J. Biol. Chem.* **1998,** *273,* 29360–29366.

Kyle, D.J.; S. Reeb; V.J. Sicotte. Infant Formula and Baby Food Containing Docosahexaenoic Acid Obtained from Dinoflagellates. Martek Biosciences Corp. U.S. Patent 5,397,591, 1995.

Kyle, D.J. Production and Use of a Single Cell Oil Which Is Highly Enriched in Docosahexaenoic Acid. *Lipid Technol.* **1996,** *2,* 109–112.

Kyle, D.J. Production and Use of a Single Cell Oil Highly Enriched in Arachidonic Acid. *Lipid Technol.* **1997a,** *9,* 116–121.

Kyle, D. Arachidonic Acid and Methods for the Production and Use Thereof. Martek Biosciences Corp. U.S. Patent 5,658,767, 1997b.

Kyle, D. Arachidonic Acid and Methods for the Production and Use Thereof. Martek Biosciences Corp. European Patent Application 1,342,787, 2003.

Kyle, D.J. Fish and the Production Thereof. Advanced BioNutrition Corp. U.S. Patent Application US/20060265766, 2006.

Kyle, D.J. Microalgal Feeds Containing Arachidonic Acid and Their Production and Use. Advanced BioNutrition Corp. U.S. Patent 7,396,548, 2008.

Metz, J.G.; P. Roessler; D. Facciotti, D.; C. Levering; F. Dittrich; M. Lassner; R. Valentine; K. Lardizabal; F. Domergue; A. Yamada; et al. Production of Polyunsaturated Fatty Acids by Polyketide Synthases in Both Prokaryotes and Eukaryotes. *Science* **2001**, *293*, 290–293.

Metz, J.G.; J. Kuner; B. Rosenzweig; J.C. Lippmeier; P. Roessler; R. Zirkle. Biochemical Characterization of Polyunsaturated Fatty Acid Synthesis in *Schizochytrium*: Release of the Products as Free Fatty Acids. *Plant Physiol. Biochem.* **2009**, *47*, 472–478.

Nakahara, T.; T. Yokocki; Y. Kamisaka; O. Suzuki. Gamma-Linolenic Acid from Genus *Mortierella*. *Single Cell Oils*; D. Kyle, C. Ratledge, Eds.; American Oil Chemists' Society Press: Champaign, IL, 1992; pp 61–97.

Parker-Barnes, J.M.; T. Das; E. Bobik; A.E. Leonard; J.M. Thurmond; L.T. Chaung; Y.S. Huang; P. Mukerji. Identification and Characterization of an Enzyme Involved in the Elongation of n-6 and n-3 Polyunsaturated Fatty Acids. *Proc. Natl. Acad. Sci. USA* **2000**, *97*, 8284–8289.

Patel, M.; K. Rajyashri. Recombinant Production Docosahexaenoic Acid (DHA) in Yeast. Avestha Gengraine Tech PVT LTD. World Patent Application WO2006064317, 2008.

Pereira, S.L.; Y.S. Huang; E.G. Bobik; A.J. Kinney; K.L. Stecca; J.C. Packer; P. Mukerji. A Novel Omega3-Fatty Acid Desaturase Involved in the Biosynthesis of Eicosapentaenoic Acid. *Biochem. J.* **2004b**, *378*, 665–671.

Ratledge, C. Microbial Lipids: Commercial Realities or Academic Curiosities. *Single Cell Oils*; D.J. Kyle, C. Ratledge, Eds.; American Oil Chemists' Society Press: Champaign, IL 1992; pp 1–15.

Ratledge, C. Microbial Production of Gamma-Linolenic Acid. *Handbook of Functional Lipids*; C. Akoh, Ed.; CRC Press LLC: Boca Raton, FL, 2006; pp 19–45.

Rooke, J.; A. Sinclair; S.A. Edwards. Feeding Tuna Oil to the Sow at Different Times during Pregnancy Has Different Effects on Piglet Long-Chain Polyunsatureated Fatty Acid Composition at Birth and Subsequent Growth. *British J. Nutr.* **2001**, *86*, 21–30.

Smit, H.; A. Ykema; E.C. Verbee; I. Verwoert; M.M. Kater. Production of Cocoa Butter Equivalents by Yeast Mutants. *Single Cell Oils*; D. Kyle, C. Ratledge, Eds.; American Oil Chemists' Society Press: Champaign, IL, 1992; pp 185–195.

Sprecher, H. An Update on the Pathways of Polyunsaturated Fatty Acid Metabolism. *Curr. Opin. Clin. Nutr. Metab. Care* **1999**, *2*, 135–138.

Tanaka, S.; T. Yaguchi; S. Shimizu; T. Sogo; S. Fujikawa. Process for Preparing Docosahexaenoic Acid and Docosapentaenoic Acid with *Ulkenia*. Suntory Ltd, Nagase & Co Ltd., Nagase Chemtex Corp. U.S. Patent 6,509,178, 2003.

Tanon, T.; D. Harvey; T.R. Larson; I.A. Graham. Identification of a Very Long Chain Polyunsaturated Fatty Acid Delta4-Desaturase from the Microalga *Pavlova lutheri*. *FEBS Lett.* **2003**, *553*, 440–444.

Weaver, K.; P. Invester; J.A. Chilton; M.D. Wilson. P. Pandev; F. Chilton. The Content of Favorable and Unfavorable Polyunsaturated Fatty Acids Found in Commonly Eaten Fish. *J. Amer. Dietetic Assn.* **2008,** *108,* 1178–1185.

Wright, T.C.; B.J. Holub; A.R. Hill; B.W. McBride. Effect of Combinations of Fish Meal and Feather Meal on Milk Fatty Acid Content and Nitrogen Utilization in Dairy Cows. *J. Dairy Sci.* **2003,** *86,* 861–869.

Yokochi, T.; T. Nakahara; M. Yamaoka; R. Kurane. Method of Producing a Polyunsaturated Fatty Acid Containing Culture and Polyunsaturated Fatty Acid Containing Oil Using Microorganisms. Agency of Industrial Science and Technology. U.S. Patent 6,461,839, 2002.

Zaslavskaia, L.A.; J.C. Lippmeier; C. Shih; D. Ehrhardt; A.R. Grossman; K.E. Apt. Trophic Conversion of an Obligate Photoautotrophic Organism through Metabolic Engineering. *Science* **2001,** *292,* 2073–2075.

Zhou, X.R.; S.S. Robert; J.R. Petrie; D.M. Frampton; M.P. Mansour; S.I. Blackburn; P.D. Nichols; A.G. Green; S.P. Singh. Isolation and Characterization of Genes from the Marine Microalga *Pavlova salina* Encoding Three Front-end Desaturases Involved in Docosahexaenoic Acid Biosynthesis. *Phytochemistry* **2007,** *68,* 785–796.

Jesus Ruben Abril, Ph.D.

Dr. Abril is Director of Ingredient Formulations and Technical Support, at Martek Biosciences, where he has worked since 1993. Dr. Abril's previous Martek posts include: Chief Analytical Chemist, Principal Scientist-Product Development, Assistant Director-Product Development, and Director of Product Development. He served a 2-year assignment in Germany in 2001-03, working in the area of lipids, mainly omega-3 fatty acids, applied to foods and animal feeds. Previous work includes: Food Processing Engineer with Nestle Corporation, in the production of instant coffee, Nescafe, and Decaf in Toluca Mexico; Professor of Food Sciences at the University of Sonora Graduate Program, in Mexico; and Head of Analytical Chemistry at the Colorado Department of Health and Environment (focus on environmental pollutants). Dr. Abril's postdoctoral work related to Environmental Chemistry of Plants of the Desert of Arizona/Sonora, completed at the University of Denver (Colorado). He also completed post-graduate studies in Food Science (M.S.) and Nutritional Biochemistry (Ph.D.) at the University of Arizona (Tucson) in 1984. In 1976, he received a Bachelor of Sciences in Biochemical Engineering from the Instituto Tecnologico y de Estudios Superiores de Monterrey in Monterrey, Mexico. Dr. Abril has written numerous publications and patents.

Akinori Ando, Ph.D.

Assistant Professor, Research Division of Microbial Sciences, Kyoto University, Japan.

Bill Barclay, Ph.D.

Bill Barclay, Ph.D. is Chief Intellectual Property Officer at Martek Biosciences Corporation, where he is responsible for developing IP-based strategies to enhance innovation and expand Martek's overall strategic perspective. Bill founded OmegaTech Inc. (Boulder, Colorado) in 1988, where he led efforts to develop a successful algae-based fermentation process for producing DHA-rich microbial oil. OmegaTech merged with Martek in 2002. As an aquatic ecologist, Bill has spent 30 years in microalgal biotechnology. He is an inventor on more than 40 granted US patents in the field of microbial technology and has authored numerous publications based on his research.

Professor Feng Chen, Ph.D.

Feng (Steven) Chen is a professor in the School of Biological Sciences at the University of Hong Kong. He is an expert in algal biotechnology. He is particularly interested in the mechanism of heterotrophic biosynthesis of micro algal metabolites, which may lead to a controllable process for manufacturing algal products on a large scale. Professor Chen has edited 4 books, filed 6 patents and published over 200 papers in SCI journals. He is serving on the editorial boards of *Process Biochemistry, Biotechnology Letters,* and *Journal of Applied Phycology.* He is an elected fellow of American Institute for Medical and Biological Engineering. Contact: Professor Steven Feng Chen; School of Biological Sciences; the University of Hong Kong; Pokfulam Road; Hong Kong;
Website: http://www.hku.hk/biosch/

Professor Zvi Cohen

Professor Zvi Cohen received his Ph.D. from the Weizmann Institute of Science in 1978. He is a member of the Blaustein Institute for Desert Research in Ben Gurion University since 1981 and is the incumbent of the Maks and Rochelle Etingin professorial chair in Desert Research. His research involves the physiology, biochemistry, molecular biology and biotechnology of PUFA production in microalgae. Professor Cohen serves on the editorial board of *Annals of Microbiology and Biotechnology Letters,* is the editor of Chemicals from Microalgae and coeditor (with Colin Ratledge) of *Single Cell Oils.*

Dr. Hui Gong

Dr. Hui Gong has been the assistant professor and aquaculture research faculty at the University of Guam since 2007. She graduated from Texas A&M University in 1999 with a Ph.D. in aquaculture nutrition. Her previous experience includes assistant production manager with a pioneer company in developing Agri-integrated aquaculture in Sonora desert, researcher in SNP marker development, and shrimp health assurance and shrimptraq database manager for a shrimp genetic breeding company. Her research interests are shrimp nutrition and genetic selection for feeding efficiency, health management in aquaculture, application of molecular tools in genetic breeding, and aquaculture development in the western Pacific region. Dr. Gong is currently serving as the principal investigator for five U.S. federally funded research projects.
Contact: Hui Gong, Ph.D.; Assistant Professor; College of Natural and Applied Sciences; University of Guam; UOG Station; Mangilao, GU 96923;
Office: (671)735-2144
Email: hgong@uguam.uog.edu
Website: http://www.wptrc.org/research_dtl.asp?rschrID=7

Dr. David J. Kyle

In 1980, Dr. Kyle received his Ph.D. from the University of Alberta, Canada, in the field of Lipid Biochemistry. He co-founded Martek Biosciences Corporation in 1985, where he developed and implemented the long-term strategic plan to commercialize the Company's DHA- and ARA-based Nutritional Products. In 2001, Dr. Kyle founded Advanced BioNutrition Corp, a biotechnology company, dedicated to providing functional nutrition and disease control elements in animal feeds to eliminate the use of antibiotics in both aquaculture and agriculture industries. In 2009, Dr. Kyle was inducted into the Space Technology Hall of Fame for his contributions to Science and Industry.

Connye Kuratko, Ph.D., R.D.

Dr. Connye Kuratko is a registered dietitian and has more than 30 years of experience in the areas of health care, public health and education. In 1991, she completed a doctoral program at Texas Tech University which emphasized diet and disease prevention, research methods, and education. She served as an Associate Professor in Pathology at the Texas Tech School of Medicine and taught nutrition and other sciences to medical and allied health students. While at the medical school, she built a research program studying the effects of omega-3 fatty acids on cancer development and gut immunity. She continues to be interested in omega-3 fatty acids and follows many aspects of nutrition research through her work as senior manager of medical affairs at Martek Biosciences in Columbia, Maryland.

Jun Ogawa, Ph.D.

Professor, Division of Applied Life Sciences, Graduate School of Agriculture, Kyoto University, Japan.

Colin Ratledge

Colin Ratledge is Emeritus Professor of Microbial Biochemistry at the University of Hull, UK, which he joined as a lecturer in 1967. He has served on most of the biotechnology committees in the UK, including being Vice President of the Society of Chemical Industry and has also served on the European Federation of Biotechnology Executive Committee. He was chairman of the Food Research Grants Group of the Biotechnology and Biological Sciences Research Council. He has been, and still is, a consultant to a number of companies both in the UK, USA and Europe for advice on microbial lipids. He has had a life-long interest in microbial oils; has written over 350 research papers, patents, reviews, and chapters in books. He is currently editor in chief of *Biotechnology Letters* and serves as an Associate Editor for *Lipids*. He is also on the editorial board of *Lipid Technology*. He is a Fellow of the Royal Society of Chemistry, Society of Biology and the Royal Society of Arts.

Alan S. Ryan

Alan S. Ryan is the Executive Director of Clinical Research at Martek Biosciences Corporation. He is involved with directing clinical research to evaluate the benefits of DHA for cognitive development and for a variety of medical conditions. Before joining Martek, Dr. Ryan was employed for 22 years at Abbott Laboratories. He was inducted into Abbott's Volwiler Society in recognition for his scientific excellence and leadership. He managed the submission of two drugs: Lovaza®, to treat hypertriglyceridemia, and NeoProfen®, to treat ductus arteriosus in premature infants. He received his Ph.D. in anthropology from the University of Michigan. After serving 27 years in the U.S. Navy Reserves, he retired as Commander.

Eiji Sakuradani, Ph.D.

Assistant Professor, Division of Applied Life Sciences, Graduate School of Agriculture, Kyoto University, Japan.

Norman Salem, Jr., Ph.D.

Norman Salem, Jr., PhD is currently the Chief Scientific Officer and Vice President of Research for Martek Biosciences Corporation. He previously served as the Chief of the Laboratory of Membrane Biochemistry & Biophysics within the Intramural Research program of the NIAAA, NIH, where he worked on studies of DHA composition, metabolism and biological function with an emphasis on the nervous system. He is an author of over 235 publications. Dr. Salem serves on the editorial board of *Lipids and Nutritional Neuroscience*. He is the Past President of the ISSFAL Society and recipient of the Supelco/Nicholas Pelick Award from the American Oil Chemists' Society. He was named recently as the recipient of the 2010 ISSFAL Lifetime Achievement Award. Dr. Salem received his B.S. degree in physics from Miami University and his Ph.D. in Neurobiology from the University of Rochester, School of Medicine.

Sakayu Shimizu, Ph.D.

Professor, Department of Bioscience and Biotechnology, Faculty of Bioenvironmental Science, Kyoto Gakuen University, Japan.

Dr. Lolke Sijtsma

After receiving his Ph.D. in microbiology, Dr. Lolke Sijtsma worked as senior scientist and project leader at ATO, currently known as Wageningen UR Food & Biobased Research. His topic of interest is microbial or enzymatic conversion of raw materials into value added products. Dr. Sijtsma coordinated several national and large multidisciplinary European research and development projects including the sustainable production of omega-3 fatty acids by marine algae. He authored and co-authored over 45 scientific papers and 6 patents. Currently, he leads a team within the Top Institute Food and Nutrition, studying the nitrogen metabolism of lactic acid bacteria using fermentation, genomics and systems biology.
E-mail: lolke.sijtsma@wur.nl
Web: www.wur.nl; www.biobasedproducts.nl

Andy Sinclair

Andy Sinclair is Professor of Human Nutrition at Deakin University, Geelong, Australia. He is immediate Past President of the Nutrition Society of Australia. His research interests include metabolism of essential fatty acids in the brain, including DPA omega 3; the effect of omega 3 deficiency and supplementation on physiological function in animals (pre- and post-natal nutrition); and the interaction between zinc and docosahexaenoic acid in neural cells. Professor Sinclair has more than 240 publications in peer-reviewed journals.

Hugo Streekstra

Since 1988 Dr. Streekstra has been working in the food industry, located in Delft, the Netherlands. He has worked in various fields, including fermentation process development, classical genetics and applied biochemistry. Currently he is employed by DSM Food Specialties, as a senior scientist in enzyme application. Hugo Streekstra started his academic training in Chemistry in 1978 at the University of Amsterdam. He graduated in 1983 with a specialization in Microbiology. This was followed by a research subject in Microbial Physiology at the same University, which led to the successful defense of a Ph.D. thesis entitled "Metabolic Uncoupling in Anaerobic *Klebsiella pneumoniae*" in 1990.

Marguerite Torrey

Marguerite Torrey is technical projects editor for *inform*, the membership magazine of the American Oil Chemists' Society (AOCS). Before joining *inform*, she served as a copyeditor/proofreader for the professional journals of AOCS. Torrey received her M.S and Ph.D. from the University of Wisconsin-Madison in Water Chemistry in the Department of Civil and Environmental Engineering. Her graduate research focused on the significance of nitrogen fixation by blue-green algae to the nitrogen budget of Lake Mendota, Madison, Wisconsin. She may be contacted at mtorrey@aocs.org. The AOCS web site is www.aocs.org.

Mary Van Elswyk

Mary Van Elswyk is a Ph.D., registered dietitian with 18 years experience researching the health benefits of DHA omega-3. She is a former Texas A&M assistant professor in nutrition who joined OmegaTech Inc. in 1997 as Director and later Vice President of Scientific Affairs. Martek Bioscience acquired OmegaTech in 2001 and she remained with Martek in various capacities including clinical, medical and scientific affairs, now serving as a consultant for regulatory affairs. She specializes in the development and substantiation of nutrition and health claims for the food and dietary supplement industry.

Mario Velasco-Escudero

Dr. Velasco-Escudero obtained his B.S. in Biology from Albright College, Pennsylvania in 1988; his M.S. in Marine Sciences from the University of Puerto Rico in 1993; and his Ph.D. in Wildlife and Fisheries Sciences, specialty in Aquaculture Nutrition from Texas A&M University in 1996. Since 2004 he directs his company, Maricultura Negocios y Tecnología S.L. (www.marnetec.com), which provides consulting services to the aquaculture sector worldwide. His previous experience is as a Laboratory and Field Technician with the Marine Systems Laboratory at the Smithsonian Institution in Washington D.C. and as a Research Scientist at The Oceanic Institute in Hawaii. Also, he has served as the Technical and R&D+i Director at PRONACA-ENACA, Guayaquil, Ecuador. His interest is developing profitable and sustainable aquaculture systems.

Todd Wills

Todd joined Martek Biosciences Food Research and Development group in 2004. His initial responsibilities at Martek were to develop successful methodologies, formulations, and delivery technologies to ensure success when fortifying foods with DHA. Todd currently provides expertise to Martek customers in formulating products and providing technical support throughout all phases of product development. Todd Wills earned an M.S. degree (2004) in Food Science from Oklahoma State. From 1997-2004, he was employed at the OSU - Food and Agriculture Product Center as HACCP coordinator. Research he conducted at OSU focused on exogenous antioxidant (Vit E) partitioning between membrane and neutral lipids in beef muscle. Todd has published in several academic journals and presented to numerous professional organizations. Contact: Todd Wills, M.S.; Scientist; Ingredient Formulations & Technical Support; Martek Biosciences; 4909 Nautilus Ct. N. #208; Boulder, CO 80301; Phone: 303-357-2815
Fax: 303-381-8181

Jim Wynn

Jim Wynn has spent the past 16 years studying the production of single cell oils. He has studied both the accumulation of lipids in general and the biosynthesis of polyunsaturated fatty acids in particular in a range of eukaryotic microorganisms. His work in this arena has taken him from the University of Hull in the North East of England, where he worked in the laboratory of Professor Colin Ratledge (one of the pioneers in the SCO field) to the East Coast of America where he worked for Martek Biosciences helping develop and commercialize both fungal and algal processes for the production of DHASCO and ARASCO. He is currently Director of Process Development and Integration at MBI a biotechnology company allied to Michigan State University with a mission to develop commercially viable processes for the manufacture of bio-based chemicals and fuels.

Quinn Q. Zhu

Since 1998 Quinn Q. Zhu has been Principal Investigator and Molecular Biology Team leader of the omega-3 program in Central Research and Development at the DuPont Company. From 1988 to 1998 Dr. Zhu served as a Post-doctorate and then Staff Scientist in the Plant Biology Laboratory (Professor Chris Lamb) at The Salk Institute for Biological Studies, San Diego, California. From 1983 to 1988 Dr. Zhu worked on Ph. D., Shanghai Institute of Plant Physiology, Chinese Academy of Science, Shanghai, China. From 1979 to 1983 Dr. Zhu worked on B.S., Biology Department, Nanjing Teachers University, Nanjing, China.

Index

Printed and bound by CPI Group (UK) Ltd, Croydon, CR0 4YY

08/05/2025

01864839-0001